Computational Mechanics of Fluid-Structure Interaction

Rajeev Kumar Jaiman · Vaibhav Joshi

Computational Mechanics of Fluid-Structure Interaction

Computational Methods for Coupled Fluid-Structure Analysis

Springer

Rajeev Kumar Jaiman
Department of Mechanical Engineering
The University of British Columbia
Vancouver, BC, Canada

Vaibhav Joshi
Department of Mechanical Engineering
Birla Institute of Technology and Science
K K Birla Goa Campus, Goa, India

ISBN 978-981-16-5357-5 ISBN 978-981-16-5355-1 (eBook)
https://doi.org/10.1007/978-981-16-5355-1

This Springer imprint is published by the registered company Springer Nature Singapore Pte Ltd.
The registered company address is: 152 Beach Road, #21-01/04 Gateway East, Singapore 189721, Singapore

Preface I

Interactions between fluid flows and flexible structures are omnipresent in nature and have numerous applications in engineering such as wind turbines, aircraft wing oscillations and flutter, solid rocket motor, offshore risers and platforms, subsea pipelines to flying bio-inspired drones and unmanned air vehicles, and flexible energy harvesting and sensing devices. The key focus has been to understand significant aspects of flow-induced deformation and fluid-elastic instabilities. The complexity of the fluid-elastic phenomena further increases with other physical effects related to two-phase flows, turbulence, and flexible multibody effects. The book aims to present general coupling formulations and algorithms to capture the interactions between these physical fields in most accurate and stable manner. In this book, fluid-structure interaction is intended to understand coupled physical phenomena that can be modeled in terms of continuum mechanics described by partial differential equations (PDEs). A particular emphasis has been placed on the *interface* treatment of fluid-structure interaction.

This book presents a broad overview of variational formulations and numerical methods for students and professionals in computationally oriented disciplines who need to solve fluid-structure interaction problems. It differs from traditional numerical analysis texts in that it focuses on the motivation and ideas behind the numerical methods presented rather than on detailed analyses of them. We attempt to convey a general understanding of the techniques available for solving problems in each major category, including proper problem formulation and interpretation of numerical results. To accommodate the diverse readers, the prerequisites for the book are fluid and solid mechanics, numerical analysis, and finite element methods. The book adopts a fairly sophisticated perspective, so a reasonable level of maturity on the part of the reader is advisable. Beyond the academic setting, we hope that the book will also be useful as a reference for engineers and scientists who may need a quick overview of numerical strategies for fluid-structure interaction. While fluid-structure interaction is still an active area of research, it will not be possible to provide a complete survey of numerical methods and formulations.

Vancouver, Canada
Goa, India

Rajeev Kumar Jaiman
Vaibhav Joshi

Preface II

The importance of fluid-structure interaction (FSI) and coupled mechanical analysis has been well acknowledged in science and engineering. Predicting the behavior of nonlinear fluid-structure interaction has tremendous importance for both industrial and natural processes. Owing to the growth of computers and advanced methods, full-scale FSI modeling has begun to be realized. This book is intended to provide a compilation of current state-of-the-start numerical methods for nonlinear fluid-structure interaction. The book aims at the continuum theory and variational formulations for the fluid-structure interaction. A particular emphasis will be placed on the treatment of fluid-structure and fluid-fluid interfaces in simulating large-scale multi-field and multidomain problems. Single- and two-phase viscous incompressible fluid flows will be considered with the increasing complexity of structures ranging from rigid body, and linear elastic and nonlinear large deformation to fully coupled flexible multibody system analysis. In particular, such coupled fluid-structure effects are omnipresent in ocean and wind environments, which can have a significant impact on the dynamical performance of engineered as well as natural systems. Understanding the impact of flow-induced loads on structural deformation and vibrations can lead to safer and cost-effective structures, especially for light and high aspect ratio structures with increased flexibility.

As one of the new aspects, this book covers the effect of two-phase flow in fluid-structure interaction for large-scale modeling of practical applications. There is a resurgence of multiphase flow FSI in numerous engineering applications, e.g., marine/offshore, biomedical, bio-locomotion, microfluidics, aerospace, and wind engineering. The interaction of multiphase flows with flexible structures poses numerous challenges. This book provides a comprehensive variational formulation to couple the two-phase fluid flow with flexible structures. The variational fluid-structure formulation includes the modeling of turbulence via dynamic large eddy simulation (LES) and hybrid RANS/LES modeling. The detailed implementation details and algorithmic steps will be unique for broad audiences who want to extend their existing finite element method codes for multiphase and fluid-structure interaction. A simple and stable approach to stabilize the phase-field Allen–Cahn equation will be covered in this book based on our recent development. The proposed variational techniques will be explored on a wide range of physical scales, non-matching spatial and temporal discretizations, the complexity associated with structure-to-fluid mass ratios (i.e., added mass effects), Reynolds number, large structural deformation, free surface, and other interacting physical fields.

The present book will cover a broad range of numerical techniques with their respective strengths and weaknesses. To our knowledge, there exists no book which covers the body-fitted FSI technique and goes in depth into the fundamental of FSI coupling for the industry standard partitioned-based methodologies. The present book will cover partitioned and monolithic methodologies for a range of applications to single- and two-phase flows. Detailed information will be given on the treatment of non-matching meshes for partitioned variational analysis. A broad range of partitioned iterative techniques and their advantages and limitations for a range of physical and geometric parameters will be provided. Finally, the present book includes the detailed formulation for flexible multibody structural interaction with the turbulent fluid flows using unstructured grids and a parallel computing environment.

Vancouver, Canada
Goa, India

Rajeev Kumar Jaiman
Vaibhav Joshi

Acknowledgements

The authors would like to acknowledge the support and contributions by several individuals, organizations, and funding agencies. The contents of chapters are based on several Ph.D. dissertations and journal papers produced under the supervision of the first author. To begin, we would like to acknowledge the kind support from our host organizations, namely the University of British Columbia (UBC), the National University of Singapore (NUS), and the University of Illinois at Urbana-Champaign. We would like to thank many current and former students and colleagues involved in our laboratory research activities. Especially, we would like to express our appreciation to those individuals whose works have been used directly in this book: Pardha S. Gurugubelli, Guojun Li, Tharindu P. Miyanawala, Yulong Li, Yun Zhi Law and among others at the Computational Multiphysics Lab at UBC and the Fluid-Structure Interaction Group (FSIG) at NUS.

Contents

Chapter 1
Introduction: A Computational Approach

This book aims at providing a survey of mathematical formulations and simulation techniques for fluid-structure interactions. As the name suggests, fluid-structure interactions involve the interplay of fluid flow and deformable/moving solid structures, aimed at understanding some physical phenomenon or designing an engineering device. Computational simulation of fluid-structure interaction is useful not just for exploring exotic or otherwise inaccessible situations, but also for exploring a wider variety of normal scenarios that could not otherwise be investigated with reasonable cost and time. In engineering design, computational simulation allows a large number of design options to be tried much more quickly, inexpensively, and safely than with traditional "build-and-test" methods using physical prototypes. A coupled fluid-structure analysis represents a special class of multiphysics problems in which it is important to study the effects of fluid flow on flexible structures and their subsequent interactions. One particular example is the study of a structure with geometric and material nonlinearities that undergo large deformation while interacting with a turbulent flow. This complex nonlinear setting is of engineering interest across a variety of mechanical fields ranging from aerospace/marine applications, such as modeling of aircraft and flying vehicles or large-scale offshore structures and long flexible risers, to biomedical applications such as modeling of blood flows in vessels. All these problems are highly nonlinear and involve strong interaction of fluid and elastic structures, and thus require a fully-coupled modeling and simulation of fluid-structure interaction.

Fluid-structure interactions can also pertain to flapping phenomena in swimming of fishes, flight of birds and insects, and fluttering of flags and leaves. The feedback process between the ambient fluid and the deformation of flexible structure forms a coupled nonlinear dynamical system, whereby the flapping motion is characterized by the mutual coupling of the fluid and the flexible body. Such flapping dynamical effects are important owing to their potential applications in designing devices for micro-energy harvesting, efficient propulsion, flow separation control, drag reduction and bio-prosthetics. The study of hydro-structural interactions and vortex-induced

R. K. Jaiman and V. Joshi, *Computational Mechanics of Fluid-Structure Interaction*,
https://doi.org/10.1007/978-981-16-5355-1_1

vibrations (VIV) is also highly relevant in which strong vortices are shed in the wake of flows passing through bluff-body structures. For example, the transport of oil and gas from ocean floor wells to oil platforms and land-based production facilities require the use of long slender pipelines called risers and subsea pipelines. Complex ocean environment can cause a pipeline to go into a self-excited oscillation and potentially cause severe fatigue damage or slam the pipeline against the seabed. The resulting damage can be very expensive to repair and may result in environmental contamination.

1.1 Some Motivating Applications and Challenges

This book is concerned with the formulations and algorithms for numerical modeling of coupled fluid-structure interaction problems that arise in several engineering applications. The mathematical theory and numerical methods of solving such problems are not well established and are still under development. In this section, some of the challenges dealing with the FSI applications are briefly discussed.

1.1.1 Marine and Offshore Engineering

Transport of oil and gas from ocean floor wells to land-based production facilities requires the use of risers and subsea pipelines that can vary in length upto several kilometers. The aspect ratios (length to diameter) of these slender structures are typically in the order of thousands and are subjected to vortex-induced vibrations (VIV) when exposed to ocean currents (Fig. 1.1). Vortex-induced vibration is a well-known fluid-structure phenomenon in which strong vortices are shed in the wake of bluff body structures. These self-excited and self-limiting vibrations are a result of unsteady fluid forces that arise from the asymmetric vortices shed behind the structure. During VIV, the phenomenon of frequency *lock-in* occurs for a certain range of fluid and structural parameters and the preferred frequency of wake deviates from its expected value determined by the Strouhal relation, while being close to the value of the natural frequency of the structure. This complex frequency *lock-in* phenomenon leads to a large amplitude of vibrations, typically around one cross-sectional unit length of the structure. Based on the marine environment and structural properties, VIV can cause significant dynamic bending stresses, drag force amplification and large deflections on the riser from a short term perspective. From the long term viewpoint for production risers, rapid accumulation of fatigue damage may lead to structural failure if the vibrations are left unchecked. Consequently, it is imperative for the safety of offshore operations that VIV response is adequately predicted on these long flexible structures. In addition, the physical understanding of vibrating flexible structures can help in developing devices and methods to suppress VIV. While experiments are expensive to perform, a high-fidelity computational framework helps to predict the

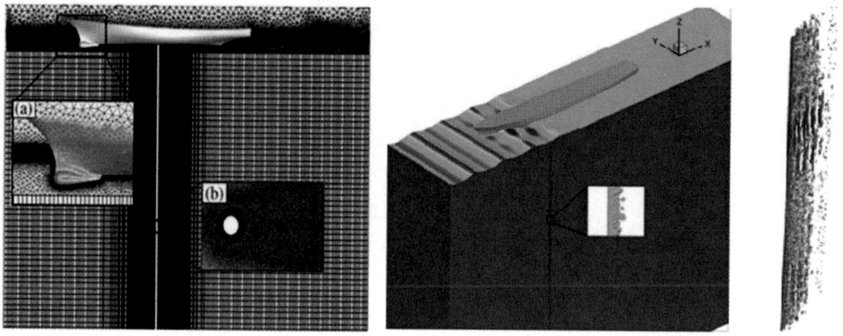

Fig. 1.1 Large-scale fluid-structure interaction of coupled vessel connected with elastic riser pipeline: unstructured finite element mesh around drillship and riser (left), representative hydrodynamics solution (middle), vortex-induced vibration of marine riser with complex vortex shedding patterns (right)

coupled fluid-structure dynamics from a first principle and rigorous physics-based understanding. Further information about these problems and predictive capability can be found in (113, 116).

1.1.2 *Energy Harvesting via Flow-Induced Vibration*

Fluid-structure interactions can exhibit complex coupled dynamics, both in useful and destructive manners. In the context of sustainable clean energy, harvesting power from flapping and vibrating structure is gaining a resurgence of interest in recent years for various engineering applications. Coupled with piezoelectric materials, the flapping and vibrating structures are capable to transform wind and ocean kinetic energy into usable energy at small-to-medium scale systems. The efficiency and performance of these devices are strongly dependent on the precise design and adaptive tuning of multiphysical properties (e.g., structures and piezoelectric materials) during nonlinear dynamic excitations. In this context, fully-coupled fluid-structure-piezoelectric modeling becomes crucial for efficient design and control of such energy harvesting systems. An elastic foil interacting with a uniform flow with its trailing edge clamped (termed as the inverted foil) is to be considered as a prototypical problem, which has a profound impact on the development of energy harvesting devices via self-sustained large-amplitude flapping (Fig. 1.2). By coupling with piezoelectric materials, the structural strain energy can be transformed into electric energy, which can be useful for low-powered onboard systems and sensors.

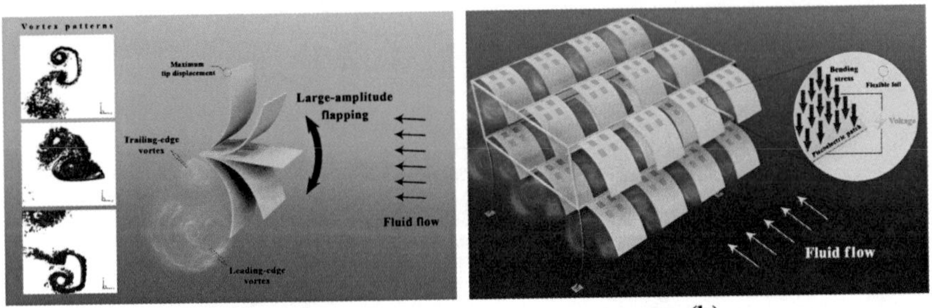

(a) (b)

Fig. 1.2 Fluid-structure interactions of inverted elastic flags for energy harvesting: large amplitude flapping of an inverted elastic flag with piezoelectric patches (left) and an array of inverted elastic flags in ocean environment (right)

1.1.3 Bio-Inspired Drones and Unmanned Air Vehicles

Compared to a conventional fixed-wing flight vehicle, the flapping flight of flying creatures (e.g., bats and owls) involves active morphing and adaptive flexible as well as serrated wing configuration, which may offer some unique benefits with regard to flight efficiency improvement, noise suppression, structural weight reduction and high maneuverability. In the past decades, there has been a large amount of research works towards the development of low-speed and small-sized vehicles, i.e. micro air vehicles (MAVs). These small-sized vehicles represent the opposite end of the flight spectrum compared to most of the military and civilian aircrafts that are currently in use. While a relatively small size of the MAV offers potential benefits with regard to maneuverability, surveillance and operational robustness, the small size with low inertia makes these systems difficult to operate with regard to mid-air control and stability. The MAVs that possess the flight agility of natural flyers remain a challenge, especially in complex terrain with very confined spaces (Fig. 1.3).

A natural extension in the design of MAVs and UAVs is to develop a hybrid aerial-aquatic vehicle that can traverse in complex multiphase conditions and perform search-and-rescue operations, surveillance and environment exploration. Such a hybrid unmanned vehicle needs to handle locomotion in air and water while overcoming the challenges associated with the transition across the air-to-water interface. Proper analysis and design of such locomotion (flying and swimming) require flexible multibody and multiphase FSI capabilities. Flapping-wing vehicle and wing kinematics should be adapted to both air and water environments. Aerial vehicle demands a large wingspan to stay aloft, while aquatic vehicle needs to minimize the surface area to reduce the hydrodynamic drag. For example, to address such conflict, bio-inspirations from the flying of a *flying-fox* bat and the swimming of a bat ray can be considered to realize a hybrid aerial-aquatic locomotion (Fig. 1.4).

$$\Gamma^s = \Gamma^s_{Rigid} \cup \Gamma^s_{Beam} \cup \Gamma^s_{Shell} \cup \Gamma^s_{Joints}$$
$$\Gamma^f = \Gamma^f_{top} \cup \Gamma^f_{bottom} \cup \Gamma^f_{LE} \cup \Gamma^f_{TE}$$

■ Rigid — Beam ▨ Shell • Joints

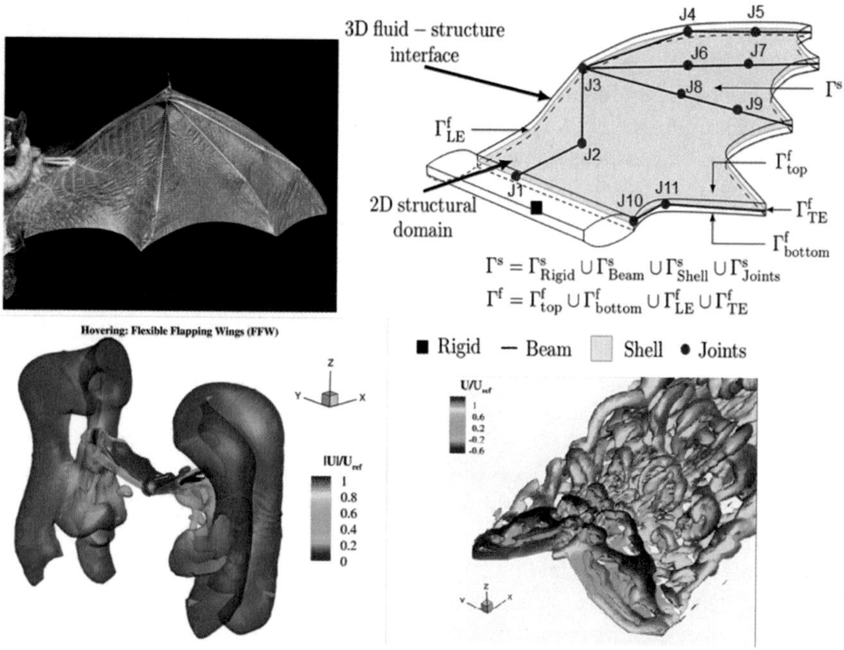

Fig. 1.3 Fluid-structure interaction of bat-inspired drones and unmanned air vehicles: a flexible multibody bat wing (top-left), illustration of flexible multibody and fluid-structure interfaces (top right), flow patterns around hovering (bottom-left) and gliding full-scale bat flight (bottom-right)

Fig. 1.4 Multiphase fluid-structure interaction of a conceptual hybrid avian-aquatic vehicle

1.1.4 Flow-Induced Vibration and Control

Bluff body flows are ubiquitous in nature and engineering structures, such as wind blowing over high rise buildings, offshore platforms and pipelines, aircraft at a high angle of attack, and suspension bridges. Coupled fluid-structure interactions can lead to a great variety of flow-induced vibrations (FIV), both in catastrophic and valuable (energy harvesting) ways in numerous engineering and scientific applications. These flow-induced vibrations of bluff bodies are particularly important for offshore structures that are increasingly deployed in the ocean environment. The offshore floating system is an integrated coupled dynamical system comprising of an offshore floating platform, risers, and moorings interacting with ocean currents and waves. These offshore structures undergo self-excited vibrations and coupled fluid-elastic instabilities, which pose a great challenge for numerical and mathematical modeling. Predicting FIV in offshore floating structures such as semi-submersible is a challenging task due to complex wake interference, vortex-induced vibrations, galloping and other fluid-elastic instabilities. These coupled instabilities associated with rhythmic oscillations are undesirable for the riser and mooring fatigue. Furthermore, the flow interference and shielding effects of tandem and side-by-side configurations significantly alter the wake dynamics and loads on the offshore floating structure. In offshore engineering, there is a growing demand to understand and to optimize the hydrodynamic loads and flow-induced motions of multicolumn floating structures (Fig. 1.5).

The development of high-fidelity tools and the discovery of new physical mechanisms can naturally lead to a host of new designs and control strategies for practical use. Using full-order FSI simulations, passive and active flow control techniques can be developed for drag reduction and the mitigation of vortex-induced vibration, for

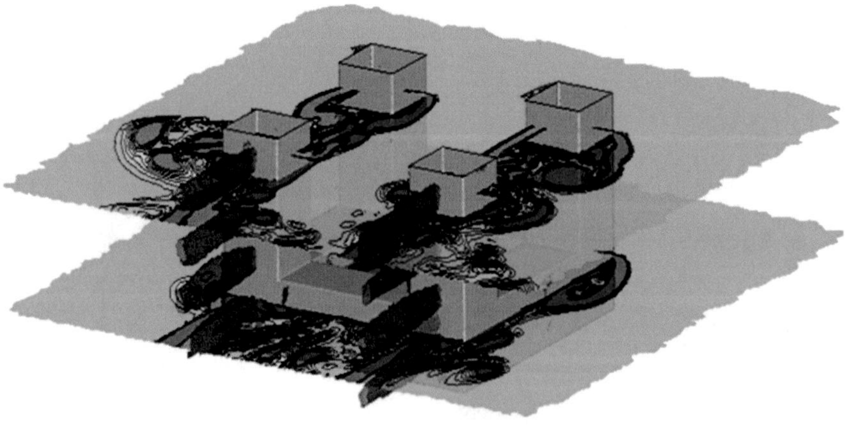

Fig. 1.5 Fluid-structure interaction of offshore platform undergoing vortex-induced vibration. Jet flow is illustrated for the wake stabilization and the suppression of flow-induced vibration

example, the wake stabilization mechanism and a new connected-C device which allows suppressing of fluid-structure interaction while maintaining low drag. This insight has a profound impact on the suppression of VIV for offshore risers and other bluff body structures. In Fig. 1.5, the blowing of jet flow in the leeward side of the three-dimensional semi-submersible model is applied, whereby the transverse displacements and lift coefficient are reduced by 23% compared to the same model without jet flow. The proposed design and arrangement are effective to minimize vortex-induced motion of semi-submersibles.

1.2 Continuum Mechanics Aspects of Fluid-Structure Interaction

Fluid-structure interaction represents a branch of mechanics that studies the interaction of fluid flows with moving or deformable structures. Throughout this book, we consider the continuum hypothesis whereas solid, liquid or gas is assumed to be continuous. The concept of a continuous medium permits us to define quantities such as density, momentum, energy, etc. at each point in the domain occupied by the material. Since these quantities are assumed to have a continuous distribution in space, we can use the framework of calculus, including the concepts of derivatives and integrals, to formulate the governing equations. It has been found that the theories of elasticity and fluid mechanics based on a continuum model yield quantitative predictions that are close to reality. This is perhaps the best justification for using the continuum hypothesis. During the modeling of continuum fluid-structure interactions, there are three main considerations and technical difficulties: (i) treatment of fluid-structure interface to satisfy the kinematic, the dynamic and the geometric boundary conditions, (ii) conflict in the representation of the dissimilar coordinate frames for the fluid and the structural domains and their motion during the fluid-structure interactions, and (iii) numerical solution of coupled nonlinear partial differential equations for the fluid and the structural systems.

1.2.1 Interface Conditions

The interplay between the continuum fluid and solid fields forms a complex nonlinear dynamical system. The coupled FSI equations comprise the initial-boundary value problems of the fluid and the structure complemented by the displacement (kinematic) and the traction (dynamic) boundary conditions at the fluid-structure interface. Furthermore, a geometric boundary conservation also needs to be satisfied at the moving fluid-structure boundary.

The first coupling condition is the kinematic condition and states that the velocity of the fluid is the same as the velocity of the structure at the fluid-structure interface.

This means that the fluid will stick to the structural boundary, which is the moving interface. This model is similar to the no-slip condition in viscous fluid dynamics. In the inviscid regime, for example, in the problems of aeroelasticity, the no-slip condition can be relaxed to a non-penetration condition that prescribes the fluid motion only in the normal direction of the fluid-structure interface, i.e., the fluid will not enter the structural boundary. The second condition is known as the dynamic condition and prescribes a balance of the forces across the fluid and the structural domain. The fluid forces on the structure include the integration of the pressure and the shear stress effects at the interface. The third condition pertains to the satisfaction of the geometric conservation across the interface. For body-fitted discretizations of the domain, the interface and the mesh are required to move together keeping the conformity of the mesh.

1.2.2 Coordinate Frames and Motion of Continuum Domains

Another major mathematical and modeling challenge for the FSI is concerned with the motion of the fluid and the structural domains and the interfaces (fluid-structure interface and fluid-fluid interfaces for multiphase flows). From a continuum mechanics perspective, there are two coordinate systems namely spatial Eulerian coordinates and material-point based Lagrangian coordinates. In structural mechanics, as the motion of the solid has to be exactly quantified, the material-point based Lagrangian coordinates is usually preferred. In a traditional sense, fluid flows are generally described in fixed domains via spatial Eulerian coordinates. In the field of computational continuum mechanics, there is a classic dilemma with regard to the selection of Lagrangian versus Eulerian descriptions wherein each frame has certain benefits depending on the behavior of deformable matter and its conditions. This essentially poses a difficulty in fluid-structure interaction, viz., structures are typically simulated using Lagrangian methods with moving material nodes and fluids using an Eulerian spatial grid. This conflict mainly lies in the constitutive response—structural stress depends on total deformation, computable from the relative positions of neighboring material nodes, whereas the fluid stress generally depends on the deformation rate, obtainable from the numerical derivatives of the velocity field on a fixed-space mesh. Furthermore, fluid flows involve phenomena which require Eulerian description like mixing, vortex stretching, vortex merging/break-up, turbulence, multiphase and inflow/outflow boundaries. On the other hand, the structural deformation is inherently Lagrangian, characterized by a relatively smaller total strains and boundary conditions that move with the deforming structure. Therefore, fluid-structure interaction is a prime example whereby the above inconsistency is problematic to solve the coupled system of equations described by the fluid and the structural domains.

One of the ways to resolve this conflict is to introduce a referential coordinate system which has advantages of both Eulerian and Lagrangian descriptions. Such a system is referred to as arbitrary Lagrangian-Eulerian (ALE) coordinate system. The methods based on ALE coordinates are Lagrangian at the interfaces and employ moving mesh strategies with Eulerian coordinates away from the interface. The challenge lies in handling the mesh motion when large deformation or motion as well as topological changes in the interfaces are involved.

1.2.3 Solution of Coupled Partial Differential Equations

Many theoretical questions are still unanswered for the governing equations of the two sub-systems involved in the fluid-structure interaction, viz., the Navier-Stokes equations for fluid dynamics and the conservation equations for nonlinear hyperelastic materials. It is therefore, impractical to comment on the results of the coupled FSI system without a complete understanding of these sub-systems. The main mathematical problem spurs from the motion of the domains and the realization of the coupling conditions. For example, the incompressible Navier-Stokes equations are mixed-parabolic and nonlinear partial differential equations (PDEs) while the structural equations are hyperbolic PDEs. The notion of the kinematic coupling condition that ties velocities of fluid and structure together, is not well-posed. By the motion of the fluid domain, which follows the deflection of the structural domain, it is required to describe a smooth and well-defined interface where the fluid-structure coupling takes place.

The FSI simulations are generally accomplished by using either monolithic or partitioned schemes. A monolithic approach assembles the fluid and structural equations into a single block and solves them simultaneously for each iteration. These schemes lack the advantage of flexibility and modularity of using existing stable fluid or structural solvers. However, they offer good numerical stability even for problems involving very strong added mass effects. In contrast, a partitioned approach solves the fluid and structural equations in a sequential manner, facilitating the coupling of the existing fluid and structural program with minimal changes. This trait of the partitioned approach, therefore, makes it an attractive option from a computational point of view.

1.3 Book Organization

The book chapters are divided into the increasing order of complexity. The present chapter starts with an overview and motivation of fluid-structure interaction. It then provides an introduction to the physical understanding of the fluid-structure system.

Chapter 2 presents the necessary mathematical background for the following chapters in the form of a brief introduction. The chapter discusses the basics of continuum mechanics known as equilibrium, mechanical forces and kinematics, which deals with purely geometrical notions such as strain, rate of deformation, etc. Balance laws of fluid and structural mechanics are presented in the Eulerian and Lagrangian descriptions.

Chapter 3 will introduce the basic models of continuum mechanics and give an overview of different material laws used to describe solids and incompressible fluids. Given enough understanding of the two continuum models, we will then formulate the coupled models for fluid-structure interactions. Both problems are coupled by means of boundary conditions on the common fluid-structure interface.

Chapter 4 introduces stabilized finite element methods for steady and unsteady convection-dominated transport problems. The difficulties of Galerkin finite elements are first recognized and the stabilized finite element techniques are introduced to address node-to-node oscillations caused by the convection and reaction effects. Most of this chapter is generic and applies to a general system of coupled partial differential equations.

Chapter 5 extends the concepts of the previous chapter to systems of nonlinear coupled partial differential equations characterizing the fluid-structure interactions. The weak form, the semi-discrete formulation and the matrix form of the fluid-structure interactions are presented. The chapter closes with a discussion on the different approaches for modeling the FSI problems, i.e., monolithic and partitioned techniques.

Chapter 6 presents the first type of strategy in solving the fluid-structure interaction system, namely, monolithic techniques. As mentioned, one of the advantages of such techniques is the numerical stability of the body-fitted fluid-structure interface. Detailed numerical implementation of a quasi-monolithic formulation is presented together with numerical convergence and verification.

While Chap. 7 presents various spatial and temporal coupling techniques for partitioned FSI, Chap. 8 extends the FSI coupling to two-phase flows with structural dynamics. Chapter 9 is devoted to the flexible multibody aspects of structural dynamics and partitioned coupling issues. The governing differential equations of the fluid-flexible multibody solver based on the Navier-Stokes and the flexible multibody equations with constraints are discussed. Afterwards, the closure problem for turbulence based on the dynamic subgrid-scale model and the delayed detached eddy simulation is described in Chap. 10.

Chapter 2
Equilibrium, Kinematics and Balance Laws

When a fluid flows past or inside a structure, loads exerted by the fluid tend to change the configuration of the structure by inducing deformations and/or displacements. In turn, this change in structural configuration affects the dynamics of the fluid flow. The aforementioned intrinsic two-way communication between the structure and the fluid leads to the fluid-structure interaction which requires fundamental knowledge of the kinematics and equilibrium of continuous matter. In this chapter, we present the basic concepts of equilibrium, kinematics and the balance laws describing the dynamics of fluids and solid. We utilize the theory of continuum mechanics to characterize the physical phenomena of a macroscopic fluid and solid system, without taking into account the knowledge of detailed compositions of internal structures at molecular levels. While fluid and solid systems can be described by fundamental physical laws governing interactions of molecules and smaller particles, such an approach is not possible for large-scale fluid-structure systems (e.g., the aerodynamics of flexible wing and hydrodynamics effects around the ship) described in the previous chapter.

Using a continuum approach for the description of the large scale dynamics, we can describe the physical behaviors of fluid and solid mechanics. Needless to say, there are limitations to this approach when we consider phenomena at micro- or nano-scales. For a solid, liquid or gas, our fundamental assumption is that the material involved can be modeled as a continuum, without considering the atomic or molecular nature. This assumption can lead to very effective continuum models at length scales much longer than typical interatomic spacings. Nevertheless, it is no longer a valid assumption if the properties are at length scales comparable to or smaller than the interatomic spacings. Later in this chapter, we present the balance laws by requiring that the continuum domain comprising a fixed set of particles obeys the conservation of mass, linear and angular momentum. An important point to note is that the choice of this continuum domain is arbitrary and a set of differential equations of field variables based on conservation laws are defined over the domain. The variations of each field variable as a function of space and time represents the

R. K. Jaiman and V. Joshi, *Computational Mechanics of Fluid-Structure Interaction*,
https://doi.org/10.1007/978-981-16-5355-1_2

solution of these differential equations subjected to appropriate boundary and initial conditions.

The continuum assumption allows us to define a volume $\Omega(t)$ for a material body B in Euclidean space at any fixed instant in time t. Each particle of the material is identified by a point $\boldsymbol{x} \in \Omega(t)$. We only observe averaged properties of the volume instead of describing each particle. The regions of space occupied by the body B are termed as configurations, i.e., $\Omega(t)$ denotes a configuration of the body at a certain instant of time. Due to the action of mechanical forces, the body will displace and/or deform and will undergo various configurations as time progresses. Using continuum mechanics principles, we wish to review the fundamental concepts and equations that describe the interplay of these averaged quantities under the moving and deforming continua $\Omega(t)$ over time.

Let us first describe the notion of a mass density field, which allows to characterize the mass of an arbitrary part of a body. Based on the continuum assumption, the mass of a body B is assumed to be distributed uniformly throughout its volume. Furthermore, the mass of body tends to zero as the volume tends to zero and any subset of B with positive volume has positive mass. Let Mass[$\Omega(t)$] and Vol[$\Omega(t)$] denote the mass and volume of an arbitrary open subset $\Omega(t)$ of B. There exists a mass density field $\rho(\boldsymbol{x}) > 0$ per unit volume such that the mass and volume of the domain $\Omega(t)$ can be defined as

$$\text{Mass}[\Omega(t)] = \int_{\Omega(t)} \rho(\boldsymbol{x}) dV_x, \qquad \text{Vol}[\Omega(t)] = \int_{\Omega(t)} dV_x \tag{2.1}$$

Therefore, the definition of the mass density field $\rho(\boldsymbol{x})$ for an arbitrary point in Ω can be written as

$$\rho(\boldsymbol{x}) = \lim_{\delta \to 0} \frac{\text{Mass}\,[\Omega_\delta(t)]}{\text{Vol}\,[\Omega_\delta(t)]} \tag{2.2}$$

where $\Omega_\delta(t)$ represents a family of volumes with the properties that Vol $[\Omega_\delta(t)] \to 0$ as $\delta \to 0$ and $\boldsymbol{x} \in \Omega_\delta(t)$ for every $\delta > 0$. By the virtue of continuum assumption, the mass distribution in a continuum is that this limit exists and is positive for each $\boldsymbol{x}$ in $\Omega(t)$. Next we present the concepts of body and surface force fields over a continuum body.

2.1 Mechanical Forces and Equilibrium

Before we can proceed to the continuum equations of the fluid-structure system, it is necessary to discuss various types of forces on the continuum mass. Mechanical interactions between parts of a physical system (body) or between a body and its environment/surroundings are described by two basic types of forces: (i) body forces, which are exerted at the interior points of a body, and (ii) surface forces, which are exerted on internal surfaces between separate parts of a body, or external surfaces between the body and its environment.

2.1.1 *Body Forces*

Any force that does not arise due to physical contact between the bodies is referred to as body force. Such force is a result of action at a distance without any physical contact, for example, gravitational force. Body forces are uniformly distributed across the mass of the body and are proportional to the mass. Let $\boldsymbol{b}(\boldsymbol{x})$ denote a body force per unit mass acting on the body in the domain $\Omega(t)$. The resultant force on $\Omega(t)$ due to a body force field is defined as

$$f_b[\Omega(t)] = \int\limits_{\Omega(t)} \rho(\boldsymbol{x})\boldsymbol{b}(\boldsymbol{x})dV_{\boldsymbol{x}} \tag{2.3}$$

and the resultant torque corresponding to the body force about a point z can be given as

$$\boldsymbol{\tau}_b[\Omega(t)] = \int\limits_{\Omega(t)} (\boldsymbol{x} - z) \times \rho(\boldsymbol{x})\boldsymbol{b}(\boldsymbol{x})dV_{\boldsymbol{x}}. \tag{2.4}$$

Body forces can be conservative or non-conservative. Some examples of conservative forces are gravity, magnetic and electrostatic forces.

2.1.2 *Surface Forces*

In contrast to body forces, surface forces are generated during physical contact between the bodies. Surface forces along any imaginary surface within the interior of a body are termed as internal surface forces (shown in Fig. 2.1), and those along the bounding surface of the body are called external surface forces. Contact forces applied to the body by the surrounding environment can be treated as external surface forces. In the context of fluid-structure interaction, fluid forces can be considered as an external force on a wetted boundary of the deformable elastic body.

2.1.2.1 Concept of Traction Field

We discuss the concept of traction field to represent the surface forces in this section. Let $\boldsymbol{n}$ represent outward unit normal field of an arbitrary oriented surface Γ in $\Omega(t)$,

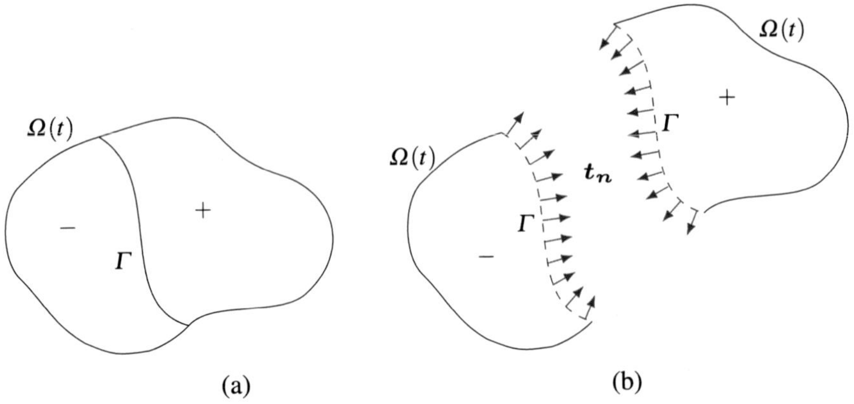

Fig. 2.1 Mechanical forces on a body: **a** body with domain $\Omega(t)$ with an imaginary internal surface, **b** internal surface forces (or tractions) along the imaginary surface

as shown in Fig. 2.1. At each point $\boldsymbol{x}$, the unit normal $\boldsymbol{n}(\boldsymbol{x})$ defines a positive side and a negative side of Γ. The force per unit area, exerted by material on the positive side on the material on the negative side is given by a function $\boldsymbol{t_n}$ which is referred as the traction or surface force field for Γ. When Γ is part of the bounding surface of $\Omega(t)$, we always choose $\boldsymbol{n}$ to be the outward unit normal field. The traction field $\boldsymbol{t_n}$ in this case represents the force per unit area, applied to the surface of $\Omega(t)$ by external forces.

Due to a traction field on an oriented surface Γ, the resultant force is defined as

$$\boldsymbol{f}_s[\Gamma] = \int_\Gamma \boldsymbol{t_n}(\boldsymbol{x}) dA_x \tag{2.5}$$

where dA_x represents an infinitesimal surface area element at $\boldsymbol{x} \in \Gamma$. Similarly, the resultant torque, about a point z, due to a traction field on Γ is defined as

$$\boldsymbol{\tau}_s[\Gamma] = \int_\Gamma (\boldsymbol{x} - z) \times \boldsymbol{t_n}(\boldsymbol{x}) dA_x. \tag{2.6}$$

The traction field $\boldsymbol{t_n}$ on a surface Γ in Ω depends only on the unit normal field $\boldsymbol{n}$. In particular, there is a function $\boldsymbol{t}$ such that

$$\boldsymbol{t_n}(\boldsymbol{x}) = \boldsymbol{t}(\boldsymbol{n}(\boldsymbol{x}), \boldsymbol{x}) \tag{2.7}$$

and the law of action and reaction can be written as

$$\boldsymbol{t}(-\boldsymbol{n}, \boldsymbol{x}) = -\boldsymbol{t}(\boldsymbol{n}, \boldsymbol{x}), \quad \forall \boldsymbol{x} \in \Omega(t). \tag{2.8}$$

2.1.2.2 The Stress Tensor

There exists a specific way in which the traction function $\boldsymbol{t}(\boldsymbol{n}, \boldsymbol{x})$ depends on $\boldsymbol{n}$. The following result is typically referred to as Cauchy's Theorem and will be of fundamental importance to the continuum formulation of fluid-structure interaction. Let $\boldsymbol{t}$ denote the traction function for a domain $\Omega(t)$. Then $\boldsymbol{t}(\boldsymbol{n}, \boldsymbol{x})$ is linear in $\boldsymbol{n}$, that is, for each $\boldsymbol{x} \in \Omega(t)$, there exists a second-order tensor $\boldsymbol{S}(\boldsymbol{x})$ such that

$$\boldsymbol{t}(\boldsymbol{n}, \boldsymbol{x}) = \boldsymbol{S}(\boldsymbol{x})\boldsymbol{n} \tag{2.9}$$

The field $\boldsymbol{S}$ is called the Cauchy stress field for $\Omega(t)$. The traction field on a surface with normal $\boldsymbol{n}(\boldsymbol{x})$ is

$$\boldsymbol{t}(\boldsymbol{n}(\boldsymbol{x}), \boldsymbol{x}) = \boldsymbol{S}(\boldsymbol{x})\boldsymbol{n}(\boldsymbol{x}). \tag{2.10}$$

In some places, we will abbreviate this relation by $\boldsymbol{t}(\boldsymbol{x}) = \boldsymbol{S}(\boldsymbol{x})\boldsymbol{n}$ or more simply $\boldsymbol{t} = \boldsymbol{S}\boldsymbol{n}$. In terms of the components in any coordinate frame with unit vectors $\{\boldsymbol{e}_i\}$ this reads $t_i = S_{ij}n_j$. The nine components of the stress tensor $\boldsymbol{S}(\boldsymbol{x})$ can be understood as the components of the three traction vectors $\boldsymbol{t}\left(\boldsymbol{e}_j, \boldsymbol{x}\right)$ on the coordinate planes at $\boldsymbol{x}$. In particular, substituting $\boldsymbol{t}\left(\boldsymbol{e}_j, \boldsymbol{x}\right) = t_i\left(\boldsymbol{e}_j, \boldsymbol{x}\right)\boldsymbol{e}_i$, we get $\boldsymbol{S}(\boldsymbol{x}) = S_{ij}(\boldsymbol{x})\boldsymbol{e}_i \otimes \boldsymbol{e}_j$ where $S_{ij}(\boldsymbol{x}) = t_i\left(\boldsymbol{e}_j, \boldsymbol{x}\right)$.

2.1.3 *Equilibrium*

We next provide the required conditions for a continuum body to be in mechanical equilibrium, i.e., at rest in a fixed frame for the Euclidean space. We also derive a local form of the conditions in terms of differential equations. Consider a continuum body in Euclidean space at rest denoted by the undeformed configuration $\widehat{\Omega}$ at a certain time instant as shown in Fig. 2.2a. Suppose the body is subjected to an external traction field $\boldsymbol{h}$ and a body force field $\boldsymbol{b}$ such that the body changes shape and comes to rest in a deformed configuration $\Omega(t)$ (Fig. 2.2b). Let the mass density field be denoted by $\rho(\boldsymbol{x})$. We implicitly assume that ρ, $\boldsymbol{h}$ and $\boldsymbol{b}$ do not depend on time.

Therefore, the resultant force on $\Omega(t)$ due to the body and the surface forces is

$$\boldsymbol{f}[\Omega(t)] = \boldsymbol{f}_b[\Omega(t)] + \boldsymbol{f}_s[\partial\Omega(t)] = \int_{\Omega(t)} \rho(\boldsymbol{x})\boldsymbol{b}(\boldsymbol{x})dV_x + \int_{\partial\Omega(t)} \boldsymbol{t}(\boldsymbol{x})dA_x \tag{2.11}$$

and the resultant torque on $\Omega(t)$, about a point z, due to corresponding forces is

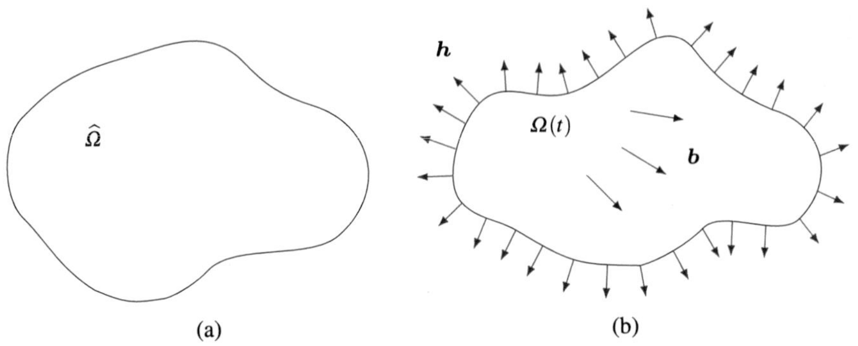

Fig. 2.2 Mechanical equilibrium of a continuum body: **a** undeformed configuration $\widehat{\Omega}$ of the body, and **b** deformed configuration $\Omega(t)$ under the action of body forces $\boldsymbol{b}$ and the surface tractions $\boldsymbol{h}$

$$\boldsymbol{\tau}[\Omega(t)] = \boldsymbol{\tau}_b[\Omega(t)] + \boldsymbol{\tau}_s[\partial\Omega(t)] \tag{2.12}$$

$$= \int_{\Omega(t)} (\boldsymbol{x} - \boldsymbol{z}) \times \rho(\boldsymbol{x})\boldsymbol{b}(\boldsymbol{x})dV_x + \int_{\partial\Omega(t)} (\boldsymbol{x} - \boldsymbol{z}) \times \boldsymbol{t}(\boldsymbol{x})dA_x \tag{2.13}$$

Necessary conditions for mechanical equilibrium can be stated as follows. If a body in a configuration $\Omega(t)$ is in mechanical equilibrium, then the resultant force and resultant torque about any fixed point, say the origin, must vanish for every open subset $\Omega(t)$ of the body B. That is

$$\boldsymbol{f}[\Omega(t)] = \int_{\Omega(t)} \rho(\boldsymbol{x})\boldsymbol{b}(\boldsymbol{x})dV_x + \int_{\partial\Omega(t)} \boldsymbol{t}(\boldsymbol{x})dA_x = \boldsymbol{0}, \qquad \forall\Omega(t) \subseteq B \tag{2.14}$$

$$\boldsymbol{\tau}[\Omega(t)] = \int_{\Omega(t)} \boldsymbol{x} \times \rho(\boldsymbol{x})\boldsymbol{b}(\boldsymbol{x})dV_x + \int_{\partial\Omega(t)} \boldsymbol{x} \times \boldsymbol{t}(\boldsymbol{x})dA_x = \boldsymbol{0}, \qquad \forall\Omega(t) \subseteq B \tag{2.15}$$

We next present a set of local equations corresponding to the equilibrium conditions. If the Cauchy stress field $\boldsymbol{S}$ is continuously differentiable, and the density field ρ and body force field $\boldsymbol{b}$ are continuous, then the equilibrium conditions are equivalent to

$$(\nabla \cdot \boldsymbol{S})(\boldsymbol{x}) + \rho(\boldsymbol{x})\boldsymbol{b}(\boldsymbol{x}) = \boldsymbol{0}, \qquad \forall\boldsymbol{x} \in \Omega(t) \tag{2.16}$$

$$\boldsymbol{S}^T(\boldsymbol{x}) = \boldsymbol{S}(\boldsymbol{x}), \qquad \forall\boldsymbol{x} \in \Omega(t) \tag{2.17}$$

or in components

$$S_{ij,j}(\boldsymbol{x}) + \rho(\boldsymbol{x})b_i(\boldsymbol{x}) = \boldsymbol{0}, \qquad \forall\boldsymbol{x} \in \Omega(t) \tag{2.18}$$

$$S_{ij}(\boldsymbol{x}) = S_{ji}(\boldsymbol{x}), \qquad \forall\boldsymbol{x} \in \Omega(t). \tag{2.19}$$

2.2 Kinematics

For a continuum, kinematics involves the quantities and results that can be found without reference to the dynamics of the continuum. Kinematics is the study of the geometry of motion without considering the influence of mass, force and stress. In this section, we present the notion of deformation of a body under the action of forces and a deformation map.

2.2.1 *Configurations and Deformations*

At any instant of time, a material body occupies an open subset $\widehat{\Omega}$ of Euclidean space $\mathbb{E}^3$. The identification of material particles with points of $\widehat{\Omega}$ defines the configuration of the body at that instant. Subjected to forces, the body in the initial or undeformed configuration $\widehat{\Omega}$ can undergo a change in the configuration to a deformed configuration $\Omega(t)$. By convention, we call $\widehat{\Omega}$ as the reference configuration and $\Omega(t)$ as the deformed configuration. We denote points in $\widehat{\Omega}$ by $\boldsymbol{X} = (X_1, X_2, X_3)^T$ and points in $\Omega(t)$ by $\boldsymbol{x} = (x_1, x_2, x_3)^T$. Since $\widehat{\Omega}$ and $\Omega(t)$ are two configurations of a single material body B, each particle in the body has two sets of coordinates: a set of material coordinates X_i for its location in $\widehat{\Omega}$, and a set of spatial coordinates x_i for its location in $\Omega(t)$. While the description in terms of the spatial variable $\boldsymbol{x}$ is called as the Eulerian formulation, the description in terms of the reference variable $\boldsymbol{X}$ is referred to as the Lagrangian formulation.

2.2.2 *Deformation Map*

The deformation of a body from a configuration $\widehat{\Omega}$ onto another configuration $\Omega(t)$ is described by a function $\boldsymbol{\varphi} : \widehat{\Omega} \rightarrow \Omega(t)$ which maps each point $\boldsymbol{X} \in \widehat{\Omega}$ to a point $\boldsymbol{x} = \boldsymbol{\varphi}(\boldsymbol{X}, t) \in \Omega(t)$ as shown in Fig. 2.3. The function $\boldsymbol{\varphi}(\boldsymbol{X}, t)$ is called the deformation map relative to the reference configuration $\widehat{\Omega}$ and this map is orientation-preserving and one-to-one. Physically, the deformation map being one-to-one means that the continuum body should not penetrate itself. Owing to the one-to-one property of the mapping $\boldsymbol{\varphi}$, we can invert it to obtain $\boldsymbol{X}$ as a function of $\boldsymbol{x}$ and t as $\boldsymbol{X} = \boldsymbol{\varphi}^{-1}(\boldsymbol{x}, t)$ for $\boldsymbol{x} \in \Omega(t)$. Thus, the motion of the body is essentially a smooth one-parameter family of deformations with time as the parameter.

As mentioned, the map $\boldsymbol{\varphi}$ satisfies certain conditions in order for it to represent the deformation of the material body, i.e., orientation preservation and being one-to-one. Mathematically, the orientation preservation property is written as $\det(\widehat{\nabla}\boldsymbol{\varphi}(\boldsymbol{X}, t)) = \det(\boldsymbol{F}) > 0$ for all $\boldsymbol{X} \in \widehat{\Omega}$, where $\det(\cdot)$ is the determinant operation, $\widehat{\nabla}$ denotes the gradient evaluated at the material coordinates $\boldsymbol{X}$ and $\boldsymbol{F} = \widehat{\nabla}\boldsymbol{\varphi}(\boldsymbol{X}, t)$ is called the deformation gradient of the mapping $\boldsymbol{\varphi}$. This ensures that the body cannot be con-

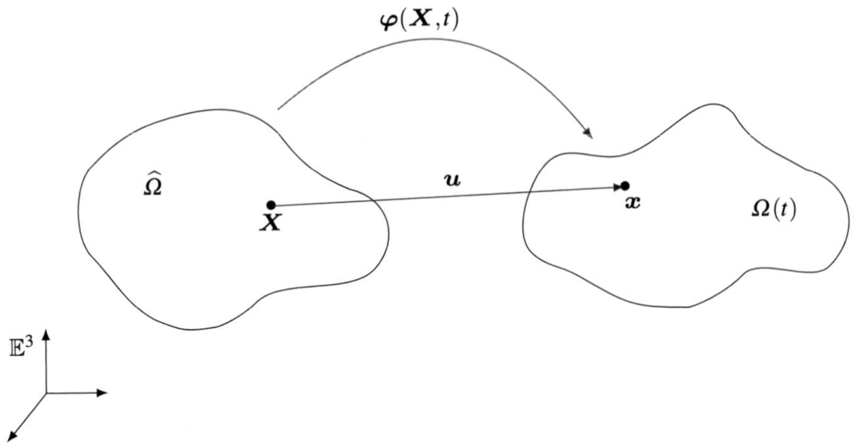

Fig. 2.3 The reference or undeformed configuration of the body $\widehat{\Omega}$ and the deformed configuration $\Omega(t)$ under the action of forces are mapped by a deformation map denoted by $\boldsymbol{\varphi}(\boldsymbol{X}, t)$

tinuously deformed to its mirror image. The one-to-one property of the function $\boldsymbol{\varphi}$ implies that two or more distinct points from $\widehat{\Omega}$ cannot simultaneously occupy the same position in $\Omega(t)$. Deformations satisfying these conditions are called admissible. Since it is both one-to-one and onto, an admissible deformation $\boldsymbol{\varphi}$ is a bijection between $\widehat{\Omega}$ and $\Omega(t)$.

The displacement of a material particle from its initial coordinates $\boldsymbol{X}$ to its final location $\boldsymbol{x}$ is given by

$$\boldsymbol{u}(\boldsymbol{X}, t) = \boldsymbol{\varphi}(\boldsymbol{X}, t) - \boldsymbol{X}, \tag{2.20}$$

where $\boldsymbol{u}$ is the displacement field associated with the mapping $\boldsymbol{\varphi}$. A natural way to quantify strain of the deformation is with the help of second-order tensor field $\boldsymbol{F}$ known as the deformation gradient, defined by

$$\boldsymbol{F} = \widehat{\nabla}\boldsymbol{\varphi}(\boldsymbol{X}, t). \tag{2.21}$$

The deformation gradient field provides the information about the local behavior of the deformation $\boldsymbol{\varphi}$. From Eq. (2.20), it can be seen that $\boldsymbol{F} = \widehat{\nabla}\boldsymbol{u} + \boldsymbol{I}$. Therefore, it follows that $\boldsymbol{F}$ in the reference configuration $\widehat{\Omega}$ is an identity tensor.

Another measure of strain is provided by the right Cauchy-Green strain tensor $\boldsymbol{C}$, which is defined as

$$\boldsymbol{C} = \boldsymbol{F}^T \boldsymbol{F}. \tag{2.22}$$

Notice that $\boldsymbol{C}$ is symmetric and positive-definite at each point in $\widehat{\Omega}$.

Strain can also be quantified by the infinitesimal strain tensor field $\boldsymbol{E}$ associated with $\boldsymbol{\varphi}$ defined as

$$\boldsymbol{E} = \mathrm{sym}(\widehat{\nabla}\boldsymbol{u}) = \frac{1}{2}\left(\widehat{\nabla}\boldsymbol{u} + (\widehat{\nabla}\boldsymbol{u})^T\right) \tag{2.23}$$

Notice that, by definition, $\boldsymbol{E}$ is symmetric at each point in $\widehat{\Omega}$. The tensor $\boldsymbol{E}$ is related to the deformation gradient $\boldsymbol{F}$ and the right Cauchy-Green tensor $\boldsymbol{C}$ using Eqs. (2.20), (2.21) and (2.22) as

$$\boldsymbol{E} = \mathrm{sym}(\boldsymbol{F} - \boldsymbol{I}) = \frac{1}{2}(\boldsymbol{C} - \boldsymbol{I}) - \frac{1}{2}(\widehat{\nabla}\boldsymbol{u})^T\widehat{\nabla}\boldsymbol{u}. \tag{2.24}$$

The tensor $\boldsymbol{E}$ is particularly useful in the case of small deformations. We say that a deformation $\boldsymbol{\varphi}$ is small if there is a number $0 \le \epsilon \ll 1$ such that $|\widehat{\nabla}\boldsymbol{u}| = \mathcal{O}(\epsilon)$ or equivalently $\partial u_i/\partial X_j = \mathcal{O}(\epsilon)$, for all points $\boldsymbol{X} \in \widehat{\Omega}$. Thus, if the norm of displacement gradient is small at all the points on the body, the deformation is said to be small. In this scenario, the higher order gradients can be neglected in Eq. (2.24) as

$$\boldsymbol{E} = \frac{1}{2}(\boldsymbol{C} - \boldsymbol{I}) + \mathcal{O}\left(\epsilon^2\right). \tag{2.25}$$

2.3 Motion Kinematics

In kinematics, we investigate the motion and deformation of bodies. Motion is defined as the continuous deformation of a body over the course of time. The motion of a body with reference domain $\widehat{\Omega}$ is described by a continuous map $\boldsymbol{\varphi}$, where for each fixed $t \ge 0$, the function $\boldsymbol{\varphi}(\boldsymbol{X}, t)$ is a deformation of $\widehat{\Omega}$. At any time $t \ge 0$, the deformation $\boldsymbol{x} = \boldsymbol{\varphi}(\boldsymbol{X}, t)$ maps the reference domain $\widehat{\Omega}$ onto a domain $\Omega(t)$, which is referred as the current or deformed configuration at time t (Fig. 2.4).

We can also define an inverse deformation $\boldsymbol{\psi}(\boldsymbol{x}, t) = \boldsymbol{\varphi}^{-1}(\boldsymbol{x}, t)$ which gives the mapping from $\Omega(t)$ to $\widehat{\Omega}$ such that

$$\boldsymbol{X} = \boldsymbol{\psi}(\boldsymbol{x}, t). \tag{2.26}$$

By properties of inverse functions, for any $t \ge 0$ we have

$$\boldsymbol{X} = \boldsymbol{\psi}(\boldsymbol{x}, t) = \boldsymbol{\psi}(\boldsymbol{\varphi}(\boldsymbol{X}, t), t) \quad \forall \boldsymbol{X} \in \widehat{\Omega} \tag{2.27}$$

and similarly

$$\boldsymbol{x} = \boldsymbol{\varphi}(\boldsymbol{X}, t) = \boldsymbol{\varphi}(\boldsymbol{\psi}(\boldsymbol{x}, t), t) \quad \forall \boldsymbol{x} \in \Omega(t). \tag{2.28}$$

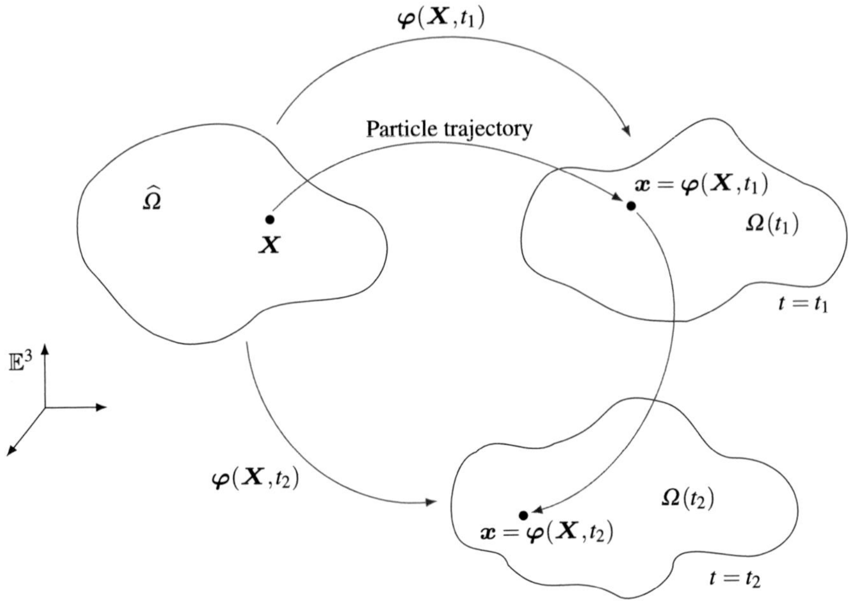

Fig. 2.4 Motion of a body with time denoted by the change in the reference configuration $\widehat{\Omega}$ via a deformation map $\varphi(X, t)$ which maps the undeformed configuration $\widehat{\Omega}$ to the deformed configuration $\Omega(t)$ at time t

For studying the motion of continuum bodies, we consider the current configuration $\Omega(t)$ whose points are defined by x. Since $x = \varphi(X, t)$, any function of $x \in \Omega(t)$ can also be expressed as a function of $X \in \widehat{\Omega}$. Similarly, there are the fields on the reference configuration $\widehat{\Omega}$ whose points are labeled by X. However, since $X = \psi(x, t)$, any function of $X \in \widehat{\Omega}$ can also be expressed as a function of $x \in \Omega(t)$. To help keep track of where a field was originally defined, and how it is currently being expressed, we introduce the following definitions. A field expressed in terms of the material points X of $\widehat{\Omega}$ is referred as a material field. On the other hand, a field which is expressed in terms of the spatial points x of $\Omega(t)$ is referred as a spatial field. The notations used in the parenthesis for the coordinates, for example, x or X will indicate whether the field is associated with the spatial or material fields respectively.

2.3.1 Spatial or Coordinate Derivatives

While dealing with different configurations, it is essential to distinguish between the material coordinates $X = (X_1, X_2, X_3)^T$ which label points in $\widehat{\Omega}$ and the spatial coordinates $x = (x_1, x_2, x_3)^T$, which label points in the current configuration $\Omega(t)$.

The symbol $\widehat{\nabla}$ is used to denote the gradient, divergence and curl of material fields with respect to the material coordinates X_i for any fixed time $t \geq 0$. Similarly, we use the symbol ∇ to denote the gradient, divergence and curl of spatial fields with respect to the spatial coordinates x_i for any fixed time $t \geq 0$. By extension, we define the Laplacian operators $\widehat{\Delta} = \widehat{\nabla} \cdot (\widehat{\nabla})$ and $\Delta = \nabla \cdot (\nabla)$.

2.3.2 *Time Derivatives*

The total time derivative of a field is defined as the rate of change of the field as measured by an observer who is tracking the motion of each particle in the body. The change in the spatial coordinates with time is given by the motion

$$\boldsymbol{x} = \boldsymbol{\varphi}(\boldsymbol{X}, t). \tag{2.29}$$

The total time derivative is denoted by an overdot over the symbol of the field. Let $\phi(\boldsymbol{X}, t)$ denote a material field. Its total time derivative will be given as

$$\dot{\phi}(\boldsymbol{X}, t) = \frac{\partial}{\partial t}\phi(\boldsymbol{X}, t). \tag{2.30}$$

On the other hand, as $\boldsymbol{x} = \boldsymbol{\varphi}(\boldsymbol{X}, t)$, the total time derivative of any spatial field $\tilde{\phi}(\boldsymbol{x}, t)$ is given by

$$\dot{\tilde{\phi}}(\boldsymbol{x}, t) = \left[\frac{\partial}{\partial t}\tilde{\phi}(\boldsymbol{\varphi}(\boldsymbol{X}, t), t)\right]\Bigg|_{\boldsymbol{X}=\boldsymbol{\psi}(\boldsymbol{x},t)} = \left[\dot{\tilde{\phi}}(\boldsymbol{X}, t)\right]\Bigg|_{\boldsymbol{X}=\boldsymbol{\psi}(\boldsymbol{x},t)}, \tag{2.31}$$

where $\tilde{\phi}(\boldsymbol{X}, t)$ denotes the material description of the spatial field $\tilde{\phi}(\boldsymbol{x}, t)$.

2.3.3 *Velocity and Acceleration Fields*

Continuing the concept of the material and spatial domains, we define the velocity and acceleration fields in this section. Let $\boldsymbol{v}(\boldsymbol{X}, t)$ denote the velocity at time t of the material particle labeled by $\boldsymbol{X}$ in $\widehat{\Omega}$. By definition of the motion $\boldsymbol{x} = \boldsymbol{\varphi}(\boldsymbol{X}, t)$, we can write the velocity in the material field as

$$\boldsymbol{v}(\boldsymbol{X}, t) = \frac{\partial}{\partial t}\boldsymbol{\varphi}(\boldsymbol{X}, t). \tag{2.32}$$

Similarly, the acceleration of the material particle can be denoted by $\boldsymbol{a}(\boldsymbol{X}, t)$. By definition of the motion, we have

$$a(X, t) = \frac{\partial^2}{\partial t^2}\varphi(X, t). \tag{2.33}$$

From the above definitions we see that the velocity and acceleration of material particles are naturally material fields. Frequently, however, we will need the spatial descriptions of these fields. We denote by $v(x, t)$ the spatial description of the material velocity field $v(X, t)$, which is given by

$$v(x, t) = \left[\frac{\partial}{\partial t}\varphi(X, t)\right]\bigg|_{X=\psi(x,t)} \tag{2.34}$$

and $a(x, t)$, which represents the spatial description of the material acceleration field $a(X, t)$ is given as

$$a(x, t) = \left[\frac{\partial^2}{\partial t^2}\varphi(X, t)\right]\bigg|_{X=\psi(x,t)}. \tag{2.35}$$

For evaluating the total time derivative of a generic scalar or vector field, let φ be a motion of a continuum body with associated spatial velocity field v, and consider an arbitrary spatial scalar field $\phi = \phi(x, t)$ and an arbitrary spatial vector field $w = w(x, t)$. Then, the total time derivatives of ϕ and w are given by

$$\dot{\phi} = \frac{\partial}{\partial t}\phi + v \cdot \nabla\phi \quad \text{and} \quad \dot{w} = \frac{\partial}{\partial t}w + v \cdot \nabla w. \tag{2.36}$$

2.4 Rate of Strain

The strain tensor represents a basic quantity for a continuum solid body assuming that a finite force leads to a finite deformation. However, in fluid flow, finite forces may give rise to infinite deformation. Therefore, in this scenario, the deformation and the deformation gradient are not of interest. It is the temporal variation of them that serves as a key quantity to model the internal forces (stresses) of the fluids. It has been mentioned that for fluid flow observations, Eulerian description is more meaningful. Thus, the measure of rate of strain is defined in the spatial coordinate system.

Let φ be a motion of a continuum fluid or solid system, and let $\Omega(t)$ and $\Omega(t')$ denote the configurations at times $t \geq 0$ and $t' \geq t$ respectively. Moreover, let Ω be a small ball of radius $\alpha > 0$ centered at a point x in $\Omega(t)$, and let Ω' be the corresponding region in $\Omega(t')$. Any quantitative measure of the rate of change of shape between Ω' and Ω in the limits $\alpha \to 0$ and $t' \to t$ is generally called rate of strain at x and t. Similarly, any quantitative measure of the rate of change of orientation of Ω' with respect to Ω is called rate of rotation or spin. Note that, in contrast to the concept of strain, the concepts of rate of strain and spin are independent of the reference configuration $\widehat{\Omega}$. For this reason, rate of strain and spin play a central role in the study of fluids.

Let $\boldsymbol{\varphi}$ be a motion of a continuum body with spatial velocity field $\boldsymbol{v}$. Then a measure of rate of strain is given by the tensor

$$\boldsymbol{L} = \text{sym}\,(\nabla \boldsymbol{v}) = \frac{1}{2}\left(\nabla \boldsymbol{v} + (\nabla \boldsymbol{v})^T\right). \tag{2.37}$$

Notice that, by definition, $\boldsymbol{L}$ is a spatial field and is symmetric for each point $\boldsymbol{x}$ in $\Omega(t)$ and time t. To any spatial velocity field $\boldsymbol{v}$, another tensor field associated with it can be defined by

$$\boldsymbol{W} = \text{skew}\,(\nabla \boldsymbol{v}) = \frac{1}{2}\left(\nabla \boldsymbol{v} - (\nabla \boldsymbol{v})^T\right), \tag{2.38}$$

where $\boldsymbol{W}$ is called as the spin tensor field associated with $\boldsymbol{v}$. By definition, $\boldsymbol{W}$ is a spatial field and is skew-symmetric for each point $\boldsymbol{x}$ in $\Omega(t)$.

2.5 Rigid Body Motions

For structural bodies undergoing small deformation (Eq. 2.25), the displacement at any point can be decomposed into a rigid body displacement and an elastic displacement. For example, in flexible multibody system, this assumption of small deformation helps to model the multiple bodies connected with joints, where they undergo small deformations although there can be large relative motions at the joints. Thus, it is essential to understand the concept of rigid body motion to formulate such complex multibody systems. Here, we will briefly summarize some of the concepts related to multibody formulation.

A motion $\boldsymbol{\varphi}$ is called rigid if

$$\boldsymbol{\varphi}(\boldsymbol{X}, t) = \boldsymbol{c}(t) + \boldsymbol{R}(t)\boldsymbol{X} \tag{2.39}$$

for some time-dependent vector $\boldsymbol{c}(t)$ and rotation tensor $\boldsymbol{R}(t)$. For such motions, we can show that the spatial velocity field can be written in the form

$$\boldsymbol{v}(\boldsymbol{x}, t) = \boldsymbol{\omega}(t) \times (\boldsymbol{x} - \boldsymbol{c}(t)) + \dot{\boldsymbol{c}}(t) \tag{2.40}$$

where $\boldsymbol{\omega}(t)$ is a time-dependent vector called the spatial angular velocity of the motion.

2.5.1 Geometric Description of Rotation

As mentioned earlier, for small deformations, the displacement can be decomposed into rigid body motion and elastic motion. Here, we discuss the concepts of rotation tensor for rigid body kinematics. Consider an inertial coordinate system $Ox_1x_2x_3$ with orthonormal basis $(\boldsymbol{i}_1, \boldsymbol{i}_2, \boldsymbol{i}_3)$ and a rotated coordinate system $AX_1X_2X_3$ with orthonormal basis $(\boldsymbol{e}_1, \boldsymbol{e}_2, \boldsymbol{e}_3)$. A rotation is the linear transformation which maps the basis $(\boldsymbol{i}_1, \boldsymbol{i}_2, \boldsymbol{i}_3)$ to $(\boldsymbol{e}_1, \boldsymbol{e}_2, \boldsymbol{e}_3)$, i.e.,

$$\boldsymbol{e}_j = \boldsymbol{D}\boldsymbol{i}_j, \tag{2.41}$$

where $\boldsymbol{D}$ is the direction cosine matrix with elements given by $D_{jk} = \boldsymbol{i}_j \cdot \boldsymbol{e}_k$. Note that rotations do not commute. The order in which the rotation operations are performed is crucial.

Now, any rotation between the two bases depicted above can be represented by three successive rotations in a plane about the three different axes, where these three rotation angles are called the Euler angles. There are many combinations of selecting the Euler angles based on the order of rotation, however the fact remains that any rotation (or the direction cosine matrix) can be represented in terms of three independent parameters only. Some of the drawbacks of Euler angles are that they can be defined in different manners depending on the order of rotation, and singularities can occur in evaluating these angles (as they are based on trigonometric functions). This leads to the method of parameterization of rotation operation based on Euler's theorem.

Euler's theorem states that any arbitrary rotation of a rigid body can be represented as a single rotation of magnitude ϕ about a unit vector $\widehat{\boldsymbol{n}}$ passing through a fixed point on the body. This is depicted in Fig. 2.5. Based on the theorem, a rotation tensor $\boldsymbol{R}$, which relates the rotation operation on a vector $\boldsymbol{a}$ to $\boldsymbol{b}$ as $\boldsymbol{b} = \boldsymbol{R}\boldsymbol{a}$ can be written in a simplified form using the rotation angle ϕ about the unit vector $\widehat{\boldsymbol{n}}$ as the Rodrigues' rotation formula:

$$\boldsymbol{R} = \boldsymbol{I} + (1 - \cos\phi)\widetilde{\boldsymbol{n}}\widetilde{\boldsymbol{n}} + \sin\phi\widetilde{\boldsymbol{n}}, \tag{2.42}$$

where $\widetilde{\boldsymbol{n}}$ is a skew-symmetric tensor known as the spin of $\widehat{\boldsymbol{n}}$, i.e.,

$$\widetilde{\boldsymbol{n}} = \begin{pmatrix} 0 & -\widehat{n}_3 & \widehat{n}_2 \\ \widehat{n}_3 & 0 & -\widehat{n}_1 \\ -\widehat{n}_2 & \widehat{n}_1 & 0 \end{pmatrix} \tag{2.43}$$

Since the orthonormal basis $(\boldsymbol{e}_1, \boldsymbol{e}_2, \boldsymbol{e}_3)$ were expressed in the inertial coordinate system in Eq. (2.41), the direction cosine matrix $\boldsymbol{D}$ is the same as the rotation matrix $\boldsymbol{R}$ evaluated in the inertial frame for the rotation of the bases.

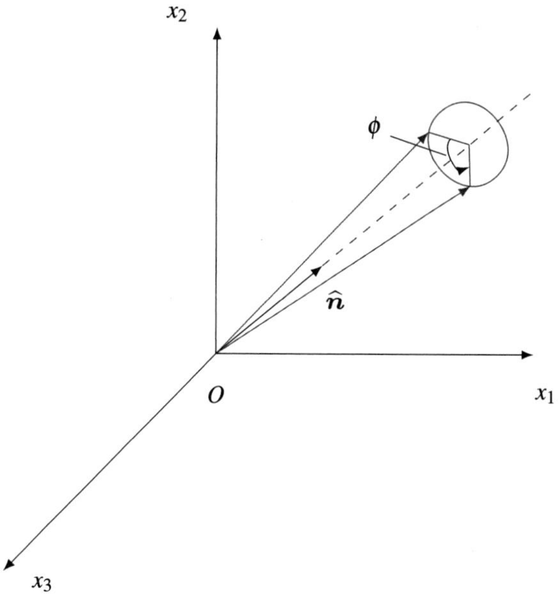

Fig. 2.5 Euler's theorem: any arbitrary rotation of a rigid body can be represented as a single rotation about a unit vector $\widehat{\boldsymbol{n}}$ passing through a fixed point on the body

2.5.2 *Parameterization of Rotation Matrix*

The rotation matrix can be parameterized based on either non-vectorial or vectorial parameterizations. The vectorial parameterizations consist of a minimal set of three parameters such as Cartesian rotation vector, Euler-Rodrigues' or Weiner-Milenković parameters, while the parameters in non-vectorial parameterizations may be minimal or redundant like three in case of Euler angles, four in Euler parameters, Cayley rotation parameters and nine in direction cosines.

2.5.2.1 Non-vectorial Parameterization

1. Euler parameters: Euler parameters are defined by the four parameters:

$$e_0 = \cos\frac{\phi}{2}, \qquad \boldsymbol{e} = \widehat{\boldsymbol{n}}\sin\frac{\phi}{2}, \tag{2.44}$$

   which gives the expression for the rotation tensor as

$$\boldsymbol{R} = \boldsymbol{I} + 2e_0\widetilde{\boldsymbol{e}} + 2\widetilde{\boldsymbol{e}}\widetilde{\boldsymbol{e}}. \tag{2.45}$$

2. Cayley rotation parameters: The rotation tensor can be expressed as a skew-symmetric tensor $\widehat{\boldsymbol{C}}$ as $\boldsymbol{R} = (\boldsymbol{I} - \widehat{\boldsymbol{C}})^{-1}(\boldsymbol{I} + \widehat{\boldsymbol{C}}) = (\boldsymbol{I} + \widehat{\boldsymbol{C}})(\boldsymbol{I} - \widehat{\boldsymbol{C}})^{-1}$, where $\widehat{\boldsymbol{C}} = \widetilde{\boldsymbol{a}}$ and is given in terms of the parameters of $\widetilde{\boldsymbol{a}}$ as

$$\boldsymbol{R} = \frac{1}{1 + \boldsymbol{a}^T \boldsymbol{a}} \left((1 + \boldsymbol{a}^T \boldsymbol{a}) \boldsymbol{I} + 2\widetilde{a} + 2\widetilde{a}\widetilde{a} \right). \tag{2.46}$$

2.5.2.2 Vectorial Parameterizations

Let the vector parameterization of rotation of angle ϕ about the normal $\widehat{\boldsymbol{n}}$ be defined as $\boldsymbol{p} = p(\phi)\widehat{\boldsymbol{n}}$. The rotation tensor is thus given by Rodrigues' rotation formula,

$$\boldsymbol{R} = \boldsymbol{I} + \xi_1(\phi)\widetilde{\boldsymbol{p}} + \xi_2(\phi)\widetilde{\boldsymbol{p}}\widetilde{\boldsymbol{p}}, \tag{2.47}$$

where $\xi_1(\phi)$ and $\xi_2(\phi)$ are even functions of the rotation angle defined as,

$$\xi_1(\phi) = \frac{\sin\phi}{p(\phi)}, \tag{2.48}$$

$$\xi_2(\phi) = \frac{1 - \cos\phi}{(p(\phi))^2}, \tag{2.49}$$

Based on the choice of the generating function $p(\phi)$, different variants of the parameterization can be obtained.

1. Cartesian rotation vector: $p(\phi) = \phi$.
2. Linear parameterization: $p(\phi) = \sin\phi$.
3. Cayley/Gibbs/Rodrigues' parameterization: $p(\phi) = 2\tan(\phi/2)$.
4. Reduced Euler-Rodrigues' parameterization: $p(\phi) = 2\sin(\phi/2)$.
5. Weiner-Milenković parameterization: $p(\phi) = 4\tan(\phi/4)$

The rotation matrix $\boldsymbol{R}$ parameterized by Wiener-Milenković parameters, which is a technique based on conformal transformation on Euler parameters, also known as conformal rotation vector (CRV) [149, 239] avoids singularities for particular values of rotation and a minimal set of parameters (three in this case) also avoids redundancies in the description [19, 74].

2.6 Change of Variables

Let $\boldsymbol{\varphi}$ be a motion of a continuum body, and for any time $t \geq 0$ consider the deformation $\boldsymbol{\varphi}(\boldsymbol{X}, t)$ which maps the reference configuration $\widehat{\Omega}$ onto the current configuration $\Omega(t)$. Formulations for fluid-structure systems require transformation of integrals defined over the regions in $\Omega(t)$ to those corresponding to $\widehat{\Omega}$. Similarly, there is also a need to transform integrals defined over surfaces in $\Omega(t)$ to integrals defined over corresponding surfaces in $\widehat{\Omega}$. This section discusses such transformations in volume and surface integrals.

2.6.1 Transformation of Volume Integrals

To begin, consider an infinitesimal volume element dV_X at an arbitrary point $\boldsymbol{X}$ in $\widehat{\Omega}$ which can be represented in terms of a triple scalar product. Let $d\boldsymbol{X}_1, d\boldsymbol{X}_2$ and $d\boldsymbol{X}_3$ be the three infinitesimal vectors based at the point $\boldsymbol{X}$. The infinitesimal volume element defined by these vectors will be

$$dV_X = (d\boldsymbol{X}_1 \times d\boldsymbol{X}_2) \cdot d\boldsymbol{X}_3 \tag{2.50}$$

The three infinitesimal vectors $d\boldsymbol{X}_k$ at $\boldsymbol{X}$ are mapped to the three infinitesimal vectors $d\boldsymbol{x}_k$ at $\boldsymbol{x} = \boldsymbol{\varphi}(\boldsymbol{X}, t)$ in $\Omega(t)$ by the deformation map $\boldsymbol{\varphi}(\boldsymbol{X}, t)$, as shown in Fig. 2.6. In particular, we have

$$d\boldsymbol{x}_k = \widehat{\nabla}\boldsymbol{\varphi}(\boldsymbol{X}, t) d\boldsymbol{X}_k = \boldsymbol{F}(\boldsymbol{X}, t) d\boldsymbol{X}_k \tag{2.51}$$

In the deformed configuration $\Omega(t)$, the infinitesimal vectors $d\boldsymbol{x}_k$ can be used to define a volume element dV_x via the triple scalar product as

$$dV_x = (d\boldsymbol{x}_1 \times d\boldsymbol{x}_2) \cdot d\boldsymbol{x}_3 \tag{2.52}$$
$$= (\boldsymbol{F}(\boldsymbol{X}, t) d\boldsymbol{X}_1 \times \boldsymbol{F}(\boldsymbol{X}, t) d\boldsymbol{X}_2) \cdot \boldsymbol{F}(\boldsymbol{X}, t) d\boldsymbol{X}_3 \tag{2.53}$$
$$= \det(\boldsymbol{F}(\boldsymbol{X}, t)) \, (d\boldsymbol{X}_1 \times d\boldsymbol{X}_2) \cdot d\boldsymbol{X}_3 \tag{2.54}$$
$$= \det(\boldsymbol{F}(\boldsymbol{X}, t)) dV_X. \tag{2.55}$$

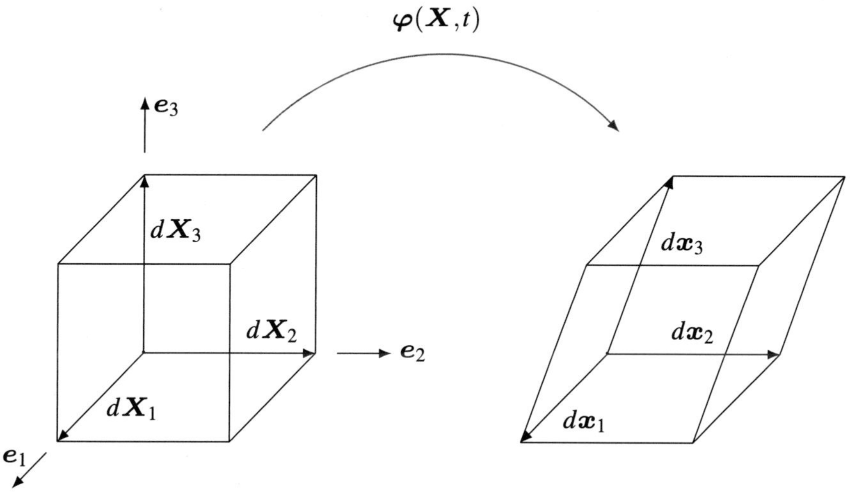

Fig. 2.6 Transformation of infinitesimal volume element due to deformation map $\boldsymbol{\varphi}(\boldsymbol{X}, t)$ from $\widehat{\Omega}$ to $\Omega(t)$

Thus, the volume integrals an be transformed from the spatial configuration $\Omega(t)$ to the reference configuration $\widehat{\Omega}$ with the help of the deformation gradient $\boldsymbol{F}(\boldsymbol{X}, t)$.

To summarize the transformation of volume integrals, let $\phi(\boldsymbol{x}, t)$ be an arbitrary spatial scalar field on $\Omega(t)$, under the deformation map $\boldsymbol{\varphi}(\boldsymbol{X}, t)$. Then,

$$\int_{\Omega(t)} \phi(\boldsymbol{x}, t) dV_{\boldsymbol{x}} = \int_{\widehat{\Omega}} \phi(\boldsymbol{X}, t) \det(\boldsymbol{F}(\boldsymbol{X}, t)) dV_{\boldsymbol{X}}, \tag{2.56}$$

where $\phi(\boldsymbol{X}, t)$ denotes the material description of the scalar field.

The field defined by $J(\boldsymbol{X}, t) = \det(\boldsymbol{F}(\boldsymbol{X}, t))$ is generally called the Jacobian field for the deformation $\boldsymbol{\varphi}(\boldsymbol{X}, t)$. It is a measure of volumetric strain caused by a deformation $\boldsymbol{\varphi}(\boldsymbol{X}, t)$ at the point $\boldsymbol{X}$ at time t. The deformation compresses or expands the material in a neighborhood of $\boldsymbol{X}$ if and only if $J(\boldsymbol{X}, t) < 1$ (volume is decreased) or $J(\boldsymbol{X}, t) > 1$ (volume is increased), respectively. When $J(\boldsymbol{X}, t) = 1$ there is no change in the material volume in the neighborhood of $\boldsymbol{X}$. In this case, the material near $\boldsymbol{X}$ is distorted in such a way that its volume is preserved. For example, it can be compressed along certain directions, and expanded along others. Notice that $J(\boldsymbol{X}, t) > 0$ for any admissible deformation.

Let us now define the time derivative of the Jacobian field. Considering $\boldsymbol{\varphi}(\boldsymbol{X}, t)$ as the motion of the continuum body with spatial velocity field $\boldsymbol{v}(\boldsymbol{x}, t)$ and the deformation gradient field given as $\boldsymbol{F}(\boldsymbol{X}, t)$, the time derivative of the Jacobian can be written as

$$\frac{\partial}{\partial t}(\det(\boldsymbol{F}(\boldsymbol{X}, t))) = \det(\boldsymbol{F}(\boldsymbol{X}, t)) \, (\nabla \cdot \boldsymbol{v}) \, (\boldsymbol{x}, t)|_{\boldsymbol{x}=\boldsymbol{\varphi}(\boldsymbol{X},t)} \,. \tag{2.57}$$

It can be inferred that the time derivative of the Jacobian depends on the Jacobian field itself and on the divergence of the spatial velocity field. This is helpful in application to the Reynolds Transport Theorem which involves the time derivative of the volume integrals which can be stated as follows: Let $\boldsymbol{\varphi}(\boldsymbol{X}, t)$ be a motion of a continuum body with associated spatial velocity field $\boldsymbol{v}(\boldsymbol{x}, t)$ and let $\Omega(t)$ be an arbitrary volume with surface denoted by $\partial\Omega(t)$ and $\boldsymbol{n}$ represents the outward unit normal field on $\partial\Omega(t)$. Then for any spatial scalar field $\Phi(\boldsymbol{x}, t)$, we have

$$\frac{d}{dt} \int_{\Omega(t)} \Phi dV_{\boldsymbol{x}} = \int_{\Omega(t)} \dot{\Phi} + \Phi \, (\nabla \cdot \boldsymbol{v}) \, dV_{\boldsymbol{x}} \tag{2.58}$$

or equivalently,

$$\frac{d}{dt} \int_{\Omega(t)} \Phi dV_{\boldsymbol{x}} = \int_{\Omega(t)} \frac{\partial}{\partial t} \Phi dV_{\boldsymbol{x}} + \int_{\partial\Omega(t)} \Phi \boldsymbol{v} \cdot \boldsymbol{n} dA_{\boldsymbol{x}}. \tag{2.59}$$

2.6.2 Transformation of Surface Integrals

In this subsection, we discuss the transformation of the surface integrals across the material and spatial domains. Consider Γ as a surface in the reference configuration $\widehat{\Omega}$ with a unit normal to the surface denoted by $\boldsymbol{N}$. Under the deformation map $\boldsymbol{\varphi}(\boldsymbol{X}, t)$, let the surface deform to $\Gamma(t)$ in $\Omega(t)$ with the unit normal denoted by $\boldsymbol{n}$. We are interested in finding the transformation of an infinitesimal area element dA_X at point $\boldsymbol{X}$ in the reference configuration under the deformation map $\boldsymbol{\varphi}(\boldsymbol{X}, t)$.

Further, we assume that the surface Γ is regular and we can describe a system of surface coordinates on Γ near a given point $\mathbf{Z}$. Thus, a region D and a corresponding map $\boldsymbol{\chi}$ can be found such that for any point $\boldsymbol{X}$ on Γ near $\boldsymbol{Z}$, $\boldsymbol{X} = \boldsymbol{\chi}(\boldsymbol{\xi})$,where $\boldsymbol{\xi} = (\xi_1, \xi_2)^T \in D$, as shown in Fig. 2.7. The infinitesimal area element dA_X and the normal associated with it $\boldsymbol{N}$ can be described based on the surface coordinate mapping $\boldsymbol{\xi}$ as

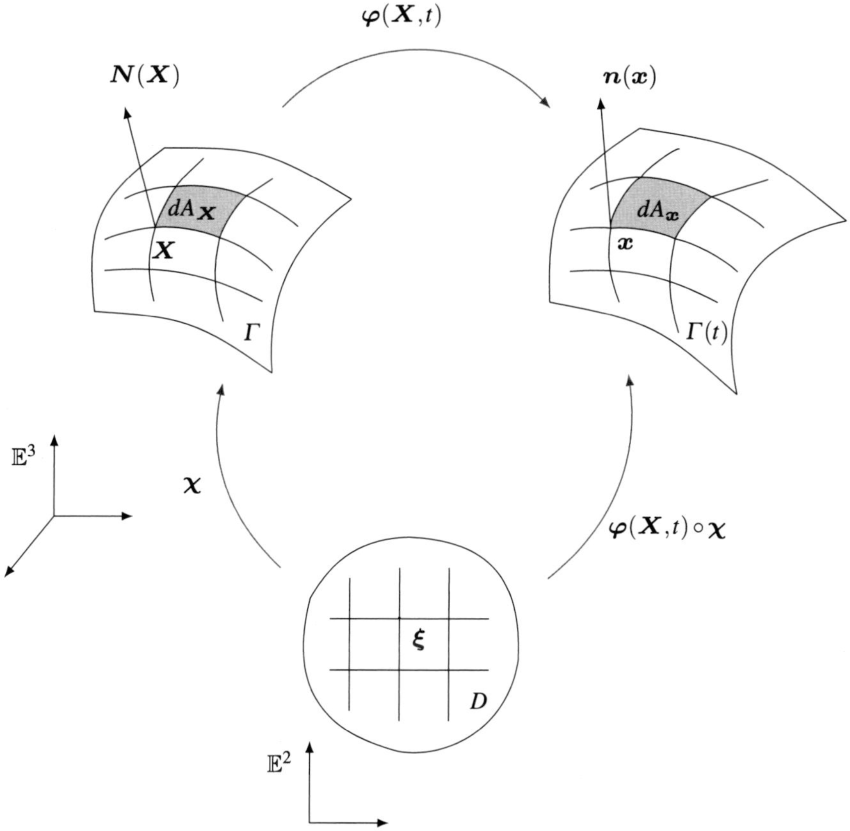

Fig. 2.7 Transformation of surface element due to deformation map $\boldsymbol{\varphi}(\boldsymbol{X}, t)$ from Γ to $\Gamma(t)$

$$N(X)dA_X = \frac{\partial}{\partial \xi_1}\chi(\xi) \times \frac{\partial}{\partial \xi_2}\chi(\xi)d\xi_1 d\xi_2. \tag{2.60}$$

In the deformed configuration $\Omega(t)$, the surface area gets transformed to $\Gamma(t)$ and surface coordinates become $z = \varphi(Z, t)$. Thus, any point x on $\Gamma(t)$ near z can be expressed as $x = \varphi(\chi(\xi), t)$, for $\xi = (\xi_1, \xi_2)^T \in D$. Similarly, the infinitesimal area element in the deformed surface can be expressed as

$$n(x)dA_x = \frac{\partial}{\partial \xi_1}\varphi(\chi(\xi), t) \times \frac{\partial}{\partial \xi_2}\varphi(\chi(\xi), t)d\xi_1 d\xi_2. \tag{2.61}$$

Furthermore,

$$\frac{\partial}{\partial \xi_1}\varphi(\chi(\xi), t) = \frac{\partial}{\partial \xi_1}\varphi^i(\chi(\xi), t)e_i \tag{2.62}$$

$$= \left[\frac{\partial}{\partial X_j}\varphi^i(X, t)\right]\Bigg|_{X=\chi(\xi)} \frac{\partial}{\partial \xi_1}\chi_j(\xi)e_i \tag{2.63}$$

$$= F(\chi(\xi), t)\frac{\partial}{\partial \xi_1}\chi(\xi), \tag{2.64}$$

and a similar expression can be written for $\frac{\partial}{\partial \xi_2}\varphi(\chi(\xi), t)$. With the help of the identity

$$Fu \times Fv = (\det(F))F^{-T}(u \times v) \tag{2.65}$$

which is true for any invertible second-order tensor F and vectors u and v, Eq. (2.61) can be written as

$$n(x)dA_x = \left[F(\chi(\xi), t)\frac{\partial}{\partial \xi_1}\chi(\xi)\right] \times \left[F(\chi(\xi), t)\frac{\partial}{\partial \xi_2}\chi(\xi)\right] d\xi_1 d\xi_2 \tag{2.66}$$

$$= \left[\det(F(\chi(\xi), t))\right] F(\chi(\xi), t)^{-T}\left[\frac{\partial}{\partial \xi_1}\chi(\xi) \times \frac{\partial}{\partial \xi_2}\chi(\xi)\right] d\xi_1 d\xi_2 \tag{2.67}$$

Using Eq. (2.60), we can thus write

$$n(x)dA_x = (\det(F(X, t)))\, F(X, t)^{-T} N(X)dA_X. \tag{2.68}$$

The above relation helps to transform the surface integrals between the reference and the deformed configurations. Let $\phi(x, t)$, $w(x, t)$ and $T(x, t)$ be an arbitrary spatial scalar field, vector field, and second-order tensor field, respectively. Consider a deformation mapping from $\widehat{\Omega}$ to $\Omega(t)$ denoted by $\varphi(X, t)$. Let the outward normal for the surface $\partial\widehat{\Omega}$ of $\widehat{\Omega}$ be denoted by $N(X)$ and the outward normal for the surface $\partial\Omega(t)$ in the deformed configuration be denoted by $n(x)$. Then, the following relations can be written:

$$\int_{\partial\Omega(t)} \phi(\boldsymbol{x},t)\boldsymbol{n}(\boldsymbol{x})dA_x = \int_{\partial\widehat{\Omega}} \phi(\boldsymbol{X},t)\boldsymbol{G}(\boldsymbol{X},t)\boldsymbol{N}(\boldsymbol{X})dA_X \tag{2.69}$$

$$\int_{\partial\Omega(t)} \boldsymbol{w}(\boldsymbol{x},t)\cdot\boldsymbol{n}(\boldsymbol{x})dA_x = \int_{\partial\widehat{\Omega}} \boldsymbol{w}(\boldsymbol{X},t)\cdot\boldsymbol{G}(\boldsymbol{X},t)\boldsymbol{N}(\boldsymbol{X})dA_X \tag{2.70}$$

$$\int_{\partial\Omega(t)} \boldsymbol{T}(\boldsymbol{x},t)\boldsymbol{n}(\boldsymbol{x})dA_x = \int_{\partial\widehat{\Omega}} \boldsymbol{T}(\boldsymbol{X},t)\boldsymbol{G}(\boldsymbol{X},t)\boldsymbol{N}(\boldsymbol{X})dA_X, \tag{2.71}$$

where $\boldsymbol{G}$ is a material second-order tensor field given as

$$\boldsymbol{G}(\boldsymbol{X},t) = (\det(\boldsymbol{F}(\boldsymbol{X},t)))\boldsymbol{F}(\boldsymbol{X},t)^{-T}. \tag{2.72}$$

We can further define the conditions for volume preservation. Let $\boldsymbol{\varphi}(\boldsymbol{X},t)$ be a motion of a continuum body with spatial velocity field $\boldsymbol{v}(\boldsymbol{x},t)$ and deformation gradient field $\boldsymbol{F}(\boldsymbol{X},t)$. Then $\boldsymbol{\varphi}(\boldsymbol{X},t)$ is volume-preserving if and only if $\det(\boldsymbol{F}(\boldsymbol{X},t)) = 1$ for all $\boldsymbol{X} \in \widehat{\Omega}$ and $t \geq 0$, or equivalently $(\nabla\cdot\boldsymbol{v})(\boldsymbol{x},t) = 0$ for all $\boldsymbol{x} \in \Omega(t)$ and $t \geq 0$.

2.7 Balance Laws in Lagrangian Form

In this section, we apply the concepts of the transformations in stating the balance laws in the material configuration, which is also known as Lagrangian form. We will limit ourselves to the conservation of mass and linear momentum, which form the core laws for the fluid-structure interactions. Given a continuum body with reference configuration $\widehat{\Omega}$ undergoing a deformation given by $\boldsymbol{\varphi}(\boldsymbol{X},t)$ to the current or deformed configuration $\Omega(t)$, any balance law which has been stated in the spatial coordinates $\boldsymbol{x} \in \Omega(t)$ can be expressed in terms of the material coordinates $\boldsymbol{X} \in \widehat{\Omega}$ by the change of variables. This results in the material or Lagrangian form of the balance laws.

2.7.1 Conservation of Mass

Consider an arbitrary configuration of the body at time t as $\Omega(t)$ with the reference configuration as $\widehat{\Omega}$ under the deformation $\boldsymbol{\varphi}(\boldsymbol{X},t)$. According to the conservation of mass, the mass of the continuum does not change during the deformation, i.e.,

$$\text{Mass}\,[\Omega(t)] = \text{Mass}[\widehat{\Omega}]. \tag{2.73}$$

Let $\boldsymbol{F}(\boldsymbol{X},t)$ denote the deformation gradient associated with the deformation $\boldsymbol{\varphi}(\boldsymbol{X},t)$. Consider $\rho(\boldsymbol{x},t)$ as the spatial mass density field in $\Omega(t)$ and let $\rho_0(\boldsymbol{X})$

represent the mass density field in the reference configuration $\widehat{\Omega}$. The conservation of mass requires that

$$\rho(\boldsymbol{X}, t)\det(\boldsymbol{F}(\boldsymbol{X}, t)) = \rho_0(\boldsymbol{X}), \quad \forall \boldsymbol{X} \in \widehat{\Omega}, t \geq 0, \tag{2.74}$$

where $\rho(\boldsymbol{X}, t)$ denotes the material description of the spatial field $\rho(\boldsymbol{x}, t)$.

2.7.2 Conservation of Linear Momentum

The law for conservation of linear momentum for the deformed configuration $\Omega(t)$ can be stated as the rate of change of linear momentum for the continuum is equal to the resultant external forces acting on it, i.e.,

$$\frac{d}{dt}\boldsymbol{l}[\Omega(t)] = \boldsymbol{f}[\Omega(t)] \tag{2.75}$$

where $\boldsymbol{l}[\Omega(t)]$ and $\boldsymbol{f}[\Omega(t)]$ are the linear momentum and the resultant external force of $\Omega(t)$, respectively and they are defined as

$$\boldsymbol{l}[\Omega(t)] = \int_{\Omega(t)} \rho(\boldsymbol{x}, t)\boldsymbol{v}(\boldsymbol{x}, t)dV_x, \tag{2.76}$$

$$\boldsymbol{f}[\Omega(t)] = \int_{\Omega(t)} \rho(\boldsymbol{x}, t)\boldsymbol{b}(\boldsymbol{x}, t)dV_x + \int_{\partial\Omega(t)} \boldsymbol{S}(\boldsymbol{x}, t)\boldsymbol{n}(\boldsymbol{x})dA_x. \tag{2.77}$$

Performing a change of variable in the integral in Eq. (2.76) to the material coordinates and using Eq. (2.74), we get

$$\boldsymbol{l}[\Omega(t)] = \int_{\Omega(t)} \rho(\boldsymbol{x}, t)\boldsymbol{v}(\boldsymbol{x}, t)dV_x \tag{2.78}$$

$$= \int_{\widehat{\Omega}} \rho(\boldsymbol{\varphi}(\boldsymbol{X}, t), t)\boldsymbol{v}(\boldsymbol{\varphi}(\boldsymbol{X}, t), t)\det(\boldsymbol{F}(\boldsymbol{X}, t))dV_X \tag{2.79}$$

$$= \int_{\widehat{\Omega}} \rho_0(\boldsymbol{X})\dot{\boldsymbol{\varphi}}(\boldsymbol{X}, t)dV_X \tag{2.80}$$

Similarly, the expression for the external forces can be written as

$$f[\Omega(t)] = \int_{\Omega(t)} \rho(\boldsymbol{x},t)\boldsymbol{b}(\boldsymbol{x},t)dV_x + \int_{\partial\Omega(t)} \boldsymbol{S}(\boldsymbol{x},t)\boldsymbol{n}(\boldsymbol{x})dA_x \tag{2.81}$$

$$= \int_{\widehat{\Omega}} \det(\boldsymbol{F}(\boldsymbol{X},t))\rho(\boldsymbol{\varphi}(\boldsymbol{X},t),t)\boldsymbol{b}(\boldsymbol{\varphi}(\boldsymbol{X},t),t)dV_X \tag{2.82}$$

$$+ \int_{\partial\widehat{\Omega}} (\det(\boldsymbol{F}(\boldsymbol{X},t)))\boldsymbol{S}(\boldsymbol{\varphi}(\boldsymbol{X},t),t)\boldsymbol{F}(\boldsymbol{X},t)^{-T}\boldsymbol{N}(\boldsymbol{X})dA_X, \tag{2.83}$$

where we have used the transformation of the surface integral. The external force consists of the body forces on $\Omega(t)$ and the external surface forces on the bounding surface $\partial\Omega(t)$. Here, let $\boldsymbol{S}(\boldsymbol{x},t)$ be the Cauchy stress field in $\Omega(t)$ associated with $\boldsymbol{\varphi}(\boldsymbol{X},t)$. We define the nominal or first Piola-Kirchhoff stress field $\boldsymbol{P}(\boldsymbol{X},t)$ as

$$\boldsymbol{P}(\boldsymbol{X},t) = (\det(\boldsymbol{F}(\boldsymbol{X},t)))\boldsymbol{S}(\boldsymbol{X},t)\boldsymbol{F}(\boldsymbol{X},t)^{-T}, \quad \forall \boldsymbol{X} \in \widehat{\Omega}, t \geq 0, \tag{2.84}$$

where $\boldsymbol{S}(\boldsymbol{X},t)$ denotes the material description of the stress field $\boldsymbol{S}(\boldsymbol{x},t)$. Furthermore, the second Piola-Kirchhoff stress field can be written as

$$\boldsymbol{\Sigma}(\boldsymbol{X},t) = \boldsymbol{F}(\boldsymbol{X},t)^{-1}\boldsymbol{P}(\boldsymbol{X},t), \quad \forall \boldsymbol{X} \in \widehat{\Omega}, t \geq 0. \tag{2.85}$$

Let $\boldsymbol{b}(\boldsymbol{X},t)$ denote the material description of the spatial body force $\boldsymbol{b}(\boldsymbol{x},t)$, then the resultant forces can be expressed as

$$f[\Omega(t)] = \int_{\widehat{\Omega}} \rho_0(\boldsymbol{X})\boldsymbol{b}(\boldsymbol{X},t)dV_X + \int_{\partial\widehat{\Omega}} \boldsymbol{P}(\boldsymbol{X},t)\boldsymbol{N}(\boldsymbol{X})dA_X. \tag{2.86}$$

From Eq. (2.75), we obtain

$$\frac{d}{dt}\int_{\widehat{\Omega}} \rho_0(\boldsymbol{X})\dot{\boldsymbol{\varphi}}(\boldsymbol{X},t)dV_X = \int_{\widehat{\Omega}} \rho_0(\boldsymbol{X})\boldsymbol{b}(\boldsymbol{X},t)dV_X + \int_{\partial\widehat{\Omega}} \boldsymbol{P}(\boldsymbol{X},t)\boldsymbol{N}(\boldsymbol{X})dA_X. \tag{2.87}$$

With the help of Divergence theorem and the fact that $\widehat{\Omega}$ is independent of time we get

$$\int_{\widehat{\Omega}} \rho_0(\boldsymbol{X})\ddot{\boldsymbol{\varphi}}(\boldsymbol{X},t)dV_X = \int_{\widehat{\Omega}} \left[\left(\widehat{\nabla}\cdot\boldsymbol{P}\right)(\boldsymbol{X},t) + \rho_0(\boldsymbol{X})\boldsymbol{b}(\boldsymbol{X},t)\right]dV_X. \tag{2.88}$$

There exists a bijection between $\widehat{\Omega}$ and $\Omega(t)$, any balance law which holds for $\Omega(t)$ must also hold for an arbitrary $\widehat{\Omega}$. Hence the conservation of linear momentum in the Lagrangian form can be written as

$$\rho_0 \ddot{\boldsymbol{\varphi}} = \widehat{\nabla} \cdot \boldsymbol{P} + \rho_0 \boldsymbol{b}, \quad \forall \boldsymbol{X} \in \widehat{\Omega}, t \geq 0. \tag{2.89}$$

We can also define the conservation of angular momentum in the Lagrangian form, which can be mathematically written as

$$\boldsymbol{P}\boldsymbol{F}^T = \boldsymbol{F}\boldsymbol{P}^T \quad \text{or equivalently} \quad \boldsymbol{\Sigma}^T = \boldsymbol{\Sigma}, \quad \forall \boldsymbol{X} \in \widehat{\Omega}, t \geq 0 \tag{2.90}$$

where $\boldsymbol{P}(\boldsymbol{X}, t)$ is the first and $\boldsymbol{\Sigma}(\boldsymbol{X}, t)$ is the second Piola-Kirchhoff stress field associated with $\boldsymbol{\varphi}(\boldsymbol{X}, t)$.

2.7.3 Application to Deformable Solid Models

In this section, we study the application of the Lagrangian form of the balance laws to model the structural equations. Various constitutive models relating the stress $\boldsymbol{P}(\boldsymbol{X}, t)$ to the deformation gradient $\boldsymbol{F} = \widehat{\nabla}\boldsymbol{\varphi}$ are reviewed, which describe the behavior of the structure. Material stresses can be expressed in terms of either first Piola-Kirchhoff stress $\boldsymbol{P}(\boldsymbol{X}, t)$, the Cauchy stress $\boldsymbol{S}(\boldsymbol{X}, t)$ or the second Piola-Kirchhoff stress $\boldsymbol{\Sigma}(\boldsymbol{X}, t)$, depending on the convenience. Note that each one of the stress can be expressed in terms of the others. Here, we discuss two classes of solids, viz., elastic and hyperelastic.

2.7.3.1 Elastic Solid

Consider a continuum body with reference material configuration $\widehat{\Omega}$ undergoing a deformation expressed by the deformation tensor $\boldsymbol{F}(\boldsymbol{X}, t)$. The structure is said to have elastic behavior if:

- The Cauchy stress field is of the form

$$\boldsymbol{S}(\boldsymbol{X}, t) = \widehat{\boldsymbol{S}}(\boldsymbol{F}(\boldsymbol{X}, t), \boldsymbol{X}), \quad \forall \boldsymbol{X} \in \widehat{\Omega}, t \geq 0 \tag{2.91}$$

  where $\widehat{\boldsymbol{S}}(\cdot)$ is a function called a stress response function for the Cauchy stress $\boldsymbol{S}(\boldsymbol{X}, t)$. This implies that the stress at a point in an elastic solid depends only on a measure of strain at that point. The stress is independent of the past history of the material deformation and the rate of strain. This can be interpreted as a generalization of Hooke's Law. It is often called a stress-strain relationship.
- The function $\widehat{\boldsymbol{S}}(\cdot)$ has the property

$$\widehat{\boldsymbol{S}}(\boldsymbol{F}, \boldsymbol{X})^T = \widehat{\boldsymbol{S}}(\boldsymbol{F}, \boldsymbol{X}), \quad \forall \boldsymbol{X} \in \widehat{\Omega}, \det(\boldsymbol{F}) > 0. \tag{2.92}$$

  This implies that the Cauchy stress field is necessarily symmetric. Thus, the conservation of angular momentum is automatically satisfied. For any admissible motion,

$\det(\boldsymbol{F}(\boldsymbol{X},t)) > 0$, thus $\widehat{\boldsymbol{S}}(\boldsymbol{F},\boldsymbol{X})$ has been considered for arguments $\boldsymbol{F}$ satisfying $\det(\boldsymbol{F}) > 0$.

Let $\rho_0(\boldsymbol{X})$ denote the mass density of an elastic body in its reference material configuration $\widehat{\Omega}$ and let $\boldsymbol{b}(\boldsymbol{X},t)$ represent the material description of a spatial body force field per unit mass $\boldsymbol{b}(\boldsymbol{x},t)$ that is $\boldsymbol{b}(\boldsymbol{X},t) = \boldsymbol{b}(\boldsymbol{\varphi}(\boldsymbol{X},t),t)$. Let $\widehat{\boldsymbol{P}}(\boldsymbol{F})$ be the stress response function for the first Piola-Kirchhoff stress tensor $\boldsymbol{P}(\boldsymbol{X},t)$. Thus, setting $\boldsymbol{P}(\boldsymbol{X},t) = \widehat{\boldsymbol{P}}(\boldsymbol{F})$ in the balance of linear momentum equation, the motion of the body can be modeled by the governing equation for all $\boldsymbol{X} \in \widehat{\Omega}$ and $t \geq 0$:

$$\rho_0\ddot{\boldsymbol{\varphi}} = \widehat{\nabla}\cdot(\widehat{\boldsymbol{P}}(\boldsymbol{F})) + \rho_0\boldsymbol{b}. \tag{2.93}$$

This is known as the Elastodynamics equation. Along with the boundary and initial conditions, the Lagrangian description of the structure can be summarized by the conservation of linear momentum equation as

$$\rho_0\ddot{\boldsymbol{\varphi}} = \widehat{\nabla}\cdot[\widehat{\boldsymbol{P}}(\boldsymbol{F})] + \rho_0\boldsymbol{b} \qquad \text{on } \widehat{\Omega}\times[0,T] \tag{2.94}$$

$$\boldsymbol{\varphi} = \boldsymbol{g} \qquad \text{on } \Gamma_D\times[0,T] \tag{2.95}$$

$$\widehat{\boldsymbol{P}}(\boldsymbol{F})\boldsymbol{N} = \boldsymbol{h} \qquad \text{on } \Gamma_N\times[0,T] \tag{2.96}$$

$$\boldsymbol{\varphi}(\cdot,0) = \boldsymbol{X} \qquad \text{on } \widehat{\Omega} \tag{2.97}$$

$$\dot{\boldsymbol{\varphi}}(\cdot,0) = \boldsymbol{v}_0 \qquad \text{on } \widehat{\Omega} \tag{2.98}$$

where, Γ_D and Γ_N denote the Dirichlet and Neumann boundaries, which are subsets of boundary of the domain $\partial\widehat{\Omega}$ such that $\Gamma_D \cup \Gamma_N = \partial\widehat{\Omega}$ and $\Gamma_D \cap \Gamma_N = \emptyset$, ρ_0 is the reference mass density, and $\boldsymbol{b}(\boldsymbol{X},t)$ is the material description of a specified spatial body force per unit mass.

2.7.3.2 Hyperelastic Solid

A continuum body with reference configuration $\widehat{\Omega}$ has hyperelastic behavior if:

- The structure is elastic with stress response functions $\widehat{\boldsymbol{P}}(\boldsymbol{F})$, $\widehat{\boldsymbol{S}}(\boldsymbol{F})$, and $\widehat{\boldsymbol{\Sigma}}(\boldsymbol{F})$ for the stress fields $\boldsymbol{P}(\boldsymbol{X},t)$, $\boldsymbol{S}(\boldsymbol{X},t)$, and $\boldsymbol{\Sigma}(\boldsymbol{X},t)$, respectively. This implies that in any motion, the various stress fields can be related to the corresponding stress response functions. For the first Piola-Kirchhoff stress field, $\boldsymbol{P}(\boldsymbol{X},t) = \widehat{\boldsymbol{P}}(\boldsymbol{F}(\boldsymbol{X},t))$.
- The response function for the first Piola-Kirchhoff stress field, $\widehat{\boldsymbol{P}}(\boldsymbol{F})$ can be written as

$$\widehat{\boldsymbol{P}}(\boldsymbol{F}) = \frac{\partial\mathscr{W}(\boldsymbol{F})}{\partial\boldsymbol{F}}, \tag{2.99}$$

  where $\mathscr{W}(\boldsymbol{F})$ is a given function called as strain energy density for the body. This relates the response function $\widehat{\boldsymbol{P}}(\boldsymbol{F})$ with the derivative of a scalar strain energy function $\mathscr{W}(\boldsymbol{F})$.

- The function $\mathscr{W}(\boldsymbol{F})$ has the property

$$\frac{\partial \mathscr{W}(\boldsymbol{F})}{\partial \boldsymbol{F}}\boldsymbol{F}^T = \boldsymbol{F}\frac{\partial \mathscr{W}(\boldsymbol{F})^T}{\partial \boldsymbol{F}}, \quad \det(\boldsymbol{F}) > 0. \tag{2.100}$$

For the stress field to be frame independent, the strain energy function $\mathscr{W}(\boldsymbol{F})$ can be represented as $\mathscr{W}(\boldsymbol{F}) = \overline{\mathscr{W}}(\boldsymbol{C})$, where $\boldsymbol{C} = \boldsymbol{F}^T\boldsymbol{F}$ is the right Cauchy-Green strain tensor. The response function $\widehat{\boldsymbol{P}}(\boldsymbol{F})$, $\widehat{\boldsymbol{S}}(\boldsymbol{F})$, and $\widehat{\boldsymbol{\Sigma}}(\boldsymbol{F})$ in terms of $\overline{\mathscr{W}}(\boldsymbol{C})$ can be written as

$$\widehat{\boldsymbol{P}}(\boldsymbol{F}) = 2\boldsymbol{F}\frac{\partial \overline{\mathscr{W}}(\boldsymbol{C})}{\partial \boldsymbol{C}}, \tag{2.101}$$

$$\widehat{\boldsymbol{S}}(\boldsymbol{F}) = 2(\det \boldsymbol{C})^{-1/2}\boldsymbol{F}\frac{\partial \overline{\mathscr{W}}(\boldsymbol{C})}{\partial \boldsymbol{C}}\boldsymbol{F}^T, \tag{2.102}$$

$$\widehat{\boldsymbol{\Sigma}}(\boldsymbol{F}) = 2\frac{\partial \overline{\mathscr{W}}(\boldsymbol{C})}{\partial \boldsymbol{C}}. \tag{2.103}$$

For a hyperelastic solid, the relation between strain and stress comes from a strain energy density function. Using the strain energy density function with phenomenological and mechanistic arguments, various types of hyperelastic models can be constructed. The constitutive expressions for $\overline{\mathscr{W}}(\boldsymbol{C})$ for some representative hyperelastic models are summarized below:

1. St. Venant-Kirchhoff model:

$$\overline{\mathscr{W}}(\boldsymbol{C}) = \frac{\lambda}{2}(\mathrm{tr}(\boldsymbol{E}))^2 + \mu\,\mathrm{tr}(\boldsymbol{E}^2) \tag{2.104}$$

$$\boldsymbol{E} = \frac{1}{2}(\boldsymbol{C} - \boldsymbol{I}), \quad \lambda, \mu > 0. \tag{2.105}$$

As can be easily shown, the first Piola-Kirchhoff stress tensor can be written as

$$\widehat{\boldsymbol{P}}(\boldsymbol{F}) = \lambda(\mathrm{tr}(\boldsymbol{E}))\boldsymbol{F} + 2\mu\boldsymbol{F}\boldsymbol{E}. \tag{2.106}$$

2. Neo-Hookean model:

$$\overline{\mathscr{W}}(\boldsymbol{C}) = a\,\mathrm{tr}(\boldsymbol{C}) + \Gamma(\sqrt{\det(\boldsymbol{C})}), \tag{2.107}$$

where

$$\Gamma(s) = cs^2 - d\ln s, \quad a, c, d > 0 \tag{2.108}$$

3. Mooney-Rivlin model:

$$\overline{\mathscr{W}}(\boldsymbol{C}) = a\,\mathrm{tr}(\boldsymbol{C}) + b\,\mathrm{tr}(\boldsymbol{C}_*) + \Gamma(\sqrt{\det(\boldsymbol{C})}) \tag{2.109}$$

$$\boldsymbol{C}_* = (\det(\boldsymbol{C}))\boldsymbol{C}^{-1}. \quad a, b > 0 \tag{2.110}$$

4. Ogden model:

$$\overline{\mathscr{W}}(\boldsymbol{C}) = \sum_{i=1}^{M} a_i\,\mathrm{tr}(\boldsymbol{C}^{\gamma_i/2}) + \sum_{j=1}^{N} b_j\,\mathrm{tr}(\boldsymbol{C}_*^{\delta_j/2}) + \Gamma(\sqrt{\det(\boldsymbol{C})}), \tag{2.111}$$

where $M, N \geq 1.\ \ a_i, b_j > 0.\ \ \gamma_i, \delta_j \geq 1.$

2.8 Balance Laws in Eulerian Form

Next, we focus on the Eulerian description of the balance laws, where we work with the current or deformed configuration $\Omega(t)$ consisting of spatial coordinates $\boldsymbol{x}$. We review the conservation of mass and linear momentum as they apply to any open subset of a continuum body.

2.8.1 *Conservation of Mass*

Let $\boldsymbol{\varphi}(\boldsymbol{X}, t)$ denote the motion of a continuum body with associated spatial velocity field $\boldsymbol{v}(\boldsymbol{x}, t)$ and spatial mass density field $\rho(\boldsymbol{x}, t)$. The conservation of mass requires

$$\frac{\partial}{\partial t}\rho + \nabla \cdot (\rho \boldsymbol{v}) = 0, \quad \forall \boldsymbol{x} \in \Omega(t), t \geq 0 \tag{2.112}$$

Equivalently, by definition of the total time derivative, we have

$$\dot{\rho} + \rho \nabla \cdot \boldsymbol{v} = 0, \quad \forall \boldsymbol{x} \in \Omega(t), t \geq 0 \tag{2.113}$$

2.8.2 *Conservation of Linear Momentum*

The conservation of linear momentum in the spatial configuration $\Omega(t)$ can be written as

$$\frac{d}{dt}\int_{\Omega(t)} \rho(\boldsymbol{x}, t)\boldsymbol{v}(\boldsymbol{x}, t) dV_x = \int_{\Omega(t)} \rho(\boldsymbol{x}, t)\boldsymbol{b}(\boldsymbol{x}, t) dV_x + \int_{\partial\Omega(t)} \boldsymbol{S}(\boldsymbol{x}, t)\boldsymbol{n}(\boldsymbol{x}) dA_x \tag{2.114}$$

where $\rho(\boldsymbol{x}, t)$ is the spatial mass density field, $\boldsymbol{v}(\boldsymbol{x}, t)$ is the spatial velocity field, $\boldsymbol{b}(\boldsymbol{x}, t)$ is the spatial body force field per unit mass and $\boldsymbol{S}(\boldsymbol{x}, t)$ is the Cauchy stress tensor field on the surface $\partial\Omega(t)$ in $\Omega(t)$.

Thus, we obtain the following relationship:

$$\rho\dot{\boldsymbol{v}} = \nabla \cdot \boldsymbol{S} + \rho\boldsymbol{b}, \quad \forall\boldsymbol{x} \in \Omega(t), t \geq 0 \tag{2.115}$$

We have the total derivative

$$\dot{\boldsymbol{v}} = \frac{\partial}{\partial t}\boldsymbol{v} + \boldsymbol{v} \cdot \nabla\boldsymbol{v}, \tag{2.116}$$

thus the balance of linear momentum equation may be written in the equivalent form

$$\rho\left[\frac{\partial}{\partial t}\boldsymbol{v} + \boldsymbol{v} \cdot \nabla\boldsymbol{v}\right] = \nabla \cdot \boldsymbol{S} + \rho\boldsymbol{b}. \tag{2.117}$$

2.8.3 *Application to Fluid Models*

We apply the Eulerian description of the conservation laws to model the fluid equations. We summarize the various fluid models and the governing equations for them in this section.

2.8.3.1 Ideal Fluid

For an ideal fluid, the Eulerian description of the balance laws on a domain $\Omega(t)$ with the boundary $\partial\Omega(t)$ for $t \geq 0$ can be written as the following governing equations:

$$\rho\left[\frac{\partial}{\partial t}\boldsymbol{v} + \boldsymbol{v} \cdot \nabla\boldsymbol{v}\right] = -\nabla p + \rho\boldsymbol{b} \quad \text{on} \quad \Omega(t) \times [0, T] \tag{2.118}$$

$$\nabla \cdot \boldsymbol{v} = 0 \quad \text{on} \quad \Omega(t) \times [0, T] \tag{2.119}$$

$$\boldsymbol{v} \cdot \boldsymbol{n} = 0 \quad \text{on} \quad \partial\Omega(t) \times [0, T] \tag{2.120}$$

$$\boldsymbol{v}(\cdot, 0) = \boldsymbol{v}_0(\cdot) \quad \text{on} \quad \Omega(0), \tag{2.121}$$

where $\boldsymbol{n}$ is the unit normal to the boundary $\partial\Omega(t)$ and $\boldsymbol{v}_0$ denotes the initial condition for the velocity at $t = 0$.

Let $\boldsymbol{\varphi}$ be a motion of a continuum body with spatial velocity field $\boldsymbol{v}$ and spin field $\boldsymbol{W} = \text{skew}\,(\nabla\boldsymbol{v})$. Then the motion $\boldsymbol{\varphi}$ is said to be irrotational if

$$\boldsymbol{W}(\boldsymbol{x}, t) = \boldsymbol{0}, \quad \forall\boldsymbol{x} \in \Omega(t), t \geq 0, \tag{2.122}$$

$$\Longrightarrow (\nabla \times \boldsymbol{v})(\boldsymbol{x}, t) = \boldsymbol{0}, \quad \forall\boldsymbol{x} \in \Omega(t), t \geq 0, \tag{2.123}$$

where $\nabla \times \boldsymbol{v}$ is called the vorticity. Furthermore, as the motion is irrotational, the velocity can be expressed as the gradient of a potential function ϕ as

$$\boldsymbol{v}(\boldsymbol{x}, t) = \nabla\phi(\boldsymbol{x}, t), \quad \forall \boldsymbol{x} \in \Omega(t), t \geq 0. \tag{2.124}$$

Now, we can express the Bernoulli's theorem for ideal fluids. Given a conservative body force per unit mass $\boldsymbol{b} = -\nabla\beta$, the conservation of linear momentum for an ideal fluid simplifies to

$$\nabla\left(\frac{\partial}{\partial t}\phi + \frac{1}{2}|\boldsymbol{v}|^2 + \frac{p}{\rho} + \beta\right) = \boldsymbol{0} \tag{2.125}$$

which can also be written as

$$\frac{\partial}{\partial t}\phi + \frac{1}{2}|\boldsymbol{v}|^2 + \frac{p}{\rho} + \beta = f(t), \tag{2.126}$$

for some function $f(t)$ depending on the motion.

2.8.3.2 Elastic Fluid

In the case of an elastic fluid, the Cauchy stress $\boldsymbol{S} = -p\boldsymbol{I}$. Thus, we obtain the following equations from the conservation laws:

$$\rho\dot{\boldsymbol{v}} = \nabla\cdot(-p\boldsymbol{I}) + \rho\boldsymbol{b} \quad \text{and} \quad \frac{\partial}{\partial t}\rho + \nabla\cdot(\rho\boldsymbol{v}) = 0, \tag{2.127}$$

where $p = \pi(\rho)$ and $\boldsymbol{b}(\boldsymbol{x}, t)$ is the spatial body force field per unit mass. Furthermore, $\nabla\cdot(-p\boldsymbol{I}) = -\nabla p = -\nabla[\pi(\rho)] = -\pi'(\rho)\nabla\rho$. Thus the spatial velocity and density fields in a body of elastic fluid with reference configuration $\widehat{\Omega}$ must satisfy the following equations for all $\boldsymbol{x} \in \Omega(t)$ and $t \geq 0$:

$$\rho\left[\frac{\partial}{\partial t}\boldsymbol{v} + \boldsymbol{v}\cdot\nabla\boldsymbol{v}\right] = -\pi'(\rho)\nabla\rho + \rho\boldsymbol{b} \tag{2.128}$$

$$\frac{\partial}{\partial t}\rho + \nabla\cdot(\rho\boldsymbol{v}) = 0 \tag{2.129}$$

These equations are known as the Elastic fluid equations.

2.8.3.3 Incompressible Newtonian Fluid

A continuum body with reference configuration $\widehat{\Omega}$ is said to be an incompressible Newtonian fluid if:

- The reference mass density field $\rho(\boldsymbol{X})$ is uniform in the sense that

$$\rho(\boldsymbol{X}) = \rho > 0 \text{ (constant)} \tag{2.130}$$

- The material is incompressible, i.e., the spatial velocity is divergence-free,

$$\nabla \cdot \boldsymbol{v} = 0 \tag{2.131}$$

- The Cauchy stress field is Newtonian, i.e., there exists a scalar field $p(\boldsymbol{x}, t)$ called the pressure and a constant fourth-order tensor $\widetilde{\boldsymbol{C}}$ such that

$$\boldsymbol{S} = -p\boldsymbol{I} + \widetilde{\boldsymbol{C}}\,(\nabla \boldsymbol{v}) \tag{2.132}$$

  where $\nabla \boldsymbol{v}(\boldsymbol{x}, t)$ is the spatial gradient of velocity. The fourth-order tensor $\widetilde{\boldsymbol{C}}$ satisfies the left minor symmetry condition $(\widetilde{\boldsymbol{C}}(\boldsymbol{A}))^T = \widetilde{\boldsymbol{C}}(\boldsymbol{A})$ for any second-order tensor $\boldsymbol{A}$, and $\text{tr}(\widetilde{\boldsymbol{C}}(\boldsymbol{A})) = 0$, $\text{tr}(\boldsymbol{A}) = 0$.

Under the frame independent property of the material frame, the possible form of $\widetilde{\boldsymbol{C}}$ will be

$$\widetilde{\boldsymbol{C}}\,(\nabla \boldsymbol{v}) = 2\mu \,\text{sym}\,(\nabla \boldsymbol{v})\,, \tag{2.133}$$

where μ is a constant called the dynamic viscosity of the fluid. Based on mechanical energy inequality arguments, $\mu \geq 0$. It is to be noted that when $\mu = 0$, the incompressible Newtonian model of fluid reduces to the ideal model.

By describing the Cauchy stress tensor for the incompressible Newtonian fluid as $\boldsymbol{S} = -p\boldsymbol{I} + 2\mu \,\text{sym}\,(\nabla \boldsymbol{v})$, the conservation of mass and linear momentum on a spatial domain $\Omega(t)$ can be written as

$$\rho \left[\frac{\partial}{\partial t}\boldsymbol{v} + \boldsymbol{v} \cdot \nabla \boldsymbol{v} \right] = \mu \Delta \boldsymbol{v} - \nabla p + \rho \boldsymbol{b}, \tag{2.134}$$

$$\nabla \cdot \boldsymbol{v} = 0. \tag{2.135}$$

These equations are known as the Navier-Stokes Equations for a Newtonian fluid. We can make some observations and simplifications. For instance, in contrast to ideal fluids, Newtonian fluids develop shear stresses. Thus, the traction $\boldsymbol{t}$ on a surface with unit normal $\boldsymbol{n}$ is not necessarily normal to the surface as $\boldsymbol{t} = \boldsymbol{S}\boldsymbol{n} = -p\boldsymbol{n} + 2\mu \,\text{sym}\,(\nabla \boldsymbol{v})\,\boldsymbol{n}$. A non-dimensional number that arises from the study of fluid flows is the Reynolds number $Re = \rho U_0 L_0/\mu$, where ρ is the mass density, U_0 is a characteristic velocity and L_0 is a characteristic length associated with the flow. Low Reynolds numbers correspond to smooth or laminar flows, while high Reynolds number flows are fluctuating and turbulent. Neglecting the acceleration term in the Navier-Stokes equations, we get linear flow equations known as Stokes equations, as follows:

$$\mu \Delta \boldsymbol{v} - \nabla p + \rho \boldsymbol{b} = \boldsymbol{0} \tag{2.136}$$

$$\nabla \cdot \boldsymbol{v} = 0 \tag{2.137}$$

These equations can be used to model nearly steady and slow flows consisting of relatively small velocity gradients and the viscous effects are very large compared to the convective acceleration term. In contrast to the full incompressible Navier-Stokes equations, the Stokes equation is relatively simple for computational analysis. However, there is still complexity of a saddle-point or incompressibility constraint and it can serve as a prototype problem for the design and analysis of variational methods.

Bibliography Notes

Further materials on motion kinematics of deformable bodies can be found in Trusdell [214, 215], Ogden [159], Chadwick [41], Gurtin [81], Gonzalez and Stuart [80]. Some of the equations and the descriptions of continuum mechanics concepts are inspired from the textbook of Gonzalez and Stuart [80].

Chapter 3
Fluid-Structure Equations with Body-Fitted Interface

In this chapter, we combine the balance laws of fluid and solid fields to formulate fluid-structure interactions. The two subdomains for fluid and solid are governed by the incompressible Navier-Stokes equations and by hyper-elastic material law. The following sections are devoted to details on the coupling of the two models. Finally, we will formulate a coupled system of equations that describes the full process, fluid, solid and interface conditions. In particular, we present how the Eulerian fluid and Lagrangian solid can be coupled via arbitrary Lagrangian-Eulerian coordinates and how to satisfy the dynamic, kinematic and geometry compatibility conditions along the body-fitted (body-conformal) interface.

3.1 Continuum Mechanics of Fluid and Solid

Before discussing the interface conditions, we first present the equations of Lagrangian solid and Eulerian fluid systems.

3.1.1 Lagrangian Solid System

The principal advantage of the Lagrangian approach lies in the treatment of boundary conditions at material surfaces. A standard initial-boundary value problem for an elastic body with reference configuration $\widehat{\Omega}$ can be written as

R. K. Jaiman and V. Joshi, *Computational Mechanics of Fluid-Structure Interaction*,
https://doi.org/10.1007/978-981-16-5355-1_3

$$\rho_0 \ddot{\boldsymbol{\varphi}} = \widehat{\nabla} \cdot [\boldsymbol{P}(\boldsymbol{X}, t)] + \rho_0 \boldsymbol{b} \quad \text{on } \widehat{\Omega} \times [0, T] \tag{3.1}$$

$$\boldsymbol{\varphi} = \boldsymbol{g} \quad \text{on } \Gamma_D \times [0, T] \tag{3.2}$$

$$\boldsymbol{P}(\boldsymbol{X}, t)\boldsymbol{N} = \boldsymbol{h} \quad \text{on } \Gamma_N \times [0, T] \tag{3.3}$$

$$\boldsymbol{\varphi}(\cdot, 0) = \boldsymbol{X} \quad \text{on } \widehat{\Omega} \tag{3.4}$$

$$\dot{\boldsymbol{\varphi}}(\cdot, 0) = \boldsymbol{v}_0 \quad \text{on } \widehat{\Omega} \tag{3.5}$$

In the above system, Γ_D and Γ_N are subsets of $\partial\widehat{\Omega}$ with the properties $\Gamma_D \cup \Gamma_N = \partial\widehat{\Omega}$ and $\Gamma_D \cap \Gamma_N = \emptyset$, ρ_0 is the reference mass density, $\boldsymbol{b}(\boldsymbol{X}, t)$ is the body force and $\boldsymbol{P}(\boldsymbol{X}, t)$ denotes the first Piola-Kirchhoff stress tensor and $\widehat{\boldsymbol{P}}(\boldsymbol{F})$ is the stress response associated with it.

Lagrangian methods, mainly designed for problems in structural mechanics, allow easy tracking of surfaces and interfaces between different materials. They are also appropriate for materials with history dependent constitutive relations (such as plasticity); but have difficulties for problems with large mesh deformations.

3.1.2 *Eulerian Fluid System*

The equations of motion for Newtonian fluids are most commonly formulated in Eulerian form, leading to the Navier-Stokes equations. In many situations of interest, Eulerian formulations permit the simulation of fluid flows using a fixed domain or control volume, which is an attractive feature. A standard initial-boundary value problem for an incompressible fluid occupying a fixed region $\Omega(t)$ can be written as

$$\rho \left[\frac{\partial}{\partial t} \boldsymbol{v} + \boldsymbol{v} \cdot \nabla \boldsymbol{v} \right] = \nabla \cdot \boldsymbol{S} + \rho \boldsymbol{b} \quad \text{on } \Omega(t) \times [0, T] \tag{3.6}$$

$$\nabla \cdot \boldsymbol{v} = 0 \quad \text{on } \Omega(t) \times [0, T] \tag{3.7}$$

$$\boldsymbol{S} \cdot \boldsymbol{n} = 0 \quad \text{on } \partial\Omega(t) \times [0, T] \tag{3.8}$$

$$\boldsymbol{v}(\cdot, 0) = \boldsymbol{v}_0(\cdot) \quad \text{on } \Omega(0), \tag{3.9}$$

where ∇ defines the gradient in the spatial coordinate system. Setting $\boldsymbol{S} = -p\boldsymbol{I} + 2\mu \,\text{sym}\,(\nabla \boldsymbol{v})$ in the balance of linear momentum equation, we obtain a closed system of equations for the spatial velocity and pressure fields in a body of Newtonian fluid.

The procedure of FSI-coupling is as follows: Let us say, we are given fluid equations in Eulerian coordinates and solid equations in a Lagrangian framework. The questions we need to pose are: How do we couple or combine the coordinate systems? Once we decide a technique, we further ask ourselves how to handle the fluid-structure interface and which amount of information needs to be transferred along the interface? This question might be posed from an application viewpoint but also from a mathematical perspective.

3.2 Kinematics of Eulerian and Lagrangian Modeling

In fluid-structure interaction, we study the kinematical motion and deformation of solid/structures due to the traction applied by fluid flow. It is important to understand the kinematical modeling and coupling of the two fields. To study the kinematics, there are two common domains in continuum mechanics: the material domain $\widehat{\Omega} \subset \mathbb{R}^{n_{\rm sd}}$, with $n_{\rm sd}$ spatial dimensions, made up of material particles X and the spatial domain $\Omega(t)$ consisting of spatial points x. Furthermore, $\Omega(0)$ is the so-called initial configuration and $\widehat{\Omega} := \Omega(0)$. In the Lagrangian coordinate system, a specific material point and its deformation are observed in time. In contrast, using Eulerian coordinates, we focus on a fixed point in space and observe what is happening at this spatial point while time is evolving. While solids/elasticity equations are usually described in the Lagrangian system, fluids (i.e., Navier-Stokes equations) are preferred in Eulerian coordinates.

A deformation of $\widehat{\Omega}$ is a smooth, one-to-one (i.e., bijective), orientation-preserving mapping

$$\varphi : \widehat{\Omega} \to \Omega(t) \quad \text{with } (X, t) \mapsto (x, t) = (\varphi(X, t), t) \tag{3.10}$$

This mapping associates each point $X \in \widehat{\Omega}$ (reference domain) to a new position $x \in \Omega(t)$ (physical domain) (Fig. 3.1).

In material/Lagrangian description, we have

$$u(X, t) = x(X, t) - X, \tag{3.11}$$

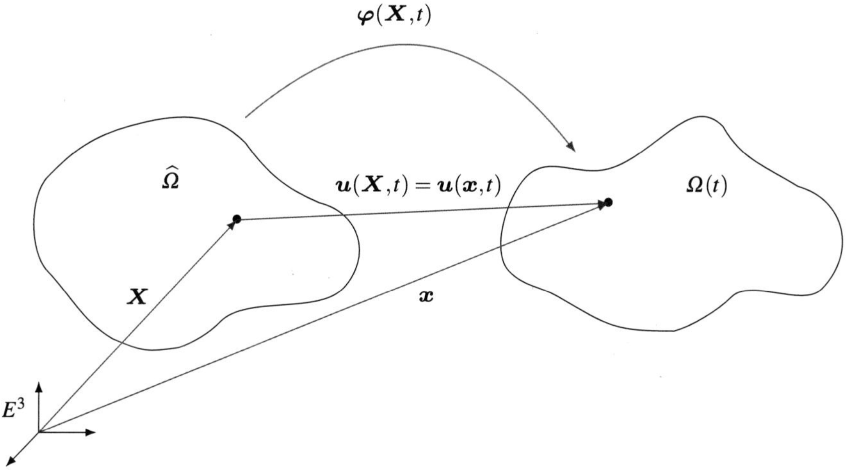

Fig. 3.1 Descriptions of displacements in Eulerian and Lagrangian coordinates. Traversing from the origin via $\widehat{\Omega}$ means $X + u(X, t) = x$ and going from the origin via $\Omega(t)$ leads to $x - u(x, t) = X$

where $\boldsymbol{u}(\boldsymbol{X}, t)$ is the material description of the displacement field and it relates a particle's position in the reference configuration $\boldsymbol{X}$ to its corresponding position in the current configuration $\boldsymbol{x}$ at time t. On the other hand, the spatial/Eulerian description can be written as

$$\boldsymbol{u}(\boldsymbol{x}, t) = \boldsymbol{x} - \boldsymbol{X}(\boldsymbol{x}, t), \tag{3.12}$$

$\boldsymbol{u}(\boldsymbol{x}, t)$ being the spatial description of the displacement field. This is formulated in terms of the current displacement, which results from its original position $\boldsymbol{X}$ plus the displacement for that position. We recapitulate that the two displacement descriptions can be transformed through the deformation map $\boldsymbol{x} = \boldsymbol{\varphi}(\boldsymbol{X}, t)$.

3.2.1 Deformation Mapping

We briefly review the deformation mapping that is required to transform and formulate equations in different coordinate systems. Specifically, we introduce concepts to further study the changes of size and shape of a body that is moved from the reference configuration $\widehat{\Omega}$ to the current domain $\Omega(t)$. For simplicity, let a body occupy $\widehat{\Omega}$ at time $t = 0$. It is described by its position vector $\boldsymbol{X} = (X_1, X_2, X_3)$ in a Cartesian coordinate system with the orthonormal basis $(\boldsymbol{E}_1, \boldsymbol{E}_2, \boldsymbol{E}_3)$. At time t, the body has evolved (and possibly changed its size and shape) and occupies the current domain $\Omega(t)$. Here, the position vector is $x_i\left(X_j, t\right)$ Consequently,

$$\boldsymbol{X} = X_i \boldsymbol{E}_i, \quad \boldsymbol{x} = x_i\left(X_j, t\right) \boldsymbol{E}_i. \tag{3.13}$$

To define the deformation gradient, let us consider an infinitesimal material vector $d\boldsymbol{X}$ in $\widehat{\Omega}$ that displaces the material particle at $\boldsymbol{X}$ such that the new position is $\boldsymbol{X} + d\boldsymbol{X}$. In the current domain, the operation yields $\boldsymbol{x} + d\boldsymbol{x}$. The key purpose of the deformation gradient $\boldsymbol{F}$ is to link $d\boldsymbol{x}$ and $d\boldsymbol{X}$. The deformation gradient has been defined and discussed to certain detail in the previous chapter. Here, we will review some of the results from the previous chapter.

We can define deformation gradient as

$$d\boldsymbol{x} = \boldsymbol{F} \cdot d\boldsymbol{X} \tag{3.14}$$

with $\boldsymbol{F} = \widehat{\nabla}\boldsymbol{x}$, i.e. , $F_{ij} = \frac{\partial x_i}{\partial X_j}$, $\widehat{\nabla}$ being the derivative with respect to the material coordinates $\boldsymbol{X}$. There exists a relationship between deformation gradient and displacement via the deformation map $\boldsymbol{x}(\boldsymbol{X}, t) = \boldsymbol{\varphi}(\boldsymbol{X}, t) = \boldsymbol{X} + \boldsymbol{u}(\boldsymbol{X}, t)$, which can be expressed as

$$\boldsymbol{F} = \widehat{\nabla}\boldsymbol{\varphi} = \boldsymbol{I} + \widehat{\nabla}\boldsymbol{u}(\boldsymbol{X}, t). \tag{3.15}$$

With the help of the transformation between the material and spatial coordinates, we can relate the different tensor fields between the undeformed or reference configuration to the deformed configuration. One example is known as the Piola transformation which is defined as

$$\boldsymbol{P}(\boldsymbol{X}, t) := J(\boldsymbol{X}, t)\boldsymbol{P}(\boldsymbol{x}, t)\boldsymbol{F}^{-T}(\boldsymbol{X}, t) \quad \text{for} \quad \boldsymbol{x} = \boldsymbol{\varphi}(\boldsymbol{X}, t). \tag{3.16}$$

The Piola transformation is later used to transform the Cauchy stress tensor from Eulerian coordinates to the Piola-Kirchhoff stress tensor into the Lagrangian framework. Nanson's formula is an important relation that can be used to go from surface areas in the current configuration to the surface areas in the reference configuration and *vice-versa*, which can be given as follows

$$\boldsymbol{n}(\boldsymbol{x})dA_{\boldsymbol{x}} = J(\boldsymbol{X}, t)\boldsymbol{F}^{-T}(\boldsymbol{X}, t)\boldsymbol{N}(\boldsymbol{X})dA_{\boldsymbol{X}} \tag{3.17}$$

where $J(\boldsymbol{X}, t) = \det(\boldsymbol{F}(\boldsymbol{X}, t))$. In view of fluid-structure interaction modeling, $J(\boldsymbol{X}, t) > 0$ allows to keep the orientation of a mapping $\boldsymbol{\varphi}(\boldsymbol{X}, t)$ and consequently the mapping is locally invertible.

3.3 Continuum Mechanics of Moving Domains

In Continuum mechanics, the Lagrangian description deals with following the particles as they move in the continuum. The material coordinates $\boldsymbol{X}$ identify the reference or undeformed configuration $\widehat{\Omega}$. As the particles move, the motion relates the material coordinates $\boldsymbol{X}$ to the spatial coordinates $\boldsymbol{x}$ in the current or deformed configuration $\Omega(t)$. The motion is given by a mapping function $\boldsymbol{\varphi}$ such that $\boldsymbol{x} = \boldsymbol{\varphi}(\boldsymbol{X}, t)$. In the coupled fluid-structure interaction problem, the deformed structure tends to change the configuration of the fluid domain too. Thus, we need to resolve the fluid dynamic equations on a moving (or, time dependent) domain. The Lagrangian description helps to easily track the free surfaces and interfaces between different materials, however, it is challenging to capture large distortions and topological changes in the domain without involving frequent remeshing procedures. On the other hand, Eulerian description is widely used for modeling of fluids. In this scenario, the computational domain is fixed and the fluid moves with respect to the grid points. Large distortions in the fluid motion can be easily modeled and captured by the Eulerian formulation, which is essential for high Reynolds number flows. But, the Eulerian description cannot handle moving boundaries, which are encountered in fluid-structure interaction problems. Various approaches have been considered in the literature to deal with this problem. One of the ways is to introduce an arbitrary reference configuration, known as arbitrary Lagrangian-Eulerian (ALE) description, which can overcome the difficulties due to reconstruction of the mesh at each time step.

3.3.1 Arbitrary Lagrangian Eulerian Description

The advantages and disadvantages of both the Lagrangian and Eulerian descriptions were highlighted above. This has shown a potential interest in developing a generalized description which combines the advantages of both the formulations (Lagrangian and Eulerian), while minimizing their drawbacks as much as possible. Such a description is classically known as the arbitrary Lagrangian Eulerian (ALE) description. Methods based on ALE description were first proposed in the finite difference and finite volume contexts.

In this description, neither the material configuration $\widehat{\Omega}$ nor the spatial configuration $\Omega(t)$ is selected as the reference configuration. A third referential configuration Ω_χ is introduced with reference coordinates denoted by $\boldsymbol{\chi}$ to identify the grid points. The three domains and the one-to-one transformations between them are depicted in Fig. 3.2. The referential domain Ω_χ is mapped into the material and spatial domains by $\widetilde{\boldsymbol{\varphi}}$ and $\widehat{\boldsymbol{\varphi}}$, respectively. The reference configuration moves arbitrarily with respect to the Eulerian and Lagrangian coordinates. The motion of the particle $\boldsymbol{\varphi}$ can then be expressed as $\boldsymbol{\varphi} = \widehat{\boldsymbol{\varphi}} \circ \widetilde{\boldsymbol{\varphi}}$, showing that the three mappings between the three descriptions $\widehat{\boldsymbol{\varphi}}$, $\widetilde{\boldsymbol{\varphi}}$ and $\boldsymbol{\varphi}$ are not independent. The mapping of $\widehat{\boldsymbol{\varphi}}$ from the referential domain to the spatial domain can be understood as the motion of the grid points in the spatial domain (Fig. 3.2).

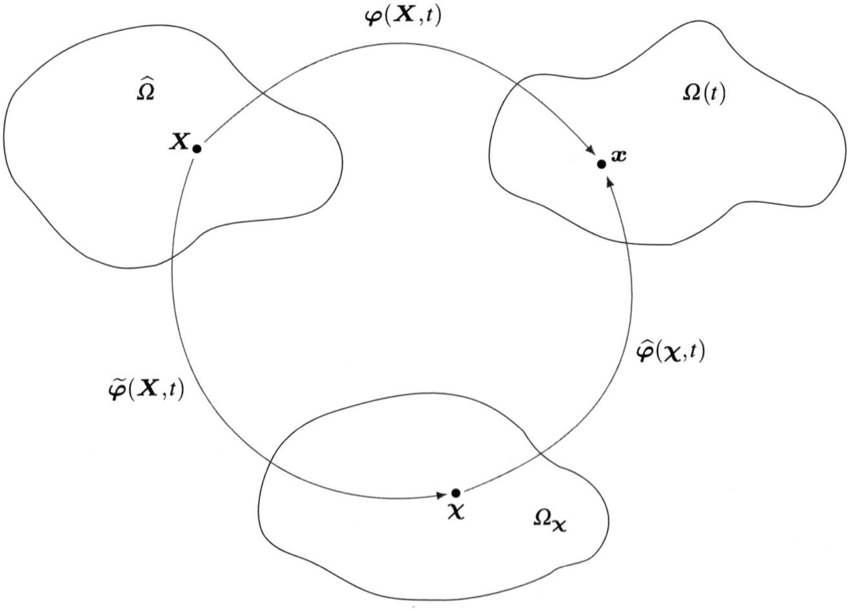

Fig. 3.2 The reference configuration Ω_χ considered in the arbitrary Lagrangian Eulerian (ALE) description and the mapping of Ω_χ with the material $\widehat{\Omega}$ and the spatial domains $\Omega(t)$

3.3.1.1 Lagrangian-to-Eulerian Map

We first define the Lagrangian-to-Eulerian map $\boldsymbol{\varphi}$

$$\boldsymbol{\varphi}(\cdot, t) : \widehat{\Omega} \to \Omega(t) = \boldsymbol{\varphi}(\widehat{\Omega}, t), \qquad \forall t \geq 0 \tag{3.18}$$

$$\boldsymbol{X} \mapsto \boldsymbol{x} = \boldsymbol{\varphi}(\boldsymbol{X}, t), \qquad \forall \boldsymbol{X} \in \widehat{\Omega} \tag{3.19}$$

where the displacement vector $\boldsymbol{u} = \boldsymbol{\varphi}(\boldsymbol{X}, t) - \boldsymbol{\varphi}(\boldsymbol{X}, 0) = \boldsymbol{\varphi}(\boldsymbol{X}, t) - \boldsymbol{X}$ and the deformation gradient is $\boldsymbol{F} = \widehat{\nabla}\boldsymbol{\varphi} = \widehat{\nabla}\boldsymbol{x}$. The gradient of the map $\boldsymbol{\varphi}(\boldsymbol{X}, t)$ can be expressed in the matrix form as

$$\frac{\partial \boldsymbol{\varphi}(\boldsymbol{X}, t)}{\partial(\boldsymbol{X}, t)} = \begin{pmatrix} \frac{\partial \boldsymbol{x}}{\partial \boldsymbol{X}} & \boldsymbol{v} \\ 0^T & 1 \end{pmatrix} = \begin{pmatrix} \boldsymbol{F} & \boldsymbol{v} \\ 0^T & 1 \end{pmatrix}, \tag{3.20}$$

where $\boldsymbol{v}$ is the material velocity given by

$$\boldsymbol{v}(\boldsymbol{X}, t) = \left.\frac{\partial \boldsymbol{x}}{\partial t}\right|_{\boldsymbol{X}}. \tag{3.21}$$

The material derivative of a scalar physical quantity during this transformation is given as

$$\dot{f}(\boldsymbol{x}, t) = \left.\frac{\partial f(\boldsymbol{x}, t)}{\partial t}\right|_{\boldsymbol{X}} \tag{3.22}$$

$$= \left.\frac{\partial f(\boldsymbol{\varphi}(\boldsymbol{X}, t), t)}{\partial t}\right|_{\boldsymbol{X}} \tag{3.23}$$

$$= \left.\frac{\partial f}{\partial t}\right|_{\boldsymbol{x}} + \nabla f \cdot \left.\frac{\partial \boldsymbol{x}}{\partial t}\right|_{\boldsymbol{X}} \tag{3.24}$$

$$= \left.\frac{\partial f}{\partial t}\right|_{\boldsymbol{x}} + \boldsymbol{v} \cdot \nabla f. \tag{3.25}$$

3.3.1.2 Referential-to-Eulerian Map

We next define the Referential-to-Eulerian map $\widehat{\boldsymbol{\varphi}}$

$$\widehat{\boldsymbol{\varphi}}(\cdot, t) : \Omega_{\chi} \to \Omega(t) = \widehat{\boldsymbol{\varphi}}\left(\Omega_{\chi}, t\right), \qquad \forall t \geq 0 \tag{3.26}$$

$$\boldsymbol{\chi} \mapsto \boldsymbol{x} = \widehat{\boldsymbol{\varphi}}(\boldsymbol{\chi}, t), \qquad \forall \boldsymbol{\chi} \in \Omega_{\chi} \tag{3.27}$$

where the displacement $\boldsymbol{u}_{\widehat{\varphi}} = \widehat{\boldsymbol{\varphi}}(\boldsymbol{\chi}, t) - \widehat{\boldsymbol{\varphi}}(\boldsymbol{\chi}, 0) = \widehat{\boldsymbol{\varphi}}(\boldsymbol{\chi}, t) - \boldsymbol{\chi}$. In this scenario, the deformation gradient can be written as $\widehat{\boldsymbol{F}} = \nabla^{\chi}\widehat{\boldsymbol{\varphi}} = \nabla^{\chi}\boldsymbol{x}$ and the gradient of the map $\widehat{\boldsymbol{\varphi}}(\boldsymbol{\chi}, t)$ can be represented as

$$\frac{\partial \widehat{\boldsymbol{\varphi}}(\boldsymbol{\chi}, t)}{\partial(\boldsymbol{\chi}, t)} = \begin{pmatrix} \frac{\partial \boldsymbol{x}}{\partial \boldsymbol{\chi}} & \boldsymbol{w} \\ \boldsymbol{0}^T & 1 \end{pmatrix} = \begin{pmatrix} \widehat{\boldsymbol{F}} & \boldsymbol{w} \\ \boldsymbol{0}^T & 1 \end{pmatrix}, \tag{3.28}$$

where $\boldsymbol{w}$ is the velocity with respect to the referential domain given by

$$\boldsymbol{w}(\boldsymbol{\chi}, t) = \left. \frac{\partial \boldsymbol{x}}{\partial t} \right|_{\boldsymbol{\chi}}. \tag{3.29}$$

The time derivative of a scalar physical quantity during this transformation can be represented as

$$\left. \frac{\partial f(\boldsymbol{x}, t)}{\partial t} \right|_{\boldsymbol{\chi}} = \left. \frac{\partial f(\widehat{\boldsymbol{\varphi}}(\boldsymbol{\chi}, t), t)}{\partial t} \right|_{\boldsymbol{\chi}} \tag{3.30}$$

$$= \left. \frac{\partial f}{\partial t} \right|_{\boldsymbol{x}} + \nabla f \cdot \left. \frac{\partial \boldsymbol{x}}{\partial t} \right|_{\boldsymbol{\chi}} \tag{3.31}$$

$$= \left. \frac{\partial f}{\partial t} \right|_{\boldsymbol{x}} + \boldsymbol{w} \cdot \nabla f. \tag{3.32}$$

3.3.1.3 Lagrangian-to-Referential Map

Similarly, we define the Lagrangian-to-Referential map $\widetilde{\boldsymbol{\varphi}}$. The map is defined as: $\widetilde{\boldsymbol{\varphi}} = \widehat{\boldsymbol{\varphi}}^{-1} \circ \boldsymbol{\varphi}$.

$$\widetilde{\boldsymbol{\varphi}}(\cdot, t) : \widehat{\Omega} \to \Omega_{\chi} = \widetilde{\boldsymbol{\varphi}}(\widehat{\Omega}, t), \qquad \forall t \geq 0 \tag{3.33}$$

$$\boldsymbol{X} \mapsto \boldsymbol{\chi} = \widetilde{\boldsymbol{\varphi}}(\boldsymbol{X}, t), \qquad \forall \boldsymbol{X} \in \widehat{\Omega} \tag{3.34}$$

The displacement function $\boldsymbol{u}_{\widetilde{\varphi}} = \widetilde{\boldsymbol{\varphi}}(\boldsymbol{X}, t) - \widetilde{\boldsymbol{\varphi}}(\boldsymbol{X}, 0) = \widetilde{\boldsymbol{\varphi}}(\boldsymbol{X}, t) - \boldsymbol{X}$, and the deformation gradient is $\widetilde{\boldsymbol{F}} = \widehat{\nabla} \widetilde{\boldsymbol{\varphi}} = \widehat{\nabla} \boldsymbol{\chi}$. The gradient of the deformation map can be expressed as

$$\frac{\partial \widetilde{\boldsymbol{\varphi}}(\boldsymbol{X}, t)}{\partial(\boldsymbol{X}, t)} = \begin{pmatrix} \frac{\partial \boldsymbol{\chi}}{\partial \boldsymbol{X}} & \widetilde{\boldsymbol{v}} \\ \boldsymbol{0}^T & 1 \end{pmatrix} = \begin{pmatrix} \widetilde{\boldsymbol{F}} & \widetilde{\boldsymbol{v}} \\ \boldsymbol{0}^T & 1 \end{pmatrix}, \tag{3.35}$$

where $\widetilde{\boldsymbol{v}}$ is the velocity with respect to the material domain given by

$$\widetilde{\boldsymbol{v}}(\boldsymbol{X}, t) = \left. \frac{\partial \boldsymbol{\chi}}{\partial t} \right|_{\boldsymbol{X}}. \tag{3.36}$$

The time derivative of a scalar field thus will be

$$\dot{f}(\chi, t) = \left.\frac{\partial f}{\partial t}\right|_{\chi} + \widetilde{\boldsymbol{v}} \cdot \nabla^{\chi} f. \tag{3.37}$$

A relation between the different velocities in the transformations mentioned can be established with the help of the dependency of the maps, i.e., $\boldsymbol{\varphi} = \widehat{\boldsymbol{\varphi}} \circ \widetilde{\boldsymbol{\varphi}}$. Differentiating this expression, we can get

$$\frac{\partial \boldsymbol{\varphi}(\boldsymbol{X}, t)}{\partial(\boldsymbol{X}, t)} = \frac{\partial \widehat{\boldsymbol{\varphi}}(\widetilde{\boldsymbol{\varphi}}(\boldsymbol{X}, t), t)}{\partial \chi} \frac{\partial \widetilde{\boldsymbol{\varphi}}(\boldsymbol{X}, t)}{\partial(\boldsymbol{X}, t)} = \frac{\partial \widehat{\boldsymbol{\varphi}}(\chi, t)}{\partial \chi} \frac{\partial \widetilde{\boldsymbol{\varphi}}(\boldsymbol{X}, t)}{\partial(\boldsymbol{X}, t)}. \tag{3.38}$$

Using Eqs. (3.20), (3.28) and (3.35), we can write

$$\begin{pmatrix} \boldsymbol{F} & \boldsymbol{v} \\ \boldsymbol{0}^T & 1 \end{pmatrix} = \begin{pmatrix} \widehat{\boldsymbol{F}} & \boldsymbol{w} \\ \boldsymbol{0}^T & 1 \end{pmatrix} \begin{pmatrix} \widetilde{\boldsymbol{F}} & \widetilde{\boldsymbol{v}} \\ \boldsymbol{0}^T & 1 \end{pmatrix}. \tag{3.39}$$

Thus, we can obtain $\boldsymbol{v} = \boldsymbol{w} + \widehat{\boldsymbol{F}}\widetilde{\boldsymbol{v}}$. Let $\boldsymbol{c} = \boldsymbol{v} - \boldsymbol{w} = \widehat{\boldsymbol{F}}\widetilde{\boldsymbol{v}}$ denote the relative velocity between the material domain and the ALE referential domain, using Eqs. (3.25) and (3.32), we can write the material time derivative as:

$$\dot{f}(\boldsymbol{x}, t) = \left.\frac{\partial f}{\partial t}\right|_{\boldsymbol{x}} + \boldsymbol{v} \cdot \nabla f \tag{3.40}$$

$$= \left.\frac{\partial f}{\partial t}\right|_{\chi} - \boldsymbol{w} \cdot \nabla f + \boldsymbol{v} \cdot \nabla f \tag{3.41}$$

$$= \left.\frac{\partial f}{\partial t}\right|_{\chi} + (\boldsymbol{v} - \boldsymbol{w}) \cdot \nabla f = \left.\frac{\partial f}{\partial t}\right|_{\chi} + \boldsymbol{c} \cdot \nabla f. \tag{3.42}$$

Therefore, $\boldsymbol{c}$ is the particle velocity relative to the ALE moving domain as seen from the Eulerian frame and $\widetilde{\boldsymbol{v}}$ is the particle velocity observed from the ALE referential domain.

ALE description provides an intermediate state in which the fluid domain is moved according to the solid. This requires a mapping between the deformed state and a reference configuration. Consider a fluid domain $\widehat{\Omega}^{\mathrm{f}}$ at rest and deformed to the configuration $\Omega^{\mathrm{f}}(t)$ as a result of deformation of the structure $\boldsymbol{\varphi}(\boldsymbol{X}, t)$ as shown in Fig. 3.3. Let this mapping of the fluid domain be denoted by $\widehat{\mathscr{A}}(\cdot, t)$.

The total time derivative of a scalar physical field under different frameworks can be summarized below. In the Lagrangian setting, the total and the partial derivatives coincide, i.e.,

$$\dot{f}(\boldsymbol{x}, t) = \left.\frac{\partial f(\boldsymbol{x}, t)}{\partial t}\right|_{\boldsymbol{X}}. \tag{3.43}$$

In the Eulerian framework, the relationship between the total time derivative and the partial time derivative reads

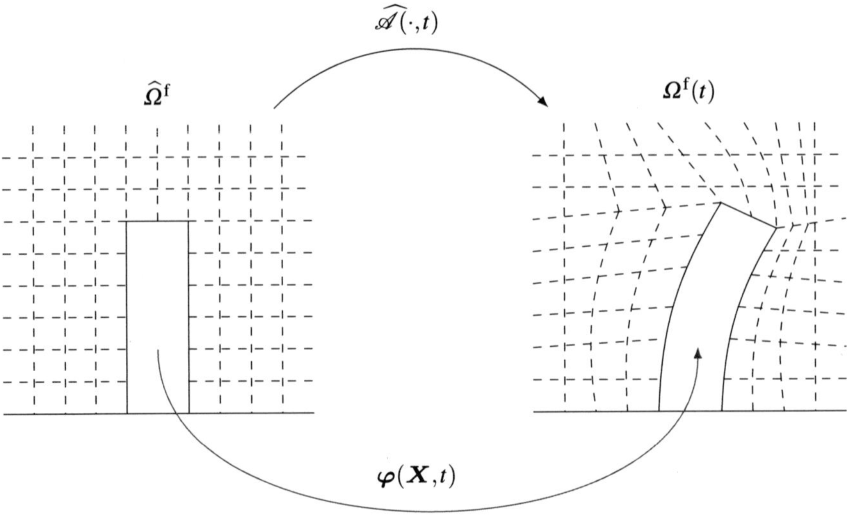

Fig. 3.3 ALE mapping of the fluid mesh from the initial configuration $\widehat{\Omega}^{\mathrm{f}}$ to the deformed configuration $\Omega^{\mathrm{f}}(t)$

$$\dot{f}(\boldsymbol{x},t) = \left.\frac{\partial f(\boldsymbol{x},t)}{\partial t}\right|_{\boldsymbol{x}} + \boldsymbol{v}\cdot\nabla f. \tag{3.44}$$

On the other hand, using the ALE description, the derivative can be written as

$$\dot{f}(\boldsymbol{x},t) = \left.\frac{\partial f(\boldsymbol{x},t))}{\partial t}\right|_{\chi} + (\boldsymbol{v}-\boldsymbol{w})\cdot\nabla f. \tag{3.45}$$

Note that the Lagrangian and Eulerian descriptions can be obtained from the ALE description based on the velocity of the computational domain $\boldsymbol{w}$. Some remarks for the ALE framework are given below:

- If $\boldsymbol{w} = \boldsymbol{v}$, i.e., the domain is moving with the fluid velocity $\boldsymbol{v}$, we get a Lagrangian description
- In a fixed Eulerian setting, $\boldsymbol{w} = 0$, i.e., the domain is fixed.
- In the ALE description, $0 \leq \boldsymbol{w} \leq \boldsymbol{v}$. In the discussions about the fluid-structure interaction coupling in the subsequent sections and chapters, we will see that $\boldsymbol{w} = \boldsymbol{v}$ at the fluid-structure interface Γ^{fs}. In the neighborhood of Γ^{fs}, $0 < \boldsymbol{w} < \boldsymbol{v}$, while $\boldsymbol{w} = 0$ (the mesh is not moving anymore) far away from the fluid-structure interface. Therefore, depending on the location in the domain, we use both Eulerian and Lagrangian frameworks with a smooth transition between them in the ALE description (Fig. 3.4).

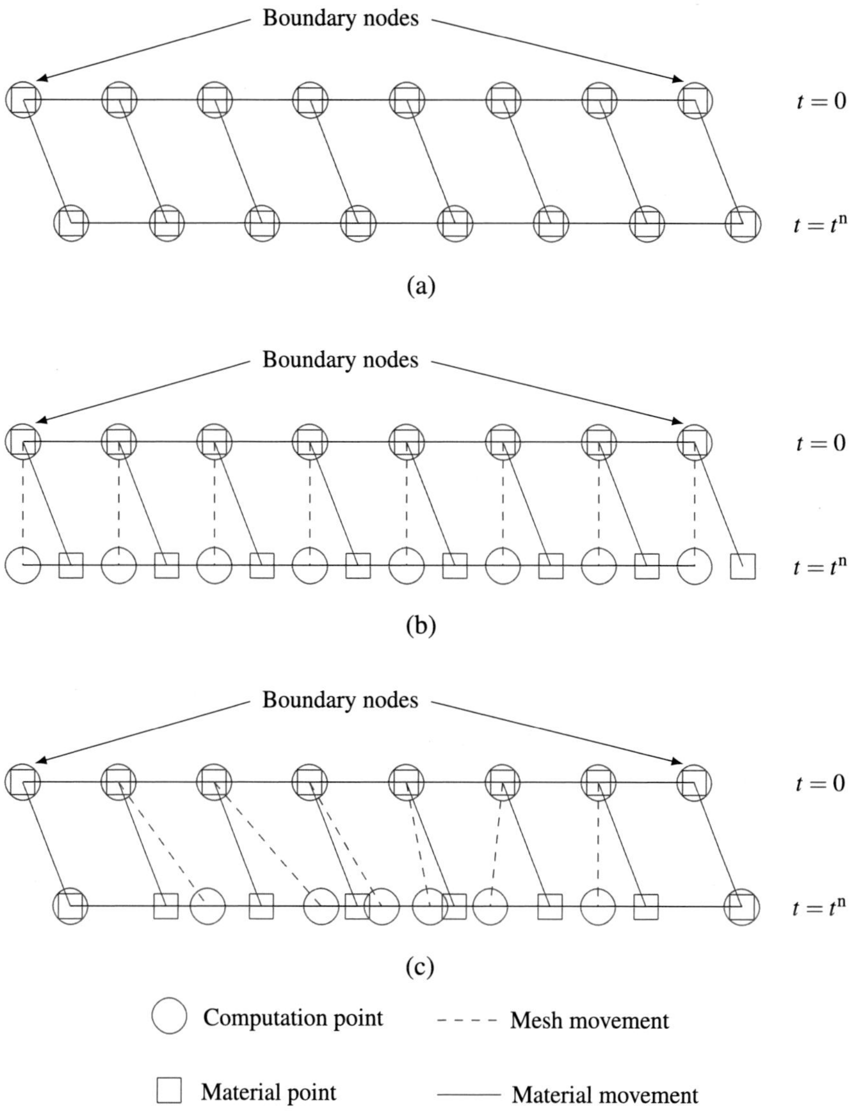

Fig. 3.4 Visualization of mesh movement compared to materials points in: **a** Lagrangian, **b** Eulerian and **c** arbitrary Lagrangian Eulerian descriptions

3.4 Coupled Fluid-Structure Equations

In this section, we will discuss the coupled equations of the fluid and the structure which form the governing equations for FSI problems. The concepts of ALE moving mesh description is applied for the movement of fluid domain along with the deforming structure. The interface conditions to be satisfied for the coupling are highlighted and the coupled FSI system is discussed which forms the basis for upcoming chapters.

3.4.1 Coordinate System

We need to consider suitable coordinate systems to solve the governing equations that describe the coupled fluid-structure interactions. In this work, we have considered three different coordinate systems. The Lagrangian or material coordinate system with material points $\boldsymbol{X}$ is considered to represent flexible structure's velocity, $\boldsymbol{v}^{\mathrm{s}}(\boldsymbol{X}, t)$ and displacement, $\boldsymbol{u}^{\mathrm{s}}(\boldsymbol{X}, t)$, at any time t. In this coordinate system, each computational node follows the corresponding material points. The Eulerian or spatial coordinate system has been considered to describe the fluid velocity, $\boldsymbol{v}^{\mathrm{f}}(\boldsymbol{x}, t)$ and pressure, $p(\boldsymbol{x}, t)$, at a spatial point $\boldsymbol{x}$ and time t. Each spatial point in this reference system is fixed in space and fluid flows through the regions formed by these nodes. However, to handle the problem of fluid-structure interactions with moving interface boundaries we introduce a third type of coordinate system for the fluid domain with reference nodes χ, known as ALE or arbitrary Lagrangian-Eulerian. In this coordinate system, the computational nodes can move relative to the spatial coordinate system. The nodes on the fluid-structure interface behave like material points and the nodes inside the fluid domain can be held fixed like spatial points or moved arbitrarily to account for the interface deformation. From now on, we will not use the notation $\widehat{(\cdot)}$ for the variables defined in the material coordinate system for simplicity. The coordinate system for the equations will be assumed based on the domain, i.e., material coordinates for the structure and ALE coordinates for the fluid. Furthermore, the notations for the fluid and the structure variables are denoted by a superscript $(\cdot)^{\mathrm{f}}$ and $(\cdot)^{\mathrm{s}}$ respectively (Fig. 3.5).

3.4.2 Coupled Formulation

Let $\Omega^{\mathrm{f}}(t) \subset \mathbb{R}^d$ be an Eulerian fluid domain at time t, where d is the space dimension. The motion of an incompressible viscous fluid in $\Omega^{\mathrm{f}}(t)$ is governed by the following Navier-Stokes equations:

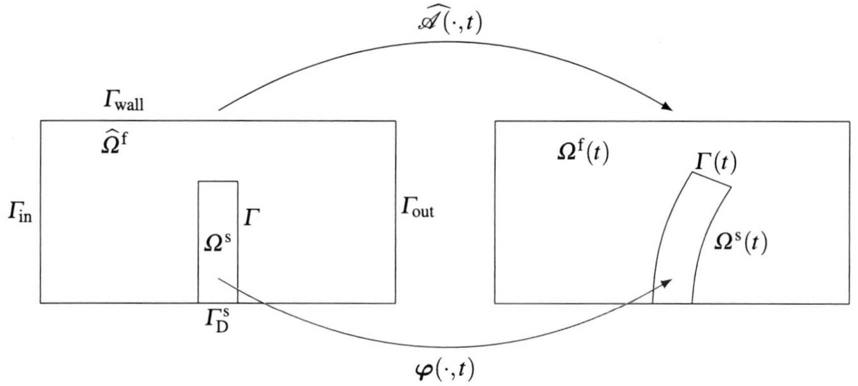

Fig. 3.5 The coupled fluid-structure system with a deformed structure and the deformed fluid domain. The deformation in the structure is mapped by the map $\boldsymbol{\varphi}(\cdot, t)$ and the deformation in the fluid domain is carried out by the ALE mapping $\widehat{\mathscr{A}}(\cdot, t)$

$$\rho^{\text{f}} \frac{\partial \boldsymbol{v}^{\text{f}}}{\partial t} + \rho^{\text{f}} \boldsymbol{v}^{\text{f}} \cdot \nabla \boldsymbol{v}^{\text{f}} = \nabla \cdot \boldsymbol{\sigma}^{\text{f}} + \rho^{\text{f}} \boldsymbol{b}^{\text{f}}, \qquad \text{on } \Omega^{\text{f}}(t) \times [0, T], \tag{3.46}$$

$$\nabla \cdot \boldsymbol{v}^{\text{f}} = 0, \qquad \text{on } \Omega^{\text{f}}(t) \times [0, T] \tag{3.47}$$

Here $\boldsymbol{v}^{\text{f}} = \boldsymbol{v}^{\text{f}}(\boldsymbol{x}, t)$ represents the fluid velocity defined for each spatial point $\boldsymbol{x}$ at time t, ρ^{f} is the fluid density, $\rho^{\text{f}} \boldsymbol{b}^{\text{f}}$ denotes the body force and $\boldsymbol{\sigma}^{\text{f}}$ is Cauchy stress tensor which is given by

$$\boldsymbol{\sigma}^{\text{f}} = -p\boldsymbol{I} + \boldsymbol{T}^{\text{f}}, \quad \boldsymbol{T}^{\text{f}} = 2\mu^{\text{f}} \boldsymbol{\epsilon}^{\text{f}}(\boldsymbol{v}^{\text{f}}), \quad \boldsymbol{\epsilon}^{\text{f}}(\boldsymbol{v}^{\text{f}}) = \frac{1}{2}\left[\nabla \boldsymbol{v}^{\text{f}} + \left(\nabla \boldsymbol{v}^{\text{f}}\right)^{T}\right], \tag{3.48}$$

where $p = p(\boldsymbol{x}, t)$ is the fluid pressure, $\boldsymbol{I}$ denotes the second-order identity tensor, $\boldsymbol{T}$ represents the shear stress tensor, μ^{f} is the fluid dynamic viscosity and $\boldsymbol{\epsilon}^{\text{f}}$ is the fluid strain rate tensor.

Let $\Omega^{\text{s}} \subset \mathbb{R}^{d}$ be the reference Lagrangian domain for a flexible elastic structure. A material point, whose initial position is given by $\boldsymbol{X} \in \Omega^{\text{s}}$, deforms to position $\boldsymbol{\varphi}^{\text{s}}(\boldsymbol{X}, t)$ at time t with the structural momentum equation

$$\rho^{\text{s}} \frac{\partial^{2} \boldsymbol{\varphi}^{\text{s}}}{\partial t^{2}} = \nabla \cdot \boldsymbol{\sigma}^{\text{s}} + \rho^{\text{s}} \boldsymbol{b}^{\text{s}}, \quad \text{on } \Omega^{\text{s}} \times [0, T], \tag{3.49}$$

where $\boldsymbol{\sigma}^{\text{s}}$ denotes the first Piola-Kirchhoff stress tensor, $\rho^{\text{s}} \boldsymbol{b}^{\text{s}}$ represents the body force vector acting on the structure, and ρ^{s} is its mass density. For a linear elastic material,

$$\boldsymbol{\sigma}^{\text{s}}(\boldsymbol{\varphi}^{\text{s}}) = \mu^{\text{s}}\left[\nabla \boldsymbol{u}^{\text{s}} + (\nabla \boldsymbol{u}^{\text{s}})^{T}\right] + \lambda^{\text{s}}(\nabla \cdot \boldsymbol{u}^{\text{s}})\boldsymbol{I}, \tag{3.50}$$

where $\boldsymbol{u}^{\mathrm{s}}(\boldsymbol{X}, t) = \boldsymbol{\varphi}^{\mathrm{s}}(\boldsymbol{X}, t) - \boldsymbol{X}$ is the displacement vector, μ^{s} and λ^{s} are the Lamé's coefficients of a material satisfying [208]

$$\lambda^{\mathrm{s}} > 0 \quad \text{and} \quad 3\lambda^{\mathrm{s}} + 2\mu^{\mathrm{s}} > 0. \tag{3.51}$$

The relationship between the Lamé's coefficients and the flexible structure's elastic properties is given by

$$\lambda^{\mathrm{s}} = \frac{E\nu^{\mathrm{s}}}{(1+\nu^{\mathrm{s}})(1-2\nu^{\mathrm{s}})} \quad \text{and} \quad \mu^{\mathrm{s}} = \frac{E}{2(1+\nu^{\mathrm{s}})}, \tag{3.52}$$

where E is Young's modulus and ν^{s} is the Poisson's ratio. For a St. Venant-Kirchhoff (SVK) material [10, 208],

$$\boldsymbol{\sigma}^{\mathrm{s}}(\boldsymbol{\varphi}^{\mathrm{s}}) = 2\mu^{\mathrm{s}}\boldsymbol{F}\boldsymbol{E} + \lambda^{\mathrm{s}}(\mathrm{tr}(\boldsymbol{E}))\,\boldsymbol{F}, \tag{3.53}$$

where $\mathrm{tr}(\cdot)$ is the tensor trace operator and $\boldsymbol{F}$ is the deformation gradient tensor which is given by

$$\boldsymbol{F} = \nabla\boldsymbol{\varphi}^{\mathrm{s}} = (\boldsymbol{I} + \nabla\boldsymbol{u}^{\mathrm{s}}), \tag{3.54}$$

and $\boldsymbol{E}$ represents the Green-Lagrangian strain tensor defined as

$$\boldsymbol{E} = \frac{1}{2}\left(\boldsymbol{F}^T\boldsymbol{F} - \boldsymbol{I}\right), \tag{3.55}$$

where the term $\boldsymbol{F}^T\boldsymbol{F}$ denotes the right Cauchy-Green deformation tensor [208].

Before we present the Eulerian-Lagrangian coupling, we rewrite Eq. (3.49) in terms of the structural velocity $\boldsymbol{v}^{\mathrm{s}}$, defined as

$$\boldsymbol{v}^{\mathrm{s}}(\boldsymbol{X}, t) = \partial_t\boldsymbol{\varphi}^{\mathrm{s}}(\boldsymbol{X}, t). \tag{3.56}$$

This modification creates the path for the implementation of the velocity continuity (i.e. kinematic equilibrium) along the fluid-structure interface. The structural governing equation (3.49) can therefore be rewritten as

$$\rho^{\mathrm{s}}\frac{\partial\boldsymbol{v}^{\mathrm{s}}}{\partial t} = \nabla\cdot\boldsymbol{\sigma}^{\mathrm{s}} + \rho^{\mathrm{s}}\boldsymbol{b}^{\mathrm{s}}, \quad \text{on } \Omega^{\mathrm{s}} \times [0, T]. \tag{3.57}$$

3.4.3 ALE Formulation

Here, we will reiterate the derivation of the Navier-Stokes equations in the ALE framework. We consider a unique one-to-one mapping function $\widehat{\boldsymbol{\varphi}}^{\mathrm{f}}$ defined as

$$x = \widehat{\varphi}^{\mathrm{f}}(\chi, t) \quad \forall t, \tag{3.58}$$

which maps each node χ in the reference coordinate system to its corresponding spatial node $\boldsymbol{x}$. The gradient of the mapping function $\widehat{\varphi}^{\mathrm{f}}$ will give us

$$\frac{\partial \widehat{\varphi}^{\mathrm{f}}(\chi, t)}{\partial(\chi, t)} = \begin{pmatrix} \frac{\partial x}{\partial \chi} & \boldsymbol{w} \\ 0^T & 1 \end{pmatrix}, \tag{3.59}$$

where $\boldsymbol{w}$ is the mesh velocity, i.e., the velocity with which spatial nodes $\boldsymbol{x}$ are moving with respect to the reference coordinate system and is written as

$$\boldsymbol{w} = \left.\frac{\partial \boldsymbol{x}}{\partial t}\right|_{\chi}. \tag{3.60}$$

Notably, both the material and mesh move with respect to the laboratory frame. We can similarly write the relationships for the inverse map.

To define the conservation laws in the ALE framework, a relation between time derivatives is needed. The total (or material) time derivatives, which are used in the conservation laws, must be transformed to referential time derivatives. The material derivative of a function $f(\chi, t)$ defined on the ALE reference coordinate system yields

$$\frac{Df}{Dt} = \left.\frac{\partial f(\chi, t)}{\partial t}\right|_{\chi} + \left.\frac{\partial f(\boldsymbol{x}, t)}{\partial \boldsymbol{x}} \frac{\partial \boldsymbol{x}}{\partial \chi} \frac{\partial \chi}{\partial t}\right|_{X}. \tag{3.61}$$

The relation can be further simplified by considering the material derivative of the spatial coordinate $\boldsymbol{x}$

$$\boldsymbol{v}^{\mathrm{f}} = \frac{D\boldsymbol{x}}{Dt} = \left.\frac{\partial \boldsymbol{x}(\chi, t)}{\partial t}\right|_{\chi} + \left.\frac{\partial \boldsymbol{x}}{\partial \chi} \frac{\partial \chi}{\partial t}\right|_{X}. \tag{3.62}$$

Therefore, substituting Eqs. (3.60) and (3.62) back into Eq. (3.61) we get

$$\frac{Df}{Dt} = \left.\frac{\partial f(\chi, t)}{\partial t}\right|_{\chi} + (\boldsymbol{v}^{\mathrm{f}} - \boldsymbol{w})\frac{\partial f(\boldsymbol{x}, t)}{\partial \boldsymbol{x}}. \tag{3.63}$$

From Eqs. (3.46 to 3.47), (3.58 to 3.59) and (3.63) we can obtain the Navier-Stokes equations in the ALE framework as

$$\rho^{\mathrm{f}} \left.\frac{\partial \boldsymbol{v}^{\mathrm{f}}}{\partial t}\right|_{\chi} + \rho^{\mathrm{f}}(\boldsymbol{v}^{\mathrm{f}} - \boldsymbol{w}) \cdot \nabla \boldsymbol{v}^{\mathrm{f}} = \nabla \cdot \boldsymbol{\sigma}^{\mathrm{f}} + \rho^{\mathrm{f}} \boldsymbol{b}^{\mathrm{f}}, \quad \text{on } \Omega^{\mathrm{f}}(t) \times [0, T], \tag{3.64}$$

$$\nabla \cdot \boldsymbol{v}^{\mathrm{f}} = 0, \quad \text{on } \Omega^{\mathrm{f}}(t) \times [0, T]. \tag{3.65}$$

3.4.4 Boundary and Initial Conditions

Let us assume that the fluid boundary $\partial\Omega^{\mathrm{f}}(t)$ can be decomposed into three disjoint portions comprising of $\Gamma_D^{\mathrm{f}}(t)$, $\Gamma_N^{\mathrm{f}}(t)$ and $\Gamma(t)$ at any time t, where Γ_D^{f} and Γ_N^{f} represent fluid Dirichlet and Neumann boundaries respectively, Γ represents the interface boundary between the fluid and structural domains at $t = 0$, i.e., $\Gamma = \partial\Omega^{\mathrm{f}}(0) \cap \partial\Omega^{\mathrm{s}}$ and $\Gamma(t)$ is the mapping of Γ from $\Omega^{\mathrm{s}} \to \Omega^{\mathrm{s}}(t)$, i.e., $\Gamma(t) = \boldsymbol{\varphi}^{\mathrm{s}}(\Gamma, t)$. Similarly, we can consider that the solid boundary $\partial\Omega^{\mathrm{s}}$ can be decomposed into Γ_D^{s}, Γ_N^{s} and Γ, where Γ_D^{s} and Γ_N^{s} represent solid Dirichlet and Neumann boundaries respectively.

The fluid and structural governing equations are coupled by the imposition of velocity and traction continuity relations along the fluid-structure interface. Mathematically, these relations can be written as

$$\boldsymbol{v}^{\mathrm{f}}(\boldsymbol{\varphi}^{\mathrm{s}}(\boldsymbol{X}, t), t) = \boldsymbol{v}^{\mathrm{s}}(\boldsymbol{X}, t) \quad \forall \boldsymbol{X} \in \Gamma, \tag{3.66}$$

$$\int_{\boldsymbol{\varphi}^{\mathrm{s}}(\boldsymbol{\gamma},t)} \boldsymbol{\sigma}^{\mathrm{f}}(\boldsymbol{x}, t) \cdot \boldsymbol{n}^{\mathrm{f}} d\Gamma + \int_{\boldsymbol{\gamma}} \boldsymbol{\sigma}^{\mathrm{s}}(\boldsymbol{X}, t) \cdot \boldsymbol{n}^{\mathrm{s}} d\Gamma = 0 \quad \forall \boldsymbol{\gamma} \subset \Gamma, \tag{3.67}$$

where $\boldsymbol{n}^{\mathrm{f}}$ and $\boldsymbol{n}^{\mathrm{s}}$ are the unit outward normals to the deformed fluid element $\boldsymbol{\varphi}^{\mathrm{s}}(\boldsymbol{\gamma}, t)$ and its corresponding undeformed solid element $\boldsymbol{\gamma}$, respectively. Here, $\boldsymbol{\gamma}$ is any part of fluid-structure interface Γ at $t = 0$. Note that it is equivalent to

$$\left(\det\left(\frac{\partial \boldsymbol{x}}{\partial \boldsymbol{X}}\right)\right)\left(\boldsymbol{\sigma}^{\mathrm{f}}(\boldsymbol{x}(\boldsymbol{X}), t)\right)\left(\frac{\partial \boldsymbol{x}}{\partial \boldsymbol{X}}\right)^{-T} \cdot \boldsymbol{n}^{\mathrm{s}} - \boldsymbol{\sigma}^{\mathrm{s}}(\boldsymbol{X}, t) \cdot \boldsymbol{n}^{\mathrm{s}} = 0, \tag{3.68}$$

for any $\boldsymbol{X} \in \Gamma$, where $\boldsymbol{x}(\boldsymbol{X}) = \boldsymbol{\varphi}^{\mathrm{s}}(\boldsymbol{X}, t)$. Refer to Nanson's formula or the derivation of area transformation.

Dirichlet and Neumann boundary conditions for the non-interface fluid boundaries are prescribed as

$$\boldsymbol{v}^{\mathrm{f}}(\boldsymbol{x}, t) = \boldsymbol{v}_D^{\mathrm{f}} \quad \text{on } \Gamma_D^{\mathrm{f}} \qquad \text{and} \qquad \boldsymbol{\sigma}^{\mathrm{f}}(\boldsymbol{x}, t) \cdot \boldsymbol{n}^{\mathrm{f}} = \boldsymbol{\sigma}_N^{\mathrm{f}} \quad \text{on } \Gamma_N^{\mathrm{f}}. \tag{3.69}$$

Here, $\boldsymbol{v}_D^{\mathrm{f}}$ and $\boldsymbol{\sigma}_N^{\mathrm{f}}$ are the prescribed fluid velocity and traction functions along Γ_D^{f} and Γ_N^{f} respectively.

Similarly, the Dirichlet and Neumann boundary conditions along the non-interface structural boundaries are

$$\boldsymbol{v}^{\mathrm{s}}(\boldsymbol{X}, t) = \boldsymbol{v}_D^{\mathrm{s}} \quad \text{on } \Gamma_D^{\mathrm{s}} \qquad \text{and} \qquad \boldsymbol{\sigma}^{\mathrm{s}}(\boldsymbol{\varphi}^{\mathrm{s}}) \cdot \boldsymbol{n}^{\mathrm{s}} = \boldsymbol{\sigma}_N^{\mathrm{s}} \quad \text{on } \Gamma_N^{\mathrm{s}}, \tag{3.70}$$

where $\boldsymbol{v}_D^{\mathrm{s}}$ denotes the solid velocity function defined along the Dirichlet boundary Γ_D^{s} and $\boldsymbol{\sigma}_N^{\mathrm{s}}$ is the prescribed traction force on the Neumann boundary Γ_N^{s}. Besides the boundary conditions defined in Eqs. (3.66–3.70), the governing equations Eqs. (3.49) and (3.64–3.65) should also be accompanied by the initial conditions to make the problem well-posed. The initial condition for the fluid velocity over domain $\Omega^{\mathrm{f}}(0)$

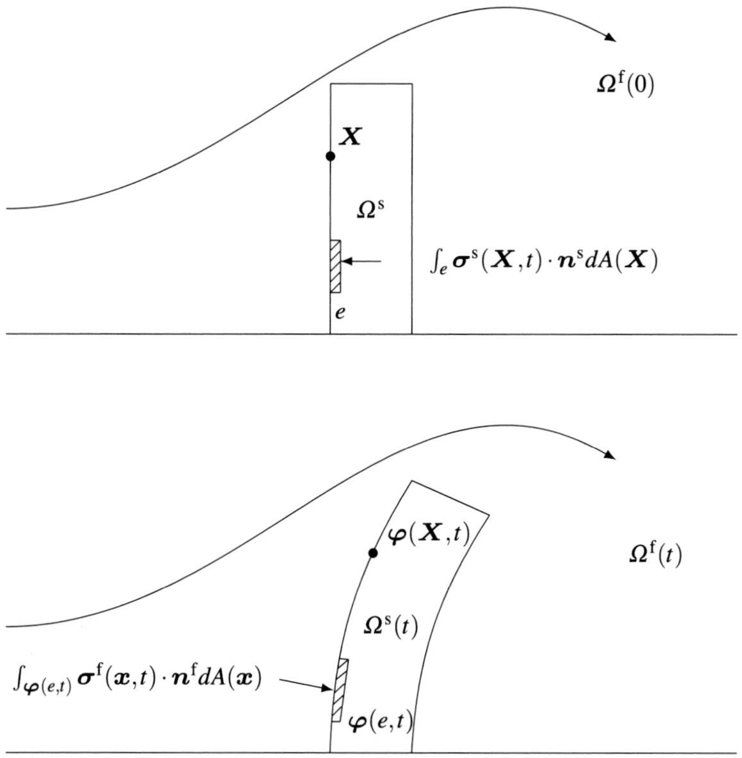

Fig. 3.6 Traction and velocity continuity conditions

are defined as

$$\boldsymbol{v}^{\mathrm{f}}(\cdot,0) = \boldsymbol{v}_0^{\mathrm{f}} \quad \text{on } \Omega^{\mathrm{f}}(0), \tag{3.71}$$

where $\boldsymbol{v}_0^{\mathrm{f}}$ is divergence-free, i.e. $\boldsymbol{v}_0^{\mathrm{f}}$ satisfies Eq. (3.47). Similarly, the initial velocity and displacement for the structural dynamics Eq. (3.49) are

$$\boldsymbol{v}^{\mathrm{s}}(\cdot,0) = \boldsymbol{v}_0^{\mathrm{s}}, \qquad \boldsymbol{\varphi}^{\mathrm{s}}(\cdot,0) = \boldsymbol{\varphi}_0^{\mathrm{s}} \quad \text{on } \Omega^{\mathrm{s}}, \tag{3.72}$$

where $\boldsymbol{v}_0^{\mathrm{s}}$ and $\boldsymbol{\varphi}_0^{\mathrm{s}}$ are the structural velocity and deformation mapping function prescribed on the computational domain Ω^{s}.

The fluid and the structural equations are coupled by the continuity of velocity and traction along the fluid-structure interface (Fig. 3.6).

3.5 Application of ALE Formulation for FSI Equations

Given any $t^{\mathrm{n}} \in \mathbb{R}$, when the time t reaches t^{n}, we define a backward in time mapping $\widehat{\boldsymbol{\varphi}}^{\mathrm{f,n}}(\cdot, t)$ which maps $\Omega^{\mathrm{f}}(t^{\mathrm{n}})$ to $\Omega^{\mathrm{f}}(t)$ for $t \leq t^{\mathrm{n}}$, and reduces to identity map when $t = t^{\mathrm{n}}$. We can rewrite by the ALE description

$$\int_{\Omega^{\mathrm{f}}(t^{\mathrm{n}})} \rho^{\mathrm{f}} \left(\frac{d}{dt} \boldsymbol{v}^{\mathrm{f}}(\widehat{\boldsymbol{\varphi}}^{\mathrm{f,n}}(\boldsymbol{x}, t), t) \Big|_{t=t^{\mathrm{n}}} + (\boldsymbol{v}^{\mathrm{f}} - \partial_t \widehat{\boldsymbol{\varphi}}^{\mathrm{f,n}}(\boldsymbol{x}, t^{\mathrm{n}})) \cdot \nabla \boldsymbol{v}^{\mathrm{f}} \right) \cdot \boldsymbol{\psi}^{\mathrm{f}} d\Omega$$
$$+ \int_{\Omega^{\mathrm{f}}(t^{\mathrm{n}})} \boldsymbol{\sigma}^{\mathrm{f}} : \nabla \boldsymbol{\psi}^{\mathrm{f}} d\Omega - \int_{\Omega^{\mathrm{f}}(t)} (\nabla \cdot \boldsymbol{v}^{\mathrm{f}}) q d\Omega$$
$$+ \int_{\Omega^{\mathrm{s}}} \rho^{\mathrm{s}} \partial_t \boldsymbol{v}^{\mathrm{s}}(\boldsymbol{X}, t^{\mathrm{n}}) \cdot \boldsymbol{\psi}^{\mathrm{s}} d\Omega + \int_{\Omega^{\mathrm{s}}} \boldsymbol{\sigma}^{\mathrm{s}} : \nabla \boldsymbol{\psi}^{\mathrm{s}} d\Omega =$$
$$\int_{\Omega^{\mathrm{f}}(t^{\mathrm{n}})} \rho^{\mathrm{f}} \boldsymbol{b}^{\mathrm{f}} \cdot \boldsymbol{\psi}^{\mathrm{f}} d\Omega + \int_{\Sigma_2} \boldsymbol{\sigma}_N^{\mathrm{f}} \cdot \boldsymbol{\psi}^{\mathrm{f}} d\Gamma + \int_{\Omega^{\mathrm{s}}} \rho^{\mathrm{s}} \boldsymbol{b}^{\mathrm{s}} \cdot \boldsymbol{\psi}^{\mathrm{s}} d\Gamma + \int_{\Sigma_4} \boldsymbol{\sigma}_N^{\mathrm{s}} \cdot \boldsymbol{\psi}^{\mathrm{s}} d\Gamma. \quad (3.73)$$

Here $\boldsymbol{\psi}^{\mathrm{f}}$, $\boldsymbol{\psi}^{\mathrm{s}}$ and q are the weighting or test functions, which will be discussed later in the context of weak/variational formulation.

The motion of the spatial coordinates on the fluid domain can be simply modeled as an elastic material in equilibrium,

$$\nabla \cdot \boldsymbol{\sigma}^{\mathrm{m}} = \mathbf{0}, \quad \text{with } \boldsymbol{\sigma}^{\mathrm{m}} = (1 + k_{\mathrm{m}}) \left[\left(\nabla \boldsymbol{u}^{\mathrm{f}} + \left(\nabla \boldsymbol{u}^{\mathrm{f}} \right)^T \right) + (\nabla \cdot \boldsymbol{u}^{\mathrm{f}}) \boldsymbol{I} \right], \quad (3.74)$$

where $\boldsymbol{u}^{\mathrm{f}}$ denotes the ALE mesh nodal displacement satisfying the boundary conditions

$$\boldsymbol{u}^{\mathrm{f}} = \boldsymbol{\varphi}(\boldsymbol{X}, t) - \boldsymbol{X} \qquad \text{on } \Gamma, \quad (3.75)$$
$$\boldsymbol{u}^{\mathrm{f}}(\boldsymbol{\chi}, t) = 0 \qquad \text{on } \partial\Omega^{\mathrm{f}}(0) \backslash \Gamma. \quad (3.76)$$

In the above equation, k_{m} represents the local element level mesh stiffness parameter chosen as a function of the element sizes to limit the distortion of the small elements located in the immediate vicinity of the fluid-structure interface. For each element $T_i \in \Omega^{\mathrm{f}}(t)$, the mesh stiffness parameter k_{m} is defined as $k_{\mathrm{m}} = \frac{\max_j |T_j| - \min_j |T_j|}{|T_i|}$ [146, 200]. The solution of Eq. (3.74) provides us the ALE mesh nodal displacements, which can be used to update the spatial nodes

$$\boldsymbol{x} = \widehat{\boldsymbol{\varphi}}^{\mathrm{f}}(\boldsymbol{\chi}, t) = \boldsymbol{\chi} + \boldsymbol{u}^{\mathrm{f}}(\boldsymbol{\chi}, t), \qquad \forall\, \boldsymbol{x} \in \Omega^{\mathrm{f}}(t) \text{ and } \boldsymbol{\chi} \in \Omega^{\mathrm{f}}(0). \quad (3.77)$$

To summarize, the interface coupling of the fluid with the structure equations must satisfy three conditions:

- continuity of velocities (no-slip);
- continuity of tractions (dynamic equilibrium);
- continuity of displacements (geometrical coupling of physical solids and fluid mesh motion).

The last condition comes from geometric compatibility while the first two conditions are of physical nature. Mathematically, the first and third condition can be classified as Dirichlet conditions and the second condition is a Neumann condition. Here, for fluid-structure interaction based on the 'arbitrary Lagrangian Eulerian' framework (ALE), the choice of an appropriate fluid mesh movement PDE is important in order to guarantee the Jacobian condition $J > 0$.

Remark: We will further discuss the weak form, the finite element formulation and the computer implementation aspects of the ALE fluid-structure equations in the upcoming chapters.

Bibliography Notes

Further materials on ALE formulation and the application to FSI system can be found [26, 56, 98, 105, 129, 132, 139].

Chapter 4
Variational and Stabilized Finite Element Methods

4.1 Introduction

Before proceeding with the variational formulation of the fluid-structure coupled system, let us look at the convection-diffusion-reaction (CDR) equation which forms a canonical equation for any continuum transport system. The present chapter discusses the variational formulation and finite element technique applied to the CDR equation and reviews various types of stabilization methods.

The CDR equation is given as:

$$\frac{\partial \varphi}{\partial t} + \boldsymbol{v} \cdot \nabla \varphi - \nabla \cdot (\boldsymbol{k} \nabla \varphi) + s\varphi = f, \tag{4.1}$$

on a d-dimensional domain $\Omega(t) \subset \mathbb{R}^d$, where φ is the unknown transport variable, $\boldsymbol{v}$, $\boldsymbol{k}$, s and f are the convection velocity, diffusivity tensor ($\boldsymbol{k} = k\boldsymbol{I}$ for isotropic diffusion, k and $\boldsymbol{I}$ being the diffusion coefficient and identity tensor respectively), reaction coefficient and source respectively.

Numerical discretization of the domain cannot capture the continuum effects properly, thus leading to numerical errors in the solution. The various effects in the CDR equation are (a) convection, which is characterized by the first-order spatial derivatives and transports any information along the characteristic curves with velocity $\boldsymbol{v}$, (b) diffusion, which is represented by the second-order spatial derivatives and spreads the information through the domain with the diffusion coefficient $\boldsymbol{k}$, and (c) reaction, denoted by $s\varphi$ which is the cause of any production or destruction in the information.

The roots of the characteristic polynomial of Eq. (4.1) in one-dimension (the sign of $D = |\boldsymbol{v}|^2 + 4ks$) can be utilized to characterize the exact solution. When $D \geq 0$, the solution is said to be in the exponential regime, with production for $s < 0$ and dissipation (or destruction) for $s \geq 0$. It constitutes the propagation regime when $D < 0$. The solution exhibits distinctive properties based on the dominant effects of the CDR equation and the boundary conditions. A boundary layer-like behavior may

R. K. Jaiman and V. Joshi, *Computational Mechanics of Fluid-Structure Interaction*,
https://doi.org/10.1007/978-981-16-5355-1_4

be observed near the regions where high gradients of the solution exist. At the outflow boundaries, exponential layers are formed, while parabolic or characteristic boundary layers are formed along the tangent to the flow direction. Any sharp discontinuity inside the computational domain can also induce such layers, called parabolic or characteristic internal layers [202].

Numerical solution with the standard Galerkin weighted residual method leads to spurious oscillations when the effects are convection dominant [93]. These oscillations are due to the inability of the discretization to resolve the high gradients. To stabilize these oscillations, several methods have been proposed in the literature called Petrov-Galerkin stabilization where the weighting function is perturbed based on a stabilization parameter τ, which tends to capture the missing subgrid scale physics. The perturbation acts as an upwinding function and removes the effect of the outflow boundary condition on the convection term. Some of the widely used techniques are streamline upwind Petrov-Galerkin (SUPG) [34], Galerkin/least-squares (GLS) [97] and subgrid scale (SGS) [48] methods. A detailed discussion about the development of these types of stabilization techniques in the literature can be found in Appendix at the end of this chapter.

In the upcoming sections, we discuss widely used linear stabilization methods (SUPG, GLS and SGS) for the discretization of the CDR equation and their numerical properties. We then derive the nonlinear expressions to impart the positivity condition at the element matrix level to form the positivity preserving variational (PPV) method [112] and generalize it to different types of mesh discretizations.

4.2 The Convection-Diffusion-Reaction Equation

The CDR equation forms a canonical form for complex transport systems like Navier-Stokes, turbulence, two-phase equations and so on. Here, we first review the semi-discrete variational form of the equation.

4.2.1 *Strong Differential Form*

Consider a d-dimensional spatial domain $\Omega(t) \subset \mathbb{R}^d$ with the Dirichlet and Neumann boundaries denoted by $\Gamma_D^{\varphi}(t)$ and $\Gamma_N^{\varphi}(t)$ respectively. The strong form of the CDR equation (with φ as the scalar variable) along with the boundary conditions can be written as

$$\frac{\partial \varphi}{\partial t} + \boldsymbol{v} \cdot \nabla \varphi - \nabla \cdot (\boldsymbol{k} \nabla \varphi) + s\varphi = f, \qquad \text{on } \Omega(t) \times [0, T], \tag{4.2}$$

$$\varphi = \varphi_D, \qquad \text{on } \Gamma_D^{\varphi}(t) \times [0, T], \tag{4.3}$$

$$\boldsymbol{k} \nabla \varphi \cdot \boldsymbol{n}^{\varphi} = \varphi_N, \qquad \text{on } \Gamma_N^{\varphi}(t) \times [0, T], \tag{4.4}$$

$$\varphi = \varphi_0, \qquad \text{on } \Omega(0), \tag{4.5}$$

where $\varphi = \varphi(\boldsymbol{x}, t)$ is the scalar unknown which depends on the spatial and temporal coordinates $\boldsymbol{x}$ and t respectively, $\boldsymbol{v}$, $\boldsymbol{k}$, s and f are the convection velocity, the diffusivity tensor, the reaction coefficient and the source term respectively. Here, we assume that the diffusivity tensor $\boldsymbol{k}$ is isotropic, i.e., $\boldsymbol{k} = k\boldsymbol{I}$, $\boldsymbol{I}$ being the identity tensor and k is a scalar positive quantity defined as the diffusion coefficient of the problem. The Dirichlet and Neumann boundary conditions are given by φ_D and φ_N respectively and $\boldsymbol{n}^\varphi$ represents the unit normal to the Neumann boundary. φ_0 denotes the initial condition on φ.

4.2.2 Semi-Discrete Variational Form

We utilize the generalized-α method [44] to discretize the equation in time. This technique allows user-defined high-frequency damping via controlling a parameter called the spectral radius ρ_∞, which is helpful for coarser discretization in space and time. The following expressions are employed for the temporal discretization:

$$\varphi^{\mathrm{n+1}} = \varphi^{\mathrm{n}} + \Delta t \partial_t \varphi^{\mathrm{n}} + \gamma \Delta t (\partial_t \varphi^{\mathrm{n+1}} - \partial_t \varphi^{\mathrm{n}}), \tag{4.6}$$

$$\partial_t \varphi^{\mathrm{n}+\alpha_{\mathrm{m}}} = \partial_t \varphi^{\mathrm{n}} + \alpha_{\mathrm{m}} (\partial_t \varphi^{\mathrm{n+1}} - \partial_t \varphi^{\mathrm{n}}), \tag{4.7}$$

$$\varphi^{\mathrm{n}+\alpha} = \varphi^{\mathrm{n}} + \alpha (\varphi^{\mathrm{n+1}} - \varphi^{\mathrm{n}}), \tag{4.8}$$

where γ, α and α_{m} are the generalized-α parameters given by

$$\alpha_{\mathrm{m}} = \frac{1}{2}\left(\frac{3 - \rho_\infty}{1 + \rho_\infty}\right), \quad \alpha = \frac{1}{1 + \rho_\infty}, \quad \gamma = \frac{1}{2} + \alpha_{\mathrm{m}} - \alpha. \tag{4.9}$$

In the present chapter, we select the parameters $\alpha = \alpha_{\mathrm{m}} = \gamma = 0.5$ which correspond to $\rho_\infty = 1$ so that $\partial_t \varphi^{\mathrm{n}+\alpha_{\mathrm{m}}} = (\varphi^{\mathrm{n}+\alpha} - \varphi^{\mathrm{n}})/(\alpha \Delta t)$. The temporally discretized CDR equation can thus be written as

$$\partial_t \varphi^{\mathrm{n}+\alpha_{\mathrm{m}}} + \boldsymbol{v} \cdot \nabla \varphi^{\mathrm{n}+\alpha} - \nabla \cdot (\boldsymbol{k} \nabla \varphi^{\mathrm{n}+\alpha}) + s \varphi^{\mathrm{n}+\alpha} = f(t^{\mathrm{n}+\alpha}), \tag{4.10}$$

$$\frac{\varphi^{\mathrm{n}+\alpha} - \varphi^{\mathrm{n}}}{\alpha \Delta t} + \boldsymbol{v} \cdot \nabla \varphi^{\mathrm{n}+\alpha} - \nabla \cdot (\boldsymbol{k} \nabla \varphi^{\mathrm{n}+\alpha}) + s \varphi^{\mathrm{n}+\alpha} = f(t^{\mathrm{n}+\alpha}). \tag{4.11}$$

We can take the temporal derivative coefficient inside the reaction term so that Eq. (4.11) can also be expressed as

$$\boldsymbol{v} \cdot \nabla \varphi^{\mathrm{n}+\alpha} - \nabla \cdot (\boldsymbol{k} \nabla \varphi^{\mathrm{n}+\alpha}) + \left(s + \frac{1}{\alpha \Delta t}\right) \varphi^{\mathrm{n}+\alpha} = \left(f + \frac{1}{\alpha \Delta t} \varphi^{\mathrm{n}}\right). \tag{4.12}$$

The above equation can be observed as a steady-state equation with modified reaction coefficient and source term. Let the modified coefficients be given by $\tilde{\boldsymbol{v}}$, $\tilde{\boldsymbol{k}}$, $\tilde{s}$ and $\tilde{f}$ defined as

$$\tilde{\boldsymbol{v}} = \boldsymbol{v}, \qquad \tilde{\boldsymbol{k}} = \boldsymbol{k}, \qquad \tilde{s} = s + \frac{1}{\alpha \Delta t}, \qquad \tilde{f} = f + \frac{1}{\alpha \Delta t}\varphi^{\mathrm{n}}. \tag{4.13}$$

Therefore, we will now discretize the following equation in the spatial domain:

$$\tilde{\boldsymbol{v}} \cdot \nabla \varphi^{\mathrm{n}+\alpha} - \nabla \cdot (\tilde{\boldsymbol{k}} \nabla \varphi^{\mathrm{n}+\alpha}) + \tilde{s}\varphi^{\mathrm{n}+\alpha} = \tilde{f} \qquad \text{on } \Omega(t). \tag{4.14}$$

The domain $\Omega(t)$ is discretized into $\mathrm{n_{el}}$ number of elements such that $\Omega(t) = \cup_{\mathrm{e}=1}^{\mathrm{n_{el}}} \Omega^{\mathrm{e}}$ and $\emptyset = \cap_{\mathrm{e}=1}^{\mathrm{n_{el}}} \Omega^{\mathrm{e}}$. The space of trial solution and test function, $\mathcal{S}_\varphi^{\mathrm{h}}$ and $\mathcal{V}_\varphi^{\mathrm{h}}$ respectively for the variational formulation are defined as

$$\mathcal{S}_\varphi^{\mathrm{h}} = \{\varphi_{\mathrm{h}} \mid \varphi_{\mathrm{h}} \in H^1(\Omega(t)), \varphi_{\mathrm{h}} = \varphi_D \text{ on } \Gamma_D^\varphi(t)\}, \tag{4.15}$$

$$\mathcal{V}_\varphi^{\mathrm{h}} = \{w_{\mathrm{h}} \mid w_{\mathrm{h}} \in H^1(\Omega(t)), w_{\mathrm{h}} = 0 \text{ on } \Gamma_D^\varphi(t)\}. \tag{4.16}$$

These function spaces form the mathematical preliminaries to the variational formulation and their properties have been discussed to some extent in Appendix at the end of this chapter. The variational statement for the discretized CDR equation is thus given as: find $\varphi_{\mathrm{h}}(\boldsymbol{x}, t^{\mathrm{n}+\alpha}) \in \mathcal{S}_\varphi^{\mathrm{h}}$ such that $\forall w_{\mathrm{h}} \in \mathcal{V}_\varphi^{\mathrm{h}}$,

$$\int_{\Omega(t)} \Big(w_{\mathrm{h}}(\tilde{\boldsymbol{v}} \cdot \nabla \varphi_{\mathrm{h}}) - w_{\mathrm{h}} \nabla \cdot (\tilde{\boldsymbol{k}} \nabla \varphi_{\mathrm{h}}) + w_{\mathrm{h}} \tilde{s} \varphi_{\mathrm{h}} \Big) d\Omega = \int_{\Omega(t)} w_{\mathrm{h}} \tilde{f} d\Omega, \tag{4.17}$$

Using the divergence theorem and the fact that $w_{\mathrm{h}} = 0$ on $\Gamma_D^\varphi(t)$, Eq. (4.17) becomes

$$\int_{\Omega(t)} \Big(w_{\mathrm{h}}(\tilde{\boldsymbol{v}} \cdot \nabla \varphi_{\mathrm{h}}) + \nabla w_{\mathrm{h}} \cdot (\tilde{\boldsymbol{k}} \nabla \varphi_{\mathrm{h}}) + w_{\mathrm{h}} \tilde{s} \varphi_{\mathrm{h}} \Big) d\Omega = \int_{\Omega(t)} w_{\mathrm{h}} \tilde{f} d\Omega + \int_{\Gamma_N^\varphi} w_{\mathrm{h}} \varphi_N d\Gamma. \tag{4.18}$$

As a result of spurious global oscillations and instability for convection- and reaction-dominated regimes in the Galerkin finite element method, various stabilization techniques have been proposed in the literature, the most widely used of which are SUPG and GLS methods. The stability is introduced through perturbing the test or weighting function so that the effect of upwinding is achieved. The standard variational formulation for such methods is: find $\varphi_{\mathrm{h}}(\boldsymbol{x}, t^{\mathrm{n}+\alpha}) \in \mathcal{S}_\varphi^{\mathrm{h}}$ such that $\forall w_{\mathrm{h}} \in \mathcal{V}_\varphi^{\mathrm{h}}$:

$$\int_{\Omega(t)} \Big(w_{\mathrm{h}}(\tilde{\boldsymbol{v}} \cdot \nabla \varphi_{\mathrm{h}}) + \nabla w_{\mathrm{h}} \cdot (\tilde{\boldsymbol{k}} \nabla \varphi_{\mathrm{h}}) + w_{\mathrm{h}} \tilde{s} \varphi_{\mathrm{h}} \Big) d\Omega$$

$$+ \sum_{\mathrm{e}=1}^{\mathrm{n_{el}}} \int_{\Omega^{\mathrm{e}}} \mathcal{L}^m w_{\mathrm{h}} \tau (\tilde{\mathcal{L}} \varphi_{\mathrm{h}} - \tilde{f}) d\Omega = \int_{\Omega(t)} w_{\mathrm{h}} \tilde{f} d\Omega + \int_{\Gamma_N^\varphi} w_{\mathrm{h}} \varphi_N d\Gamma, \tag{4.19}$$

Table 4.1 Differential operators on the weighting function for stabilization methods

Method	Stabilization operator ($\mathscr{L}^m$)
SUPG	$\mathscr{L}_{adv} = \tilde{\boldsymbol{v}} \cdot \nabla$
GLS	$\tilde{\mathscr{L}} = \tilde{\boldsymbol{v}} \cdot \nabla - \nabla \cdot (\tilde{\boldsymbol{k}}\nabla) + \tilde{s}$
SGS	$-\tilde{\mathscr{L}}^* = \tilde{\boldsymbol{v}} \cdot \nabla + \nabla \cdot (\tilde{\boldsymbol{k}}\nabla) - \tilde{s}$

where $\mathscr{L}^m$ is the operator on the weighting function given in Table 4.1 and the expression for the stabilization parameter τ is [193]

$$\tau = \left[\left(\frac{1}{\alpha \Delta t}\right)^2 + 9\left(\frac{4\tilde{k}}{h^2}\right)^2 + \left(\frac{2|\tilde{\boldsymbol{v}}|}{h}\right)^2 + \tilde{s}^2\right]^{-1/2}, \tag{4.20}$$

where h is the characteristic element length and $|\tilde{\boldsymbol{v}}|$ is the magnitude of the convection velocity. The formula for τ has been extensively studied in the literature with several variations, and can be estimated through error analysis. The generality of the expression is a topic of discussion in Sect. 4.3.2.4. The residual of the CDR equation is defined as

$$\mathscr{R}(\varphi_{\mathrm{h}}) = \tilde{\boldsymbol{v}} \cdot \nabla\varphi_{\mathrm{h}} - \nabla \cdot (\tilde{\boldsymbol{k}}\nabla\varphi_{\mathrm{h}}) + \tilde{s}\varphi_{\mathrm{h}} - \tilde{f} = \tilde{\mathscr{L}}\varphi_{\mathrm{h}} - \tilde{f}, \tag{4.21}$$

where $\tilde{\mathscr{L}}$ is the differential operator corresponding to the differential Eq. (4.14).

4.3 The Positivity Preserving Variational (PPV) Method

We discuss how we can impart the positivity property to the variational formulation in this section. We design the nonlinear stabilization terms to reduce the oscillations which remain even after the linear stabilization of SUPG or GLS discussed in the previous section.

4.3.1 Linear Stabilization

We begin by analyzing the linear stabilization methods (SUPG, GLS and SGS) with respect to the effect of the sign of the reaction coefficient ($\tilde{s}$), owing to the destruction or production effects. Fourier analysis of the discretized methods (GLS and SGS) [112] showed that the SGS method performs well when $\tilde{s}$ is negative, but, it loses accuracy when $\tilde{s} >> 0$ due to excessive dissipation [87]. On the other hand, the GLS method is not as diffusive as SGS when $\tilde{s} \geq 0$, but it suffers from phase error when $\tilde{s} < 0$. Thus, we select a combination of these methods, so that the formulation is benefited in both production and destruction regimes. Note that the effect of the diffusion term is assumed negligible in the differential operator owing to the use of linear and multilinear finite elements. The second term in Eq. (4.19) can thus be modified as

$$\sum_{\mathrm{e}=1}^{\mathrm{n_{el}}} \int_{\Omega^{\mathrm{e}}} \mathscr{L} w_{\mathrm{h}} \tau (\tilde{\mathscr{L}} \varphi_{\mathrm{h}} - \tilde{f}) d\Omega, \tag{4.22}$$

where $\mathscr{L} = \tilde{\boldsymbol{v}} \cdot \nabla - \nabla \cdot (\tilde{\boldsymbol{k}} \nabla) + |\tilde{s}|$. Here, the absolute value function $\mathscr{F}(x) = |x|$ is defined as

$$\mathscr{F}(x) = \begin{cases} x, & \text{if } x > 0, \\ 0, & \text{if } x = 0, \\ -x, & \text{if } x < 0. \end{cases} \tag{4.23}$$

As can be easily observed, the above method will behave as GLS for $\tilde{s} \geq 0$ and as SGS when $\tilde{s} < 0$. We next design the nonlinear stabilization terms to maintain the positivity property in the variational procedure.

4.3.2 *Positivity and Nonlinear Stabilization*

As mentioned earlier, SUPG and GLS methods are successful in the reduction of the spurious oscillations which are observed in the Galerkin formulation, there still exist some overshoots and undershoots in the solution near the regions of high gradients or a sharp discontinuity in the solution. This is a result of the fact that neither of two methods is monotone or positivity preserving [45]. Physical solutions in nature should preclude such oscillations, maintaining the sign of their neighbor. Therefore, we look into some of the key criteria to enforce the positivity property at the discrete level.

4.3.2.1 The Positivity Condition

To describe the positivity condition, we consider the Galerkin formulation for the convection-dominated problems. The positivity preserving condition for the Eq. (4.2) can be defined by considering a simplified form of the CDR equation with only convection effects:

$$\partial_t \varphi + \boldsymbol{v} \cdot \nabla \varphi = 0. \tag{4.24}$$

The finite element approximation of an explicit scheme for a one-dimensional element between $i-1$ and i nodes can be written as

$$\frac{\varphi_i^{\mathrm{n}+1}-\varphi_i^{\mathrm{n}}}{\Delta t} = -\int\limits_{\Omega(t)} w_{\mathrm{h}}\left(u\frac{\partial \varphi_{\mathrm{h}}}{\partial x}\right)d\Omega = -\int\limits_{\Omega(t)} N^T u\frac{\partial N}{\partial x}d\Omega \begin{bmatrix}\varphi_{i-1}^{\mathrm{n}}\\ \varphi_i^{\mathrm{n}}\end{bmatrix}$$
$$= -\frac{u}{2h}\begin{bmatrix}-1 & 1\\ -1 & 1\end{bmatrix}\begin{bmatrix}\varphi_{i-1}^{\mathrm{n}}\\ \varphi_i^{\mathrm{n}}\end{bmatrix}, \tag{4.25}$$

where N is the row-vector of shape functions based on linear Lagrange polynomials for one-dimensional elements satisfying the partition of unity property and u is the convection velocity in one-dimension. Therefore, after assembly of the elements for a uniform grid, the finite element based stencil can be written as follows

$$\frac{\varphi_i^{\mathrm{n}+1}-\varphi_i^{\mathrm{n}}}{\Delta t} = -\frac{u}{2h}\left(\varphi_{i+1}^{\mathrm{n}}-\varphi_{i-1}^{\mathrm{n}}\right), \tag{4.26}$$

which has a similar structure as that of the central difference scheme. Any scheme that can be written in the form

$$\frac{\varphi_i^{\mathrm{n}+1}-\varphi_i^{\mathrm{n}}}{\Delta t} = C^+(\varphi_{i+1}^{\mathrm{n}}-\varphi_i^{\mathrm{n}}) - C^-(\varphi_i^{\mathrm{n}}-\varphi_{i-1}^{\mathrm{n}}), \tag{4.27}$$

satisfies the positivity preserving property if the coefficients C^+ and C^- satisfy [85]

$$C^+ \geq 0, \qquad C^- \geq 0, \qquad C^+ + C^- \leq 1. \tag{4.28}$$

Note that the Galerkin approximation in Eq. (4.26) does not satisfy this condition. To impart the positivity property, the scheme has to be modified by addition of the stabilization terms as suggested by [85] for finite difference approximations.

We observed the case with an explicit scheme. Now, let us see what happens in the case of an implicit scheme. An implicit scheme can be expressed as a system of linear equations of the form $\boldsymbol{A}\boldsymbol{\varphi} = \boldsymbol{b}$, where $\boldsymbol{A}$ and $\boldsymbol{b}$ are the left-hand side matrix and the right-hand side vector or force vector respectively, with $\boldsymbol{\varphi}$ being the vector of unknowns. The positivity preservation property can be generalized to the implicit matrix form of the scheme by transforming the matrix $\boldsymbol{A} = \{a_{ij}\}$ to an M-matrix which ensures the positivity and convergence. The elements of the M-matrix have the following properties [130]

$$a_{ii} > 0, \forall i, \tag{4.29}$$
$$a_{ij} \leq 0, \forall j \neq i, \tag{4.30}$$
$$\sum_j a_{ij} = 0, \forall i. \tag{4.31}$$

These make up the sufficient conditions for imposing the positivity property to the matrix $\boldsymbol{A}$. The transformation of $\boldsymbol{A}$ into an M-matrix is carried out by the addition of the discrete upwind matrix $\boldsymbol{D} = \{d_{ij}\}$ to $\boldsymbol{A}$, the elements of which satisfy [131]:

$$d_{ij} = d_{ji} = -\max\{0, a_{ij}, a_{ji}\}, \tag{4.32}$$

$$d_{ii} = -\sum_{j \neq i} d_{ij}. \tag{4.33}$$

Remark 4.1 Galerkin as well as linear stabilization methods such as SUPG and GLS are not positivity preserving methods [45]. Addition of the linear stabilization term in SUPG and GLS does not satisfy the required conditions in Eqs. (4.29–4.31) and therefore they are not positivity preserving. The PPV method takes the combination of the GLS-SGS approach in the linear stabilization (which is not positivity preserving) and further adds a nonlinear stabilization term, which is designed in the next section, to impose the positivity property.

4.3.2.2 Enforcement of Positivity Condition In One-Dimension

Let us design the nonlinear stabilization term to impart the positivity property in the linear stabilization method based on combined GLS-SGS methodology in Sect. 4.3.1. We begin by considering the steady-state CDR equation in one-dimension:

$$u\frac{\mathrm{d}\varphi}{\mathrm{d}x} - k\frac{\mathrm{d}^2\varphi}{\mathrm{d}x^2} + s\varphi = f, \quad u \geq 0. \tag{4.34}$$

Assume that only Dirichlet boundary conditions are specified and $f = 0$ for simplicity. The variational formulation gives: find $\varphi_\mathrm{h}(\boldsymbol{x}) \in \mathscr{S}^\mathrm{h}_\varphi$ such that $\forall w_\mathrm{h} \in \mathscr{V}^\mathrm{h}_\varphi$:

$$\int_\Omega \left(w_\mathrm{h} u \frac{\mathrm{d}\varphi_\mathrm{h}}{\mathrm{d}x} + \frac{\mathrm{d}w_\mathrm{h}}{\mathrm{d}x} k \frac{\mathrm{d}\varphi_\mathrm{h}}{\mathrm{d}x} + w_\mathrm{h} s \varphi_\mathrm{h} \right) d\Omega + \sum_{\mathrm{e}=1}^{\mathrm{n_{el}}} \int_{\Omega^e} \mathscr{L} w_\mathrm{h} \tau (\tilde{\mathscr{L}} \varphi_\mathrm{h}) d\Omega = 0, \tag{4.35}$$

where $\mathscr{L}$ is the differential operator of the PPV method, i.e., in one-dimension, $\mathscr{L} = u\frac{\mathrm{d}}{\mathrm{d}x} + |s|$ (Here, the higher-order diffusion term is neglected due to the linear finite elements), which gives

$$\int_\Omega \left(w_\mathrm{h} u \frac{\mathrm{d}\varphi_\mathrm{h}}{\mathrm{d}x} + \frac{\mathrm{d}w_\mathrm{h}}{\mathrm{d}x} k \frac{\mathrm{d}\varphi_\mathrm{h}}{\mathrm{d}x} + w_\mathrm{h} s \varphi_\mathrm{h} \right) d\Omega$$
$$+ \sum_{\mathrm{e}=1}^{\mathrm{n_{el}}} \int_{\Omega^e} \left(u\frac{\mathrm{d}w_\mathrm{h}}{\mathrm{d}x} + |s| w_\mathrm{h} \right) \tau \left(u\frac{\mathrm{d}\varphi_\mathrm{h}}{\mathrm{d}x} + s\varphi_\mathrm{h} \right) d\Omega = 0. \tag{4.36}$$

Let $\boldsymbol{A}_c^{\mathrm{e}} = [a_{c\{ij\}}^{\mathrm{e}}]$, $\boldsymbol{A}_d^{\mathrm{e}} = [a_{d\{ij\}}^{\mathrm{e}}]$ and $\boldsymbol{A}_r^{\mathrm{e}} = [a_{r\{ij\}}^{\mathrm{e}}]$ be the local element matrices for the first three terms in Eq. (4.36) corresponding to the convection, diffusion and reaction effects respectively. Discretizing the one-dimensional domain using two-node linear elements and $w_i = N_i$ being the shape functions at the corresponding nodes of the finite element, the matrices can be expressed as

$$a_{c\{ij\}}^{\mathrm{e}} = \int_{\Omega^{\mathrm{e}}} N_i u \frac{\mathrm{d}N_j}{\mathrm{d}x} d\Omega, \qquad a_{d\{ij\}}^{\mathrm{e}} = \int_{\Omega^{\mathrm{e}}} \frac{\mathrm{d}N_i}{\mathrm{d}x} k \frac{\mathrm{d}N_j}{\mathrm{d}x} d\Omega, \qquad a_{r\{ij\}}^{\mathrm{e}} = \int_{\Omega^{\mathrm{e}}} N_i s N_j d\Omega,$$

which are computed to be

$$\boldsymbol{A}_c^{\mathrm{e}} = \frac{u}{2}\begin{bmatrix} -1 & 1 \\ -1 & 1 \end{bmatrix}, \qquad \boldsymbol{A}_d^{\mathrm{e}} = \frac{k}{h}\begin{bmatrix} 1 & -1 \\ -1 & 1 \end{bmatrix}, \qquad \boldsymbol{A}_r^{\mathrm{e}} = \frac{sh}{6}\begin{bmatrix} 2 & 1 \\ 1 & 2 \end{bmatrix}, \tag{4.37}$$

where h is the characteristic length of the element. Similarly, each expression in Eq. (4.36) can be expressed in the matrix form. The combined element level matrix $\boldsymbol{A}^{\mathrm{e}}$ is thus given by

$$\boldsymbol{A}^{\mathrm{e}} = \begin{bmatrix} -\frac{u}{2} + \frac{k}{h} + \frac{sh}{3} + \frac{\tau u^2}{h} - \frac{u\tau s}{2} - \frac{u\tau|s|}{2} + \frac{\tau s|s|h}{3} & \frac{u}{2} - \frac{k}{h} + \frac{sh}{6} - \frac{\tau u^2}{h} - \frac{u\tau s}{2} + \frac{u\tau|s|}{2} + \frac{\tau s|s|h}{6} \\ -\frac{u}{2} - \frac{k}{h} + \frac{sh}{6} - \frac{\tau u^2}{h} + \frac{u\tau s}{2} - \frac{u\tau|s|}{2} + \frac{\tau s|s|h}{6} & \frac{u}{2} + \frac{k}{h} + \frac{sh}{3} + \frac{\tau u^2}{h} + \frac{u\tau s}{2} + \frac{u\tau|s|}{2} + \frac{\tau s|s|h}{3} \end{bmatrix}. \tag{4.38}$$

The discrete upwind matrix corresponding to $\boldsymbol{A}^{e}$ can be written as

$$\begin{aligned} \boldsymbol{D}^{\mathrm{e}} &= \begin{bmatrix} \max\left\{\frac{|u-u\tau s+u\tau|s||}{2} - \frac{k+\tau u^2}{h} + \frac{(s+\tau s|s|)h}{6}, 0\right\} & -\max\left\{\frac{|u-u\tau s+u\tau|s||}{2} - \frac{k+\tau u^2}{h} + \frac{(s+\tau s|s|)h}{6}, 0\right\} \\ -\max\left\{\frac{|u-u\tau s+u\tau|s||}{2} - \frac{k+\tau u^2}{h} + \frac{(s+\tau s|s|)h}{6}, 0\right\} & \max\left\{\frac{|u-u\tau s+u\tau|s||}{2} - \frac{k+\tau u^2}{h} + \frac{(s+\tau s|s|)h}{6}, 0\right\} \end{bmatrix} \\ &= \frac{k^{\mathrm{add}}}{h}\begin{bmatrix} 1 & -1 \\ -1 & 1 \end{bmatrix}, \end{aligned} \tag{4.39}$$

where k^{add} is

$$\boxed{k^{\mathrm{add}} = \max\left\{\frac{|u - \tau u s + \tau u|s||h}{2} - (k + \tau u^2) + \frac{(s + \tau s|s|)h^2}{6}, 0\right\}} \tag{4.40}$$

The discrete upwind matrix $\boldsymbol{D}^{\mathrm{e}}$ converts $\boldsymbol{A}^{\mathrm{e}}$ to an M-matrix when added to it, thus imparting the positivity property. It can be observed as addition of diffusion to the linear stabilized form. This procedure reduces the order of accuracy of the method to first-order. Therefore, to counteract the loss of accuracy, a nonlinear regulation of this added diffusion is required. This regulation is provided by a nonlinear residual-based solution-dependent parameter which adds diffusion only to the elements where a non-regular solution or oscillations are present. As a result, the loss of accuracy is avoided in the regions where smooth solution is present as the nonlinear term has a small contribution owing to the negligible residual.

The discrete PPV formulation for steady-state CDR equation in one-dimension is: find $\varphi_h(x) \in \mathscr{S}_\varphi^h$ such that $\forall w_h \in \mathscr{V}_\varphi^h$:

$$\int_\Omega \left(w_h (u \frac{d\varphi_h}{dx}) + k \frac{dw_h}{dx}\frac{d\varphi_h}{dx} + w_h s \varphi_h \right) d\Omega + \sum_{e=1}^{n_{el}} \int_{\Omega^e} \mathscr{L} w_h \tau (\tilde{\mathscr{L}} \varphi_h - f) d\Omega$$
$$+ \sum_{e=1}^{n_{el}} \int_{\Omega^e} \chi \frac{|\mathscr{R}(\varphi_h)|}{|\nabla \varphi_h|} k^{add} \frac{dw_h}{dx}\frac{d\varphi_h}{dx} d\Omega = \int_\Omega w_h f d\Omega + \int_{\Gamma_N^\varphi} w_h \varphi_N d\Gamma, \quad (4.41)$$

where $\mathscr{R}(\varphi_h)$ is the residual of the given equation and χ is a scaling parameter defined using the problem constants (u, k, s and h) which non-dimensionalizes $\frac{|\mathscr{R}(\varphi_h)|}{|\nabla \varphi_h|}$. The added term has been the standard form to add discontinuity capturing term to linear stabilized equation in the literature. The next step is to derive an expression for χ. Let us look into the variation of $\frac{|\mathscr{R}(\varphi_h)|}{|\nabla \varphi_h|}$ with s, u, k and h inside a finite element. To accomplish this, we consider a pure reaction problem ($s\varphi = f$) and follow a similar procedure as that in [155]. We obtain a lumped mass matrix after adding the discrete upwind matrix to the left-hand side of the element-level matrix. For such problems, it is known that Gibbs oscillations can be prevented by the lumped mass technique. Therefore, we evaluate $\frac{|\mathscr{R}(\varphi_h)|}{|\nabla \varphi_h|}$ assuming that the solution tends to that obtained by lumped mass approach. For this purpose, consider a step function as the source $f(x)$

$$f(x) = \begin{cases} 0, & \forall x \in [0, 0.25 + \eta h], \\ q, & \text{elsewhere}, \end{cases} \quad (4.42)$$

where h is the element length, η is the location of the discontinuity in the function $f(x)$. The lumped mass matrix solution can then be written as

$$\varphi = \left(\frac{q}{s}\right)\left\{0, \ldots, 0, \frac{(1-\eta)^2}{2}, \frac{(2-\eta^2)}{2}, 1, \ldots, 1\right\}. \quad (4.43)$$

Let us observe how $\frac{|\mathscr{R}(\varphi_h)|}{|\nabla \varphi_h|}$ behaves inside the element containing the discontinuity at η. For this particular element,

$$\mathscr{R}(\varphi_h) = s\varphi_h - f = q\left(\frac{(1-\xi)}{2}\frac{(1-\eta)^2}{2} + \frac{(1+\xi)}{2}\frac{(2-\eta^2)}{2}\right) - f, \quad (4.44)$$

$$\nabla \varphi_h = \frac{q}{2sh}(2\eta - 2\eta^2 + 1), \quad (4.45)$$

which gives

$$\frac{|\mathscr{R}(\varphi_h)|}{|\nabla \varphi_h|} = \left|\frac{\mathscr{R}(\varphi_h)}{\nabla \varphi_h}\right| = \left|\frac{sh}{2}\frac{(1-\xi)(1-\eta)^2 + (1+\xi)(2-\eta^2)}{2\eta - 2\eta^2 + 1} - \frac{2fsh}{q(2\eta - 2\eta^2 + 1)}\right|. \quad (4.46)$$

Generalizing for all the elements,

$$\frac{|\mathscr{R}(\varphi_{\rm h})|}{|\nabla\varphi_{\rm h}|} = \frac{|s\varphi_{\rm h} - f|}{|\nabla\varphi_{\rm h}|} = \left|\frac{sh}{2}p(\eta,\xi)\right|, \tag{4.47}$$

where $p(\eta,\xi)$ is a function depending on the location of the discontinuity η inside an element and the natural coordinate ξ and is given as

$$p(\eta,\xi) = \begin{cases} \frac{(1-\xi)(1-\eta)^2+(1+\xi)(2-\eta^2)}{2\eta-2\eta^2+1} - \frac{4f}{q(2\eta-2\eta^2+1)}, & \Omega^{\rm e} \text{ with discontinuity,} \\ (1+\xi) - \frac{4f}{q(1-\eta)^2}, & \Omega^{\rm e} \text{ adjoining discontinuity,} \\ 0, & \text{elsewhere.} \end{cases} \tag{4.48}$$

Thus, the nonlinear stabilization term can be expressed as

$$\int_{\Omega^{\rm e}} \chi \frac{|\mathscr{R}(\varphi_{\rm h})|}{|\nabla\varphi_{\rm h}|} k^{\rm add} \frac{{\rm d}w_{\rm h}}{{\rm d}x}\frac{{\rm d}\varphi_{\rm h}}{{\rm d}x} d\Omega = \int_{\Omega^{\rm e}} \chi \left|\frac{sh}{2}p(\eta,\xi)\right| k^{\rm add} \frac{{\rm d}w_{\rm h}}{{\rm d}x}\frac{{\rm d}\varphi_{\rm h}}{{\rm d}x} d\Omega. \tag{4.49}$$

Let the local element matrix after the discretization of the above term be denoted by $\boldsymbol{K}^{\rm e}_{\rm add}$. It can be written as

$$\boldsymbol{K}^{\rm e}_{\rm add} = \frac{k^{\rm add}\chi}{h} \int_{\Omega^{\rm e}} \left|\frac{sh}{2}p(\eta,\xi)\right| d\Omega \begin{bmatrix} 1 & -1 \\ -1 & 1 \end{bmatrix}. \tag{4.50}$$

After some algebraic manipulations, the above integral can be integrated using the property of the absolute function (Eq. 4.23) as

$$\int_{\Omega^{\rm e}} \left|\frac{sh}{2}p(\eta,\xi)\right| d\Omega = \begin{cases} \frac{|s|h^2}{2} r(\eta), & \Omega^{\rm e} \text{ with discontinuity,} \\ \frac{|s|h^2}{2}, & \Omega^{\rm e} \text{ adjoining discontinuity,} \\ 0, & \text{elsewhere,} \end{cases} \tag{4.51}$$

where $r(\eta)$ is a function of the location of the discontinuity inside the element $(0 \leq \eta \leq 1)$ given as [155]

$$r(\eta) = \left[\frac{1+2\eta(1-\eta)[1-2\eta(1-\eta)]}{1+2\eta(1-\eta)}\right]. \tag{4.52}$$

After the evaluation of the ratio $\frac{|\mathscr{R}(\varphi_{\rm h})|}{|\nabla\varphi_{\rm h}|}$, we next look into defining the parameter χ. As the change of variable to the local coordinates in the integration introduces Jacobian of the element ($h/2$ in one-dimensional case) which is geometry-dependent, we need to re-scale χ such that it becomes geometry-independent. This leads to scaling the integral as $\chi = 2/(|s|h)$ so that the regulatory term depends purely on the function

$r(\eta)$, which is geometry-independent. To extend this concept to various regimes, consider the following cases:

Case 1: For the convection-diffusion problem with $f = 0$,

$$\frac{|\mathscr{R}(\varphi_h)|}{|\nabla\varphi_h|} = \frac{|u\nabla\varphi_h|}{|\nabla\varphi_h|} = |u|. \tag{4.53}$$

Therefore, we define $\chi = 1/|u|$.

Case 2: In the diffusion-reaction equation, we construct the same expression for the integral as Eq. (4.49) which gives $\chi = 2/(|s|h)$. For linear finite elements, the Laplacian is exactly zero. But, for multilinear elements it is negligible. Therefore, we conclude that for the reaction-dominated problems ($u \to 0$), $\chi \to 2/(|s|h)$ and for the convection-dominated problems ($s \to 0$), $\chi \to 1/|u|$. Thus, χ can be defined as a combined expression as follows:

$$\boxed{\chi = \frac{2}{|s|h + 2|u|}} \tag{4.54}$$

Remark 4.2 Note that we have not made any assumption on the sign of the reaction coefficient s. The definition of the absolute function takes care of the change in the sign of the reaction coefficient in the evaluated integral.

The above derivation completes the design of the nonlinear PPV technique where we impart the positivity condition to the element-level matrix in one-dimension. The bounds for the positivity condition (Eq. 4.28) for the PPV formulation have been shown for some cases in [112]. We next intend to generalize the procedure to multi-dimensional CDR equation.

4.3.2.3 Extension to Multi-Dimensions

The discrete form of the PPV formulation for the steady-state CDR equation in multi-dimensions can be written as: find $\varphi_h(\boldsymbol{x}) \in \mathscr{S}^h_\varphi$ such that $\forall w_h \in \mathscr{V}^h_\varphi$:

$$\begin{aligned}
&\int_\Omega \Big(w_h(\boldsymbol{v}\cdot\nabla\varphi_h) + \nabla w_h \cdot (\boldsymbol{k}\nabla\varphi_h) + w_h s \varphi_h \Big) d\Omega + \sum_{e=1}^{n_{el}} \int_{\Omega^e} \mathscr{L} w_h \tau (\tilde{\mathscr{L}}\varphi_h - f) d\Omega \\
&+ \sum_{e=1}^{n_{el}} \int_{\Omega^e} \chi \frac{|\mathscr{R}(\varphi_h)|}{|\nabla\varphi_h|} k_s^{add} \nabla w_h \cdot \left(\frac{\boldsymbol{v}\otimes\boldsymbol{v}}{|\boldsymbol{v}|^2} \right) \cdot \nabla\varphi_h d\Omega \\
&+ \sum_{e=1}^{n_{el}} \int_{\Omega^e} \chi \frac{|\mathscr{R}(\varphi_h)|}{|\nabla\varphi_h|} k_c^{add} \nabla w_h \cdot \left(\boldsymbol{I} - \frac{\boldsymbol{v}\otimes\boldsymbol{v}}{|\boldsymbol{v}|^2} \right) \cdot \nabla\varphi_h d\Omega = \int_\Omega w_h f d\Omega + \int_{\Gamma_N^\varphi} w_h \varphi_N d\Gamma,
\end{aligned} \tag{4.55}$$

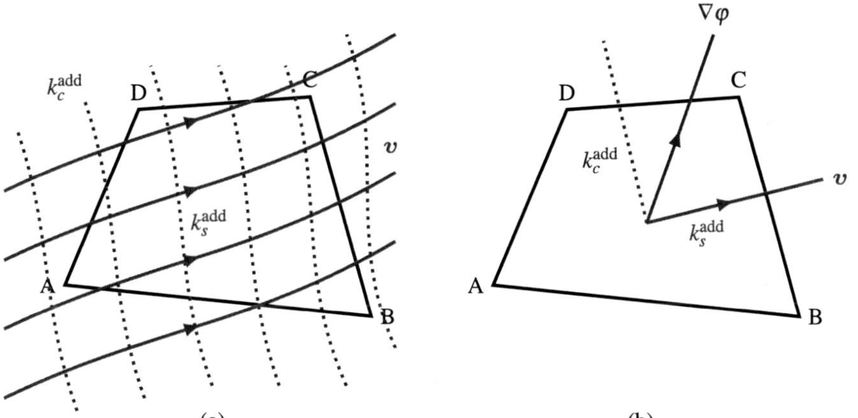

Fig. 4.1 Sketch of the residual-based stabilization inside a generic element with ABCD sub-domain boundary: **a** four-node quadrilateral element with convection velocity and the added diffusions k_s^{add} and k_c^{add} in the streamline and crosswind directions respectively; **b** added diffusions and solution gradient ($\nabla\varphi$) shown for designing positivity. Solid (red) and dashed (blue) lines represent the streamline and crosswind directions of the convection field respectively while solid (blue) line is the solution gradient vector

where k_s^{add} and k_c^{add} are the added diffusion in the streamline and the crosswind directions respectively. We directly extend the diffusion formula derived for the satisfaction of the positivity condition in one-dimension (Eq. 4.40) to multi-dimensions by working with two principle directions, viz., streamline and crosswind (Fig. 4.1). Implementation of this procedure requires only minor modifications to the existing stabilized finite element codes.

Remark 4.3 The PPV technique adds diffusion consistently in both the streamline and crosswind directions. For the case of velocity perpendicular to the solution gradient, positivity and local boundedness is thus maintained due to the sufficient diffusion (see Fig. 4.1b). The effectiveness of the method has been thoroughly tested by conducting several numerical experiments in Sect. 4.5.

To elaborate, let us consider the streamline direction first. To maintain positivity, the following expression adds diffusion in the streamline direction:

$$\text{Stabilization term}_s = \sum_{\mathrm{e}=1}^{\mathrm{n_{el}}} \int_{\Omega^{\mathrm{e}}} \chi \frac{|\mathscr{R}(\varphi_{\mathrm{h}})|}{|\nabla\varphi_{\mathrm{h}}|} k_s^{add} \nabla w_{\mathrm{h}} \cdot \left(\frac{\boldsymbol{v}\otimes\boldsymbol{v}}{|\boldsymbol{v}|^2}\right) \cdot \nabla\varphi_{\mathrm{h}} d\Omega, \quad (4.56)$$

where k_s^{add} is the diffusion in the streamline direction, given by

$$k_s^{add} = \max\left\{\frac{||\boldsymbol{v}| - \tau|\boldsymbol{v}|s + \tau|\boldsymbol{v}||s||h}{2} - (k + \tau|\boldsymbol{v}|^2) + \frac{(s + \tau s|s|)h^2}{6}, 0\right\}. \quad (4.57)$$

On the other hand, in the crosswind direction, we add the following term:

$$\text{Stabilization term}_c = \sum_{\mathrm{e}=1}^{\mathrm{n_{el}}} \int_{\Omega^{\mathrm{e}}} \chi \frac{|\mathscr{R}(\varphi_{\mathrm{h}})|}{|\nabla \varphi_{\mathrm{h}}|} k_c^{\mathrm{add}} \nabla w_{\mathrm{h}} \cdot \left(\boldsymbol{I} - \frac{\boldsymbol{v} \otimes \boldsymbol{v}}{|\boldsymbol{v}|^2} \right) \cdot \nabla \varphi_{\mathrm{h}} d\Omega, \tag{4.58}$$

where k_c^{add} is the diffusion in the crosswind direction, given by

$$k_c^{\mathrm{add}} = \max \left\{ \frac{||\boldsymbol{v}| + \tau |\boldsymbol{v}||s||h}{2} - k + \frac{(s + \tau s|s|)h^2}{6}, 0 \right\}. \tag{4.59}$$

Here, $|\boldsymbol{v}| = \sqrt{v_i v_i}$ is the magnitude of the multi-dimensional velocity vector.
Remark 4.4 The one-dimensional formula for k^{add} has been directly extended to multi-dimensions. Note that there is no sense of crosswind direction in one-dimension. Therefore, k^{add} in Eq. (4.40) is added along the streamline. While in multi-dimensions, the expression for streamline diffusion is same as that in one-dimension (Eq. 4.40), and for the crosswind diffusion (Eq. 4.59), we omit the terms related to the linear stabilization and the convection velocity. This becomes clearer if you assume the linear stabilization to be in the streamline direction (SUPG). The terms in Eq. (4.40) depicting this streamline stabilization are $| - \tau |\boldsymbol{v}|s|h/2$ and $\tau |\boldsymbol{v}|^2$. However, in the crosswind direction, these terms will not be present, giving Eq. (4.59). This case is elaborated as follows. Suppose that we have SUPG stabilization in one-dimension, then the added term in the linear stabilization is

$$\sum_{\mathrm{e}=1}^{\mathrm{n_{el}}} \int_{\Omega^{\mathrm{e}}} u \frac{\mathrm{d}w_{\mathrm{h}}}{\mathrm{d}x} \tau \left(u \frac{\mathrm{d}\varphi_{\mathrm{h}}}{\mathrm{d}x} + s\varphi_{\mathrm{h}} - f \right) d\Omega. \tag{4.60}$$

The left-hand side of the stabilized expression after the finite element discretization can be written in the matrix form at the local element level as

$$\begin{bmatrix} \frac{\tau u^2}{h} - \frac{u\tau s}{2} & -\frac{\tau u^2}{h} - \frac{u\tau s}{2} \\ -\frac{\tau u^2}{h} + \frac{u\tau s}{2} & \frac{\tau u^2}{h} + \frac{u\tau s}{2} \end{bmatrix} \tag{4.61}$$

When we write the k^{add} expression for the SUPG method, these extra terms will correspond to the diffusion in the streamline direction. Therefore, the terms relating to the expressions above are omitted from the crosswind diffusion in Eq. (4.59).

Remark 4.5 The gradients in the solution in any direction are taken care of by the PPV technique, unlike earlier methods in the literature which added null nonlinear stabilization term when the convection velocity is perpendicular to the solution gradient.

Remark 4.6 The PPV technique is slightly more diffusive than that proposed in [45, 155, 156]. Neglecting the effects of the added nonlinear diffusion in the streamline direction, the method can be simplified to that of [45] by using a factor of 0.7 rather than 1 in the first term of the k_c^{add} for the convection-diffusion equation:

$$k_c^{\mathrm{add}} = \max\left\{\frac{0.7|\boldsymbol{v}|h}{2} - k, 0\right\}. \tag{4.62}$$

Taking the constant as 1 rather than 0.7 maintains positivity throughout the computational domain. With 0.7 as the factor, a non-regular solution near the boundary for the 90° convection case is observed (Fig. 4.12f).

Remark 4.7 There are only two assumptions that have been made in the design of the method: (i) the diffusion is isotropic and (ii) higher-order derivatives in the variational form and the differential operator of the stabilization term are neglected pertaining to the use of linear and multilinear finite elements.

Remark 4.8 Another numerical challenge is the evaluation of the Jacobian for the implicit and nonlinear schemes due to the presence of the absolute value function. Here, we have utilized Picard fixed-point iterative procedure for the nonlinear iterations, so that $|\mathscr{R}(\varphi_{\mathrm{h}})|/|\nabla\varphi_{\mathrm{h}}|$ is evaluated using the known values of φ_{h} at the previous nonlinear iteration.

4.3.2.4 Characteristic Element Length Scale

A lack of general definition of the element length scale poses a challenge to the error analysis of the residual-based stabilization of the Galerkin formulation. It is straightforward to select the characteristic element length in one-dimension. But, the task is very challenging in multi-dimensions, especially for highly irregular and anisotropic elements. Under generalized isoparametric transformation, the formula for the stabilization parameter τ was shown to be [193]

$$\tau = \left[\left(\frac{1}{\alpha\Delta t}\right)^2 + \boldsymbol{v}\cdot\boldsymbol{G}\boldsymbol{v} + C_I k^2 \boldsymbol{G} : \boldsymbol{G} + s^2\right]^{-1/2}, \tag{4.63}$$

where C_I is a constant derived from inverse estimates [83] and $\boldsymbol{G}$ is the element contravariant metric tensor for isoparametric mapping given by

$$\boldsymbol{G} = \left(\frac{\partial\boldsymbol{\xi}}{\partial\boldsymbol{x}}\right)^T \frac{\partial\boldsymbol{\xi}}{\partial\boldsymbol{x}}, \tag{4.64}$$

$\boldsymbol{\xi}$ and $\boldsymbol{x}$ being the vectors of natural and global coordinates, respectively. In the case of one-dimension, the above formula for τ reduces to Eq. (4.20). Thus, in one-dimension,

$$\boldsymbol{v}\cdot\boldsymbol{G}\boldsymbol{v} = \left(\frac{2u}{h}\right)^2. \tag{4.65}$$

Similarly, in multi-dimensions, we can define the element length such that one-dimensional equivalent streamline length scale is obtained. Such length scale measures have been used in the literature [49, 50, 151]:

$$h_u = \frac{2|\boldsymbol{v}|}{\sqrt{\boldsymbol{v} \cdot \boldsymbol{G}\boldsymbol{v}}}, \tag{4.66}$$

where h_u is the streamline element length and $|\boldsymbol{v}| = \sqrt{v_i v_i}$. It can be observed that the above expressions for the length scale do not consider the anisotropic character of the element, especially when there is a large directionality mismatch between the equation coefficients and the wave vector associated with subscale physics. This is the reason for the oscillations in the numerical solution for some cases when the discretization is anisotropic. A generalized Fourier analysis of the CDR equation was carried out in [173] to study this phenomenon. We highlight some key points from that analysis.

The origin of the stabilization parameter has been attributed to the multiscale phenomenon in which the unresolved subgrid scales are modeled via residual-based stabilization. The expression for the stabilization parameter τ is derived from the Fourier representation of the Green's function for a subscale problem at the element level, where the subscale term vanishes on the element boundary. The formula for τ for the subscale problem can be written using the inverse Fourier transform as [173]

$$\tau = \left[(K_i K_j k^r_{ij} + s)^2 + (K_j v^r_j)^2 \right]^{-1/2}, \tag{4.67}$$

where K_i and K_j are the components of the wave vector $\mathbf{K}$ from the Fourier analysis, v^r_j and k^r_{ij} are the convection velocity and the diffusion coefficient of the transformed CDR equation in the local coordinates, i.e., $v^r_i = v_j J^{-T}_{ij}$ and $k^r_{lm} = J^{-T}_{mi} k_{ij} J^{-T}_{lj}$ respectively, and J is the Jacobian of a finite element. By neglecting the directionality of the wave vector and writing the terms in the above expression for τ, we get

$$K_i K_j k^r_{ij} \approx ||\mathbf{K}||^2 \sqrt{k^r_{ij} k^r_{ij}}, \tag{4.68}$$

$$K_j v^r_j \approx ||\mathbf{K}|| ||\boldsymbol{v}^r||, \tag{4.69}$$

where $|| \cdot ||$ is the invariant of the vector. The above expression is similar to Eq. (4.63) when $s = 0$. However, neglecting the directionality of the wave vector has consequences on an anisotropic grid with variable coefficients. Therefore, a heuristic approach was proposed in [173] to find the direction where the maximum instability occurs based on the element Peclet and Damköhler numbers. A similar definition of the element length scale is adopted here and demonstrated for anisotropic and unstructured grids in Sect. 4.5.3.

For generality to transient problems, all the preceding developments can be extended to the time-dependent case by considering the transient term and deriving the diffusion expressions similar to the above procedure.

4.4 Convergence and Stability Analysis

In this section, a systematic analysis is carried out of the different stabilization methods for solving the one-dimensional CDR equation. Furthermore, convergence, accuracy and stability properties of the PPV method are also assessed. We assess these properties across a wide range of characteristic non-dimensional parameters. To begin, different stabilization methods are compared by analyzing the variation of the non-dimensionalized L^2 error with $Da = sh/u$ (convection-reaction case) and with $\psi = sh^2/k$ (diffusion-reaction case). Then, the algorithmic damping and phase velocity (or algorithmic frequency) ratios are quantified for the PPV technique based on Fourier analysis to understand the stability properties. Finally, accuracy of the various methods for the steady-state convection-reaction, diffusion-reaction and CDR equations is analyzed by performing a mesh convergence study. The non-dimensionalized L^2 error is computed for a fixed number of sample nodal points throughout the analysis as

$$\text{Error} = \frac{||\varphi_{\text{numerical}} - \varphi_{\text{exact}}||_2}{||\varphi_{\text{exact}}||_2}, \tag{4.70}$$

where φ_{exact} is the analytical value for the equation and $||\cdot||_2$ is the L^2 norm.

4.4.1 Dependence of Error on the Non-Dimensional Parameters

For the error analysis, we consider the one-dimensional problem in a computational domain of length $L = 1$ by varying the non-dimensional numbers while keeping the element length size to 0.1 (10 number of elements) and 0.025 (40 number of elements).

For the convection-reaction case, the left-hand side Dirichlet condition is set to 1 with $f = 0$ and for the diffusion-reaction case, the left-hand side and right-hand side nodal values are 8 and 3 respectively with $f = 0$. From Fig. 4.2a, c, a monotonically decreasing error is observed in the convection-dominated regime ($Da < 1$) for all the techniques, while for high Da, the error from the methods apart from PPV behave like an asymptote to a constant value. On the other hand, PPV minimizes the error even in the reaction-dominated regime ($Da > 1$). Moreover, the error decreases with the increase in Da. This can be attributed to the reduction in the undershoots and

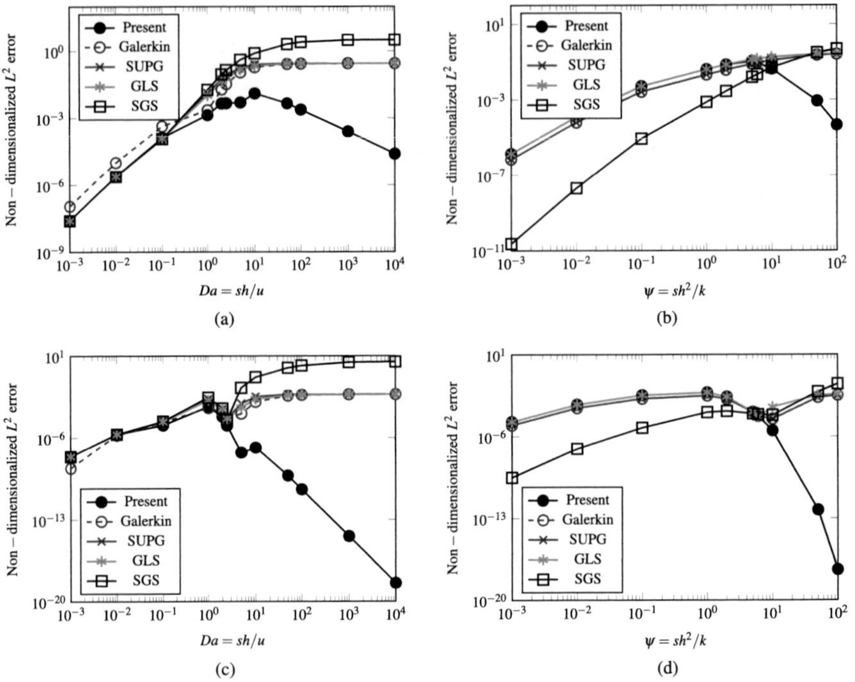

Fig. 4.2 Dependence of non-dimensionalized L^2 error on non-dimensional numbers as a function of: Da for convection-reaction equation with number of elements (**a**) and (**c**); and ψ for diffusion-reaction problem (**b**) and (**d**). Number of elements is kept constant at 10 for (**a**, **b**) and 40 for (**c**, **d**) with a domain length $L = 1$

overshoots of the solution near the boundary layers in the case of PPV. The maximum error occurs around $Da \approx 10$ and $Da \approx 1$ for the coarse and fine meshes respectively.

Similarly, from Fig. 4.2b, d, SGS method behaves quite well for diffusion-dominated regime ($\psi < 1$) with the monotonically decreasing error as ψ decreases. An asymptotic error is observed to be reached at higher ψ for all the methods apart from PPV. The error decreases monotonically for PPV in the reaction-dominated regime ($\psi > 1$). Note that the performance of the PPV is very similar to that of SUPG and GLS in the diffusion-dominated regime. The maximum error for PPV occurs around $\psi \approx 5$ and $\psi \approx 1$ for the coarse and fine meshes respectively. To summarize, the improvement in the error norm for PPV with increasing Da and ψ for the reaction-dominated effects is mainly due to the preservation of the positivity property and reduction of oscillations near the boundary layer.

In the middle of the regime of the non-dimensional numbers in Fig. 4.2, an increase in the computed error can be observed. This can be explained as follows. An increase in Da (or ψ) leads to a sharper boundary layer in the exact solution of the equation. Due to the numerical discretization using finite nodes and fixed number of elements, the error depends on the ability of each node to resolve the boundary layer. At lower

Da (or ψ), the reaction effects are small, showing less numerical error. For higher Da (or ψ), the ability of the nodes to resolve the boundary layer is tested, which introduces some error as the node may fall near the edge of the layer. For much higher non-dimensional numbers, the error decreases further due to the positivity preservation property of PPV as the solution approaches to the nodally exact one and the edge of the boundary layer recedes away from the node. The high error in the middle range of non-dimensional numbers can be reduced via an adaptive refinement as shown in the finer mesh (Fig. 4.2c, d) where we observe a similar trend.

4.4.2 One- and Two-Dimensional Fourier Analysis

We continue assessing the different methods by looking into the stability and accuracy properties in this section. To accomplish this, we conduct Fourier analysis on the discretized equation without the nonlinear stabilization term and quantify the dispersion and diffusion errors.

The analysis has been carried out for the GLS and SGS methods in [112] for one- and two-dimensions. We briefly highlight the analysis here for PPV technique. Assume that the exact solution of the one-dimensional transient CDR equation can be written as a Fourier series $\varphi = Ae^{\nu t + iKx}$, where K is the spatial wave number and ν denotes the evolution of the solution with time. ν can be expressed as $\nu = -\xi - i\omega$, ξ and ω being the damping coefficient and frequency respectively and $i = \sqrt{-1}$. Substituting these expressions in the continuous CDR equation, one obtains

$$\xi = s + kK^2, \tag{4.71}$$

$$\omega = uK. \tag{4.72}$$

The group velocity for the continuous equation is given as $v_g = \partial\omega/\partial K = u$. Here, u is the convection velocity of the one-dimensional CDR equation.

Similarly, the discrete solution can be written in terms of a Fourier series as $\varphi_{\mathrm{h}}(x_j, t) = Ae^{\nu^{\mathrm{h}} t + iKjh}$, where $\nu^{\mathrm{h}} = -\xi^{\mathrm{h}} - i\omega^{\mathrm{h}}$, where ξ^{h} and ω^{h} are the discrete counterparts of ξ and ω termed as algorithmic damping coefficient and algorithmic frequency, respectively. It is assumed here that the temporal discretization has a negligible error for simplicity. The Fourier analysis of the PPV method in one-dimension leads to the following amplification factor:

$$\nu^{\mathrm{h}}_{\mathrm{PPV}} = \frac{2(\cos(Kh)-1)(k+u^2\tau) - \frac{sh^2}{3}(\cos(Kh)+2)(1+|s|\tau) - ihu\sin(Kh)(1-s\tau+|s|\tau))}{\frac{h^2}{3}(\cos(Kh)+2)(1+|s|\tau) - ihu\tau\sin(Kh)}. \tag{4.73}$$

The above expression reduces to that of GLS method when $s \geq 0$ and to that of SGS method when $s < 0$. Detailed steps of the derivation can be found in [112].

The accuracy of the PPV formulation can be analyzed by expanding ξ^{h} and ω^{h} as follows:

$$\omega^{\mathrm{h}}_{\mathrm{PPV}} = uK + ukK^3\tau - \left(\frac{uK^5}{180}\right)h^4 + \mathcal{O}(\tau\Delta t, \Delta t^2, \tau^2, h^6), \tag{4.74}$$

$$\xi^{\mathrm{h}}_{\mathrm{PPV}} = s + kK^2 - |s|kK^2\tau + \left(\frac{kK^4}{12}\right)h^2 + \mathcal{O}(\tau\Delta t, \Delta t^2, \tau^2, h^4). \tag{4.75}$$

The accuracy of the algorithmic damping coefficient (ξ^{h}) and the algorithmic frequency (ω^{h}) with regard to h depend on the stabilization parameter τ when $u \neq 0$. Depending on the dominant phenomenon among convection, diffusion and reaction effects, τ varies with h, according to Eq. (4.20). The algorithmic damping ratio (ξ^{h}/ξ), the algorithmic frequency ratio ($\omega^{\mathrm{h}}/\omega$) and the group velocity ratio (v^{h}_g/c) which is the ratio of the group velocity of the numerical scheme to that of the continuous differential equation, are also evaluated. It is seen that the PPV technique has relatively less damping than the SGS when $s \geq 0$, very similar to that of GLS. When $s < 0$, PPV adds appropriate damping rather than GLS, which has a larger negative damping. In this range, the PPV behaves similar to the SGS method. To summarize, the behavior of PPV will be the same as that of GLS for positive reaction and that of SGS for negative reaction. Fourier analysis of the fully discrete form of the CDR equation can be found in [112].

Extending the analysis to two dimensions, we assume the exact solution as a Fourier series as $\varphi = Ae^{\nu t + i\mathbf{K}\cdot\mathbf{x}}$, where $\mathbf{K} = K\cos\alpha_k\mathbf{i} + K\sin\alpha_k\mathbf{j}$ and $\mathbf{x}$ are the wave vector and the position vector respectively. It is assumed that the velocity vector $\boldsymbol{v} = u\mathbf{i} + v\mathbf{j}$ has the same direction as that of the wave vector, for simplicity in the calculations. The damping coefficient and frequency for the continuous CDR equation in two-dimensions are given by

$$\xi = s + kK^2, \tag{4.76}$$

$$\omega = |\boldsymbol{v}|K. \tag{4.77}$$

The detailed Fourier analysis in two-dimensions is presented in [112]. In two-dimensions, the amplification factor for the PPV method can be derived as

$$\begin{aligned}\nu^{\mathrm{h}}_{\mathrm{PPV}} = &\frac{1}{\gamma_k h_x^2 M_x M_y(1+|s|\tau) - i\tau h_x(u\gamma_k M_y \sin\alpha_x + vM_x\sin\alpha_y)} \\ &\times\Bigg[\frac{2M_x(\cos\alpha_y - 1)}{\gamma_k}(k+\tau v^2) + 2M_y(\cos\alpha_x - 1)\gamma_k(k+\tau u^2) - 2\tau uv\sin\alpha_x\sin\alpha_y \\ &-s\gamma_k h_x^2 M_x M_y(1+|s|\tau) - ih_x u\gamma_k M_y\sin\alpha_x(1 - s\tau + |s|\tau) \\ &-ih_x vM_x\sin\alpha_y(1 - s\tau + |s|\tau)\Bigg],\end{aligned} \tag{4.78}$$

where h_x, γ_k, α_x, α_y, M_x and M_y are defined in [112]. Note that the above expression can be simplified to its one-dimensional counterpart when the wave vector is considered along the coordinate directions. The phase velocity ratio is defined as $\omega^{\mathrm{h}}/\omega$ and the damping ratio as ξ^{h}/ξ. Figure 4.3 shows the polar plots of the phase velocity ratio and the damping ratio. Similar inferences can be made from the two-

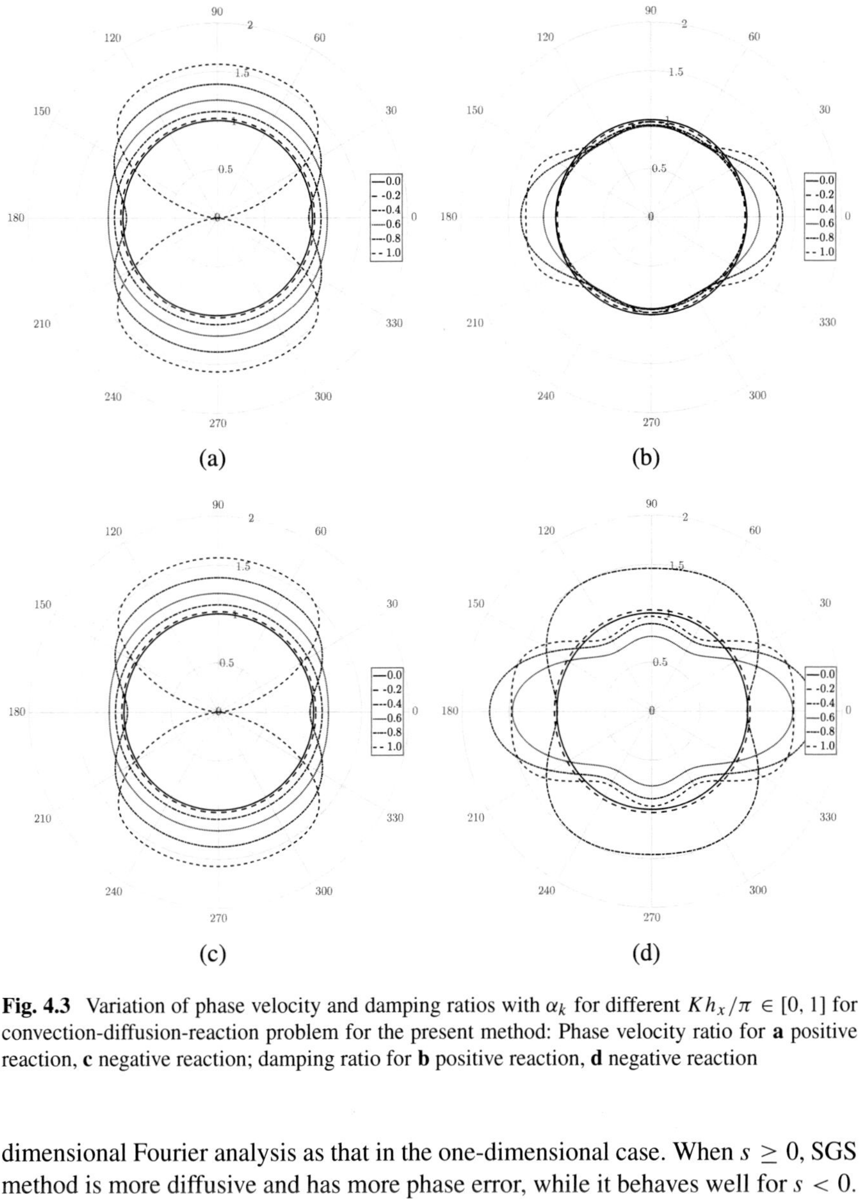

Fig. 4.3 Variation of phase velocity and damping ratios with α_k for different $Kh_x/\pi \in [0, 1]$ for convection-diffusion-reaction problem for the present method: Phase velocity ratio for **a** positive reaction, **c** negative reaction; damping ratio for **b** positive reaction, **d** negative reaction

dimensional Fourier analysis as that in the one-dimensional case. When $s \geq 0$, SGS method is more diffusive and has more phase error, while it behaves well for $s < 0$.

4.4.3 Mesh Convergence Study

In this section, we focus on the accuracy of the various stabilization schemes, in particular, with regard to the convergence rate. We employ the Taylor series to expand the one-dimensional finite element discretization to obtain the truncation error. Let us first consider the linear stabilized (GLS) discretized form of steady-state CDR equation:

$$\left(-\tfrac{u}{2h}-\tfrac{k}{h^2}+\tfrac{s}{6}-\tfrac{u^2\tau}{h^2}+\tfrac{u\tau s}{2h}-\tfrac{u\tau s}{2h}+\tfrac{s^2\tau}{6}\right)\varphi_{\mathrm{h},j-1}+\left(\tfrac{2k}{h^2}+\tfrac{2s}{3}+\tfrac{2u^2\tau}{h^2}+\tfrac{2s^2\tau}{3}\right)\varphi_{\mathrm{h},j}$$
$$+\left(\tfrac{u}{2h}-\tfrac{k}{h^2}+\tfrac{s}{6}-\tfrac{u^2\tau}{h^2}-\tfrac{u\tau s}{2h}+\tfrac{u\tau s}{2h}+\tfrac{s^2\tau}{6}\right)\varphi_{\mathrm{h},j+1}=0. \tag{4.79}$$

Expanding $\varphi_{\mathrm{h},j+1}$ and $\varphi_{\mathrm{h},j-1}$ around $\varphi_{\mathrm{h},j}$ and rearranging, we obtain the modified equation which is solved by the numerical method [236] as

$$u\frac{\mathrm{d}\varphi}{\mathrm{d}x}-k\frac{\mathrm{d}^2\varphi}{\mathrm{d}x^2}+s\varphi+\left(-u^2\frac{\mathrm{d}^2\varphi}{\mathrm{d}x^2}+s^2\varphi\right)\tau+\frac{k}{12}\frac{\mathrm{d}^4\varphi}{\mathrm{d}x^4}h^2$$
$$+\left(-\frac{u^2}{12}\frac{\mathrm{d}^4\varphi}{\mathrm{d}x^4}+\frac{s^2}{6}\frac{\mathrm{d}^2\varphi}{\mathrm{d}x^2}\right)\tau h^2+\mathscr{O}(\tau^2,h^4)=0. \tag{4.80}$$

The local truncation error can be obtained by subtracting the modified equation above with the exact differential equation, as

$$\epsilon_{\mathrm{GLS}}=\left(-u^2\tfrac{\mathrm{d}^2\varphi}{\mathrm{d}x^2}+s^2\varphi\right)\tau+\tfrac{k}{12}\tfrac{\mathrm{d}^4\varphi}{\mathrm{d}x^4}h^2+\left(-\tfrac{u^2}{12}\tfrac{\mathrm{d}^4\varphi}{\mathrm{d}x^4}+\tfrac{s^2}{6}\tfrac{\mathrm{d}^2\varphi}{\mathrm{d}x^2}\right)\tau h^2+\mathscr{O}(\tau^2,h^4). \tag{4.81}$$

Now, let us observe the behavior of the truncation error under different regimes. When $k=0$, i.e., in the convection-reaction regime, the coefficient of τ becomes null using rearrangement and substitution from the convection-reaction equation. Now, the leading error is of the order τh^2. As τ is of the order h when the mesh is refined, GLS is formally third-order accurate in this regime. On the other hand, note that the Galerkin method is fourth-order accurate since $\tau=0$ and $k=0$. When $u=0$ in the diffusion-reaction regime, GLS is second-order accurate as τ is of order h^2 with the mesh refinement.

In a similar fashion, the truncation error for the SGS method can be derived as

$$\epsilon_{\mathrm{SGS}}=\left(-u^2\frac{\mathrm{d}^2\varphi}{\mathrm{d}x^2}-2us\frac{\mathrm{d}\varphi}{\mathrm{d}x}-s^2\varphi\right)\tau+\frac{k}{12}\frac{\mathrm{d}^4\varphi}{\mathrm{d}x^4}h^2$$
$$+\left(-\frac{u^2}{12}\frac{\mathrm{d}^4\varphi}{\mathrm{d}x^4}-\frac{us}{3}\frac{\mathrm{d}^3\varphi}{\mathrm{d}x^3}-\frac{s^2}{6}\frac{\mathrm{d}^2\varphi}{\mathrm{d}x^2}\right)\tau h^2+\mathscr{O}(\tau^2,h^4). \tag{4.82}$$

For this stabilization method, a third-order of accuracy is observed in the convection-reaction regime which is due to the cancellation of the coefficient of τ. However, in the diffusion-reaction regime, a peculiar fourth-order of accuracy for the SGS method can be seen. This can be explained if we manipulate the terms in the error such that the second-order terms get cancelled: assume that the mesh refinement could be interpreted as an increase in the diffusion effects in the equation, thus, the stabilization parameter can be approximated as

$$\tau \approx \frac{h^2}{12k}. \tag{4.83}$$

Substituting this term in $-s^2\tau\varphi$ above and manipulating $\frac{k}{12}\frac{\mathrm{d}^4\varphi}{\mathrm{d}x^4}h^2$, we observe that these terms indeed get cancelled. Furthermore, in this case, the order of τ is h^2 leading to the order of truncation as τh^2 making the scheme fourth-order accurate.

The truncation error for the PPV technique in the steady-state problem is given by

$$\begin{aligned}\epsilon_{\text{PPV}} = &-\chi\frac{|\mathscr{R}(\varphi)|}{|\nabla\varphi|}k^{\text{add}}\frac{\mathrm{d}^2\varphi}{\mathrm{d}x^2} + \left(-u^2\frac{\mathrm{d}^2\varphi}{\mathrm{d}x^2} - us\frac{\mathrm{d}\varphi}{\mathrm{d}x} + u|s|\frac{\mathrm{d}\varphi}{\mathrm{d}x} + s|s|\varphi\right)\tau\\ &+\left(\frac{k}{12}\frac{\mathrm{d}^4\varphi}{\mathrm{d}x^4} - \frac{1}{12}\chi\frac{|\mathscr{R}(\varphi)|}{|\nabla\varphi|}k^{\text{add}}\frac{\mathrm{d}^4\varphi}{\mathrm{d}x^4}\right)h^2\\ &+\left(-\frac{u^2}{12}\frac{\mathrm{d}^4\varphi}{\mathrm{d}x^4} - \frac{us}{6}\frac{\mathrm{d}^3\varphi}{\mathrm{d}x^3} + \frac{u|s|}{6}\frac{\mathrm{d}^3\varphi}{\mathrm{d}x^3} + \frac{s|s|}{6}\frac{\mathrm{d}^2\varphi}{\mathrm{d}x^2}\right)\tau h^2 + \mathscr{O}(\tau^2, h^4).\end{aligned} \tag{4.84}$$

The PPV technique is observed to be at least second-order accurate in all the regimes. A third- and fourth-order of accuracy is noted for the convection-reaction and the propagation regime respectively. This is due to the behavior of PPV as the parent linear stabilization method, i.e., GLS when $s \geq 0$ and SGS when $s < 0$; when the residual of the equation tends to be negligible in the asymptotic convergence regime.

The observations made from the truncation analysis about the accuracy are confirmed by the mesh convergence plots in Fig. 4.4. The mesh convergence is carried out by uniformly refining the element length h, while keeping the parameters u, k and s as constants. The mesh convergence is shown in Fig. 4.4a–d for the convection-reaction, the diffusion-reaction, the CDR with destruction and production cases respectively. For the CR equation, $u = 1, s = 50, f = 0$ with left-hand node satisfying the Dirichlet condition of $\varphi = 1$; for the DR equation, $k = 0.01, s = 50, f = 0$, left-hand and right-hand nodes satisfy $\varphi = 8$ and $\varphi = 3$ respectively. $u = 4, k = 0.01$ and $f = 1$ with Dirichlet condition of $\varphi = 0$ is imposed at both the extreme nodes for the CDR equation with $s = 60$ and $s = -6$ for destruction and production effects. The number of elements is increased by a factor of 2 from 10 to 2560 in the study. Some of the inferences from the plots are as follows. First, the measured slope of the PPV method is higher than 2 for the steady-state convection-reaction equation, as shown in Fig. 4.4a. Second, the GLS, SUPG and PPV methods have second-order accuracy for the diffusion-reaction and the CDR equations. Third, the SGS shows an order of accuracy up to 4 for the diffusion-reaction and the CDR equations, as shown in Fig. 4.4b–d.

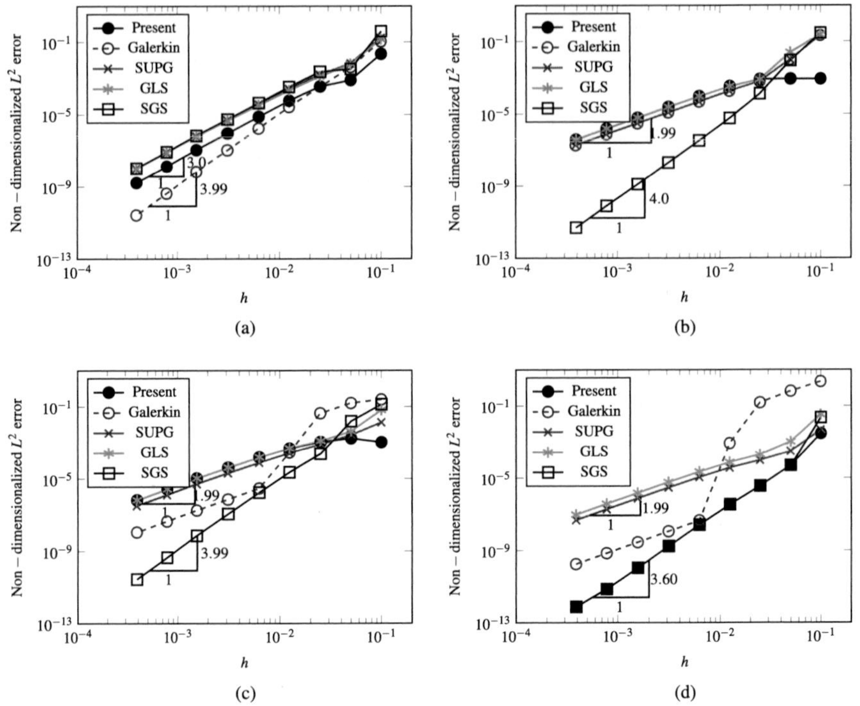

Fig. 4.4 Mesh convergence study for various finite element methods through dependence of non-dimensionalized L^2 error as a function of uniform mesh refinement h: **a** convection-reaction; **b** diffusion-reaction; **c** CDR (destruction) and **d** CDR (production) problem. Number of elements is increased by decreasing h

4.5 Numerical Results

In this section, we continue assessing the different stabilization methods by conducting numerical tests for the CDR equation in one- and two-dimensions across a wide range of parameters. Moreover, we demonstrate the generality of the PPV method for unstructured and anisotropic two-dimensional meshes.

4.5.1 One-Dimensional Cases

We begin the numerical tests with one-dimensional steady state CDR equation. A computational domain of length $L = 1$ is discretized into 10 linear finite elements. The parameters for the various test cases are given below:

- Canonical forms

- Convection-diffusion equation: $Pe = 1$ with $f = 1$ and $(0, 0)$ (Fig. 4.5a)
- Convection-reaction equation: $Da = 10$ with $f = 1$ and inlet condition $= 0$ (Fig. 4.5b)
- Diffusion-reaction equation: $\psi = 10$ with $f = 1$ and $(0, 0)$ (Fig. 4.5c)

- With source term ($f \neq 0$)

- $Pe = 0.01$ and $Da = 100$ with $(0, 0)$ (Fig. 4.6a)
- $Pe = 1$ and $Da = 10$ with $(0, 0)$ (Fig. 4.6b)
- $Pe = 10$ and $Da = 1$ with $(0, 0)$ (Fig. 4.6c)

- Without source term ($f = 0$)

- $Pe = 0.1$ and $Da = 10$ with $(0, 1)$ (Fig. 4.7a)
- $Pe = 1$ and $Da = 10$ with $(8, 3)$ (Fig. 4.7b)
- $Pe = 10$ and $Da = 10$ with $(0, 1)$ and $(1, 0)$ (Fig. 4.7c,d)
- $Pe = 10$ and $Da = 0.1$ with $(1, 0)$ (Fig. 4.7e)
- $Pe = 10$ and $Da = 0.2$ with $(8, 3)$ (Fig. 4.7f)
- $Pe = 10$ and $Da = 1$ with $(8, 3)$ (Fig. 4.7g)
- $Pe = 1$ and $Da = 60$ with $(8, 3)$ (Fig. 4.7h)

- Production regime ($s < 0$)

- Exponential regime ($u^2 + 4ks \geq 0$)

  · $Pe = 1$ and $Da = -0.4$ with $(0, 1)$ (Fig. 4.8a)
  · $Pe = 1.5$ and $Da = -0.5$ with $f = 1$ and $(0, 0)$ (Fig. 4.8b)
  · $Pe = 9$ and $Da = -0.15$ with $(0, 1)$ and $(1, 0)$ (Fig. 4.8c, d)

- Propagation regime ($u^2 + 4ks < 0$)

  · $Pe = 0.05$ and $Da = -3$ with $(0, 1)$ (Fig. 4.9a)
  · $Pe = 0.1$ and $Da = -6$ with $(0, 1)$ (Fig. 4.9b)
  · $Pe = 0.15$ and $Da = -9$ with $(0, 1)$ and $(1, 0)$ (Fig. 4.9c, d)

where $Pe = uh/2k$, $Da = sh/u$ and $\psi = sh^2/k$ are the characteristic dimensionless quantities and the values in brackets refer to the left and right Dirichlet boundary node values respectively.

The solutions of the various cases are shown in Figs. 4.5, 4.6, 4.7, 4.8 and 4.9. The comparison of the different methods with the PPV gives a glance into its ability to preserve positivity and preclude oscillations. The other methods (SUPG, GLS and SGS) lack the enforcement of positivity and local boundedness property. The PPV technique behaves quite well in the propagation regime similar to the behavior of the SGS method (Fig. 4.9) with minimal phase error, which has been analyzed in [112]. SGS method is highly diffusive for high Da and Pe numbers (Fig. 4.7c, h). On the other hand, the Galerkin solution is very oscillatory in Fig. 4.8c, d and is partially shown.

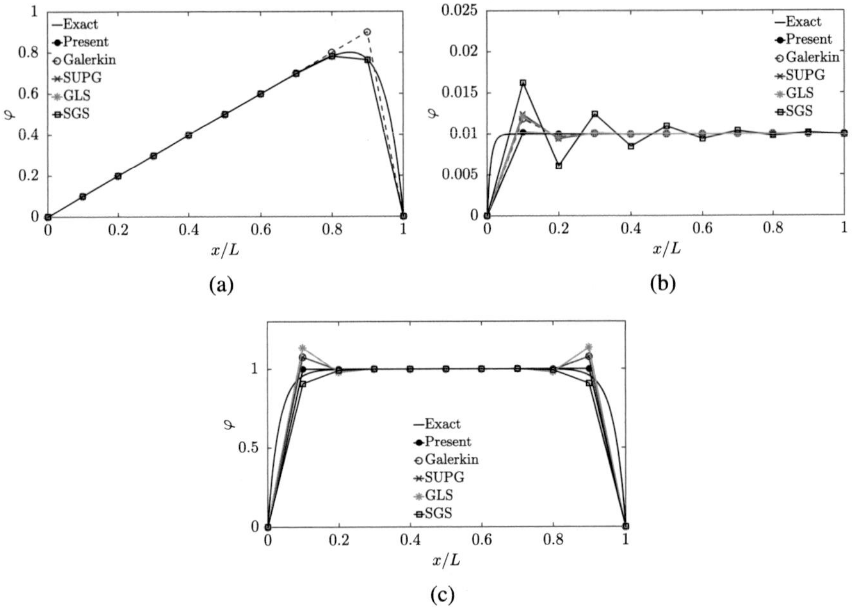

Fig. 4.5 Variation of scalar field φ of the steady CDR equation in one-dimension in the three canonical regimes: **a** convection-diffusion ($Pe = 1$, $f = 1$, $(0, 0)$); **b** convection-reaction ($Da = 10$, $f = 1$, inlet Dirichlet condition = 0); **c** diffusion-reaction ($\psi = 10$, $f = 1$, $(0, 0)$)

For the one-dimensional transient problem, we consider pure convection of two rectangular pulses, which was also considered in [155]. The domain is discretized using 200 linear two-node elements and the time step Δt is chosen as 0.002s with convection velocity $u = 1$ and $k = s = f = 0$. The initial condition for the problem is

$$\varphi(x, t = 0) = \begin{cases} 1, & \forall x \in [0.1, 0.2] \cup [0.3, 0.4], \\ 0, & \text{else}. \end{cases} \tag{4.85}$$

Fig. 4.10 shows the evolution of the rectangular pulses at different time instances $t = 0$s, $t = 0.1$s and $t = 0.5$s. No oscillations are observed during the convection phenomenon. The solution is smooth and the symmetry of the profile of the pulse is maintained at all times.

4.5.2 Two-Dimensional Cases

Next, we conduct numerical tests for two-dimensional steady state CDR equation. The computational domain is discretized using bilinear structured quadrilateral ele-

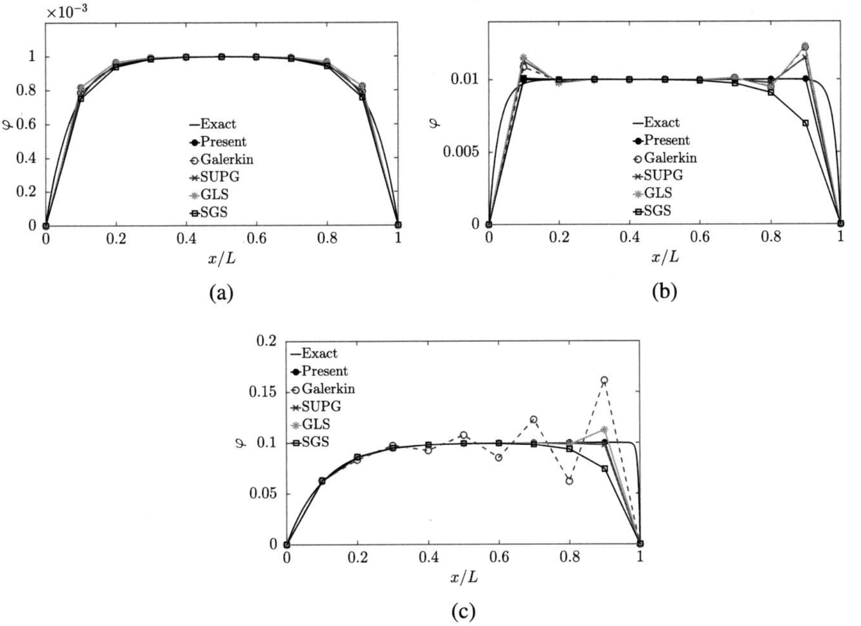

Fig. 4.6 Solution of the scalar field φ of the steady CDR equation in one-dimension by setting the source term, $f \neq 0$: **a** $Pe = 0.01$, $Da = 100$, $f = 1$, $(0, 0)$; **b** $Pe = 1.0$, $Da = 10$, $f = 1$, $(0, 0)$; **c** $Pe = 10$, $Da = 1.0$, $f = 1$, $(0, 0)$

ments. Two different meshes are considered: 20×20 and 80×80. The schematic of the cases with the boundary conditions are shown in Fig. 4.11 with the parameters given below:

1. Convection-diffusion problem with unit convection velocity skewed to the mesh at $\delta = 0°$, $30°$, $45°$, $60°$ and $90°$ with $|\boldsymbol{v}| = 1$, $k = 10^{-8}$, $s = 0$ and $f = 0$ (Fig. 4.12)
2. Convection-diffusion problem with constant source term. $\boldsymbol{v} = (1, 0)$, $k = 10^{-8}$, $s = 0$ and $f = 1$ (Fig. 4.13a)
3. Diffusion-reaction problem with constant source term. $\boldsymbol{v} = (0, 0)$, $k = 10^{-8}$, $s = 1$ and $f = 1$ (Fig. 4.13b)
4. Convection-diffusion problem with varying velocity and constant source term. $\boldsymbol{v} = (y, -x)$, $k = 10^{-8}$, $s = 0$ and $f = 1$ (Fig. 4.13c)
5. Convection-diffusion with discontinuous source term. $\boldsymbol{v} = (1, 0)$, $k = 10^{-8}$, $s = 0$ and $f(x \leq 0.5, y) = 1$, $f(x > 0.5, y) = -1$ (Fig. 4.13d)

The results for the two-dimensional steady state cases are shown in Figs. 4.12 and 4.13. The results are observed to be slightly more diffusive for the PPV method as discussed in Remark 4.6. The internal and exponential layers in the numerical solution are accurately captured with minimal oscillations, even in the case of 90° skewed convection. We observe some negative variation in the result for the refined case in

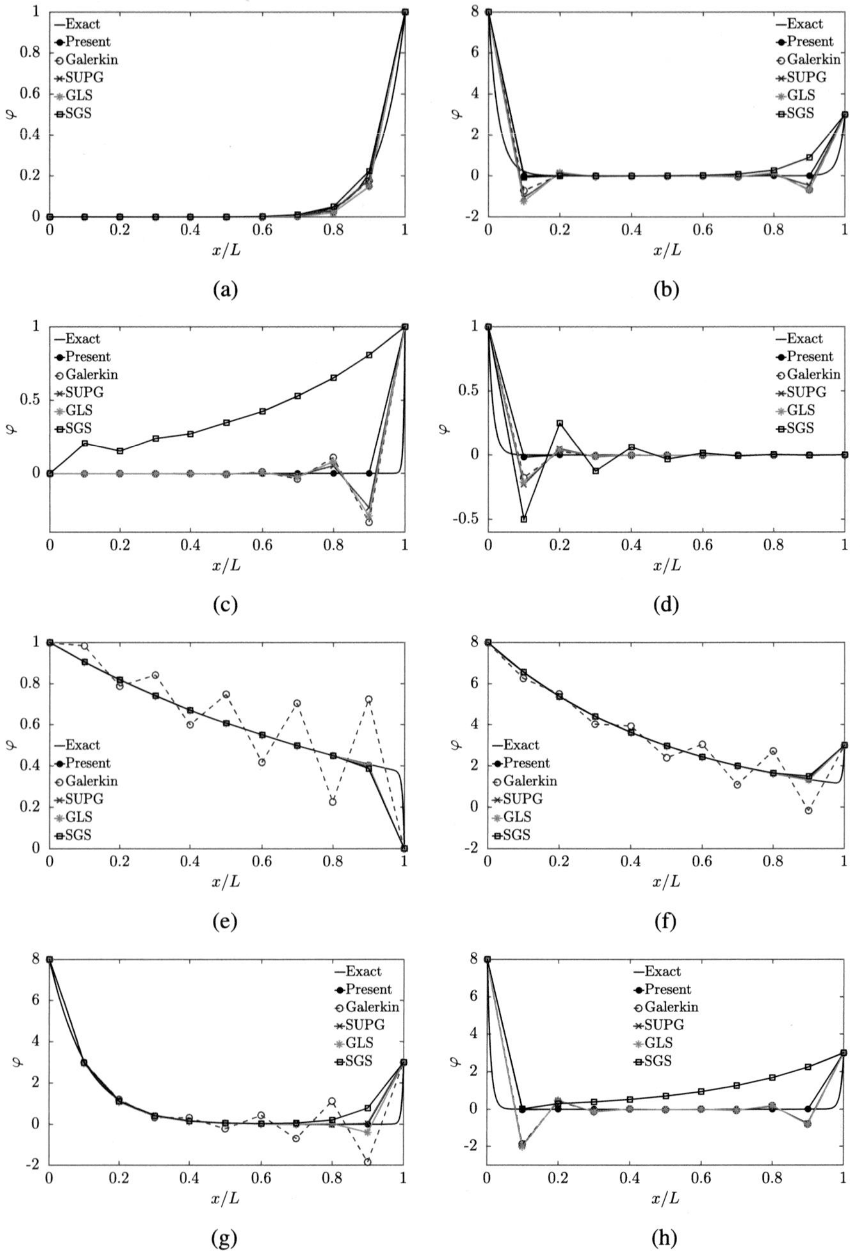

Fig. 4.7 Solution of the scalar field φ of the steady CDR equation in one-dimension by setting the source term, $f = 0$: **a** $Pe = 0.1$, $Da = 10$, (0, 1); **b** $Pe = 1.0$, $Da = 10$, (8, 3); **c** $Pe = 10$, $Da = 10$, (0, 1); **d** $Pe = 10$, $Da = 10$, (1, 0); **e** $Pe = 10$, $Da = 0.1$, (1, 0); **f** $Pe = 10$, $Da = 0.2$, (8, 3); **g** $Pe = 10$, $Da = 1.0$, (8, 3); **h** $Pe = 1.0$, $Da = 60$, (8, 3)

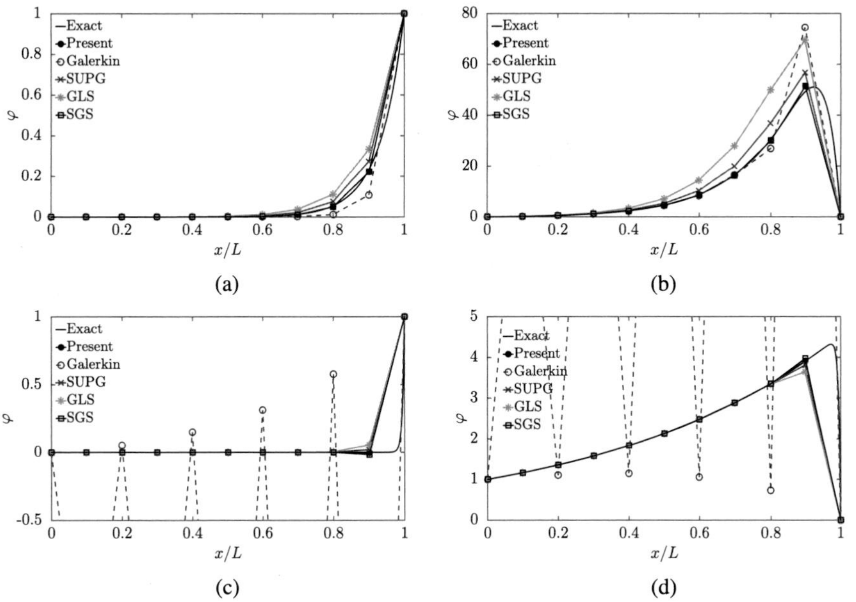

Fig. 4.8 Solution of the steady CDR equation in one-dimension in the exponential regime: **a** $Pe = 1.0$, $Da = -0.4$, $f = 0$, (0, 1); **b** $Pe = 1.5$, $Da = -0.5$, $f = 1$, (0, 0); **c** $Pe = 9.0$, $Da = -0.15$, $f = 0$, (0, 1); **d** $Pe = 9.0$, $Da = -0.15$, $f = 0$, (1, 0). The Galerkin solution is very oscillatory and hence is shown partially in Fig. 4.8c,d

Fig. 4.13d in the region where the source term $f < 0$. This variation is generally manifested by all the discontinuity capturing methods based on Petrov-Galerkin formulation [46].

The transient two-dimensional test case consists of a convection of a circular bubble in a domain of [0, 3] × [0, 3] at an angle of 45° with the X-axis [38, 156] with $\boldsymbol{v} = (0.5, 0.5)$, $k = 10^{-30}$, $s = 0$ and $f = 0$. The domain was discretized using structured 300 × 300 quadrilateral elements. The time step size was selected as $\Delta t = 0.005$s. The initial condition is set as

$$\varphi(\mathbf{r}, t = 0) = \mathrm{H}(\mathrm{R} - |\mathbf{r} - \mathbf{r}^c|), \tag{4.86}$$

where H() is the Heaviside function defined as

$$\mathrm{H}(y) = \frac{1 + \mathrm{sgn}(y)}{2} = \begin{cases} 0, & y < 0, \\ 0.5, & y = 0, \\ 1, & y > 0, \end{cases} \tag{4.87}$$

where $\mathrm{sgn}(y)$ is defined by $\frac{y}{|y|}$, $\mathbf{r}^c$ is the center of the circular bubble (0.5, 0.5), $\mathbf{r}$ is the position vector in the computational domain and $\mathrm{R} = 0.25$. The evolution of

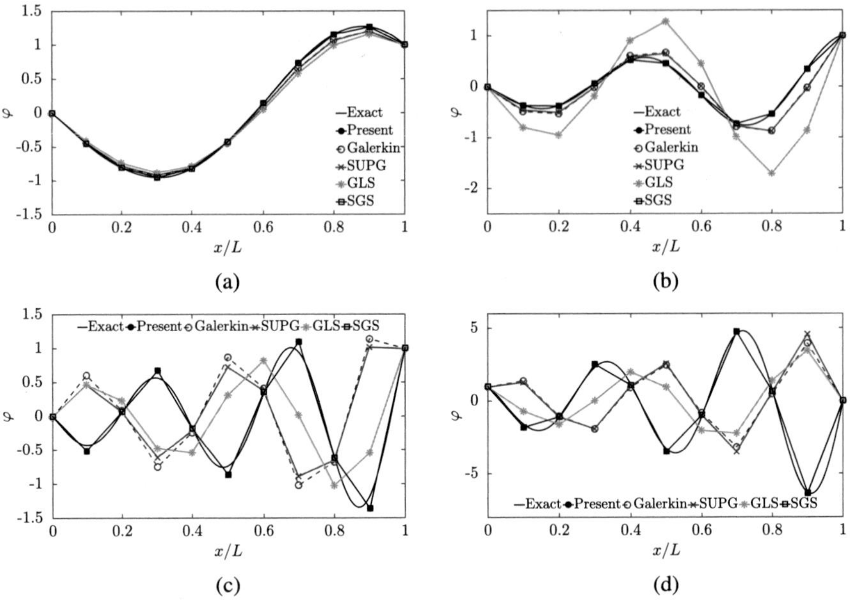

Fig. 4.9 Solution of the steady CDR equation in one-dimension in the propagation regime: **a** $Pe = 0.05$, $Da = -3.0$, $f = 0$, $(0, 1)$; **b** $Pe = 0.1$, $Da = -6.0$, $f = 0$, $(0, 1)$; **c** $Pe = 0.15$, $Da = -9.0$, $f = 0$, $(0, 1)$; **d** $Pe = 0.15$, $Da = -9.0$, $f = 0$, $(1, 0)$

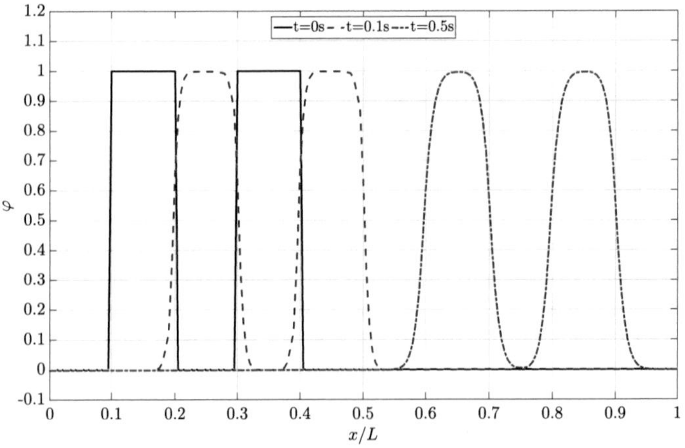

Fig. 4.10 Solution of the transient CDR equation in one-dimension: transient convection of two rectangular pulses, $u = 1$, $k = s = f = 0$

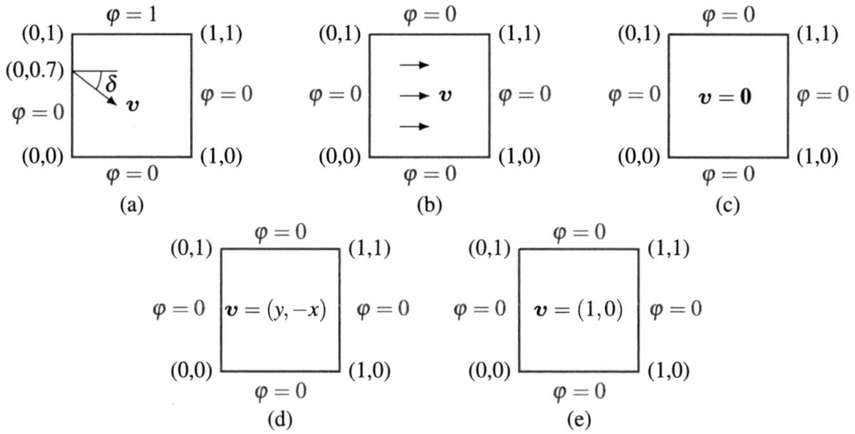

Fig. 4.11 Schematic diagrams of the two-dimensional steady-state test cases: **a** convection-diffusion problem with unit skewed convection velocity, $s = 0,\ f = 0$; **b** convection-diffusion problem with constant source term, $s = 0,\ f = 1$; **c** diffusion-reaction problem with constant source term, $s = 1,\ f = 1$; **d** convection-diffusion problem with varying velocity and constant source term, $s = 0,\ f = 1$; **e** convection-diffusion problem with discontinuous source term, $s = 0$, $f(x \leq 0.5, y) = 1,\ f(x > 0.5, y) = -1$. The diffusion coefficient (k) is taken constant with a value of 10^{-8} for all the cases

the bubble in time is shown in Fig. 4.14. The results show that the amplitude of the bubble is maintained in the convection process with no oscillations. The solution is visually monotone as the mesh resolves the important characteristics of the solution properly.

The errors in all the two-dimensional cases are quantified by comparing the extreme values obtained from the numerical solution with that of the exact solution in Table 4.2.

4.5.3 Non-Uniform Unstructured and Anisotropic Meshes

Finally, we assess the generality and robustness of the stabilization methods by conducting the numerical tests on non-uniform unstructured and anisotropic discretizations. We consider three configurations: unstructured isotropic, structured anisotropic and unstructured anisotropic quadrilateral meshes. Three cases on a computational domain of $\Omega = [0, 1] \times [0, 1]$ are carried out:

1. Skewed convection-diffusion problem at 60° with $|\boldsymbol{v}| = 1, k = 10^{-8}, s = 0$ and $f = 0$ (Fig. 4.11a)
2. Convection-diffusion problem with constant source term with $\boldsymbol{v} = (1, 0), k = 10^{-8}, s = 0$ and $f = 1$ (Fig. 4.11b)

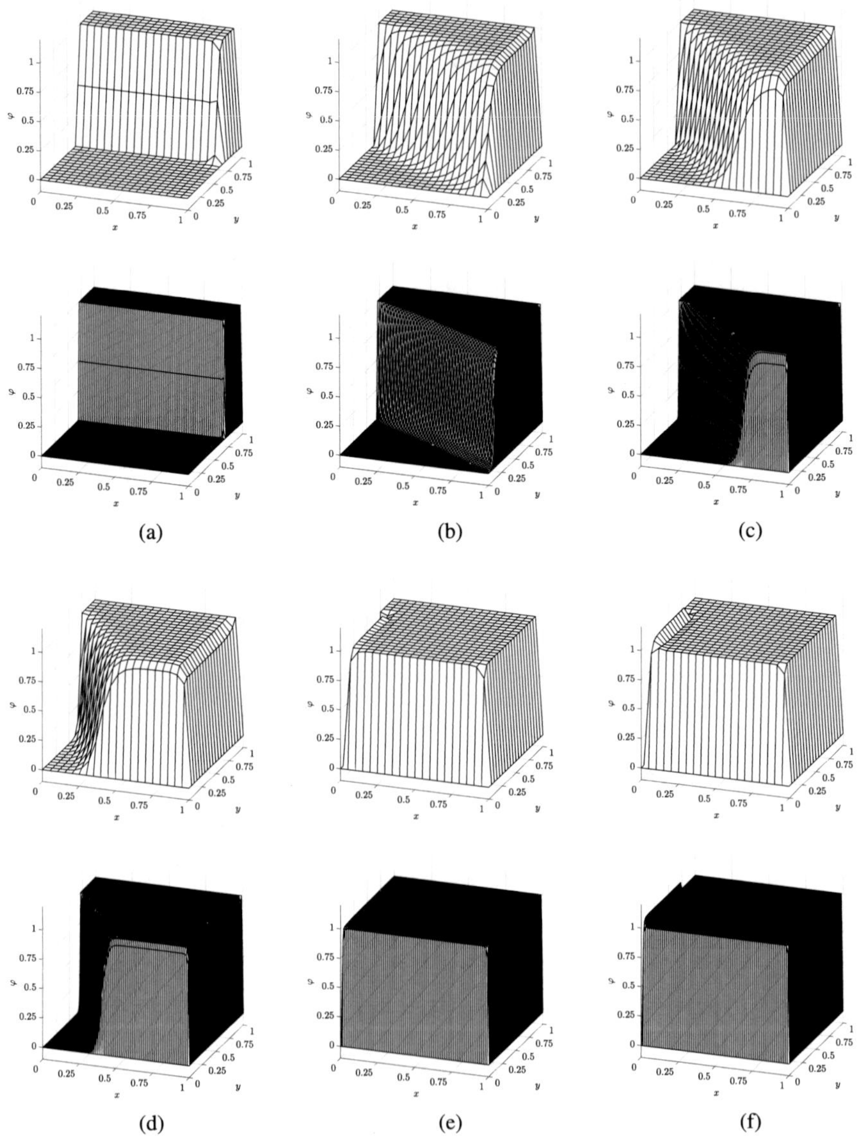

Fig. 4.12 Solution of the steady-state CDR equation in two-dimensions at 20 × 20 and 80 × 80 mesh resolutions: Convection-diffusion problem with unit convection velocity skewed at **a** 0°; **b** 30°; **c** 45°; **d** 60°; **e** 90°; and **f** 90° (Codina [45])

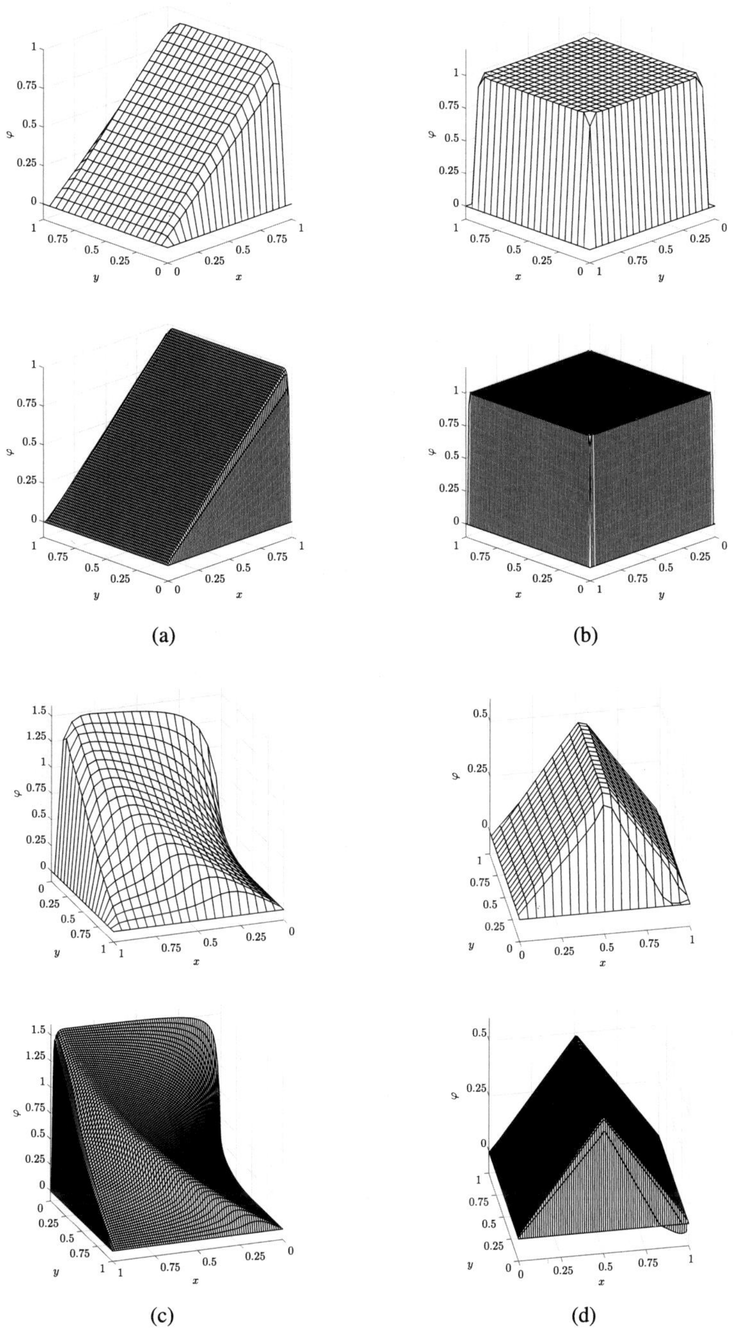

Fig. 4.13 Solution of the steady-state CDR equation in two-dimensions at 20 × 20 and 80 × 80 mesh resolutions: **a** convection-diffusion problem with constant source term; **b** diffusion-reaction problem with constant source term; **c** convection-diffusion problem with varying velocity and constant source term; **d** convection-diffusion problem with discontinuous source term

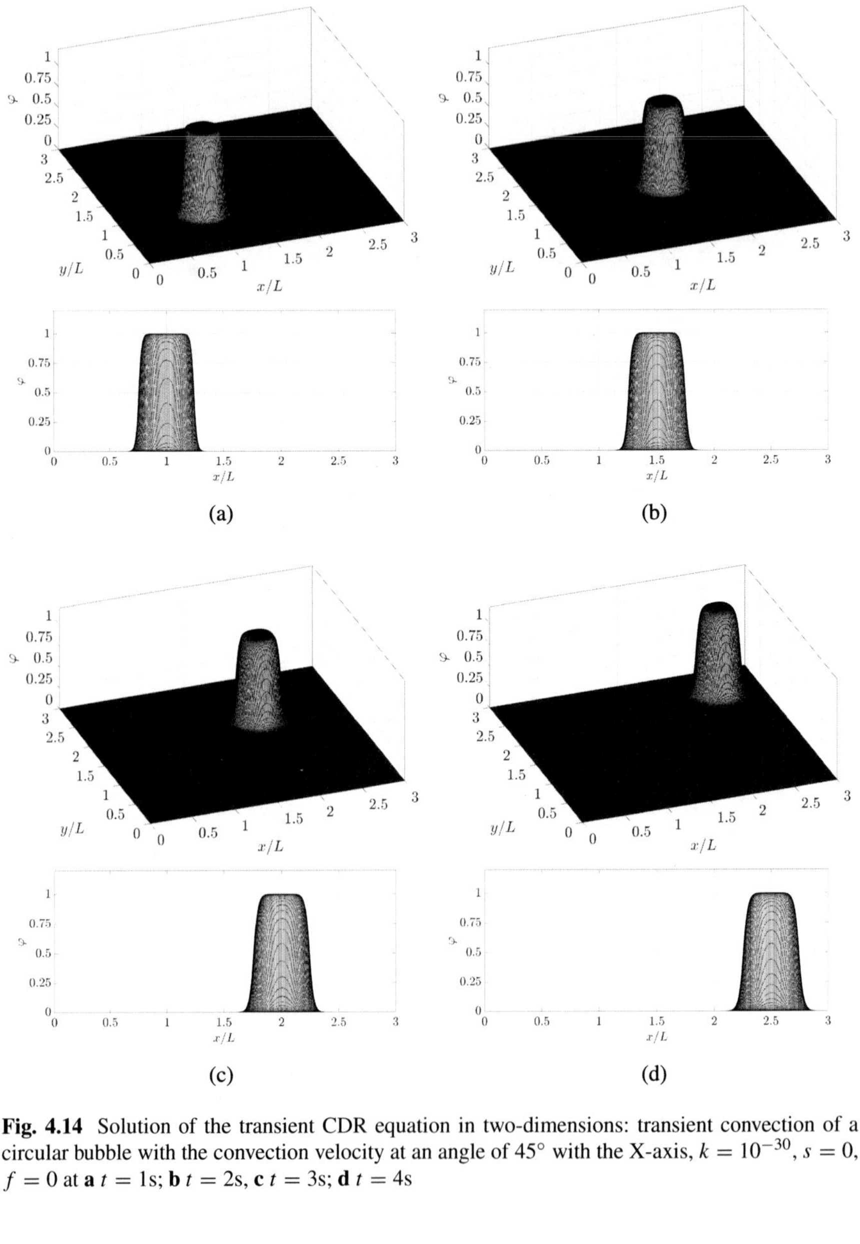

Fig. 4.14 Solution of the transient CDR equation in two-dimensions: transient convection of a circular bubble with the convection velocity at an angle of 45° with the X-axis, $k = 10^{-30}$, $s = 0$, $f = 0$ at **a** $t = 1$s; **b** $t = 2$s, **c** $t = 3$s; **d** $t = 4$s

Table 4.2 Quantification of the minimum and maximum values of the variable φ in different test cases in two-dimensions

Case	$\max(\varphi)$	$\max(\varphi_{\text{exact}})$	$\min(\varphi)$	$\min(\varphi_{\text{exact}})$
Fig. 4.12a (20 × 20)	1.00002	1.0	-4.06514×10^{-6}	0.0
Fig. 4.12a (80 × 80)	1.00001	1.0	-3.40949×10^{-6}	0.0
Fig. 4.12b (20 × 20)	1.00165	1.0	0.0	0.0
Fig. 4.12b (80 × 80)	1.00165	1.0	-2.60600×10^{-6}	0.0
Fig. 4.12c (20 × 20)	1.00077	1.0	0.0	0.0
Fig. 4.12c (80 × 80)	1.00077	1.0	0.0	0.0
Fig. 4.12d (20 × 20)	1.00102	1.0	0.0	0.0
Fig. 4.12d (80 × 80)	1.00102	1.0	-5.67374×10^{-7}	0.0
Fig. 4.12e (20 × 20)	1.01578	1.0	0.0	0.0
Fig. 4.12e (80 × 80)	1.01634	1.0	0.0	0.0
Fig. 4.13a (20 × 20)	0.94999	0.95	0.0	0.0
Fig. 4.13a (80 × 80)	0.98749	0.9875	0.0	0.0
Fig. 4.13b (20 × 20)	1.00527	1.0	0.0	0.0
Fig. 4.13b (80 × 80)	1.00532	1.0	0.0	0.0
Fig. 4.13c (20 × 20)	1.50854	<1.5708	0.0	0.0
Fig. 4.13c (80 × 80)	1.55723	<1.5708	0.0	0.0
Fig. 4.13d (20 × 20)	0.50833	0.5	−0.01208	0.0
Fig. 4.13d (80 × 80)	0.50208	0.5	−0.04734	0.0
Fig. 4.14a	1.0	1.0	0.0	0.0
Fig. 4.14b	1.0	1.0	0.0	0.0
Fig. 4.14c	0.99999	1.0	0.0	0.0
Fig. 4.14d	0.99999	1.0	0.0	0.0

3. Diffusion-reaction problem with no convection with $\boldsymbol{v} = (0, 0)$, $k = 10^{-8}$, $s = 1$, $f = 1$ (Fig. 4.11c)

As mentioned earlier, three types of mesh configurations are selected for the assessment:

- Mesh A: Unstructured quadrilateral in 20 × 20 grid (Fig. 4.15a)
- Mesh B: Structured anisotropic quadrilateral in 10 × 100 grid (Fig. 4.15b)
- Mesh C: Unstructured anisotropic quadrilateral in 20 × 100 grid (Fig. 4.15c)

The unstructured meshes are formed by perturbation of the coordinates of the nodes inside the domain of the structured mesh via a pseudo-random process given by

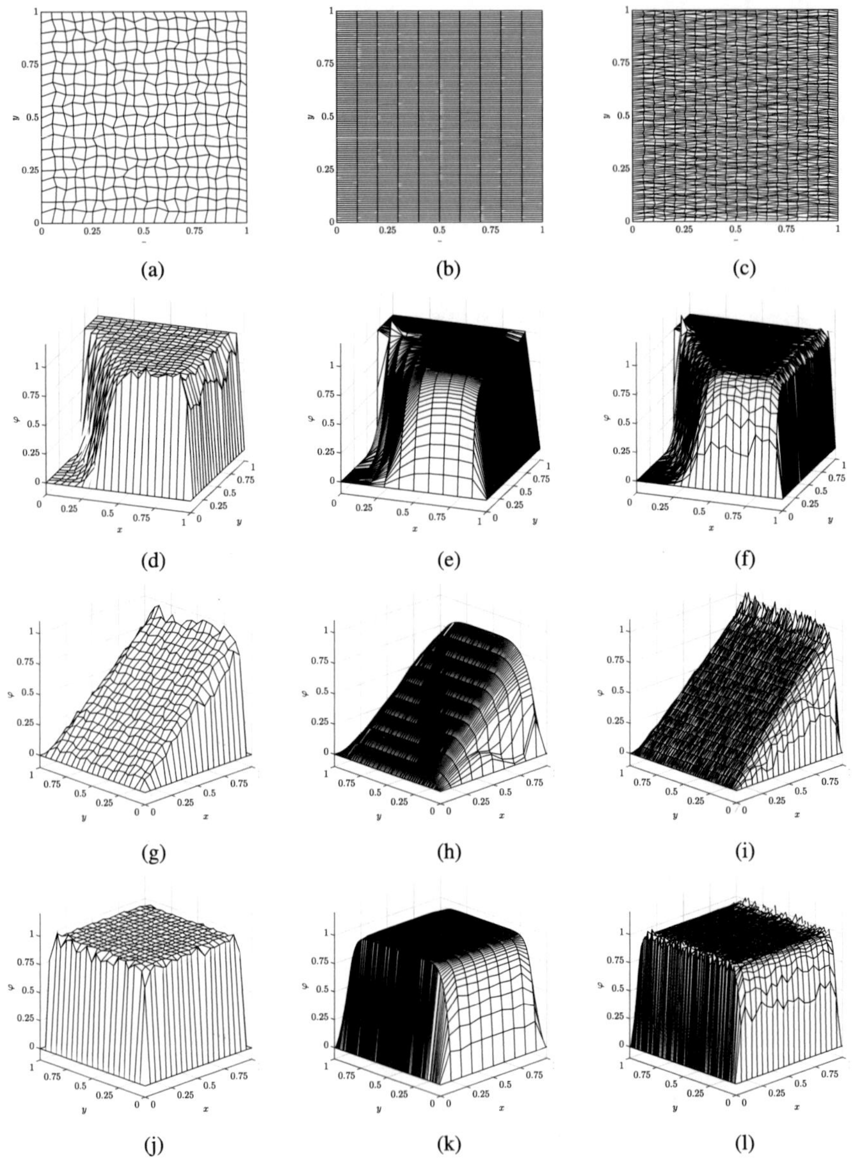

Fig. 4.15 Three representative unstructured and anisotropic grids: **a** mesh A, **b** mesh B, **c** mesh C. Results of the steady-state CDR equation in two-dimensions for skewed convection-diffusion problem at 60° (second row); convection-diffusion problem with constant source term (third row); diffusion-reaction problem (fourth row)

Table 4.3 Quantification of the minimum and maximum values of the variable φ in different test cases in two-dimensions: unstructured and anisotropic meshes

Case	$\max(\varphi)$	$\max(\varphi_{\text{exact}})$	$\min(\varphi)$	$\min(\varphi_{\text{exact}})$
Fig. 4.15d	1.0756	1.0	0.0	0.0
Fig. 4.15e	1.1111	1.0	-1.1959×10^{-4}	0.0
Fig. 4.15f	1.1313	1.0	0.0	0.0
Fig. 4.15g	1.0012	< 1.0	0.0	0.0
Fig. 4.15h	0.8999	< 1.0	0.0	0.0
Fig. 4.15i	1.0787	< 1.0	0.0	0.0
Fig. 4.15j	1.0305	1.0	0.0	0.0
Fig. 4.15k	1.0	1.0	0.0	0.0
Fig. 4.15l	1.0728	1.0	0.0	0.0

Table 4.4 The range of non-dimensional parameters: Unstructured and anisotropic meshes

Case	$Pe = \lvert\boldsymbol{v}\rvert h/(2k)$	$Da = sh/\lvert\boldsymbol{v}\rvert$	$\psi = sh^2/k$
Fig. 4.15d	$1.395 \times 10^6 - 3.875 \times 10^6$	0	0
Fig. 4.15e	4.33×10^6	0	0
Fig. 4.15f	$1.755 \times 10^6 - 2.595 \times 10^6$	0	0
Fig. 4.15g	$1.18 \times 10^6 - 3.895 \times 10^6$	0	0
Fig. 4.15h	5×10^6	0	0
Fig. 4.15i	$1.35 \times 10^5 - 3.01 \times 10^6$	0	0
Fig. 4.15j	0	∞	$1.354 \times 10^5 - 6.972 \times 10^5$
Fig. 4.15k	0	∞	1×10^6
Fig. 4.15l	0	∞	$1.648 \times 10^5 - 3.634 \times 10^5$

$$x_u = x_s + h_x \delta_x \text{rand}, \tag{4.88}$$

$$y_u = x_s + h_y \delta_y \text{rand}, \tag{4.89}$$

where (x_u, y_u) and (x_s, y_s) denote the nodal coordinates of the unstructured and structured meshes respectively, h_x and h_y are the mesh size for the structured mesh in x- and y-directions respectively and δ_x and δ_y are the mesh perturbation parameters in the respective directions with rand representing uniformly distributed random numbers in the interval (0, 1). For mesh A, the mesh resolution $h_x = h_y = 1/20$ and $\delta_x = \delta_y = 0.5$ is selected and for mesh C, we consider $h_x = 1/20$, $h_y = 1/100$, $\delta_x = 0.2$ and $\delta_y = 1.0$. Figure 4.15a–c show the representative meshes for the various cases with the numerical solution superimposed for each case in the same figure.

The results show that the considered length scale behaves reasonably well. The spurious oscillations are reduced for the convection- and reaction-dominated cases. The internal layers are captured with reasonable accuracy in the skewed convection case with unstructured quadrilaterals (Fig. 4.15d) but under-resolved and highly anisotropic grids tend to have small oscillations at the extreme values of the non-dimensional numbers for the other cases. Note that using the length scale in Eq. (4.66) leads to high oscillations in the anisotropic cases. These oscillations tend to reduce by using the length scale which is derived considering the directionality of the wave vector in Eq. (4.67). The error is quantified by comparing the extreme values of the numerical solution with that of the exact solution in Table 4.3. Furthermore, the range of the non-dimensional parameters for the cases considered are summarized based on the minimum and maximum length scales obtained from Eq. (4.67) in Table 4.4.

This completes the discussion regarding the application of the variational stabilized finite element methods to a canonical problem of convection-diffusion-reaction equation. Error analyses are performed to quantify the accuracy, convergence and stability properties of various techniques and several numerical tests are conducted to assess the mentioned methods. Next, we continue our journey to more complex equations of fluid-structure interaction and their variational finite element discretization.

Appendix

Function Spaces

We review some of the mathematical preliminaries and definitions which are helpful in formulating the weak formulation in a systematic and formal way. Consider a spatial domain $\Omega \subset \mathbb{R}^{n_{sd}}$, where $n_{sd} = 1, 2$ or 3 based on the spatial dimensions. Let Γ denote the boundary to the domain. A mapping function from the domain $\bar{\Omega} = \Omega \cup \Gamma$ to $\mathbb{R}$, $f : \bar{\Omega} \to \mathbb{R}$ is said to be of class $C^m(\Omega)$ if all the derivatives of the function up to the order m exist and are continuous functions. While solving most of higher-order differential equations, one encounters a boundary where the first derivative becomes discontinuous leading to undefined higher derivatives. Therefore, in the variational formulation, we employ the integral form of the differential equations to reduce the burden of evaluating those higher derivatives. The topic of function spaces gives a mathematical preliminary to such space of functions which obey certain restrictions which can be helpful for circumventing the issue of undefined higher derivatives.

Lebesgue space: A function in the $L^p(\Omega)$ space is considered to be Lebesgue integrable over the domain Ω to the power $p \in [1, \infty)$, i.e.,

$$L^p(\Omega) = \left\{ f \middle| \int_{\Omega} |f(x)|^p d\Omega < \infty \right\}, \tag{4.90}$$

and is equipped with the norm

$$||f||_{L^p(\Omega)} = \left(\int_{\Omega} |f|^p d\Omega \right)^{1/p}. \tag{4.91}$$

One particular case of Lebesgue space is the $L^2(\Omega)$ space which consists of functions that are square integrable over the domain. In such a case, the norm is $||f||_{L^2(\Omega)} = (f, f)^{1/2}$, where the inner product is defined as

$$(f, g) = \int_{\Omega} fg d\Omega. \tag{4.92}$$

Therefore, $L^2(\Omega)$ is equipped with an inner product with a norm that makes it a complete metric space, a type of **Hilbert space**.

Sobolev space: Functions in the Sobolev space are such that they belong to L^p space and their derivatives up to a certain order α also belong to L^p, i.e.,

$$W^{k,p}(\Omega) = \left\{ f \in L^p(\Omega) \middle| \frac{\partial^{|\alpha|} f}{\partial x_1^{\alpha_1} \partial x_2^{\alpha_2} ... \partial x_{n_{sd}}^{\alpha_{n_{sd}}}} \in L^p(\Omega) \ \forall |\alpha| \leq k \right\}, \tag{4.93}$$

where k is a non-negative integer, $\alpha = (\alpha_1, \alpha_2, ..., \alpha_{n_{sd}}) \in \mathbb{N}^{n_{sd}}$ and $|\alpha| = \alpha_1 + \alpha_2 + \cdots + \alpha_{n_{sd}}$.

Note that $W^{0,2}(\Omega) = L^2(\Omega)$. If we consider $p = 2$, the Sobolev space becomes a Hilbert space, i.e., $H^k(\Omega) = W^{k,2}(\Omega)$. For $k = 1$, the Hilbert space is defined as

$$H^1(\Omega) = \left\{ f \in L^2(\Omega) \middle| \frac{\partial f}{\partial x_i} \in L^2(\Omega), \ i = 1, 2, ..., n_{sd} \right\}, \tag{4.94}$$

with the inner product and the norm respectively,

$$(f, g)_1 = \int_{\Omega} \left(fg + \sum_{i=1}^{n_{sd}} \frac{\partial f}{\partial x_i} \frac{\partial g}{\partial x_i} \right) d\Omega, \qquad ||f||_1 = \sqrt{(f, f)_1}. \tag{4.95}$$

The above function spaces can be easily extended to vector spaces. For any two vectors, $\boldsymbol{u}, \boldsymbol{v} : \Omega \to \mathbb{R}^n$, the Hilbert space $H^k(\Omega)$ has the norm as,

$$||\boldsymbol{u}||_k = \left(\sum_{i=1}^{n} ||u_i||_k^2 \right)^{1/2}. \tag{4.96}$$

Stabilization Methods in the Literature

The concept of the stabilization methods is to provide some kind of upwinding phenomenon in the numerical scheme to add stability and accuracy. The upwinding function, introduced in [93] was noticed to cause an excessive crosswind diffusion and was limited only in the streamline direction. A streamline upwind Petrov-Galerkin (SUPG) method was suggested by [34] which introduces upwinding in the convection, transient as well as source terms. The SUPG method does not prevent overshooting and undershooting of the solution near the sharp boundary and internal layers in multi-dimensions. The numerical solution may be oscillatory or unbounded owing to the loss of conservation at the discrete level and the existing discontinuities. Satisfying the discrete maximum principle (DMP) preserves this positivity behavior by averting any maxima or minima inside the computational domain. The finite element local matrix was constructed to be non-negative and satisfy this DMP condition in [152], but the scheme was restricted to linear triangular elements. In [176], a monotone method with conservation properties was proposed where interpolation of the element geometry and flow rates led to the approximation of the convection term. This method was also limited to bilinear quadrilateral elements.

Apart from the linear methods mentioned above, nonlinear methods satisfying the DMP condition and circumventing the Godunov's theorem [135] (which limits the accuracy of a linear monotone method to first-order) for finite difference approximation were put forward. In the finite element context, a nonlinear discontinuity capturing term was introduced in the direction of the solution gradient [99] which satisfied the DMP. Some other modifications to the variational finite element formulation were proposed in [58, 70]. To improve convergence and stability of the method with accuracy, methods based on least-squares framework such as Galerkin/least-squares (GLS) and gradient Galerkin/least-squares (GGLS) method were introduced in [97] and [68] respectively. Nonlinearity to the GLS method was introduced by a discontinuity capturing term in [193], with a simplified formula for the stabilization parameter. Further development in [45] suggested the use of crosswind diffusion instead of isotropic diffusion for the nonlinear term, providing relatively accurate and improved convergence in the results. A review and comparison of the various linear stabilization methods such as SUPG, space-time Galerkin/least-squares (ST-GLS), subgrid-scale (SGS), characteristic Galerkin (CG) and Taylor-Galerkin (TG) was carried out in [47].

The origin of the stabilization parameter was attributed to the subgrid-scale phenomena in [94, 96]. The concept lies in the fact that across the spectrum of scales to be captured by the discretized method, the large scales are resolved by the method and the small unresolvable scales are modeled by Fourier representation of the Green's function of the subscale problem. The residual-based variational multiscale (RBVMS) method arises from the concept mentioned above with its application to the Navier-Stokes equations in [4, 22]. The stabilization parameter was further derived using the Taylor series and the concept of flow balance in a finite domain using the finite increment calculus (FIC) method in [162–165], which gave another interpretation to its origins. Furthermore, the concept of flux corrected schemes (FEM-FCT) and total variational diminishing schemes (FEM-TVD) in the finite element context proposed in [27, 131, 244] added an anti-diffusive flux to the first-order accurate method (discrete upwinding), to capture the discontinuities and shocks. These type of methods are referred as high resolution methods for finite-difference/finite-volume approximations [85].

The stabilization methodology was extended to the transport equation consisting of the reaction effects leading to SUPG with diffusion for reaction-dominated regions (SUPG+DRD), SUPG with centered Petrov-Galerkin (SU+C)PG and Galerkin/least-squares/gradient least-squares (GLS/GLS) methods in [84, 102, 210] respectively. The transient convection-diffusion equation was shown to behave as a transport equation with reaction effects in [101]. For the CDR equation, the stabilization parameter τ was derived while satisfying the DMP in [47], however the method did not generate a non-negative matrix in multi-dimensions. Methods based on the subgrid-scale phenomena were proposed for the reaction-dominated regimes as algebraic subgrid-scale method (SGS), unusual stabilized finite element method (USFEM) and its variants and methods employing the Green's function approach in [48, 69] and [86, 87] respectively. Stabilization for the CDR equation on distorted meshes employing the VMS technique was carried out in [173]. Nonlinear finite element methods using the discrete upwind operator to satisfy the DMP condition for convection- and reaction-dominated regimes were proposed in [112, 155, 156].

Chapter 5
Fluid-Structure Interaction: Variational Formulation

5.1 Introduction

In the previous chapter, we studied some of the basics of the finite element variational formulation for a transient convection-diffusion-reaction equation, which forms a canonical form for the nonlinear fluid-structure interaction equations. In this chapter, we apply the variational formulation to the separate equations governing the fluid and structural domains which were discussed in Sect. 3.4. We present the variational form of the Navier-Stokes equations written in ALE framework [Eqs. (3.64)–(3.65)] for the fluid and the structural dynamics equation [Eq. (3.49)] written in the Lagrangian coordinate system along with the boundary conditions mentioned in Sect. 3.4.4. We carry forward the notations defined in Sect. 3.4 in this chapter.

5.2 Weak Variational Form for Fluid-Structure Interaction

As mentioned in the previous chapters, fluid-structure interaction problems involve the coupling of two fields- flow and structure. In the current context, we will consider an incompressible Newtonian fluid with a three-dimensional structure for the variational formulation. However, the variational framework is general in its application and can be applied to any differential equation.

5.2.1 Trial and Test Function Spaces

For writing the weak formulations of the equations discussed above, we need to define the appropriate function spaces (see Appendix at the end of the previous chapter for more details) to select the trial solution and test or weighting functions which are dependent on the spatial coordinates. For the fluid equation, we define the following spaces

R. K. Jaiman and V. Joshi, *Computational Mechanics of Fluid-Structure Interaction*,
https://doi.org/10.1007/978-981-16-5355-1_5

$$\mathscr{V}_{\boldsymbol{\psi}^{\mathrm{f}}} = \{\boldsymbol{\psi}^{\mathrm{f}} \in H^1(\Omega^{\mathrm{f}}(t)) | \boldsymbol{\psi}^{\mathrm{f}} = \mathbf{0} \text{ on } \Gamma_D^{\mathrm{f}}\}, \tag{5.1}$$

$$\mathscr{V}_q = \{q \in L^2(\Omega^{\mathrm{f}}(t))\}, \tag{5.2}$$

$$\mathscr{S}_{\boldsymbol{v}^{\mathrm{f}}} = \{\boldsymbol{v}^{\mathrm{f}} \in H^1(\Omega^{\mathrm{f}}(t)) | \boldsymbol{v}^{\mathrm{f}} = \boldsymbol{v}_D^{\mathrm{f}} \text{ on } \Gamma_D^{\mathrm{f}}\}, \tag{5.3}$$

$$\mathscr{S}_p = \{p \in L^2(\Omega^{\mathrm{f}}(t))\}, \tag{5.4}$$

where $\mathscr{V}_{\boldsymbol{\psi}^{\mathrm{f}}}$ and $\mathscr{V}_q$ denote the test function spaces for the momentum and continuity equations respectively and $\mathscr{S}_{\boldsymbol{v}^{\mathrm{f}}}$ and $\mathscr{S}_p$ denote the spaces from where we select the trial solution for velocity and pressure respectively.

Similarly, for the structural equation, we define the following function spaces

$$\mathscr{V}_{\boldsymbol{\psi}^{\mathrm{s}}} = \{\boldsymbol{\psi}^{\mathrm{s}} \in H^1(\Omega^{\mathrm{s}}) | \boldsymbol{\psi}^{\mathrm{s}} = \mathbf{0} \text{ on } \Gamma_D^{\mathrm{s}}\}, \tag{5.5}$$

$$\mathscr{S}_{\boldsymbol{v}^{\mathrm{s}}} = \{\boldsymbol{v}^{\mathrm{s}} \in H^1(\Omega^{\mathrm{s}}) | \boldsymbol{v}^{\mathrm{s}} = \boldsymbol{v}_D^{\mathrm{s}} \text{ on } \Gamma_D^{\mathrm{s}}\}, \tag{5.6}$$

where $\mathscr{V}_{\boldsymbol{\psi}^{\mathrm{s}}}$ and $\mathscr{S}_{\boldsymbol{v}^{\mathrm{s}}}$ denote the test function and trial solution spaces for the structural velocity respectively. For the mesh equation, the following spaces are defined

$$\mathscr{V}_{\boldsymbol{\psi}^{\mathrm{m}}} = \{\boldsymbol{\psi}^{\mathrm{m}} \in H^1(\Omega^{\mathrm{f}}) | \boldsymbol{\psi}^{\mathrm{m}} = \mathbf{0} \text{ on } \Gamma_D^{\mathrm{m}}\}, \tag{5.7}$$

$$\mathscr{S}_{\boldsymbol{u}^{\mathrm{f}}} = \{\boldsymbol{u}^{\mathrm{f}} \in H^1(\Omega^{\mathrm{f}}) | \boldsymbol{u}^{\mathrm{f}} = \boldsymbol{u}_D^{\mathrm{f}} \text{ on } \Gamma_D^{\mathrm{m}}\}. \tag{5.8}$$

Now that we have the appropriate spaces to select the weighting functions and trial solution, we are ready to form the weak form of the flow and structural equations.

5.2.2 *Weak Formulation for FSI*

In this section, we derive the weak or variational form for the coupled fluid-structure equations. Let us begin with the incompressible Navier-Stokes equations [Eqs. (3.64–3.65)]. The weak form is written by multiplying the equation by a weighting function and then integrating it over the whole domain. Thus, for the flow equations,

$$\int_{\Omega^{\mathrm{f}}(t)} \left(\rho^{\mathrm{f}} \frac{\partial \boldsymbol{v}^{\mathrm{f}}}{\partial t}\bigg|_{\chi} + \rho^{\mathrm{f}} (\boldsymbol{v}^{\mathrm{f}} - \boldsymbol{w}) \cdot \nabla \boldsymbol{v}^{\mathrm{f}} - \nabla \cdot \boldsymbol{\sigma}^{\mathrm{f}} \right) \cdot \boldsymbol{\psi}^{\mathrm{f}} d\Omega = \int_{\Omega^{\mathrm{f}}(t)} \rho^{\mathrm{f}} \boldsymbol{b}^{\mathrm{f}} \cdot \boldsymbol{\psi}^{\mathrm{f}} d\Omega, \tag{5.9}$$

$$\int_{\Omega^{\mathrm{f}}(t)} (\nabla \cdot \boldsymbol{v}^{\mathrm{f}}) q d\Omega = 0, \tag{5.10}$$

Notice that there are higher order derivatives for velocity in the term $\nabla \cdot \boldsymbol{\sigma}^{\mathrm{f}}$ which makes the requirement for the velocity as $\boldsymbol{v}^{\mathrm{f}} \in H^2(\Omega^{\mathrm{f}}(t))$. Therefore, using Green's identity and Gauss' divergence theorem, we relax this requirement.

$$\int_{\Omega^{\mathrm{f}}(t)} \left(\rho^{\mathrm{f}} \frac{\partial \boldsymbol{v}^{\mathrm{f}}}{\partial t}\bigg|_{\chi} + \rho^{\mathrm{f}} (\boldsymbol{v}^{\mathrm{f}} - \boldsymbol{w}) \cdot \nabla \boldsymbol{v}^{\mathrm{f}} \right) \cdot \boldsymbol{\psi}^{\mathrm{f}} d\Omega + \int_{\Omega^{\mathrm{f}}(t)} \boldsymbol{\sigma}^{\mathrm{f}} : \nabla \boldsymbol{\psi}^{\mathrm{f}} d\Omega$$
$$= \int_{\Omega^{\mathrm{f}}(t)} \rho^{\mathrm{f}} \boldsymbol{b}^{\mathrm{f}} \cdot \boldsymbol{\psi}^{\mathrm{f}} d\Omega + \int_{\Gamma_D^{\mathrm{f}}} \boldsymbol{\sigma}^{\mathrm{f}} \cdot \boldsymbol{\psi}^{\mathrm{f}} d\Gamma + \int_{\Gamma_N^{\mathrm{f}}} \boldsymbol{\sigma}_N^{\mathrm{f}} \cdot \boldsymbol{\psi}^{\mathrm{f}} d\Gamma + \int_{\Gamma(t)} (\boldsymbol{\sigma}^{\mathrm{f}} \cdot \boldsymbol{n}^{\mathrm{f}}) \cdot \boldsymbol{\psi}^{\mathrm{f}} d\Gamma, \tag{5.11}$$

$$\int_{\Omega^{\mathrm{f}}(t)} (\nabla \cdot \boldsymbol{v}^{\mathrm{f}}) q d\Omega = 0. \tag{5.12}$$

Now, the continuity requirement on the velocity is $H^1(\Omega^{\mathrm{f}}(t))$ and the definitions of the function spaces in Sect. 5.2.1 can be employed. Recalling that $\boldsymbol{\psi}^{\mathrm{f}} = \boldsymbol{0}$ on Γ_D^{f},

$$\int_{\Omega^{\mathrm{f}}(t)} \left(\rho^{\mathrm{f}} \frac{\partial \boldsymbol{v}^{\mathrm{f}}}{\partial t}\bigg|_{\chi} + \rho^{\mathrm{f}} (\boldsymbol{v}^{\mathrm{f}} - \boldsymbol{w}) \cdot \nabla \boldsymbol{v}^{\mathrm{f}} \right) \cdot \boldsymbol{\psi}^{\mathrm{f}} d\Omega + \int_{\Omega^{\mathrm{f}}(t)} \boldsymbol{\sigma}^{\mathrm{f}} : \nabla \boldsymbol{\psi}^{\mathrm{f}} d\Omega$$
$$= \int_{\Omega^{\mathrm{f}}(t)} \rho^{\mathrm{f}} \boldsymbol{b}^{\mathrm{f}} \cdot \boldsymbol{\psi}^{\mathrm{f}} d\Omega + \int_{\Gamma_N^{\mathrm{f}}} \boldsymbol{\sigma}_N^{\mathrm{f}} \cdot \boldsymbol{\psi}^{\mathrm{f}} d\Gamma + \int_{\Gamma(t)} (\boldsymbol{\sigma}^{\mathrm{f}} \cdot \boldsymbol{n}^{\mathrm{f}}) \cdot \boldsymbol{\psi}^{\mathrm{f}} d\Gamma, \tag{5.13}$$

$$\int_{\Omega^{\mathrm{f}}(t)} (\nabla \cdot \boldsymbol{v}^{\mathrm{f}}) q d\Omega = 0. \tag{5.14}$$

We can combine the weak form of the momentum and continuity equations to write the variational statement as: find $[\boldsymbol{v}^{\mathrm{f}}, p] \in \mathscr{S}_{\boldsymbol{v}^{\mathrm{f}}} \times \mathscr{S}_p$ such that $\forall [\boldsymbol{\psi}^{\mathrm{f}}, q] \in \mathscr{V}_{\boldsymbol{\psi}^{\mathrm{f}}} \times \mathscr{V}_q$,

$$\int_{\Omega^{\mathrm{f}}(t)} \left(\rho^{\mathrm{f}} \frac{\partial \boldsymbol{v}^{\mathrm{f}}}{\partial t}\bigg|_{\chi} + \rho^{\mathrm{f}} (\boldsymbol{v}^{\mathrm{f}} - \boldsymbol{w}) \cdot \nabla \boldsymbol{v}^{\mathrm{f}} \right) \cdot \boldsymbol{\psi}^{\mathrm{f}} d\Omega + \int_{\Omega^{\mathrm{f}}(t)} \boldsymbol{\sigma}^{\mathrm{f}} : \nabla \boldsymbol{\psi}^{\mathrm{f}} d\Omega$$
$$+ \int_{\Omega^{\mathrm{f}}(t)} (\nabla \cdot \boldsymbol{v}^{\mathrm{f}}) q d\Omega = \int_{\Omega^{\mathrm{f}}(t)} \rho^{\mathrm{f}} \boldsymbol{b}^{\mathrm{f}} \cdot \boldsymbol{\psi}^{\mathrm{f}} d\Omega + \int_{\Gamma_N^{\mathrm{f}}} \boldsymbol{\sigma}_N^{\mathrm{f}} \cdot \boldsymbol{\psi}^{\mathrm{f}} d\Gamma + \int_{\Gamma(t)} (\boldsymbol{\sigma}^{\mathrm{f}} \cdot \boldsymbol{n}^{\mathrm{f}}) \cdot \boldsymbol{\psi}^{\mathrm{f}} d\Gamma. \tag{5.15}$$

Proceeding in a similar way for the structural equation, we can write the variational statement as: find $\boldsymbol{v}^{\mathrm{s}} \in \mathscr{S}_{\boldsymbol{v}^{\mathrm{s}}}$ such that $\forall \boldsymbol{\psi}^{\mathrm{s}} \in \mathscr{V}_{\boldsymbol{\psi}^{\mathrm{s}}}$,

$$\int_{\Omega^{\mathrm{s}}} \left(\rho^{\mathrm{s}} \frac{\partial \boldsymbol{v}^{\mathrm{s}}}{\partial t} \right) \cdot \boldsymbol{\psi}^{\mathrm{s}} d\Omega + \int_{\Omega^{\mathrm{s}}} \boldsymbol{\sigma}^{\mathrm{s}} : \nabla \boldsymbol{\psi}^{\mathrm{s}} d\Omega$$
$$= \int_{\Omega^{\mathrm{s}}} \rho^{\mathrm{s}} \boldsymbol{b}^{\mathrm{s}} \cdot \boldsymbol{\psi}^{\mathrm{s}} d\Omega + \int_{\Gamma_N^{\mathrm{s}}} \boldsymbol{\sigma}_N^{\mathrm{s}} \cdot \boldsymbol{\psi}^{\mathrm{s}} d\Gamma + \int_{\Gamma} (\boldsymbol{\sigma}^{\mathrm{s}} \cdot \boldsymbol{n}^{\mathrm{s}}) \cdot \boldsymbol{\psi}^{\mathrm{s}} d\Gamma. \tag{5.16}$$

Similarly for the mesh equation, we get: find $\boldsymbol{u}^{\mathrm{f}} \in \mathscr{S}_{\boldsymbol{u}^{\mathrm{f}}}$ such that $\forall \boldsymbol{\psi}^{\mathrm{m}} \in \mathscr{V}_{\boldsymbol{\psi}^{\mathrm{m}}}$,

$$\int_{\Omega^{\mathrm{f}}} \nabla \boldsymbol{\psi}^{\mathrm{m}} : \boldsymbol{\sigma}^{\mathrm{m}} d\Omega = 0, \tag{5.17}$$

where $\boldsymbol{\sigma}^{\mathrm{m}} = (1 + k_{\mathrm{m}})[\nabla \boldsymbol{u}^{\mathrm{f}} + (\nabla \boldsymbol{u}^{\mathrm{f}})^T + (\nabla \cdot \boldsymbol{u}^{\mathrm{f}})\boldsymbol{I}]$.

Remark 5.1 The velocity continuity condition at the fluid-structure interface is satisfied by moving the fluid domain such that the velocity of the fluid domain at the fluid-structure boundary is the same as the structural velocity at the boundary. After solving for the fluid displacement field $\boldsymbol{u}^{\mathrm{f}}$, the spatial coordinates are updated as $\boldsymbol{x} = \boldsymbol{\chi} + \boldsymbol{u}^{\mathrm{f}}(\boldsymbol{\chi}, t)$ for all $\boldsymbol{x} \in \Omega^{\mathrm{f}}(t)$ and $\boldsymbol{\chi} \in \Omega^{\mathrm{f}}(0)$.

5.3 Semi-Discrete Temporal Discretization

As the independent variables in the FSI equations are the spatial and temporal coordinates, we need to discretize the domain in both space and time. One way of doing this is to define the trial and test function spaces in the weak form such that they are space and time dependent. Such methods lead to what are called space-time finite element methods. In the present case, we will utilize a simpler approach of discretizing the temporal variable by Taylor series (similar to finite difference methods) such that the trial and test function spaces have only spatial dependence. An example of such discrete time stepping is the generalized-α time integration [44, 111] technique for integrating in time from time step t^{n} to $t^{\mathrm{n+1}}$. While one can use any other time integration technique, we will present the formulation using the generalized-α method for the rest of this chapter.

5.3.1 *Generalized-α Time Integration*

Here, we discuss the generalized-α [44, 111] predictor-corrector method for the temporal discretization. It enables a user-controlled high frequency damping via a single parameter called the spectral radius ρ_∞, which allows for a coarser discretization in time. We solve the equation at the time interval $\mathrm{n} + \alpha$ while integrating the equation from n to $\mathrm{n} + 1$ in time. Suppose the equation can be written as $G(\partial_t u^{\mathrm{n}+\alpha_{\mathrm{m}}}, u^{\mathrm{n}+\alpha}, \phi^{\mathrm{n}+\alpha}) = 0$, where $G(\cdot)$ can be a nonlinear function of the variable ϕ and its first and second-order derivatives, i.e., $u = \partial_t \phi$ and $\partial_t u = \partial_{tt}^2 \phi$, respectively. The expressions for the variables and their derivatives in the equation can be written as

$$\phi^{\mathrm{n+1}} = \phi^{\mathrm{n}} + \Delta t u^{\mathrm{n}} + \Delta t^2 \big(\beta \partial_t u^{\mathrm{n+1}} + (1/2 - \beta)\partial_t u^{\mathrm{n}}\big), \tag{5.18}$$

$$u^{\mathrm{n+1}} = u^{\mathrm{n}} + \Delta t\big(\gamma \partial_t u^{\mathrm{n+1}} + (1-\gamma)\partial_t u^{\mathrm{n}}\big), \tag{5.19}$$

$$\partial_t u^{\mathrm{n}+\alpha_{\mathrm{m}}} = \partial_t u^{\mathrm{n}} + \alpha_{\mathrm{m}}\big(\partial_t u^{\mathrm{n+1}} - \partial_t u^{\mathrm{n}}\big), \tag{5.20}$$

$$u^{\mathrm{n}+\alpha} = u^{\mathrm{n}} + \alpha(u^{\mathrm{n+1}} - u^{\mathrm{n}}), \tag{5.21}$$

$$\phi^{\mathrm{n}+\alpha} = \phi^{\mathrm{n}} + \alpha(\phi^{\mathrm{n+1}} - \phi^{\mathrm{n}}), \tag{5.22}$$

where Δt is the time step size, α_{m}, α, β and γ are defined as

$$\alpha_{\mathrm{m}} = \frac{1}{2}\left(\frac{3-\rho_\infty}{1+\rho_\infty}\right), \quad \alpha = \frac{1}{1+\rho_\infty}, \quad \beta = \frac{1}{4}(1+\alpha_{\mathrm{m}}-\alpha)^2, \quad \gamma = \frac{1}{2} + \alpha_{\mathrm{m}} - \alpha. \tag{5.23}$$

Note that the above time integration method simplifies to the following methods under the given conditions:

- Crank-Nicolson method: $\rho_\infty = 1$
- Gear's two-step method: $\rho_\infty = 0$
- Implicit Euler (Backward Euler) method: $\alpha = \alpha_{\mathrm{m}} = \gamma = 1$.

The predictor-corrector algorithm within a time step from n to n + 1 consists of the following steps:

1. Predict the variables at n + 1.
2. Evaluate the variables at the intermediate time interval $\mathrm{n} + \alpha$.
3. Linearize the nonlinear equation $G(\partial_t u^{\mathrm{n}+\alpha_{\mathrm{m}}}, u^{\mathrm{n}+\alpha}, \phi^{\mathrm{n}+\alpha}) = 0$ with the help of Newton-Raphson method.
4. Calculate the increments in the variables by solving the linear system of equations.
5. Update the variables at $\mathrm{n} + \alpha$.
6. Correct the variables at n + 1 and proceed to the next nonlinear iteration.

5.3.2 Semi-Discrete Temporal Discretization Applied to FSI

For the fluid equations, the generalized-α expressions for the fluid velocity $\boldsymbol{v}^{\mathrm{f}}$ and acceleration $\partial_t \boldsymbol{v}^{\mathrm{f}}$ can be written as:

$$\boldsymbol{v}^{\mathrm{f,n+1}} = \boldsymbol{v}^{\mathrm{f,n}} + \Delta t(\gamma^{\mathrm{f}} \partial_t \boldsymbol{v}^{\mathrm{f,n+1}} + (1-\gamma^{\mathrm{f}})\partial_t \boldsymbol{v}^{\mathrm{f,n}}), \tag{5.24}$$

$$\partial_t \boldsymbol{v}^{\mathrm{f,n}+\alpha_{\mathrm{m}}^{\mathrm{f}}} = \partial_t \boldsymbol{v}^{\mathrm{f,n}} + \alpha_{\mathrm{m}}^{\mathrm{f}}(\partial_t \boldsymbol{v}^{\mathrm{f,n+1}} - \partial_t \boldsymbol{v}^{\mathrm{f,n}}), \tag{5.25}$$

$$\boldsymbol{v}^{\mathrm{f,n}+\alpha^{\mathrm{f}}} = \boldsymbol{v}^{\mathrm{f,n}} + \alpha^{\mathrm{f}}(\boldsymbol{v}^{\mathrm{f,n+1}} - \boldsymbol{v}^{\mathrm{f,n}}). \tag{5.26}$$

On the other hand, the expressions for the structural equation involving the structural velocity $\boldsymbol{v}^{\mathrm{s}}$, position $\boldsymbol{\varphi}^{\mathrm{s}}$ and acceleration $\partial_t \boldsymbol{v}^{\mathrm{s}}$ are

$$\boldsymbol{\varphi}^{\mathrm{s,n+1}} = \boldsymbol{\varphi}^{\mathrm{s,n}} + \Delta t \boldsymbol{v}^{\mathrm{s,n}} + \Delta t^2 \bigg(\Big(\frac{1}{2} - \beta^{\mathrm{s}}\Big)\partial_t \boldsymbol{v}^{\mathrm{s,n}} + \beta^{\mathrm{s}} \partial_t \boldsymbol{v}^{\mathrm{s,n+1}} \bigg), \tag{5.27}$$

$$\boldsymbol{v}^{\mathrm{s,n+1}} = \boldsymbol{v}^{\mathrm{s,n}} + \Delta t \bigg((1-\gamma^{\mathrm{s}})\partial_t \boldsymbol{v}^{\mathrm{s,n}} + \gamma^{\mathrm{s}} \partial_t \boldsymbol{v}^{\mathrm{s,n+1}} \bigg), \tag{5.28}$$

$$\partial_t \boldsymbol{v}^{\mathrm{s,n}+\alpha_{\mathrm{m}}^{\mathrm{s}}} = \partial_t \boldsymbol{v}^{\mathrm{s,n}} + \alpha_{\mathrm{m}}^{\mathrm{s}} (\partial_t \boldsymbol{v}^{\mathrm{s,n+1}} - \partial_t \boldsymbol{v}^{\mathrm{s,n}}), \tag{5.29}$$

$$\boldsymbol{v}^{\mathrm{n}+\alpha^{\mathrm{s}}} = \boldsymbol{v}^{\mathrm{s,n}} + \alpha^{\mathrm{s}} (\boldsymbol{v}^{\mathrm{s,n+1}} - \boldsymbol{v}^{\mathrm{s,n}}), \tag{5.30}$$

$$\boldsymbol{\varphi}^{\mathrm{n}+\alpha^{\mathrm{s}}} = \boldsymbol{\varphi}^{\mathrm{s,n}} + \alpha^{\mathrm{s}} (\boldsymbol{\varphi}^{\mathrm{s,n+1}} - \boldsymbol{\varphi}^{\mathrm{s,n}}), \tag{5.31}$$

The equations written in the semi-discrete form are solved at the time level $t^{\mathrm{n}+\alpha}$ and the values at $t^{\mathrm{n+1}}$ are updated via the predictor-corrector algorithm.

5.4 Finite Element Space Discretization for FSI

The next step is to select finite element spaces $\mathscr{V}^{\mathrm{h}}$ and $\mathscr{S}^{\mathrm{h}}$ from the space of test and trial functions $\mathscr{V}$ and $\mathscr{S}$ respectively. We discretize the domain Ω into non-intersecting finite elements $\Omega = \cup_{e=1}^{\mathrm{n_{el}}} \Omega^e$. The discrete spaces for trial and test functions for the fluid equations are defined as

$$\mathscr{V}_{\boldsymbol{\psi}^{\mathrm{f}}}^{\mathrm{h}} = \{\boldsymbol{\psi}_{\mathrm{h}}^{\mathrm{f}} \in H^1(\Omega^{\mathrm{f}}(t^{\mathrm{n+1}})) | \boldsymbol{\psi}_{\mathrm{h}}^{\mathrm{f}}|_{\Omega^e} \in \mathbb{P}_m(\Omega^e) \forall e \text{ and } \boldsymbol{\psi}_{\mathrm{h}}^{\mathrm{f}} = \boldsymbol{0} \text{ on } \Gamma_D^{\mathrm{f}}\}, \tag{5.32}$$

$$\mathscr{V}_{q}^{\mathrm{h}} = \{q_{\mathrm{h}} \in L^2(\Omega^{\mathrm{f}}(t^{\mathrm{n+1}})) | q_{\mathrm{h}}|_{\Omega^e} \in \mathbb{P}_m(\Omega^e) \forall e\}, \tag{5.33}$$

$$\mathscr{S}_{\boldsymbol{v}^{\mathrm{f}}}^{\mathrm{h}} = \{\boldsymbol{v}_{\mathrm{h}}^{\mathrm{f}} \in H^1(\Omega^{\mathrm{f}}(t^{\mathrm{n+1}})) | \boldsymbol{v}_{\mathrm{h}}^{\mathrm{f}}|_{\Omega^e} \in \mathbb{P}_m(\Omega^e) \forall e \text{ and } \boldsymbol{v}_{\mathrm{h}}^{\mathrm{f}} = \boldsymbol{v}_D^{\mathrm{f}} \text{ on } \Gamma_D^{\mathrm{f}}\}, \tag{5.34}$$

$$\mathscr{S}_{p}^{\mathrm{h}} = \{p_{\mathrm{h}} \in L^2(\Omega^{\mathrm{f}}(t^{\mathrm{n+1}})) | p_{\mathrm{h}}|_{\Omega^e} \in \mathbb{P}_m(\Omega^e) \forall e\}, \tag{5.35}$$

where $\mathbb{P}_m(\Omega^e)$ is the space of polynomials of degree $\leq m$.

The finite element variational statement for the element Ω^e for the fluid equations can thus be written as: find $[\boldsymbol{v}_{\mathrm{h}}^{\mathrm{f,n}+\alpha^{\mathrm{f}}}, p_{\mathrm{h}}^{\mathrm{n+1}}] \in \mathscr{S}_{\boldsymbol{v}^{\mathrm{f}}}^{\mathrm{h}} \times \mathscr{S}_{p}^{\mathrm{h}}$ such that $\forall [\boldsymbol{\psi}_{\mathrm{h}}^{\mathrm{f}}, q_{\mathrm{h}}] \in \mathscr{V}_{\boldsymbol{\psi}^{\mathrm{f}}}^{\mathrm{h}} \times \mathscr{V}_{q}^{\mathrm{h}}$,

$$\begin{aligned} &\int_{\Omega^{\mathrm{f}}(t^{\mathrm{n+1}})} \left(\rho^{\mathrm{f}} \partial_t \boldsymbol{v}_{\mathrm{h}}^{\mathrm{f,n}+\alpha_{\mathrm{m}}^{\mathrm{f}}}\big|_{\chi} + \rho^{\mathrm{f}} (\boldsymbol{v}_{\mathrm{h}}^{\mathrm{f,n}+\alpha^{\mathrm{f}}} - \boldsymbol{w}) \cdot \nabla \boldsymbol{v}_{\mathrm{h}}^{\mathrm{f,n}+\alpha^{\mathrm{f}}} \right) \cdot \boldsymbol{\psi}_{\mathrm{h}}^{\mathrm{f}} d\Omega \\ &+ \int_{\Omega^{\mathrm{f}}(t^{\mathrm{n+1}})} \boldsymbol{\sigma}_{\mathrm{h}}^{\mathrm{f,n}+\alpha^{\mathrm{f}}} : \nabla \boldsymbol{\psi}_{\mathrm{h}}^{\mathrm{f}} d\Omega + \int_{\Omega^{\mathrm{f}}(t^{\mathrm{n+1}})} (\nabla \cdot \boldsymbol{v}_{\mathrm{h}}^{\mathrm{f,n}+\alpha^{\mathrm{f}}}) q_{\mathrm{h}} d\Omega \\ &= \int_{\Omega^{\mathrm{f}}(t^{\mathrm{n+1}})} \rho^{\mathrm{f}} \boldsymbol{b}_{\mathrm{h}}^{\mathrm{f,n}+\alpha^{\mathrm{f}}} \cdot \boldsymbol{\psi}_{\mathrm{h}}^{\mathrm{f}} d\Omega + \int_{\Gamma_N^{\mathrm{f}}} \boldsymbol{\sigma}_N^{\mathrm{f,n}+\alpha^{\mathrm{f}}} \cdot \boldsymbol{\psi}_{\mathrm{h}}^{\mathrm{f}} d\Gamma + \int_{\Gamma(t^{\mathrm{n+1}})} (\boldsymbol{\sigma}_{\mathrm{h}}^{\mathrm{f,n}+\alpha^{\mathrm{f}}} \cdot \boldsymbol{n}^{\mathrm{f}}) \cdot \boldsymbol{\psi}_{\mathrm{h}}^{\mathrm{f}} d\Gamma. \end{aligned} \tag{5.36}$$

The above equation consists of the Galerkin formulation without any stabilization terms for the fluid domain. For it to be stable for convection-dominated regimes, one needs compatible polynomial spaces for velocity-pressure coupling or stabilization terms (see Appendix at the end of this chapter). Here, we employ Petrov-Galerkin stabilization for stability in convection-dominated regimes. The extra stabilization terms are introduced as:

$$\begin{aligned}
&\int_{\Omega^{\mathrm{f}}(t^{\mathrm{n}+1})} \left(\rho^{\mathrm{f}} \partial_t \boldsymbol{v}_{\mathrm{h}}^{\mathrm{f,n}+\alpha_{\mathrm{m}}^{\mathrm{f}}}\Big|_{\chi} + \rho^{\mathrm{f}} (\boldsymbol{v}_{\mathrm{h}}^{\mathrm{f,n}+\alpha^{\mathrm{f}}} - \boldsymbol{w}) \cdot \nabla \boldsymbol{v}_{\mathrm{h}}^{\mathrm{f,n}+\alpha^{\mathrm{f}}}\right) \cdot \boldsymbol{\psi}_{\mathrm{h}}^{\mathrm{f}} d\Omega \\
&+ \int_{\Omega^{\mathrm{f}}(t^{\mathrm{n}+1})} \boldsymbol{\sigma}_{\mathrm{h}}^{\mathrm{f,n}+\alpha^{\mathrm{f}}} : \nabla \boldsymbol{\psi}_{\mathrm{h}}^{\mathrm{f}} d\Omega + \sum_{e=1}^{\mathrm{n_{el}}} \int_{\Omega^e} \tau_m \big(\rho^{\mathrm{f}} (\boldsymbol{v}_{\mathrm{h}}^{\mathrm{f,n}+\alpha^{\mathrm{f}}} - \boldsymbol{w}) \cdot \nabla \boldsymbol{\psi}_{\mathrm{h}}^{\mathrm{f}} + \nabla q_{\mathrm{h}}\big) \cdot \boldsymbol{R}_m d\Omega^e \\
&+ \int_{\Omega^{\mathrm{f}}(t^{\mathrm{n}+1})} (\nabla \cdot \boldsymbol{v}_{\mathrm{h}}^{\mathrm{f,n}+\alpha^{\mathrm{f}}}) q_{\mathrm{h}} d\Omega + \sum_{e=1}^{\mathrm{n_{el}}} \int_{\Omega^e} \nabla \cdot \boldsymbol{\psi}_{\mathrm{h}}^{\mathrm{f}} \tau_c \boldsymbol{R}_c d\Omega^e \\
&= \int_{\Omega^{\mathrm{f}}(t^{\mathrm{n}+1})} \rho^{\mathrm{f}} \boldsymbol{b}_{\mathrm{h}}^{\mathrm{f,n}+\alpha^{\mathrm{f}}} \cdot \boldsymbol{\psi}_{\mathrm{h}}^{\mathrm{f}} d\Omega + \int_{\Gamma_N^{\mathrm{f}}} \boldsymbol{\sigma}_N^{\mathrm{f,n}+\alpha^{\mathrm{f}}} \cdot \boldsymbol{\psi}_{\mathrm{h}}^{\mathrm{f}} d\Gamma + \int_{\Gamma(t^{\mathrm{n}+1})} (\boldsymbol{\sigma}_{\mathrm{h}}^{\mathrm{f,n}+\alpha^{\mathrm{f}}} \cdot \boldsymbol{n}^{\mathrm{f}}) \cdot \boldsymbol{\psi}_{\mathrm{h}}^{\mathrm{f}} d\Gamma,
\end{aligned} \tag{5.37}$$

where the second term in the second line represents the stabilization term for the momentum equation and the second term in the third line depicts the same for the continuity equation. $\boldsymbol{R}_m$ and $\boldsymbol{R}_c$ are the residual of the momentum and continuity equations respectively. The stabilization parameters τ_m and τ_c are the least-squares metrics added to the element-level integrals [34, 67, 193, 209] defined as

$$\tau_m = \left[\left(\frac{2\rho^{\mathrm{f}}}{\Delta t}\right)^2 + (\rho^{\mathrm{f}})^2 (\boldsymbol{v}_{\mathrm{h}}^{\mathrm{f,n}+\alpha^{\mathrm{f}}} - \boldsymbol{w}) \cdot \boldsymbol{G} (\boldsymbol{v}_{\mathrm{h}}^{\mathrm{f,n}+\alpha^{\mathrm{f}}} - \boldsymbol{w}) + C_I (\mu^{\mathrm{f}})^2 \boldsymbol{G} : \boldsymbol{G}\right]^{-1/2}, \tag{5.38}$$

$$\tau_c = \frac{1}{8 \mathrm{tr}(\boldsymbol{G}) \tau_m}, \tag{5.39}$$

where C_I is a constant derived from inverse estimates [83], tr() denotes the trace and $\boldsymbol{G}$ is the contravariant metric tensor given by

$$\boldsymbol{G} = \left(\frac{\partial \boldsymbol{\xi}}{\partial \boldsymbol{x}}\right)^T \frac{\partial \boldsymbol{\xi}}{\partial \boldsymbol{x}}, \tag{5.40}$$

where $\boldsymbol{x}$ and $\boldsymbol{\xi}$ are the physical and parametric coordinates respectively.

Coming to the structural equation, we define the finite spaces as follows:

$$\mathcal{V}_{\boldsymbol{\psi}^{\mathrm{s}}}^{\mathrm{h}} = \{\boldsymbol{\psi}_{\mathrm{h}}^{\mathrm{s}} \in H^1(\Omega^{\mathrm{s}}) \big| \boldsymbol{\psi}_{\mathrm{h}}^{\mathrm{s}}|_{\Omega^e} \in \mathbb{P}_m(\Omega^e) \forall e \text{ and } \boldsymbol{\psi}_{\mathrm{h}}^{\mathrm{s}} = \boldsymbol{0} \text{ on } \Gamma_D^{\mathrm{s}}\}, \tag{5.41}$$

$$\mathcal{S}_{\boldsymbol{v}^{\mathrm{s}}}^{\mathrm{h}} = \{\boldsymbol{v}_{\mathrm{h}}^{\mathrm{s}} \in H^1(\Omega^{\mathrm{s}}) \big| \boldsymbol{v}_{\mathrm{h}}^{\mathrm{s}}|_{\Omega^e} \in \mathbb{P}_m(\Omega^e) \forall e \text{ and } \boldsymbol{v}_{\mathrm{h}}^{\mathrm{s}} = \boldsymbol{v}_D^{\mathrm{s}} \text{ on } \Gamma_D^{\mathrm{s}}\}, \tag{5.42}$$

and the finite element variational statement is written as: find $\boldsymbol{v}_{\mathrm{h}}^{\mathrm{s},\mathrm{n}+\alpha^{\mathrm{s}}} \in \mathcal{S}_{\boldsymbol{v}^{\mathrm{s}}}^{\mathrm{h}}$ such that $\forall \boldsymbol{\psi}_{\mathrm{h}}^{\mathrm{s}} \in \mathcal{V}_{\boldsymbol{\psi}^{\mathrm{s}}}^{\mathrm{h}}$,

$$\int_{\Omega^{\mathrm{s}}} \left(\rho^{\mathrm{s}} \partial_t \boldsymbol{v}_{\mathrm{h}}^{\mathrm{s},\mathrm{n}+\alpha_{\mathrm{m}}^{\mathrm{s}}}\right) \cdot \boldsymbol{\psi}_{\mathrm{h}}^{\mathrm{s}} d\Omega + \int_{\Omega^{\mathrm{s}}} \boldsymbol{\sigma}_{\mathrm{h}}^{\mathrm{s},\mathrm{n}+\alpha^{\mathrm{s}}} : \nabla \boldsymbol{\psi}_{\mathrm{h}}^{\mathrm{s}} d\Omega$$
$$= \int_{\Omega^{\mathrm{s}}} \rho^{\mathrm{s}} \boldsymbol{b}_{\mathrm{h}}^{\mathrm{s},\mathrm{n}+\alpha^{\mathrm{s}}} \cdot \boldsymbol{\psi}_{\mathrm{h}}^{\mathrm{s}} d\Omega + \int_{\Gamma_N^{\mathrm{s}}} \boldsymbol{\sigma}_N^{\mathrm{s},\mathrm{n}+\alpha^{\mathrm{s}}} \cdot \boldsymbol{\psi}_{\mathrm{h}}^{\mathrm{s}} d\Gamma + \int_{\Gamma} (\boldsymbol{\sigma}_{\mathrm{h}}^{\mathrm{s},\mathrm{n}+\alpha^{\mathrm{s}}} \cdot \boldsymbol{n}^{\mathrm{s}}) \cdot \boldsymbol{\psi}_{\mathrm{h}}^{\mathrm{s}} d\Gamma. \quad (5.43)$$

Similarly for the mesh equation, we define the finite element spaces and the variational statement is written as: find $\boldsymbol{u}_{\mathrm{h}}^{\mathrm{f},\mathrm{n}+1} \in \mathcal{S}_{\boldsymbol{u}^{\mathrm{f}}}^{\mathrm{h}}$ such that $\forall \boldsymbol{\psi}_{\mathrm{h}}^{\mathrm{m}} \in \mathcal{V}_{\boldsymbol{\psi}^{\mathrm{m}}}^{\mathrm{h}}$,

$$\int_{\Omega^{\mathrm{f}}} \nabla \boldsymbol{\psi}_{\mathrm{h}}^{\mathrm{m}} : \boldsymbol{\sigma}_{\mathrm{h}}^{\mathrm{m},\mathrm{n}+1} d\Omega = 0. \quad (5.44)$$

5.5 Matrix Form of the Linear System of Equations

After the assembly across all the finite elements and mapping to the global nodes for the FSI equations and utilizing linearization strategies like Newton-Raphson, the finite element formulation can be finally written in the matrix form. For the fluid equations, the matrix form looks like the following

$$\begin{pmatrix} \boldsymbol{K}^{\mathrm{f}} & -\boldsymbol{G}^{\mathrm{f}} \\ (\boldsymbol{G}^{\mathrm{f}})^T & \boldsymbol{C}^{\mathrm{f}} \end{pmatrix} \begin{pmatrix} \Delta \underline{\boldsymbol{v}}_{\mathrm{h}}^{\mathrm{f},\mathrm{n}+\alpha^{\mathrm{f}}} \\ \Delta \underline{p}_{\mathrm{h}}^{\mathrm{n}+1} \end{pmatrix} = \begin{pmatrix} \boldsymbol{F}^{\mathrm{f}} \\ \boldsymbol{0} \end{pmatrix}, \quad (5.45)$$

where $\boldsymbol{K}^{\mathrm{f}}$ consists of the transient, convection, viscous and convective stabilization terms, $\boldsymbol{G}^{\mathrm{f}}$ is the gradient operator, $(\boldsymbol{G}^{\mathrm{f}})^T$ is the divergence operator for the continuity equation and $\boldsymbol{C}^{\mathrm{f}}$ consists of the stabilization terms corresponding to pressure-pressure coupling. The increments in the velocity and pressure for the Newton-Raphson iterative procedure are given by $\Delta \underline{\boldsymbol{v}}_{\mathrm{h}}^{\mathrm{f},\mathrm{n}+\alpha^{\mathrm{f}}}$ and $\Delta \underline{p}_{\mathrm{h}}^{\mathrm{n}+1}$ respectively.

Similarly, for the structural and mesh equation, we get the following matrix forms:

$$(\boldsymbol{K}^{\mathrm{s}})\left(\Delta \underline{\boldsymbol{v}}_{\mathrm{h}}^{\mathrm{s},\mathrm{n}+\alpha^{\mathrm{s}}}\right) = (\boldsymbol{F}^{\mathrm{s}}), \quad (5.46)$$

$$(\boldsymbol{K}^{\mathrm{m}})\left(\Delta \underline{\boldsymbol{u}}_{\mathrm{h}}^{\mathrm{f},\mathrm{n}+1}\right) = (\boldsymbol{0}), \quad (5.47)$$

where $\boldsymbol{K}^{\mathrm{s}}$ and $\boldsymbol{K}^{\mathrm{m}}$ are the linearized matrices for the respective equations.

5.6 Solution Procedure

After the temporal and spatial discretization of the FSI equations, they can be solved by either monolithic or partitioned solution strategies. These strategies are discussed in the current section. In the monolithic strategy, the whole problem is considered as a single field and all the components are advanced simultaneously in time. On the other hand, *partitioning* is the process of spatial separation of the discrete model into the interacting components which are called *partitions*. On the other hand, the temporal discretization can also be decomposed into what is called *splitting* within a particular time step size (See Fig. 5.1). The time splitting can be additive or multiplicative.

To understand this concept, consider a very simplified problem of two fields X and Y coupled with each other and advancing in time as shown in Fig. 5.2. Fields X and Y solve for the variables $x(t)$ and $y(t)$ respectively and are coupled as follows:

$$5\dot{x} + 2x - 3y = f(t), \tag{5.48}$$

$$\dot{y} + 8y - 7x = g(t), \tag{5.49}$$

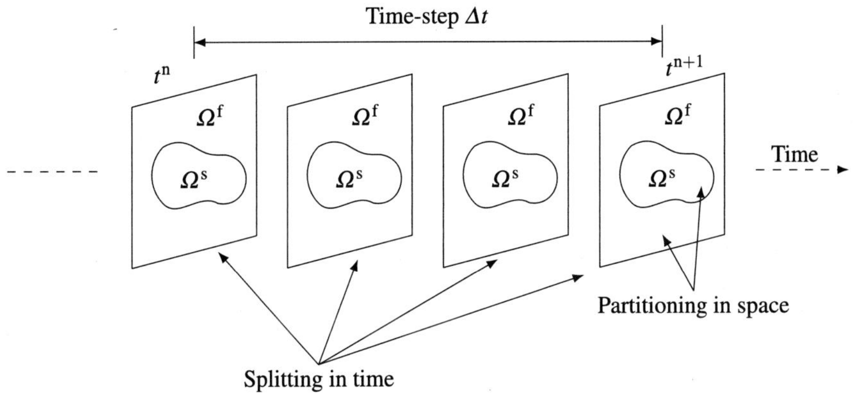

Fig. 5.1 Partition in spatial and splitting in the temporal discretizations

Fig. 5.2 Coupled fields X and Y advancing in time t

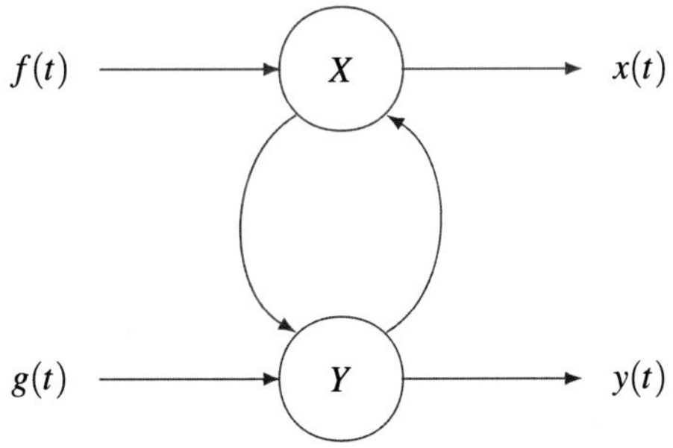

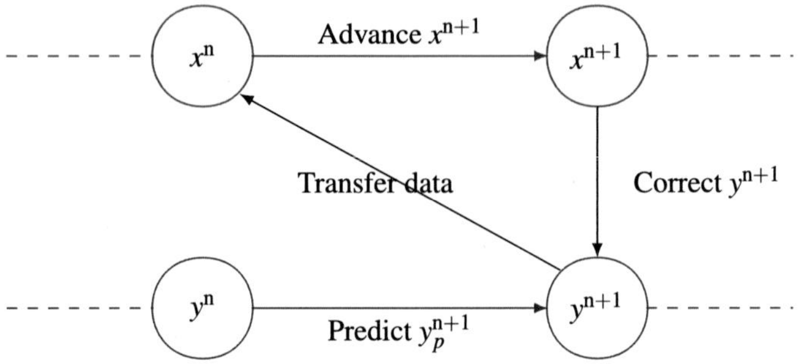

Fig. 5.3 Partitioned staggered technique for coupled fields X and Y advancing in time t

where $f(t)$ and $g(t)$ are the applied forces on the system. Discretizing the system in time using backward Euler,

$$\begin{pmatrix} 5+2\Delta t & -3\Delta t \\ -7\Delta t & 1+8\Delta t \end{pmatrix} \begin{pmatrix} x^{n+1} \\ y^{n+1} \end{pmatrix} = \begin{pmatrix} \Delta t f(t^{n+1}) + 5x^n \\ \Delta t g(t^{n+1}) + y^n \end{pmatrix} \tag{5.50}$$

whre Δt is the time step size. If one solves Eq. (5.50) directly in a time step, the technique is termed as monolithic.

For a partitioned procedure, both the fields are advanced in time separately. It consists of a predictor-corrector algorithm as follows in a particular time step:

1. Predictor step: $y_P^{n+1} = y^n$ or $y_P^{n+1} = y^n + \Delta t \dot{y}^n$
2. Transfer of data: Send the predicted value to the second field
3. Advance the second field:

$$x^{n+1} = \frac{1}{5+2\Delta t}(\Delta t f(t^{n+1}) + 5x^n + 3\Delta t y_P^{n+1}) \tag{5.51}$$

4. Corrector step:

$$y^{n+1} = \frac{1}{1+8\Delta t}(\Delta t g(t^{n+1}) + y^n + 7\Delta t x^{n+1}) \tag{5.52}$$

The above algorithm is called *staggered partitioned* procedure (Fig. 5.3). For linear problems, it is observed that staggering does not harm stability and accuracy of the problem, given the condition that the predictor is chosen properly. But for more general nonlinear problems, stability could become an issue. On the other hand, accuracy is degraded compared to the same problem solved by a monolithic scheme. This can be resolved by iterating the staggered procedure within a time step so that the scheme now forms a predictor-multicorrector format (Fig. 5.4). For nonlinear problems, these iterations help in the convergence and capture of the nonlinearities. However, these multiple iterations add to the computational cost and a monolithic

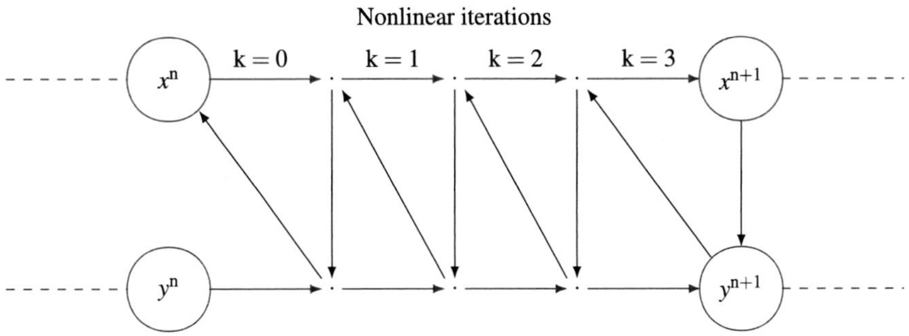

Fig. 5.4 Partitioned iterative staggered technique for coupled fields X and Y advancing in time t

scheme could be more advantageous for some problems in that case. Therefore, there is a tradeoff based on the type of problem being solved. Further details can be found in [64, 129]. Next, we briefly look into how these strategies differ for an FSI problem, before going into the details in the upcoming chapters.

5.6.1 Monolithic Solution for FSI

In the monolithic technique, the fluid and structural velocities are advanced in time simultaneously and both the fluid and structure are considered as a combined field. Note that when written in a combined field form, the dynamic equilibrium at the fluid-structure interface is satisfied naturally and we don't need to worry about the transfer of forces along the interface. This technique will be discussed in more detail in Chap. 6.

5.6.2 Partitioned Solution for FSI

In the partitioned procedure for FSI problems, we solve the fluid and structural fields separately and transfer the interface data between the fields. As the equations involved are nonlinear in nature, we opt for the predictor-multicorrector partitioned staggered procedure presented above. The matrices involved in the solution are given in Eqs. (5.45), (5.46) and (5.47) for the fluid, structure and mesh equations, respectively. The steps involved in a particular nonlinear iteration of the algorithm are given below:

1. Solve Eq. (5.46) for the predictor structural velocity ($\boldsymbol{v}^{\mathrm{s,n+1}}_{(\mathrm{k+1})}$) based on the fluid forces evaluated at the previous time step, $\boldsymbol{f}^{\mathrm{s,n}}_{(\mathrm{k})}$
2. Transfer the structural velocity at the fluid-structure interface to the fluid solver by maintaining the velocity continuity condition at the interface and satisfying the ALE mesh compatibility. This is carried out as follows:

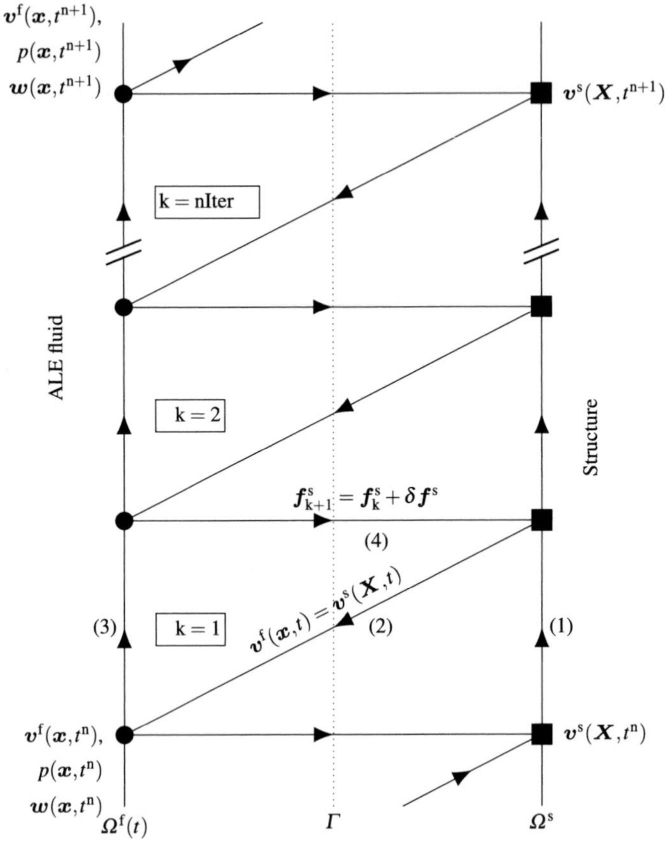

Fig. 5.5 Partitioned iterative staggered technique for coupled fluid-structure interaction system

- Equate mesh displacement to the structural displacement at the interface

$$\boldsymbol{u}^{\mathrm{m,n+1}}_{\mathrm{(k+1)}} = \boldsymbol{u}^{\mathrm{s,n+1}}_{\mathrm{(k+1)}}, \text{ on } \Gamma \tag{5.53}$$

- Equate fluid velocity to mesh velocity at the interface to satisfy conservation property of moving mesh elements in the fluid domain

$$\boldsymbol{v}^{\mathrm{f,n+\alpha^f}}_{\mathrm{(k+1)}} = \boldsymbol{w}^{\mathrm{n+\alpha^f}}_{\mathrm{(k+1)}}, \text{ on } \Gamma, \tag{5.54}$$

where

$$\boldsymbol{w}^{\mathrm{n+\alpha^f}}_{\mathrm{(k+1)}} = \frac{\boldsymbol{u}^{\mathrm{m,n+1}}_{\mathrm{(k+1)}} - \boldsymbol{u}^{\mathrm{m,n}}_{\mathrm{(k+1)}}}{\Delta t} = \boldsymbol{v}^{\mathrm{s},n+\alpha^{\mathrm{s}}}_{\mathrm{(k+1)}}, \text{ on } \Gamma \tag{5.55}$$

3. Solve Eq. (5.47) for the displacements of the fluid nodes and calculate mesh velocity field; and solve Eq. (5.45) for getting the updated velocity ($\boldsymbol{v}^{\mathrm{f,n+1}}_{(\mathrm{k+1})}$) and pressure ($p^{\mathrm{n+1}}_{(\mathrm{k+1})}$) and evaluate the fluid forces
4. Transfer the corrected fluid forces to the structural solver to satisfy the dynamic equilibrium at the fluid-structure interface, i.e., $\boldsymbol{f}^{\mathrm{s,n+\alpha^s}}_{(\mathrm{k+1})} = \boldsymbol{f}^{\mathrm{s,n+\alpha^s}}_{(\mathrm{k})} + \delta \boldsymbol{f}^{\mathrm{s}}$

The above steps form a nonlinear iteration. The fluid-structure coupling then advances in time at the end of the nonlinear iteration steps once the convergence criteria has been achieved. This scheme is shown in Fig. 5.5. The various challenges with the transfer of forces along the interface and the satisfaction of the equilibrium conditions have been discussed in detail in Chap. 7. Several techniques to ensure accurate and conservative transfer of data across the fluid-structure interface are also discussed.

To summarize, we began with the fluid-structure continuum equations and expressed the variational finite element form for the stabilized fluid system, structural system and mesh equation. We then briefly discussed the two main strategies for solving the coupled FSI system while noting some of the major differences between them. In the next few chapters, we deal with these two approaches in further detail.

Appendix

The Pressure-Velocity Coupling

Consider the Stokes equations for incompressible Newtonian flows in a domain $\Omega^{\mathrm{f}} \subset \mathbb{R}^d$ with a boundary Γ consisting of the Dirichlet and Neumann boundaries Γ_D and Γ_N respectively:

$$-\mu^{\mathrm{f}}\nabla^2 \boldsymbol{v}^{\mathrm{f}} + \nabla p = \rho^{\mathrm{f}}\boldsymbol{b}^{\mathrm{f}}, \quad \text{on } \Omega^{\mathrm{f}}, \tag{5.56}$$

$$\nabla \cdot \boldsymbol{v}^{\mathrm{f}} = 0, \quad \text{on } \Omega^{\mathrm{f}}, \tag{5.57}$$

$$\boldsymbol{v}^{\mathrm{f}} = \boldsymbol{v}^{\mathrm{f}}_D, \quad \text{on } \Gamma^{\mathrm{f}}_D, \tag{5.58}$$

$$-\nabla \boldsymbol{v}^{\mathrm{f}} \cdot \boldsymbol{n} + p\boldsymbol{n} = \boldsymbol{h}^{\mathrm{f}}, \quad \text{on } \Gamma^{\mathrm{f}}_N, \tag{5.59}$$

where $\boldsymbol{v}^{\mathrm{f}}$ and p are the fluid velocity and pressure respectively, $\boldsymbol{v}^{\mathrm{f}}_D$ and $\boldsymbol{h}^{\mathrm{f}}$ denote the Dirichlet and Neumann conditions respectively. The weak statement can be written as: find $[\boldsymbol{v}^{\mathrm{f}}, p] \in \mathcal{S}_{\boldsymbol{v}^{\mathrm{f}}} \times \mathcal{S}_p$ such that $\forall[\boldsymbol{\psi}^{\mathrm{f}}, q] \in \mathcal{V}_{\boldsymbol{\psi}^{\mathrm{f}}} \times \mathcal{V}_q$,

$$\int_{\Omega^{\mathrm{f}}} \nabla \boldsymbol{\psi}^{\mathrm{f}} : \left(\mu^{\mathrm{f}}\nabla \boldsymbol{v}^{\mathrm{f}} - p\boldsymbol{I}\right) d\Omega = \int_{\Omega^{\mathrm{f}}} \boldsymbol{\psi}^{\mathrm{f}} \cdot \rho^{\mathrm{f}}\boldsymbol{b}^{\mathrm{f}} d\Omega, \tag{5.60}$$

$$\int_{\Omega^{\mathrm{f}}} q(\nabla \cdot \boldsymbol{v}^{\mathrm{f}}) d\Omega = 0, \tag{5.61}$$

which in the compact form can be written as

$$a(\boldsymbol{\psi}^{\mathrm{f}}, \boldsymbol{v}^{\mathrm{f}}) + b(\boldsymbol{\psi}^{\mathrm{f}}, p) = (\boldsymbol{\psi}^{\mathrm{f}}, \rho^{\mathrm{f}} \boldsymbol{b}^{\mathrm{f}}), \tag{5.62}$$

$$b(\boldsymbol{v}^{\mathrm{f}}, q) = 0, \tag{5.63}$$

where

$$a(\boldsymbol{\psi}^{\mathrm{f}}, \boldsymbol{v}^{\mathrm{f}}) = \int_{\Omega^{\mathrm{f}}} \nabla \boldsymbol{\psi}^{\mathrm{f}} : \mu^{\mathrm{f}} \nabla \boldsymbol{v}^{\mathrm{f}} d\Omega, \qquad b(\boldsymbol{\psi}^{\mathrm{f}}, p) = \int_{\Omega^{\mathrm{f}}} \nabla \boldsymbol{\psi}^{\mathrm{f}} : (-p\boldsymbol{I}) d\Omega. \tag{5.64}$$

Remark 5.2 Equations (5.62) and (5.63) represent the saddle point of the Lagrangian functional

$$\mathscr{I}(\boldsymbol{\psi}^{\mathrm{f}}, q) = \frac{1}{2} a(\boldsymbol{\psi}^{\mathrm{f}}, \boldsymbol{\psi}^{\mathrm{f}}) + b(\boldsymbol{\psi}^{\mathrm{f}}, q) - (\boldsymbol{\psi}^{\mathrm{f}}, \rho^{\mathrm{f}} \boldsymbol{b}^{\mathrm{f}}). \tag{5.65}$$

A saddle point problem is to find a solution $(\boldsymbol{v}^{\mathrm{f}}, p)$ such that the functional is minimized with respect to the first argument and maximized with respect to the second one, i.e., $\mathscr{I}(\boldsymbol{v}^{\mathrm{f}}, q) \leq \mathscr{I}(\boldsymbol{v}^{\mathrm{f}}, p) \leq \mathscr{I}(\boldsymbol{\psi}^{\mathrm{f}}, p)$.

Similarly, for each element domain Ω^e, the variational finite element discretization is written as: find $[\boldsymbol{v}_{\mathrm{h}}^{\mathrm{f}}, p_{\mathrm{h}}] \in \mathscr{S}_{v^{\mathrm{f}}}^{\mathrm{h}} \times \mathscr{S}_{p}^{\mathrm{h}}$ such that $\forall [\boldsymbol{\psi}_{\mathrm{h}}^{\mathrm{f}}, q_{\mathrm{h}}] \in \mathscr{V}_{\psi^{\mathrm{f}}}^{\mathrm{h}} \times \mathscr{V}_{q}^{\mathrm{h}}$,

$$\int_{\Omega^e} \nabla \boldsymbol{\psi}_{\mathrm{h}}^{\mathrm{f}} : \left(\mu^{\mathrm{f}} \nabla \boldsymbol{v}_{\mathrm{h}}^{\mathrm{f}} - p_{\mathrm{h}} \boldsymbol{I} \right) d\Omega = \int_{\Omega^e} \boldsymbol{\psi}_{\mathrm{h}}^{\mathrm{f}} \cdot \rho^{\mathrm{f}} \boldsymbol{b}^{\mathrm{f}} d\Omega, \tag{5.66}$$

$$\int_{\Omega^e} q_{\mathrm{h}} (\nabla \cdot \boldsymbol{v}_{\mathrm{h}}^{\mathrm{f}}) d\Omega = 0. \tag{5.67}$$

Using the polynomial approximation as $v_{\mathrm{h}(j)}^{\mathrm{f}} = \sum_{i=1}^{nen} N_i v_{i(j)}^{\mathrm{f}}$ for each velocity component j and $p_{\mathrm{h}} = \sum_{i=1}^{nen} M_i p_i$ and using the Galerkin formulation, the final linear matrix system after assembling the contributions from each element can be written as

$$\begin{pmatrix} \boldsymbol{A} & -\boldsymbol{B} \\ \boldsymbol{B}^T & \boldsymbol{0} \end{pmatrix} \begin{pmatrix} \underline{\boldsymbol{v}}^{\mathrm{f}} \\ \underline{p} \end{pmatrix} = \begin{pmatrix} \boldsymbol{f}_m \\ \boldsymbol{0} \end{pmatrix}, \tag{5.68}$$

where $\underline{\boldsymbol{v}}^{\mathrm{f}}$ and $\underline{p}$ are the unknown vectors for velocity components and pressure respectively at each finite element node, $\boldsymbol{f}_m$ is the right-hand side vector and $\boldsymbol{A}$, $\boldsymbol{B}$ are the block matrices in the left-hand side. Equation (5.68) is the matrix representation of the saddle point problem. To solve the problem, the left-hand side matrix has to be non-singular, or invertible, which can be satisfied mainly by the following methods.

Mixed Finite Element Methods

The polynomial spaces for the velocity and pressure should be compatible with each other based on the requirement of $\mathrm{null}(\boldsymbol{B}) = 0$. A necessary, but not sufficient condition to ensure this criteria is that $\dim \mathscr{V}_q^{\mathrm{h}} \leq \dim \mathscr{V}_{\boldsymbol{\psi}^{\mathrm{f}}}^{\mathrm{h}}$. The sufficient condition is called Ladyzhenskaya-Babuška-Brezzi (LBB) or inf-sup condition:

Theorem 5.1 *There exists a $\beta > 0$ for the bilinear form $b(\boldsymbol{\psi}_{\mathrm{h}}^{\mathrm{f}}, q_{\mathrm{h}})$ such that*

$$\inf_{q_{\mathrm{h}} \in \mathscr{V}_q^{\mathrm{h}}} \sup_{\boldsymbol{\psi}_{\mathrm{h}}^{\mathrm{f}} \in \mathscr{V}_{\boldsymbol{\psi}^{\mathrm{f}}}^{\mathrm{h}}} \frac{b(\boldsymbol{\psi}_{\mathrm{h}}^{\mathrm{f}}, q_{\mathrm{h}})}{||\boldsymbol{\psi}_{\mathrm{h}}^{\mathrm{f}}||_{\mathscr{V}_{\boldsymbol{\psi}_{\mathrm{h}}^{\mathrm{f}}}} ||q_{\mathrm{h}}||_{\mathscr{V}_p^{\mathrm{h}}}} \geq \beta, \tag{5.69}$$

where β is independent of the mesh size h.

The polynomial spaces for velocity and pressure which satisfy the LBB condition are said to be compatible and LBB stable. For such cases, there exists a unique $\boldsymbol{v}_{\mathrm{h}}^{\mathrm{f}} \in \mathscr{S}_{\boldsymbol{\psi}_{\mathrm{h}}^{\mathrm{f}}}^{\mathrm{h}}$ and $p_{\mathrm{h}} \in \mathscr{S}_q^{\mathrm{h}}$ for the weak form of the equation.

Stabilization Methods

There is a way to circumvent the LBB condition. Till now, we have seen that the only major requirement is for the left-hand side matrix to be non-singular which is not possible due to the zero block matrix. To enforce this positive definiteness property to the global left-hand side matrix, we can modify the weak form of the equation such that we do not get the zero block matrix. These type of methods can be categorized into *Petrov-Galerkin* techniques. In contrast to Galerkin method where we take the weighting function space to be the same as the trial solution space, in the Petrov-Galerkin methods, we perturb the weighting function by certain amount. More details about these methods can be found in [56].

Chapter 6
Quasi-Monolithic Fluid-Structure Formulation

6.1 Introduction

In this chapter, we deal with the first type of strategy in solving the fluid-structure interaction system, namely, monolithic techniques. As mentioned in the previous chapter, one of the advantages of such techniques is the numerical stability of the body-fitted fluid-structure interface, while partitioned techniques suffer from instabilities [39, 66]. Partitioned techniques [147] tend to involve nonlinear iterations per time step for maintaining stability, which increases the computational cost. Thus, monolithic techniques offer an alternative for techniques which can be computationally efficient and stable for low structure-to-fluid mass ratios.

The body-fitted arbitrary Lagrangian-Eulerian (ALE) based methods employing a monolithic approach [3, 78, 88, 90, 179] assemble the fluid and structural equations in a single block matrix and solve them simultaneously at each iteration. This strategy offers numerical stability for problems involving strong added mass effects, however lacks the flexibility and modularity of using the existing stable codes for solving fluid and structural equations.

We present a quasi-monolithic technique for fluid-structure interaction which is inspired by the one-fluid formulation in [217] for studying multiphase problems and the combined fluid-particle formulation [91]. A similar technique called the one-field formulation was considered in [138] for fluid-structure coupling which was only first-order accurate in time. One of the salient features of the formulation is that the kinematic and dynamic equilibrium conditions along the interface are absorbed in the formulation and are satisfied implicitly, unlike partitioned techniques [63, 64, 147] in which interface conditions have to be enforced explicitly and tractions evaluated along the interface [218]. The ALE mesh equation is decoupled from the rest of the fluid-structure system, thus making the formulation, quasi-monolithic. Such decoupling allows to linearize the Navier-Stokes equations without losing the stability and accuracy leading to solving the coupled system only once per time step, thus reducing the computational cost.

R. K. Jaiman and V. Joshi, *Computational Mechanics of Fluid-Structure Interaction*,
https://doi.org/10.1007/978-981-16-5355-1_6

We begin with the variational description of the combined fluid-structure system. We then discuss the quasi-monolithic and the fully stabilized quasi-monolithic formulations for the FSI problems with their implementations and verify them with standard FSI benchmarks available in the literature. Finally, we demonstrate the quasi-monolithic framework for a flapping dynamics problem of a flexible foil.

6.2 Weak Variational Form

In this section, we write the weak variational form of the fluid-structure system in a combined-field form.

6.2.1 Combined Fluid-Structure Formulation

In order to write the weak form of the combined fluid-structure formulation, let us first introduce the trial function spaces $\mathscr{S}_{\boldsymbol{v}}$ and $\mathscr{S}_p$ along with their corresponding test function spaces $\mathscr{V}_{\boldsymbol{\psi}}$ and $\mathscr{V}_q$ for velocity and pressure, respectively. The trial function spaces for the combined field are defined as

$$
\begin{aligned}
\mathscr{S}_{\boldsymbol{v}} = \big\{ & (\boldsymbol{v}^{\mathrm{f}}, \boldsymbol{v}^{\mathrm{s}}) \in H^1(\Omega^{\mathrm{f}}(t)) \times H^1(\Omega^{\mathrm{s}}) \big| \\
& \boldsymbol{v}^{\mathrm{f}}(\boldsymbol{\varphi}^{\mathrm{s}}(\boldsymbol{X}, t)) = \boldsymbol{v}^{\mathrm{s}}(\boldsymbol{X}, t) \text{ on } \Gamma, \\
& \boldsymbol{v}^{\mathrm{f}}(\boldsymbol{x}, t) = \boldsymbol{v}_D^{\mathrm{f}} \text{ on } \Gamma_D^{\mathrm{f}} \text{ and } \boldsymbol{v}^{\mathrm{s}}(\boldsymbol{X}, t) = \boldsymbol{v}_D^{\mathrm{s}} \text{ on } \Gamma_D^{\mathrm{s}} \big\}, \\
\mathscr{S}_p = & \{ p \in L^2(\Omega^{\mathrm{f}}(t)) \}.
\end{aligned}
$$

Note that the trial function spaces and the corresponding test function spaces differ only at the Dirichlet boundary. Thus, the test function spaces can be defined as

$$
\begin{aligned}
\mathscr{V}_{\boldsymbol{\psi}} = \big\{ & (\boldsymbol{\psi}^{\mathrm{f}}, \boldsymbol{\psi}^{\mathrm{s}}) \in H^1(\Omega^{\mathrm{f}}(t)) \times H^1(\Omega^{\mathrm{s}}) \big| \\
& \boldsymbol{\psi}^{\mathrm{f}}(\boldsymbol{\varphi}^{\mathrm{s}}(\boldsymbol{X}, t)) = \boldsymbol{\psi}^{\mathrm{s}}(\boldsymbol{X}) \text{ on } \Gamma, \\
& \boldsymbol{\psi}^{\mathrm{f}}(\boldsymbol{x}) = 0 \text{ on } \Gamma_D^{\mathrm{f}} \text{ and } \boldsymbol{\psi}^{\mathrm{s}}(\boldsymbol{X}) = 0 \text{ on } \Gamma_D^{\mathrm{s}} \big\}, \\
\mathscr{V}_q = & \{ q \in L^2(\Omega^{\mathrm{f}}(t)) \}.
\end{aligned}
$$

One may observe that the test functions of fluid and structural velocities are defined on different domains $\Omega^{\mathrm{f}}(t)$ and Ω^{s} which satisfy the matching condition along the interface,

$$
\boldsymbol{\psi}^{\mathrm{f}}(\boldsymbol{x}) = \boldsymbol{\psi}^{\mathrm{s}}(\boldsymbol{X}) \quad \forall \boldsymbol{X} \in \Gamma, \tag{6.1}
$$

where $\boldsymbol{x} = \boldsymbol{\varphi}^{\mathrm{s}}(\boldsymbol{X}, t)$ with $\boldsymbol{\varphi}^{\mathrm{s}}$ being the mapping function. The weak form of the Navier-Stokes Eqs. (3.64) and (3.65) in the ALE framework is

$$\int_{\Omega^{\rm f}(t)} \rho^{\rm f} \left(\left.\frac{\partial \boldsymbol{v}^{\rm f}}{\partial t}\right|_{\chi} + (\boldsymbol{v}^{\rm f} - \boldsymbol{w}) \cdot \nabla \boldsymbol{v}^{\rm f} \right) \cdot \boldsymbol{\psi}^{\rm f} d\boldsymbol{x} + \int_{\Omega^{\rm f}(t)} \boldsymbol{\sigma}^{\rm f} : \nabla \boldsymbol{\psi}^{\rm f} d\boldsymbol{x} =$$

$$\int_{\Omega^{\rm f}(t)} \rho^{\rm f} \boldsymbol{b}^{\rm f} \cdot \boldsymbol{\psi}^{\rm f} d\boldsymbol{x} + \int_{\Gamma_N^{\rm f}} \boldsymbol{\sigma}_N^{\rm f} \cdot \boldsymbol{\psi}^{\rm f} d\Gamma + \int_{\Gamma(t)} \left(\boldsymbol{\sigma}^{\rm f}(\boldsymbol{x}, t) \cdot \boldsymbol{n}^{\rm f} \right) \cdot \boldsymbol{\psi}^{\rm f} d\Gamma, \tag{6.2}$$

$$\int_{\Omega^{\rm f}(t)} (\nabla \cdot \boldsymbol{v}^{\rm f}) q d\boldsymbol{x} = 0. \tag{6.3}$$

The weak forms presented in Eqs. (6.2) and (6.3) are obtained by multiplying Eq. (3.64) with $\boldsymbol{\psi}^{\rm f}$ and Eq. (3.65) with q using a dot product, then integrating them over the fluid domain $\Omega^{\rm f}(t)$. Similarly, we can define the weak form of the structural equilibrium Eq. (3.57) as

$$\int_{\Omega^{\rm s}} \rho^{\rm s} \frac{\partial \boldsymbol{v}^{\rm s}}{\partial t} \cdot \boldsymbol{\psi}^{\rm s} d\boldsymbol{X} + \int_{\Omega^{\rm s}} \boldsymbol{\sigma}^{\rm s} : \nabla \boldsymbol{\psi}^{\rm s} d\boldsymbol{X} =$$

$$\int_{\Omega^{\rm s}} \rho^{\rm s} \boldsymbol{b}^{\rm s} \cdot \boldsymbol{\psi}^{\rm s} d\boldsymbol{X} + \int_{\Gamma_N^{\rm s}} \boldsymbol{\sigma}_N^{\rm s} \cdot \boldsymbol{\psi}^{\rm s} d\Gamma + \int_{\Gamma} (\boldsymbol{\sigma}^{\rm s}(\boldsymbol{X}, t) \cdot \boldsymbol{n}^{\rm s}) \cdot \boldsymbol{\psi}^{\rm s}(\boldsymbol{X}) d\Gamma. \tag{6.4}$$

Furthermore, the weak form for the traction continuity condition given in Eq. (3.67) will be

$$\int_{\Gamma(t)} \left(\boldsymbol{\sigma}^{\rm f}(\boldsymbol{x}, t) \cdot \boldsymbol{n}^{\rm f} \right) \cdot \boldsymbol{\psi}^{\rm f}(\boldsymbol{x}) d\Gamma + \int_{\Gamma} (\boldsymbol{\sigma}^{\rm s}(\boldsymbol{X}, t) \cdot \boldsymbol{n}^{\rm s}) \cdot \boldsymbol{\psi}^{\rm s}(\boldsymbol{X}) d\Gamma = 0, \tag{6.5}$$

where $\boldsymbol{\psi}^{\rm f}$ and $\boldsymbol{\psi}^{\rm s}$ are required to satisfy $\boldsymbol{\psi}^{\rm f}(\boldsymbol{\varphi}^{\rm s}(\cdot)) = \boldsymbol{\psi}^{\rm s}(\cdot)$ on Γ. A detailed derivation of the above weak form in Eq. (6.5) from its strong form in Eq. (3.67) can be found in [138]. Now we can combine Eqs. (6.2)-(6.4) using Eqs. (6.5) and (6.1) to construct a single unique relation for the combined fluid-structure domain, which is given as: find $[\boldsymbol{v}^{\rm f}, \boldsymbol{v}^{\rm s}, p] \in \mathscr{S}_{\boldsymbol{v}} \times \mathscr{S}_p$ such that $\forall [\boldsymbol{\psi}^{\rm f}, \boldsymbol{\psi}^{\rm s}, q] \in \mathscr{V}_{\boldsymbol{\psi}} \times \mathscr{V}_q$:

$$\int_{\Omega^{\rm f}(t)} \rho^{\rm f} \left(\left.\frac{\partial \boldsymbol{v}^{\rm f}}{\partial t}\right|_{\chi} + (\boldsymbol{v}^{\rm f} - \boldsymbol{w}) \cdot \nabla \boldsymbol{v}^{\rm f} \right) \cdot \boldsymbol{\psi}^{\rm f}(\boldsymbol{x}) d\boldsymbol{x} + \int_{\Omega^{\rm f}(t)} \boldsymbol{\sigma}^{\rm f} : \nabla \boldsymbol{\psi}^{\rm f} d\boldsymbol{x}$$

$$- \int_{\Omega^{\rm f}(t)} (\nabla \cdot \boldsymbol{v}^{\rm f}) q d\boldsymbol{x}$$

$$+ \int_{\Omega^{\rm s}} \rho^{\rm s} \frac{\partial \boldsymbol{v}^{\rm s}}{\partial t} \cdot \boldsymbol{\psi}^{\rm s} d\boldsymbol{X} + \int_{\Omega^{\rm s}} \boldsymbol{\sigma}^{\rm s} : \nabla \boldsymbol{\psi}^{\rm s} d\boldsymbol{X} =$$

$$\int_{\Omega^{\rm f}(t)} \rho^{\rm f} \boldsymbol{b}^{\rm f} \cdot \boldsymbol{\psi}^{\rm f} d\boldsymbol{x} + \int_{\Gamma_N^{\rm f}} \boldsymbol{\sigma}_N^{\rm f} \cdot \boldsymbol{\psi}^{\rm f} d\Gamma + \int_{\Omega^{\rm s}} \rho^{\rm s} \boldsymbol{b}^{\rm s} \cdot \boldsymbol{\psi}^{\rm s} d\boldsymbol{X} + \int_{\Gamma_N^{\rm s}} \boldsymbol{\sigma}_N^{\rm s} \cdot \boldsymbol{\psi}^{\rm s} d\Gamma. \tag{6.6}$$

As can be observed, the velocity continuity (kinematic condition) at the interface is enforced in the function space and the traction continuity (dynamic equilibrium) is implicitly built in the formulation. Thus, the crucial idea is to solve the fluid and the structural domains as a single unique domain $\Omega = \Omega^{\mathrm{f}} \cup \Omega^{\mathrm{s}}$.

6.2.2 Finite Element Space Discretization

In contrast to many FSI schemes available in the literature [109, 172] which apply a finite-volume discretization for the fluid equations in combination with a finite-element discretization for the structural equations, a spatial discretization with finite elements is applied throughout this formulation. We adopt the Galerkin finite element formulation to discretize the weak formulation in Eq. (6.6). As mentioned in Appendix at the end of the previous chapter, the pressure-velocity coupling can be solved either by satisfying the LBB condition or using stabilization methods. Stabilization methods give an additional advantage of stability in convection-dominated regimes.

Here, we apply both approaches to the quasi-monolithic formulation. Section 6.3 deals with the formulation where we approximate the fluid velocity, pressure and structural velocity using $\mathbb{P}_2/\mathbb{P}_1/\mathbb{P}_2$ isoparametric elements, to satisfy the inf-sup or LBB condition for well-posedness. On the other hand, Sect. 6.5 discusses the stabilized quasi-monolithic technique. $\mathscr{T}_{\mathrm{h}}^{\mathrm{f}}(t)$ and $\mathscr{T}_{\mathrm{h}}^{\mathrm{s}}$ represent the triangular mesh on domains $\Omega^{\mathrm{f}}(t)$ and Ω^{s}, respectively. To increase the computational efficiency, we assume that the edges of the isoparametric elements are straight unless they represent either the interface or curved boundaries.

6.3 Quasi-Monolithic Formulation

In this section, we present the second-order time discretization of the combined fluid-structure formulation given in Sect. 6.2.1. The explicit construction of the interface at the start of each time step decouples the fluid-structure interface and ALE mesh from the computation of fluid-structure variables $(\boldsymbol{v}^{\mathrm{f}}, p, \boldsymbol{v}^{\mathrm{s}})$. Additionally, the decoupling of the ALE mesh enables us to determine the convective velocity of the fluid flow explicitly and linearize the nonlinear Navier-Stokes equations. Hence, the quasi-monolithic formulation does not require nonlinear iterations per time step.

6.3.1 Second-Order in Time Discretization

Let $\mathbb{P}_2(\Omega_{\mathrm{h}})$ denote the standard second-order Lagrange finite element space on domain $\Omega_{\mathrm{h}} = \Omega_{\mathrm{h}}^{\mathrm{f}} \cup \Omega_{\mathrm{h}}^{\mathrm{s}}$. First, we employ the second-order extrapolation to describe the deformation vector $\boldsymbol{\varphi}_{\mathrm{h}}^{\mathrm{s,n}}$ of the structure as

$$\boldsymbol{\varphi}_{\mathrm{h}}^{\mathrm{s,n}}(\boldsymbol{X}_i) = \boldsymbol{\varphi}_{\mathrm{h}}^{\mathrm{s,n-1}}(\boldsymbol{X}_i) + \frac{3\Delta t}{2}\boldsymbol{v}_{\mathrm{h}}^{\mathrm{s,n-1}}(\boldsymbol{X}_i) - \frac{\Delta t}{2}\boldsymbol{v}_{\mathrm{h}}^{\mathrm{s,n-2}}(\boldsymbol{X}_i) \qquad \forall \boldsymbol{X}_i \in \mathscr{T}_{\mathrm{h}}^{\mathrm{s}}. \quad (6.7)$$

We then use $\boldsymbol{\varphi}_{\mathrm{h}}^{\mathrm{s,n}}(\boldsymbol{X}_i)$ to update the ALE mesh displacement vector $\boldsymbol{u}_{\mathrm{h}}^{\mathrm{f,n}}$ for the interface nodes through Eq. (3.75). Now that we have both the boundary conditions Eqs. (3.75) and (3.76) required for solving the ALE Eq. (3.74), we solve Eq. (3.74) employing $\mathbb{P}_1$ finite element space. The assumption made in Sect. 6.2.2 that all the isoparametric element edges are straight unless they are on the interface or curved boundaries enables us to use $\mathbb{P}_1$ finite element discretization instead of the $\mathbb{P}_2$. As a result of this, the size of the system of linear equations required for solving the ALE mesh displacement vector, $\boldsymbol{u}_{\mathrm{h}}^{\mathrm{f,n}}$, for finite element space with $\mathbb{P}_1$ discretization would be smaller than for the $\mathbb{P}_2$ discretization space without losing the accuracy of the coupled fluid-structure solver.

We now use the solution of $\boldsymbol{u}_{\mathrm{h}}^{\mathrm{f,n}}$ computed on the $\mathbb{P}_1$ finite element space to update the location of triangular mesh $\mathscr{T}_{\mathrm{h},t^{\mathrm{n}}}^{\mathrm{f}}$ vertices using the Eq. (3.77). Since the interior edges are straight, we can position the non-vertex computational node at the center of the edge. In this way, we are able to determine the ALE mesh displacement for all the $\mathbb{P}_2$ finite element mesh $\mathscr{T}_{\mathrm{h},t^{\mathrm{n}}}^{\mathrm{f}}$ computational nodes even by solving the ALE Eq. (3.74) on a $\mathbb{P}_1$ finite element mesh.

Let us now define $\mathbb{P}_{\mathrm{m}}/\mathbb{P}_{\mathrm{m}-1}/\mathbb{P}_{\mathrm{m}}$ solution space $\mathscr{V}_{\mathrm{h}}$ as

$$\begin{aligned}\mathscr{V}_{\mathrm{h}}(t^{\mathrm{n}}, \boldsymbol{\varphi}_{\mathrm{h}}^{\mathrm{s,n}}, \alpha, \boldsymbol{b}) = \Big\{&(\boldsymbol{v}_{\mathrm{h}}^{\mathrm{f}}, p_{\mathrm{h}}, \boldsymbol{v}_{\mathrm{h}}^{\mathrm{s}}) : \boldsymbol{v}_{\mathrm{h}}^{\mathrm{f}} \in \mathbb{P}_{\mathrm{m}}(\Omega_{\mathrm{h},t^{\mathrm{n}}}^{\mathrm{f}}), p_{\mathrm{h}} \in \mathbb{P}_{\mathrm{m}-1}(\Omega_{\mathrm{h},t^{\mathrm{n}}}^{\mathrm{f}}), \boldsymbol{v}_{\mathrm{h}}^{\mathrm{s}} \in \mathbb{P}_{\mathrm{m}}(\Omega_{\mathrm{h}}^{\mathrm{s}}),\\ &\forall\, \boldsymbol{X}_i \in \mathscr{T}_{\mathrm{h}}^{\Gamma},\ \boldsymbol{v}_{\mathrm{h}}^{\mathrm{f}}(\boldsymbol{\varphi}_{\mathrm{h}}^{\mathrm{s,n}}(\boldsymbol{X}_i)) = \alpha \boldsymbol{v}_{\mathrm{h}}^{\mathrm{s}}(\boldsymbol{X}_i) + \boldsymbol{b}\Big\}, \end{aligned} \quad (6.8)$$

where $\mathscr{T}_{\mathrm{h}}^{\Gamma}$ is the mesh on the fluid-structure interface Γ. There are two parameters α and $\boldsymbol{b}$ in the definition of Eq. (6.8). These two parameters enable us to enforce the following condition on the fluid-structure interface. The test function space is chosen to be $\mathscr{V}_{\mathrm{h}}(t^{\mathrm{n}}, \boldsymbol{\varphi}_{\mathrm{h}}^{\mathrm{s,n}}, 1, \boldsymbol{0})$. By setting $\alpha = \frac{3}{4}$ and $\boldsymbol{b} = \frac{1}{2}\boldsymbol{v}_{\mathrm{h}}^{\mathrm{s,n-1}} - \frac{1}{4}\boldsymbol{v}_{\mathrm{h}}^{\mathrm{s,n-2}}$, and then finding $(\boldsymbol{v}_{\mathrm{h}}^{\mathrm{f,n}}, p_{\mathrm{h}}, \boldsymbol{v}_{\mathrm{h}}^{\mathrm{s,n}}) \in \mathscr{V}_{\mathrm{h}}(t^{\mathrm{n}}, \boldsymbol{\varphi}_{\mathrm{h}}^{\mathrm{s,n}}, \alpha, \boldsymbol{b})$ we can enforce the following condition:

$$\boldsymbol{v}_{\mathrm{h}}^{\mathrm{f,n}}(\boldsymbol{\varphi}_{\mathrm{h}}^{\mathrm{s,n}}(\boldsymbol{X}_i)) = \frac{3}{4}\boldsymbol{v}_{\mathrm{h}}^{\mathrm{s,n}}(\boldsymbol{X}_i) + \frac{1}{2}\boldsymbol{v}_{\mathrm{h}}^{\mathrm{s,n-1}}(\boldsymbol{X}_i) - \frac{1}{4}\boldsymbol{v}_{\mathrm{h}}^{\mathrm{s,n-2}}(\boldsymbol{X}_i). \quad (6.9)$$

The above equation is required to be valid for any $\boldsymbol{X}_i$ representing a grid point on $\mathscr{T}_{\mathrm{h}}^{\Gamma}$. Equation (6.9) is required to prove the energy based stability of the quasi-monolithic formulation which has been shown in [139]. Fortunately, the accuracy is also maintained since Eq. (6.9) is a second-order approximation of the interface condition Eq. (3.66).

We now define a second-order time-accurate extrapolation function as

$$\check{\boldsymbol{v}}_{\mathrm{h}}^{\mathrm{f}}(\boldsymbol{\Psi}_{\mathrm{h}}^{\mathrm{n}}(\boldsymbol{x}, t^{\mathrm{n}})) = 2\boldsymbol{v}_{\mathrm{h}}^{\mathrm{f,n-1}}(\boldsymbol{\Psi}_{\mathrm{h}}^{\mathrm{n}}(\boldsymbol{x}, t^{\mathrm{n-1}})) - \boldsymbol{v}_{\mathrm{h}}^{\mathrm{f,n-2}}(\boldsymbol{\Psi}_{\mathrm{h}}^{\mathrm{n}}(\boldsymbol{x}, t^{\mathrm{n-2}})) \quad (6.10)$$

where $\boldsymbol{\Psi}_{\mathrm{h}}^{\mathrm{n}}(\cdot, t^{\mathrm{n}-j})$ is the backward mapping function for the spatial grid points on the mesh $\mathscr{T}_{\mathrm{h},t^{\mathrm{n}}}^{\mathrm{f}}$ to $\mathscr{T}_{\mathrm{h},t^{\mathrm{n}-j}}^{\mathrm{f}}$ and mesh velocity $\boldsymbol{w}_{\mathrm{h}}^{\mathrm{n}}(\boldsymbol{x})$ is defined as

$$\begin{aligned} \boldsymbol{w}_{\mathrm{h}}^{\mathrm{n}}(\boldsymbol{x}) &= \sum_{i=1}^{G} \psi_i^{\mathrm{f,n}}(\boldsymbol{x}) \frac{1}{\Delta t}\left((\boldsymbol{x}_i^{\mathrm{n}} - \boldsymbol{x}_i^{\mathrm{n}-1}) + \frac{1}{2}\left(\boldsymbol{x}_i^{\mathrm{n}-1} - \boldsymbol{x}_i^{\mathrm{n}-2}\right) - \frac{1}{2}\left(\boldsymbol{x}_i^{\mathrm{n}-2} - \boldsymbol{x}_i^{\mathrm{n}-3}\right)\right) \\ &= \sum_{i=1}^{G} \psi_i^{\mathrm{f,n}}(\boldsymbol{x}) \frac{1}{\Delta t}\left(\boldsymbol{x}_i^{\mathrm{n}} - \frac{1}{2}\boldsymbol{x}_i^{\mathrm{n}-1} - \boldsymbol{x}_i^{\mathrm{n}-2} + \frac{1}{2}\boldsymbol{x}_i^{\mathrm{n}-3}\right), \end{aligned} \tag{6.11}$$

where G is the total number of grid points and $\psi_i^{\mathrm{f,n}}$ denotes the shape function in the fluid domain. Here, $\boldsymbol{w}_{\mathrm{h}}^{\mathrm{n}}$ is a second-order approximation of the fluid mesh velocity $\partial_t \boldsymbol{\Psi}_{\mathrm{h}}^{\mathrm{n}}(\boldsymbol{x}, t^{\mathrm{n}})$ as can be seen from the expression:

$$f'(t^{\mathrm{n}}) = \frac{1}{\Delta t}\left(f(t^{\mathrm{n}}) - \frac{1}{2} f(t^{\mathrm{n}-1}) - f(t^{\mathrm{n}-2}) + \frac{1}{2} f(t^{\mathrm{n}-3})\right) + \mathscr{O}(\Delta t^2).$$

We next show that

$$\boldsymbol{w}_{\mathrm{h}}^{\mathrm{n}}(\boldsymbol{x}_j^{\mathrm{n}}) = \check{\boldsymbol{v}}_{\mathrm{h}}^{\mathrm{f}}(\boldsymbol{x}_j^{\mathrm{n}}) \quad \text{on} \quad \Gamma_{\mathrm{h},t^{\mathrm{n}}}. \tag{6.12}$$

To prove the above equation, we rewrite Eq. (6.11) as

$$\begin{aligned} \boldsymbol{w}_{\mathrm{h}}^{\mathrm{n}}(\boldsymbol{x}_j^{\mathrm{n}}) =& \frac{1}{\Delta t}\left((\boldsymbol{x}_j^{\mathrm{n}} - \boldsymbol{x}_j^{\mathrm{n}-1}) + \frac{1}{2}\left(\boldsymbol{x}_j^{\mathrm{n}-1} - \boldsymbol{x}_j^{\mathrm{n}-2}\right) - \frac{1}{2}\left(\boldsymbol{x}_j^{\mathrm{n}-2} - \boldsymbol{x}_j^{\mathrm{n}-3}\right)\right) \\ =& \frac{\boldsymbol{\varphi}_{\mathrm{h}}^{\mathrm{s,n}}(\boldsymbol{X}_j) - \boldsymbol{\varphi}_{\mathrm{h}}^{\mathrm{s,n}-1}(\boldsymbol{X}_j)}{\Delta t} + \frac{\boldsymbol{\varphi}_{\mathrm{h}}^{\mathrm{s,n}-1}(\boldsymbol{X}_j) - \boldsymbol{\varphi}_{\mathrm{h}}^{\mathrm{s,n}-2}(\boldsymbol{X}_j)}{2\Delta t} \\ & - \frac{\boldsymbol{\varphi}_{\mathrm{h}}^{\mathrm{s,n}-2}(\boldsymbol{X}_j) - \boldsymbol{\varphi}_{\mathrm{h}}^{\mathrm{s,n}-3}(\boldsymbol{X}_j)}{2\Delta t} \ \forall\, \boldsymbol{X}_j \in \Gamma_{\mathrm{h},t^{\mathrm{n}}}. \end{aligned} \tag{6.13}$$

By substituting the definition of $\boldsymbol{\varphi}^{\mathrm{s,n}}(\boldsymbol{X})$ from Eq. (6.7) into the above Eq. (6.13) and simplifying, we get

$$\begin{aligned} \boldsymbol{w}_{\mathrm{h}}^{\mathrm{n}}(\boldsymbol{x}_j^{\mathrm{n}}) =& 2\left(\frac{3}{4}\boldsymbol{v}_{\mathrm{h}}^{\mathrm{s,n}-1}(\boldsymbol{X}_j) + \frac{1}{2}\boldsymbol{v}_{\mathrm{h}}^{\mathrm{s,n}-2}(\boldsymbol{X}_j) - \frac{1}{4}\boldsymbol{v}_{\mathrm{h}}^{\mathrm{s,n}-3}(\boldsymbol{X}_j)\right) \\ & - \left(\frac{3}{4}\boldsymbol{v}_{\mathrm{h}}^{\mathrm{s,n}-2}(\boldsymbol{X}_j) + \frac{1}{2}\boldsymbol{v}_{\mathrm{h}}^{\mathrm{s,n}-3}(\boldsymbol{X}_j) - \frac{1}{4}\boldsymbol{v}_{\mathrm{h}}^{\mathrm{s,n}-4}(\boldsymbol{X}_j)\right). \end{aligned}$$

With the aid of Eq. (6.9), we can finally obtain $\boldsymbol{w}_{\mathrm{h}}^{\mathrm{n}}(\boldsymbol{x}_j^{\mathrm{n}}) = \check{\boldsymbol{v}}_{\mathrm{h}}^{\mathrm{f}}(\boldsymbol{x}_j^{\mathrm{n}})$ from the above equation. A detailed energy based proof on the stability of the quasi-monolithic formulation can be found in [139]. We next proceed to write the complete fluid-structure formulation.

6.3.2 Complete Scheme

In this subsection, we present the fully discretized finite element form of the quasi-monolithic formulation. The variational statement reads as:

$$\text{find}\quad (\boldsymbol{v}_{\mathrm{h}}^{\mathrm{f,n}}, p_{\mathrm{h}}^{\mathrm{f,n}}, \boldsymbol{v}_{\mathrm{h}}^{\mathrm{s,n}}) \in \mathscr{V}_{\mathrm{h}}\left(t^{\mathrm{n}}, \boldsymbol{\varphi}_{\mathrm{h}}^{\mathrm{s,n}}, \frac{3}{4}, \frac{1}{2}\boldsymbol{v}_{\mathrm{h}}^{\mathrm{s,n-1}} - \frac{1}{4}\boldsymbol{v}_{\mathrm{h}}^{\mathrm{s,n-2}}\right)$$

with $\boldsymbol{v}_{\mathrm{h}}^{\mathrm{f,n}}|_{\Gamma_D^{\mathrm{f}}} = \boldsymbol{v}_D^{\mathrm{f}}$ and $\boldsymbol{v}_{\mathrm{h}}^{\mathrm{s,n}}|_{\Gamma_D^{\mathrm{s}}} = \boldsymbol{v}_D^{\mathrm{s}}$ such that for any finite element triple

$$(\boldsymbol{\psi}^{\mathrm{f}}, q^{\mathrm{f}}, \boldsymbol{\psi}^{\mathrm{s}}) \in \mathscr{V}_{\mathrm{h}}(t^{\mathrm{n}}, \boldsymbol{\varphi}_{\mathrm{h}}^{\mathrm{s,n}}, 1, \mathbf{0}) \tag{6.14}$$

with $\boldsymbol{\psi}_{\mathrm{h}}^{\mathrm{f}}|_{\Gamma_D^{\mathrm{f}}} = 0$ and $\boldsymbol{\psi}_{\mathrm{h}}^{\mathrm{s}}|_{\Gamma_D^{\mathrm{s}}} = 0$,

$$\left.\begin{aligned}
&\int_{\Omega_{\mathrm{h}}^{\mathrm{f}}(t^{\mathrm{n}})} \left[\frac{\rho^{\mathrm{f}}}{\Delta t}\left(\frac{3}{2}\boldsymbol{v}_{\mathrm{h}}^{\mathrm{f,n}}(\boldsymbol{x}) - 2\boldsymbol{v}_{\mathrm{h}}^{\mathrm{f,n-1}}(\boldsymbol{\Psi}_{\mathrm{h}}^{\mathrm{n}}(\boldsymbol{x}, t^{\mathrm{n-1}})) + \frac{1}{2}\boldsymbol{v}_{\mathrm{h}}^{\mathrm{f,n-2}}(\boldsymbol{\Psi}_{\mathrm{h}}^{\mathrm{n}}(\boldsymbol{x}, t^{\mathrm{n-2}}))\right)\right.\\
&\qquad\left. + \left(\check{\boldsymbol{v}}_{\mathrm{h}}^{\mathrm{f}} - \boldsymbol{w}_{\mathrm{h}}^{\mathrm{n}}\right)\cdot\nabla\boldsymbol{v}_{\mathrm{h}}^{\mathrm{f,n}} + \frac{1}{2}\left(\nabla\cdot\check{\boldsymbol{v}}_{\mathrm{h}}^{\mathrm{f}}\right)\boldsymbol{v}_{\mathrm{h}}^{\mathrm{f,n}}\right]\cdot\boldsymbol{\psi}^{\mathrm{f}}d\boldsymbol{x}\\
&\qquad + \int_{\Omega_{\mathrm{h}}^{\mathrm{f}}(t^{\mathrm{n}})} \mu^{\mathrm{f}}\left(\nabla\boldsymbol{v}_{\mathrm{h}}^{\mathrm{f,n}} + (\nabla\boldsymbol{v}_{\mathrm{h}}^{\mathrm{f,n}})^T\right) : \nabla\boldsymbol{\psi}^{\mathrm{f}}d\boldsymbol{x}
\end{aligned}\right\} A$$

$$\left. - \int_{\Omega_{\mathrm{h}}^{\mathrm{f}}(t^{\mathrm{n}})} p_{\mathrm{h}}^{\mathrm{f,n}}(\nabla\cdot\boldsymbol{\psi}^{\mathrm{f}})d\boldsymbol{x}\right\} B$$

$$\left. - \int_{\Omega_{\mathrm{h}}^{\mathrm{f}}(t^{\mathrm{n}})} q^{\mathrm{f}}(\nabla\cdot\boldsymbol{v}_{\mathrm{h}}^{\mathrm{f,n}})d\boldsymbol{x}\right\} C$$

$$\left.\begin{aligned}
&+\int_{\Omega_{\mathrm{h}}^{\mathrm{s}}} \frac{\rho^{\mathrm{s}}}{\Delta t}\left(\frac{3}{2}\boldsymbol{v}_{\mathrm{h}}^{\mathrm{s,n}} - 2\boldsymbol{v}_{\mathrm{h}}^{\mathrm{s,n-1}} + \frac{1}{2}\boldsymbol{v}_{\mathrm{h}}^{\mathrm{s,n-2}}\right)\cdot\boldsymbol{\psi}^{\mathrm{s}}d\boldsymbol{X}\\
&+\frac{1}{2}\int_{\Omega_{\mathrm{h}}^{\mathrm{s}}} \left(\boldsymbol{\sigma}^{\mathrm{s}}(\boldsymbol{\varphi}_{\mathrm{h}}^{\mathrm{s,n-1}}) + \boldsymbol{\sigma}^{\mathrm{s}}(\boldsymbol{\varphi}_{\mathrm{h}}^{\mathrm{s,n+1}})\right) : \nabla\boldsymbol{\psi}^{\mathrm{s}}d\boldsymbol{X}
\end{aligned}\right\} D$$

$$\left. = \int_{\Omega_{\mathrm{h}}^{\mathrm{f}}(t^{\mathrm{n}})} \rho^{\mathrm{f}}\boldsymbol{b}^{\mathrm{f}}\cdot\boldsymbol{\psi}^{\mathrm{f}}d\boldsymbol{x} + \int_{(\Gamma_H^{\mathrm{f}})_{\mathrm{h}}} \boldsymbol{\sigma}_N^{\mathrm{f}}\cdot\boldsymbol{\psi}^{\mathrm{f}}d\Gamma + \int_{\Omega_{\mathrm{h}}^{\mathrm{s}}} \rho^{\mathrm{s}}\boldsymbol{b}^{\mathrm{s}}\cdot\boldsymbol{\psi}^{\mathrm{s}}d\boldsymbol{X} + \int_{(\Gamma_N^{\mathrm{s}})_{\mathrm{h}}} \boldsymbol{\sigma}_N^{\mathrm{s}}\cdot\boldsymbol{\psi}^{\mathrm{f}}d\Gamma,\right\} E \tag{6.15}$$

where A contains the transient, convective and diffusive contributions, B and C are the pressure and continuity terms respectively, D represents the momentum equation for the structure and E consists of the external body force and boundary conditions. The technique of adding the term $\frac{1}{2}\left(\nabla\check{\boldsymbol{v}}_{\mathrm{h}}^{\mathrm{f}}\right)\boldsymbol{v}_{\mathrm{h}}^{\mathrm{f,n}}$ in part A of Eq. (6.15) is to stabilize the convective term, which is discussed in detail in [207].

For linearly elastic materials, we can represent $\boldsymbol{\sigma}^{\mathrm{s}}(\boldsymbol{\varphi}^{\mathrm{s,n+1}})$ from the part D in Eq. (6.15) using Eq. (6.7) as

$$\boldsymbol{\sigma}^{\mathrm{s}}(\boldsymbol{\varphi}^{\mathrm{s,n+1}}) = \boldsymbol{\sigma}^{\mathrm{s}}\left(\boldsymbol{\varphi}^{\mathrm{s,n}} + \frac{\Delta t}{2}\left(3\boldsymbol{v}^{\mathrm{s,n}} - \boldsymbol{v}^{\mathrm{s,n-1}}\right)\right). \tag{6.16}$$

Similarly for nonlinear material, we can approximate the last term of part D from Eq. (6.15) as

$$\int_{\Omega_{\mathrm{h}}^{\mathrm{s}}} \boldsymbol{\sigma}^{\mathrm{s}}(\boldsymbol{\varphi}_{\mathrm{h}}^{\mathrm{s,n+1}}) : \nabla\boldsymbol{\psi}^{\mathrm{s}} d\boldsymbol{X} \tag{6.17}$$

by the sum of a known forcing term and a linear term for the unknown $\boldsymbol{v}_{\mathrm{h}}^{\mathrm{s,n}}$,

$$\int_{\Omega_{\mathrm{h}}^{\mathrm{s}}} \boldsymbol{\sigma}^{\mathrm{s}}\left(\boldsymbol{\varphi}_{\mathrm{h}}^{\mathrm{s,n}} - \frac{\Delta t}{2}\boldsymbol{v}_{\mathrm{h}}^{\mathrm{s,n-1}}\right) : \nabla\boldsymbol{\psi}^{\mathrm{s}} d\boldsymbol{X} + \frac{3\Delta t}{2} A_s\left(\nabla\left(\boldsymbol{\varphi}_{\mathrm{h}}^{\mathrm{s,n}} - \frac{\Delta t}{2}\boldsymbol{v}_{\mathrm{h}}^{\mathrm{s,n-1}}\right); \nabla\boldsymbol{v}_{\mathrm{h}}^{\mathrm{s,n}}, \nabla\boldsymbol{\psi}^{\mathrm{s}}\right). \tag{6.18}$$

We have ignored a remainder term of size $\mathscr{O}(\Delta t^2)$ in the above linearization. For a St. Venant-Kirchhoff material (see for example [10, 138]), the quantity A_s can be written as

$$A_s(\boldsymbol{F}; \boldsymbol{g}, \boldsymbol{H}) = \int_{\Omega^{\mathrm{s}}} \Big(\lambda^{\mathrm{s}}(\mathrm{tr}\boldsymbol{E})\boldsymbol{H} : \boldsymbol{g} + \frac{\lambda^{\mathrm{s}}}{4}\mathrm{tr}(\boldsymbol{H}^T\boldsymbol{F} + \boldsymbol{F}^T\boldsymbol{H})\mathrm{tr}(\boldsymbol{g}^T\boldsymbol{F} + \boldsymbol{F}^T\boldsymbol{g})$$
$$+ \mu^{\mathrm{s}}\boldsymbol{E} : (\boldsymbol{H}^T\boldsymbol{g} + \boldsymbol{g}^T\boldsymbol{H}) + \frac{\mu^{\mathrm{s}}}{2}(\boldsymbol{H}^T\boldsymbol{F} + \boldsymbol{F}^T\boldsymbol{H}) : (\boldsymbol{g}^T\boldsymbol{F} + \boldsymbol{F}^T\boldsymbol{g})\Big), \tag{6.19}$$

where $\boldsymbol{E} = \frac{1}{2}\left(\boldsymbol{F}\boldsymbol{F}^T - \boldsymbol{I}\right)$ and $\boldsymbol{H} : \boldsymbol{g} = \mathrm{tr}(\boldsymbol{H}^T\boldsymbol{g})$. $A_s(\boldsymbol{F}; \boldsymbol{g}, \boldsymbol{H})$ is a bilinear functional of $\boldsymbol{g}$ and $\boldsymbol{H}$. Therefore, $\boldsymbol{g}$ and $\boldsymbol{H}$ for Eq. (6.18) will be $\nabla\boldsymbol{v}_{\mathrm{h}}^{\mathrm{s,n}}$ and $\nabla\boldsymbol{\psi}^{\mathrm{s}}$, respectively.

6.3.3 Algorithm

The presented variational formulation discussed above is expressed in the algorithmic format in this section. Suppose a mesh $\mathscr{T}_{\mathrm{h}}^{\mathrm{s}}$ for the structural reference domain Ω^{s} shares a part of its boundary grid points with $\mathscr{T}^{\mathrm{h}}$ along Γ. Assume that $\boldsymbol{v}_{\mathrm{h}}^{\mathrm{f,n-1}}$, $\boldsymbol{v}_{\mathrm{h}}^{\mathrm{s,n-1}}$, and $\boldsymbol{\varphi}_{\mathrm{h}}^{\mathrm{s,n-1}}$ are known for the mesh $\mathscr{T}_{\mathrm{h},t^{\mathrm{n}-1}}^{\mathrm{f}}$ which is defined on the domain $\Omega_{\mathrm{h}}^{\mathrm{f}}(t^{\mathrm{n-1}})$. Here, $\Omega_{\mathrm{h}}^{\mathrm{f}}(t^{\mathrm{n-1}})$ denotes the numerical approximation for the fluid domain at time $t^{\mathrm{n-1}}$. It should be noted that we have considered $\mathbb{P}_{\mathrm{m}}/\mathbb{P}_{\mathrm{m-1}}/\mathbb{P}_{\mathrm{m}}$ elements. To ensure the optimal rate of approximation on an isoparametric finite element mesh, all the constituent elements are considered as straight-edged standard Lagrangian elements. The basic steps to be performed in the quasi-monolithic combined fluid-structure formulation are summarized below:

Algorithm 1 Second-order quasi-monolithic formulation for fluid-structure interactions

1. Start from known solutions of the field variables $\boldsymbol{v}_{\mathrm{h}}^{\mathrm{f,n-1}}$, $\boldsymbol{v}_{\mathrm{h}}^{\mathrm{f,n-2}}$, $\boldsymbol{v}_{\mathrm{h}}^{\mathrm{s,n-1}}$, $\boldsymbol{v}_{\mathrm{h}}^{\mathrm{s,n-2}}$, $\boldsymbol{\varphi}_{\mathrm{h}}^{\mathrm{s,n-1}}$ at time $t^{\mathrm{n-1}}$
2. Advance from $t^{\mathrm{n-1}}$ to t^{n}
   (a) Define the structural position $\boldsymbol{\varphi}_{\mathrm{h}}^{\mathrm{s,n}}$ using Eq. (6.7)
   (b) Determine positions of vertices of triangles on the fluid domain $\Omega_{\mathrm{h}}^{\mathrm{f}}(t^{\mathrm{n}})$ by solving Eq. (3.74) using $\mathbb{P}_1$ elements on $\Omega_{\mathrm{h}}^{\mathrm{f}}(0)$
   (c) Use result from (b) to determine the updated mesh of $\Omega_{\mathrm{h}}^{\mathrm{f}}(t^{\mathrm{n}})$
   (d) Evaluate $\check{\boldsymbol{v}}_{\mathrm{h}}^{\mathrm{f}}$ and $\boldsymbol{w}_{\mathrm{h}}^{\mathrm{n}}$ by Eqs. (6.10) and (6.11)
   (e) Solve for the updated field properties $\boldsymbol{v}_{\mathrm{h}}^{\mathrm{f,n}}$, $p_{\mathrm{h}}^{\mathrm{n}}$, and $\boldsymbol{v}_{\mathrm{h}}^{\mathrm{s,n}}$ at current time t^{n} using Eq. (6.15)

At each time step, the scheme is solved only once and no nonlinear iterations are necessary for this mathematically nonlinear problem.

6.4 Extension to Multiple Flexible-Bodies

One of the key features of the quasi-monolithic with explicit interface updating formulation given in Sect. 6.3 is that it can be easily extended for FSI problems involving multiple flexible structures. For the case of multiple flexible bodies, let us consider that the solid domain Ω^{s} is defined as $\Omega^{\mathrm{s}} = \Omega_1^{\mathrm{s}} \cup \Omega_2^{\mathrm{s}} \cdots \Omega_j^{\mathrm{s}}$, where j is the number of flexible bodies interacting with the fluid Ω^{f}. Each of these flexible structures should satisfy the dynamic equilibrium Eq. (3.49) i.e.

$$\rho_j^{\mathrm{s}} \frac{\partial^2 \boldsymbol{\varphi}_j^{\mathrm{s}}}{\partial t^2} = \nabla \cdot \boldsymbol{\sigma}_j^{\mathrm{s}} + \rho_j^{\mathrm{s}} \boldsymbol{b}_j^{\mathrm{s}}, \qquad \text{on} \quad \Omega_j^{\mathrm{s}}, \tag{6.20}$$

The interface traction and velocity continuity conditions in Eqs. (3.66-3.67) will remain the same since $\Gamma = \Gamma_1 \cup \Gamma_2 \cup \cdots \Gamma_j$, where Γ_j is the interface between Ω^{f} and Ω_j^{s}. We can write down the weak-form of the dynamic equilibrium Eq. (6.20) in the similar manner to Eq. (6.4) as

$$\int_{\Omega_j^{\mathrm{s}}} \rho_j^{\mathrm{s}} \frac{\partial \boldsymbol{v}_j^{\mathrm{s}}}{\partial t} \cdot \boldsymbol{\psi}^{\mathrm{s}} d\boldsymbol{X} + \int_{\Omega_j^{\mathrm{s}}} \boldsymbol{\sigma}_j^{\mathrm{s}} : \nabla \boldsymbol{\psi}^{\mathrm{s}} d\boldsymbol{X} =$$

$$\int_{\Omega_j^{\mathrm{s}}} \rho_j^{\mathrm{s}} \boldsymbol{b}_j^{\mathrm{s}} \cdot \boldsymbol{\psi}^{\mathrm{s}} d\boldsymbol{X} + \int_{\Gamma_{N,j}^{\mathrm{s}}} \boldsymbol{\sigma}_{N,j}^{\mathrm{s}} \cdot \boldsymbol{\psi}^{\mathrm{s}} d\Gamma + \int_{\Gamma_j} \left(\boldsymbol{\sigma}_j^{\mathrm{s}}(\boldsymbol{X}, t) \cdot \boldsymbol{n}^{\mathrm{s}} \right) \cdot \boldsymbol{\psi}^{\mathrm{s}}(\boldsymbol{X}) d\Gamma, \tag{6.21}$$

where $\Gamma^{\mathrm{s}}_{N,j}$ represents the Neumann boundary of Ω^{s}_{j} and $\boldsymbol{\sigma}^{\mathrm{s}}_{N,j}$ is the traction function along $\Gamma^{\mathrm{s}}_{N,j}$. The subscript j in ρ^{s}, $\boldsymbol{v}^{\mathrm{s}}$, $\boldsymbol{\sigma}^{\mathrm{s}}$ and $\boldsymbol{b}^{\mathrm{s}}$ represent the structural density, velocity, first Piola-Kirchhoff stress tensor and the body force acting on the jth flexible structure. Now, we can merge Eqs. (6.2, 6.3, 6.21, 3.66–3.67) to construct the weak-form of the combined fluid-structure formulation for multiple flexible structures. The resulting weak-form is

$$\int_{\Omega^{\mathrm{f}}(t)} \rho^{\mathrm{f}}\left(\partial_t \boldsymbol{v}^{\mathrm{f}}(\boldsymbol{x},t) + (\boldsymbol{v}^{\mathrm{f}} - \boldsymbol{w})\cdot\nabla\boldsymbol{v}^{\mathrm{f}}\right)\cdot\boldsymbol{\psi}^{\mathrm{f}}(\boldsymbol{x})d\boldsymbol{x} + \int_{\Omega^{\mathrm{f}}(t)} \boldsymbol{\sigma}^{\mathrm{f}} : \nabla\boldsymbol{\psi}^{\mathrm{f}} d\boldsymbol{x} - \int_{\Omega^{\mathrm{f}}(t)} (\nabla\cdot\boldsymbol{v}^{\mathrm{f}}) q d\boldsymbol{x} + \sum_j \int_{\Omega^{\mathrm{s}}_j} \rho^{\mathrm{s}}_j \partial_t \boldsymbol{v}^{\mathrm{s}}_j \cdot \boldsymbol{\psi}^{\mathrm{s}} d\boldsymbol{X} + \sum_j \int_{\Omega^{\mathrm{s}}_j} \boldsymbol{\sigma}^{\mathrm{s}}_j : \nabla\boldsymbol{\psi}^{\mathrm{s}} d\boldsymbol{X} =$$

$$\int_{\Omega^{\mathrm{f}}(t)} \rho^{\mathrm{f}}\boldsymbol{b}^{\mathrm{f}}\cdot\boldsymbol{\psi}^{\mathrm{f}} d\boldsymbol{x} + \int_{\Gamma^{\mathrm{f}}_N} \boldsymbol{\sigma}^{\mathrm{f}}_N \cdot \boldsymbol{\psi}^{\mathrm{f}} d\Gamma + \sum_j \int_{\Omega^{\mathrm{s}}_j} \rho^{\mathrm{s}}_j \boldsymbol{b}^{\mathrm{s}}_j \cdot \boldsymbol{\psi}^{\mathrm{s}} d\boldsymbol{X} + \sum_j \int_{\Gamma^{\mathrm{s}}_{N,j}} \boldsymbol{\sigma}^{\mathrm{s}}_{N,j} \cdot \boldsymbol{\psi}^{\mathrm{s}} d\Gamma. \tag{6.22}$$

Let us consider that $\mathscr{T}^{\mathrm{s}}_{\mathrm{h},j}$ represents the finite element discretization of the flexible structure Ω^{s}_j. The explicit extrapolation of the structural deformation vector for each Lagrangian node $(\boldsymbol{X}_j)$ on structural meshes $\mathscr{T}^{\mathrm{s}}_{\mathrm{h},i}$ will be

$$\boldsymbol{\varphi}^{\mathrm{s,n}}_{\mathrm{h},j}(\boldsymbol{X}_i) = \boldsymbol{\varphi}^{\mathrm{s,n-1}}_{\mathrm{h},j}(\boldsymbol{X}_i) + \frac{3\Delta t}{2}\boldsymbol{v}^{\mathrm{s,n-1}}_{\mathrm{h},j}(\boldsymbol{X}_i) - \frac{\Delta t}{2}\boldsymbol{v}^{\mathrm{s,n-2}}_{\mathrm{h},j}(\boldsymbol{X}_i) \qquad \forall \boldsymbol{X}_i \in \mathscr{T}^{\mathrm{s}}_{\mathrm{h},j}, \tag{6.23}$$

where $\boldsymbol{\varphi}^{\mathrm{s,n}}_{\mathrm{h},j}$ is the jth flexible structure's deformation vector at time step t^{n}. The fully discretized quasi-monolithic combined fluid-structure formulation for multiple flexible bodies using second-order backward difference approximation (BDF2) in time, is given as

$$\int_{\Omega^{\mathrm{f}}_{\mathrm{h}}(t^{\mathrm{n}})} \Bigg[\frac{\rho^{\mathrm{f}}}{\Delta t}\left(\frac{3}{2}\boldsymbol{v}^{\mathrm{f,n}}_{\mathrm{h}}(\boldsymbol{x}) - 2\boldsymbol{v}^{\mathrm{f,n-1}}_{\mathrm{h}}(\boldsymbol{\Psi}^{\mathrm{n}}_{\mathrm{h}}(\boldsymbol{x},t^{\mathrm{n-1}})) + \frac{1}{2}\boldsymbol{v}^{\mathrm{f,n-2}}_{\mathrm{h}}(\boldsymbol{\Psi}^{\mathrm{n}}_{\mathrm{h}}(\boldsymbol{x},t^{\mathrm{n-2}}))\right) + (\check{\boldsymbol{v}}^{\mathrm{f}}_{\mathrm{h}} - \boldsymbol{w}^{\mathrm{n}}_{\mathrm{h}})\cdot\nabla\boldsymbol{v}^{\mathrm{f,n}}_{\mathrm{h}} + \frac{1}{2}\left(\nabla\check{\boldsymbol{v}}^{\mathrm{f}}_{\mathrm{h}}\right)\boldsymbol{v}^{\mathrm{f,n}}_{\mathrm{h}}\Bigg]\cdot\boldsymbol{\psi}^{\mathrm{f}} d\boldsymbol{x}$$

$$+ \int_{\Omega^{\mathrm{f}}_{\mathrm{h}}(t^{\mathrm{n}})} \mu^{\mathrm{f}}\left(\nabla\boldsymbol{v}^{\mathrm{f,n}}_{\mathrm{h}} + (\nabla\boldsymbol{v}^{\mathrm{f,n}}_{\mathrm{h}})^T\right) : \nabla\boldsymbol{\psi}^{\mathrm{f}} d\boldsymbol{x} - \int_{\Omega^{\mathrm{f}}_{\mathrm{h}}(t^{\mathrm{n}})} p^{\mathrm{f,n}}_{\mathrm{h}}(\nabla\cdot\boldsymbol{\psi}^{\mathrm{f}}) d\boldsymbol{x} - \int_{\Omega^{\mathrm{f}}_{\mathrm{h}}(t^{\mathrm{n}})} q^{\mathrm{f}}(\nabla\cdot\boldsymbol{v}^{\mathrm{f,n}}_{\mathrm{h}}) d\boldsymbol{x}$$

$$+ \sum_j \int_{\Omega^{\mathrm{s}}_{\mathrm{h},j}} \frac{\rho^{\mathrm{s}}_j}{\Delta t}\left(\frac{3}{2}\boldsymbol{v}^{\mathrm{s,n}}_{\mathrm{h},j} - 2\boldsymbol{v}^{\mathrm{s,n-1}}_{\mathrm{h},j} + \frac{1}{2}\boldsymbol{v}^{\mathrm{s,n-2}}_{\mathrm{h},j}\right)\cdot\boldsymbol{\psi}^{\mathrm{s}} d\boldsymbol{X}$$

$$
\begin{aligned}
&+\sum_j \frac{1}{2} \int_{\Omega^{\mathrm{s}}_{\mathrm{h},j}} \left(\boldsymbol{\sigma}^{\mathrm{s}}(\boldsymbol{\varphi}^{\mathrm{s,n-1}}_{\mathrm{h},j}) + \boldsymbol{\sigma}^{\mathrm{s}}(\boldsymbol{\varphi}^{\mathrm{s,n+1}}_{\mathrm{h},j}) \right) : \nabla \boldsymbol{\psi}^{\mathrm{s}} d\boldsymbol{X} \\
&\qquad = \int_{\Omega^{\mathrm{f}}_{\mathrm{h}}(t^{\mathrm{n}})} \rho^{\mathrm{f}} \boldsymbol{b}^{\mathrm{f}} \cdot \boldsymbol{\psi}^{\mathrm{f}} d\boldsymbol{x} + \int_{(\Gamma^{\mathrm{f}}_{N})_{\mathrm{h}}} \boldsymbol{\sigma}^{\mathrm{f}}_{N} \cdot \boldsymbol{\psi}^{\mathrm{f}} d\Gamma \\
&+\sum_j \int_{\Omega^{\mathrm{s}}_{\mathrm{h},j}} \rho^{\mathrm{s}}_j \boldsymbol{b}^{\mathrm{s}}_j \cdot \boldsymbol{\psi}^{\mathrm{s}} d\boldsymbol{X} + \sum_j \int_{(\Gamma^{\mathrm{s}}_{N,j})_{\mathrm{h}}} \boldsymbol{\sigma}^{\mathrm{s}}_{N,j} \cdot \boldsymbol{\psi}^{\mathrm{s}} d\Gamma.
\end{aligned}
\tag{6.24}
$$

The algorithm of the quasi-monolithic combined fluid-structure formulation for multiple flexible structures is identical to that for single flexible structure presented in Sect. 6.3.3. The only difference is that multiple solid meshes $\mathcal{T}^{\mathrm{s}}_{\mathrm{h},i}$ corresponding to each flexible structure are considered.

6.5 Fully Stabilized Quasi-Monolithic Formulation

One of the limitations of the Galerkin finite element discretization used for discretizing $\Omega^{\mathrm{f}}(t)$ and Ω^{s} of the quasi-monolithic formulation presented in Sect. 6.3 is the presence of non-physical spurious oscillations for convection-dominant problems [56]. As discussed in Chapter 4, these oscillations are circumvented by replacing the traditional Galerkin method with Petrov-Galerkin stabilization methods which utilize weighting functions that have more weightage for the upstream section of the flow [34, 92]. Such streamwise upwind techniques can be interpreted as combination of traditional Galerkin method and a stabilization term calculated at the interior of an element. This elemental level stabilization term introduces artificial numerical diffusion which stabilizes the spurious oscillations. The weak-form of the combined fluid-structure formulation given in Eq. (6.6) can be written in the Galerkin/Least square (GLS) stabilization form as

$$
\left.
\begin{aligned}
&\int_{\Omega^{\mathrm{f}}_{\mathrm{h}}(t)} \rho^{\mathrm{f}} \left(\partial_t \boldsymbol{v}^{\mathrm{f}} + (\boldsymbol{v}^{\mathrm{f}} - \boldsymbol{w}) \cdot \nabla \boldsymbol{v}^{\mathrm{f}} \right) \cdot \boldsymbol{\psi}^{\mathrm{f}} d\Omega + \int_{\Omega^{\mathrm{f}}(t)} \boldsymbol{\sigma}^{\mathrm{f}} : \nabla \boldsymbol{\psi}^{\mathrm{f}} d\Omega \\
&\qquad - \int_{\Omega^{\mathrm{f}}(t)} (\nabla \cdot \boldsymbol{v}^{\mathrm{f}}) q d\Omega
\end{aligned}
\right\} A
$$

$$
\left.
\begin{aligned}
&+\sum_{e=1}^{n_{\mathrm{el}}} \int_{\Omega^{\mathrm{e}}} \tau_m \left[\rho^{\mathrm{f}} (\boldsymbol{v}^{\mathrm{f}} - \boldsymbol{w}) \cdot \nabla \boldsymbol{\psi}^{\mathrm{f}} + \nabla q \right] \cdot \\
&\left[\rho^{\mathrm{f}} \partial_t \boldsymbol{v}^{\mathrm{f}} + \rho^{\mathrm{f}} (\boldsymbol{v}^{\mathrm{f}} - \boldsymbol{w}) \cdot \nabla \boldsymbol{v}^{\mathrm{f}} - \nabla \cdot \boldsymbol{\sigma}^{\mathrm{f}} - \rho^{\mathrm{f}} \boldsymbol{b}^{\mathrm{f}} \right] d\Omega^{\mathrm{e}}
\end{aligned}
\right\} B
$$

$$
\left.+\sum_{e=1}^{n_{\mathrm{el}}}\int_{\Omega^{\mathrm{e}}} \nabla\cdot\boldsymbol{\psi}^{\mathrm{f}}\tau_c\nabla\cdot\boldsymbol{v}^{\mathrm{f}}d\Omega^{\mathrm{e}}\right\} C
$$

$$
\left.+\int_{\Omega^{\mathrm{s}}} \rho^{\mathrm{s}}\partial_t\boldsymbol{v}^{\mathrm{s}}\cdot\boldsymbol{\psi}^{\mathrm{s}}d\Omega+\int_{\Omega^{\mathrm{s}}}\boldsymbol{\sigma}^{\mathrm{s}}:\nabla\boldsymbol{\psi}^{\mathrm{s}}d\Omega=\right\} D
$$

$$
\left.\int_{\Omega_{\mathrm{h}}^{\mathrm{f}}(t)} \rho^{\mathrm{f}}\boldsymbol{b}^{\mathrm{f}}\cdot\boldsymbol{\psi}^{\mathrm{f}}d\Omega+\int_{(\Gamma_N^{\mathrm{f}})_{\mathrm{h}}}\boldsymbol{\sigma}_N^{\mathrm{f}}\cdot\boldsymbol{\psi}^{\mathrm{f}}d\Gamma+\int_{\Omega_{\mathrm{h}}^{\mathrm{s}}}\rho^{\mathrm{s}}\boldsymbol{b}^{\mathrm{s}}\cdot\boldsymbol{\psi}^{\mathrm{s}}d\Omega+\int_{(\Gamma_N^{\mathrm{s}})_{\mathrm{h}}}\boldsymbol{\sigma}_N^{\mathrm{s}}\cdot\boldsymbol{\psi}^{\mathrm{s}}d\Gamma.\right\} E \tag{6.25}
$$

One can observe that terms A, D and E combine to form the Galerkin weak-form in Eq. (6.6). The terms B and C represent the local element level GLS stabilization terms for the momentum and continuity equations, respectively. They damp the spurious oscillations and circumvent the inf-sup condition. Unlike the finite element mesh discretization in Sect. 6.2.2 where fluid velocity, pressure and structural velocity were approximated using $\mathbb{P}_2/\mathbb{P}_1/\mathbb{P}_2$ isoparametric elements to justify the inf-sup condition or LBB condition for well-posedness, the above stabilized combined fluid-structure weak-form in Eq. (6.25) is approximated using six-node wedge or eight-node hexahedral element with equal order for both fluid velocity and pressure. Equal order approximation for fluid velocity and pressure simplifies the computational framework significantly. The stabilizing parameters τ_m and τ_c in the terms B and C respectively are the least squares metrics [34, 67, 97, 193]. The convective stabilization parameter τ_m is defined as [3]

$$
\tau_m=\left[\left(\frac{2\rho^{\mathrm{f}}}{\Delta t}\right)^2+(\rho^{\mathrm{f}})^2\left(\boldsymbol{v}^{\mathrm{f}}-\boldsymbol{w}\right)\cdot\boldsymbol{G}\cdot\left(\boldsymbol{v}^{\mathrm{f}}-\boldsymbol{w}\right)+12(\mu^{\mathrm{f}})^2\boldsymbol{G}:\boldsymbol{G}\right]^{-\frac{1}{2}}, \tag{6.26}
$$

where $\boldsymbol{G}$ is the elemental contravariant metric tensor which is defined as

$$
\boldsymbol{G}=\left(\frac{\partial\boldsymbol{\xi}}{\partial\boldsymbol{x}}\right)^T\frac{\partial\boldsymbol{\xi}}{\partial\boldsymbol{x}}, \tag{6.27}
$$

where $\boldsymbol{\xi}$ is local element level coordinate system and it depends on the element shape. The least squares metric τ_c for the continuity equation is defined as

$$
\tau_c=\frac{1}{8\,\mathrm{tr}\,(\boldsymbol{G})\,\tau_m}. \tag{6.28}
$$

The stabilization in the variational form provides stability to the velocity field in convection dominated regimes of the fluid domain and circumvents the LBB condition which is required to be satisfied by any standard mixed Galerkin method. The element metric tensor $\boldsymbol{G}$ deals with different element topology for different mesh discretizations and has been greatly studied in the literature.

The fully discretized quasi-monolithic stabilized fluid-structure formulation for multiple structures using BDF2 can be written as

$$
\begin{aligned}
&\int\limits_{\Omega^{\mathrm{f}}_{\mathrm{h}}(t^{\mathrm{n}})} \Bigg[\frac{\rho^{\mathrm{f}}}{\Delta t} \left(\frac{3}{2} \boldsymbol{v}^{\mathrm{f,n}}_{\mathrm{h}}(\boldsymbol{x}) - 2\boldsymbol{v}^{\mathrm{f,n-1}}_{\mathrm{h}}(\boldsymbol{\Psi}^{\mathrm{n}}_{\mathrm{h}}(\boldsymbol{x}, t^{\mathrm{n-1}})) + \frac{1}{2}\boldsymbol{v}^{\mathrm{f,n-2}}_{\mathrm{h}}(\boldsymbol{\Psi}^{\mathrm{n}}_{\mathrm{h}}(\boldsymbol{x}, t^{\mathrm{n-2}})) \right) \\
&\qquad + \left(\check{\boldsymbol{v}}^{\mathrm{f}}_{\mathrm{h}} - \boldsymbol{w}^{\mathrm{n}}_{\mathrm{h}} \right) \cdot \nabla \boldsymbol{v}^{\mathrm{f,n}}_{\mathrm{h}} + \frac{1}{2} \left(\nabla \check{\boldsymbol{v}}^{\mathrm{f}}_{\mathrm{h}} \right) \boldsymbol{v}^{\mathrm{f,n}}_{\mathrm{h}} \Bigg] \cdot \boldsymbol{\psi}^{\mathrm{f}} d\boldsymbol{x} \\
&+ \int\limits_{\Omega^{\mathrm{f}}_{\mathrm{h}}(t^{\mathrm{n}})} \mu^{\mathrm{f}} \left(\nabla \boldsymbol{v}^{\mathrm{f,n}}_{\mathrm{h}} + (\nabla \boldsymbol{v}^{\mathrm{f,n}}_{\mathrm{h}})^{T} \right) : \nabla \boldsymbol{\psi}^{\mathrm{f}} d\boldsymbol{x} - \int\limits_{\Omega^{\mathrm{f}}_{\mathrm{h}}(t^{\mathrm{n}})} p^{\mathrm{f,n}}_{\mathrm{h}} (\nabla \cdot \boldsymbol{\psi}^{\mathrm{f}}) d\boldsymbol{x} \\
&- \int\limits_{\Omega^{\mathrm{f}}_{\mathrm{h}}(t^{\mathrm{n}})} q^{\mathrm{f}} (\nabla \cdot \boldsymbol{v}^{\mathrm{f,n}}_{\mathrm{h}}) d\boldsymbol{x} \\
&+ \sum_{e=1}^{n_{\mathrm{el}}} \int\limits_{\Omega^{\mathrm{e}}_{\mathrm{h}}} \tau_m \left[\rho^{\mathrm{f}} \left(\check{\boldsymbol{v}}^{\mathrm{f}}_{\mathrm{h}} - \boldsymbol{w}^{\mathrm{n}}_{\mathrm{h}} \right) \cdot \nabla \boldsymbol{\psi}^{\mathrm{f}} + \nabla q \right] \cdot \\
&\left[\frac{\rho^{\mathrm{f}}}{\Delta t} \left(1.5 \boldsymbol{v}^{\mathrm{f,n}}_{\mathrm{h}} - 2\boldsymbol{v}^{\mathrm{f,n-1}}_{\mathrm{h}} + 0.5 \boldsymbol{v}^{\mathrm{f,n-2}}_{\mathrm{h}} \right) + \rho^{\mathrm{f}} \left(\check{\boldsymbol{v}}^{\mathrm{f}}_{\mathrm{h}} - \boldsymbol{w}^{\mathrm{n}}_{\mathrm{h}} \right) \cdot \nabla \boldsymbol{v}^{\mathrm{f,n}}_{\mathrm{h}} - \nabla \cdot \boldsymbol{\sigma}^{\mathrm{f,n}}_{\mathrm{h}} - \rho^{\mathrm{f}} \boldsymbol{b}^{\mathrm{f}} \right] d\Omega^{\mathrm{e}} \\
&+ \sum_{e=1}^{n_{\mathrm{el}}} \int\limits_{\Omega^{\mathrm{e}}_{\mathrm{h}}} \nabla \cdot \boldsymbol{\psi}^{\mathrm{f}} \tau_c \nabla \cdot \boldsymbol{v}^{\mathrm{f,n}}_{\mathrm{h}} d\Omega^{\mathrm{e}} \\
&+ \sum_{j} \int\limits_{\Omega^{\mathrm{s}}_{\mathrm{h},j}} \frac{\rho^{\mathrm{s}}_{j}}{\Delta t} \left(\frac{3}{2} \boldsymbol{v}^{\mathrm{s,n}}_{\mathrm{h},j} - 2\boldsymbol{v}^{\mathrm{s,n-1}}_{\mathrm{h},j} + \frac{1}{2} \boldsymbol{v}^{\mathrm{s,n-2}}_{\mathrm{h},j} \right) \cdot \boldsymbol{\psi}^{\mathrm{s}} d\boldsymbol{X} \\
&+ \sum_{j} \frac{1}{2} \int\limits_{\Omega^{\mathrm{s}}_{\mathrm{h},j}} \left(\boldsymbol{\sigma}^{\mathrm{s}}(\boldsymbol{\varphi}^{\mathrm{s,n-1}}_{\mathrm{h},j}) + \boldsymbol{\sigma}^{\mathrm{s}}(\boldsymbol{\varphi}^{\mathrm{s,n+1}}_{\mathrm{h},j}) \right) : \nabla \boldsymbol{\psi}^{\mathrm{s}} d\boldsymbol{X} \\
&= \int\limits_{\Omega^{\mathrm{f}}_{\mathrm{h},t^{\mathrm{n}}}} \rho^{\mathrm{f}} \boldsymbol{b}^{\mathrm{f}} \cdot \boldsymbol{\psi}^{\mathrm{f}} d\boldsymbol{x} + \int\limits_{(\Gamma^{\mathrm{f}}_{N})_{\mathrm{h}}} \boldsymbol{\sigma}^{\mathrm{f}}_{N} \cdot \boldsymbol{\psi}^{\mathrm{f}} d\Gamma \\
&+ \sum_{j} \int\limits_{\Omega^{\mathrm{s}}_{\mathrm{h},j}} \rho^{\mathrm{s}}_{j} \boldsymbol{b}^{\mathrm{s}}_{j} \cdot \boldsymbol{\psi}^{\mathrm{s}} d\boldsymbol{X} + \sum_{j} \int\limits_{(\Gamma^{\mathrm{s}}_{N,j})_{\mathrm{h}}} \boldsymbol{\sigma}^{\mathrm{s}}_{N,j} \cdot \boldsymbol{\psi}^{\mathrm{s}} d\Gamma.
\end{aligned}
\tag{6.29}
$$

The implementation of the above fully stabilized quasi-monolithic combined fluid-structure formulation differs slightly from the implementation in Sect. 6.3. Instead of Eq. (6.10) we define an alternative second-order time accurate explicit function given by

$$\check{\boldsymbol{v}}_{\mathrm{h}}^{\mathrm{f}}(\boldsymbol{\Psi}_{\mathrm{h}}^{\mathrm{n}}(\boldsymbol{x}, t^{\mathrm{n}})) = 2.25\boldsymbol{v}_{\mathrm{h}}^{\mathrm{f,n-1}}(\boldsymbol{\Psi}_{\mathrm{h}}^{\mathrm{n}}(\boldsymbol{x}, t^{\mathrm{n-1}})) - 1.5\boldsymbol{v}_{\mathrm{h}}^{\mathrm{f,n-2}}(\boldsymbol{\Psi}_{\mathrm{h}}^{\mathrm{n}}(\boldsymbol{x}, t^{\mathrm{n-2}})) + 0.25\boldsymbol{v}_{\mathrm{h}}^{\mathrm{f,n-3}}(\boldsymbol{\Psi}_{\mathrm{h}}^{\mathrm{n}}(\boldsymbol{x}, t^{\mathrm{n-3}})). \tag{6.30}$$

Similarly, we also define an alternate function for $\boldsymbol{w}_{\mathrm{h}}^{\mathrm{n}}$ as

$$\boldsymbol{w}_{\mathrm{h}}^{\mathrm{n}}(\boldsymbol{x}) = \sum_{i=1}^{G} \psi_i^{\mathrm{f,n}}(\boldsymbol{x}) \frac{1}{\Delta t}\left(\frac{3}{2}\boldsymbol{x}_i^{\mathrm{n}} - 2\boldsymbol{x}_i^{\mathrm{n-1}} + \frac{1}{2}\boldsymbol{x}_i^{\mathrm{n-2}}\right). \tag{6.31}$$

Redefining $\check{\boldsymbol{v}}_{\mathrm{h}}^{\mathrm{f}}$ and $\boldsymbol{w}_{\mathrm{h}}$ in Eqs. (6.30–6.31) enables us to implement the exact interface continuity i.e.

$$\boldsymbol{v}_{\mathrm{h}}^{\mathrm{f,n}} = \boldsymbol{v}_{\mathrm{h}}^{\mathrm{s,n}} \tag{6.32}$$

instead of the second-order approximation in time given by Eq. (6.10). Unlike the velocity continuity in Eq. (3.66) which requires us to enforce the condition explicitly, we can satisfy the velocity continuity in Eq. (6.32) implicitly by treating the fluid and its corresponding solid node on the interface as a single unique node. Thereby, we can decrease the size of the algebraic system of equations required per time step compared to the implementation presented in Sect. 6.3.

6.5.1 Algorithm

Unlike the quasi-monolithic combined fluid-structure formulation in Sect. 6.3 where we have considered $\mathbb{P}_{\mathrm{m}}/\mathbb{P}_{\mathrm{m-1}}/\mathbb{P}_{\mathrm{m}}$ elements to satisfy the inf-sup or LBB condition, here we use equal order elements for both fluid pressure and velocity. The incremental velocity and pressure are computed via the matrix-free implementation of the restarted Generalized Minimal RESidual (GMRES) [185]. The basic steps to be performed in the fully-stabilized quasi-monolithic combined fluid-structure formulation are summarized below:

Algorithm 2 Second-order fully stabilized quasi-monolithic formulation for fluid-structure interactions

1. Start from known solutions of the field variables $\boldsymbol{v}_{\mathrm{h}}^{\mathrm{f,n-1}}$, $\boldsymbol{v}_{\mathrm{h}}^{\mathrm{f,n-2}}$, $\boldsymbol{v}_{\mathrm{h}}^{\mathrm{s,n-1}}$, $\boldsymbol{v}_{\mathrm{h}}^{\mathrm{s,n-2}}$, $\boldsymbol{\varphi}_{\mathrm{h}}^{\mathrm{s,n-1}}$ at time $t^{\mathrm{n-1}}$
2. Advance from $t^{\mathrm{n-1}}$ to t^{n}
   (a) Define the structural position $\boldsymbol{\varphi}_{\mathrm{h}}^{\mathrm{s,n}}$ using Eq. (6.7)
   (b) Determine positions of vertices of triangles on the fluid domain $\Omega_{\mathrm{h}}^{\mathrm{f}}(t^{\mathrm{n}})$ by solving Eq. (3.74) on $\Omega_{\mathrm{h}}^{\mathrm{f}}(0)$
   (c) Use result from (b) to determine the updated mesh of $\Omega_{\mathrm{h}}^{\mathrm{f}}(t^{\mathrm{n}})$
   (d) Evaluate $\check{\boldsymbol{v}}_{\mathrm{h}}^{\mathrm{f}}$ and $\boldsymbol{w}_{\mathrm{h}}^{\mathrm{n}}$ by Eqs. (6.30) and (6.31)
   (e) Determine the element level stabilization parameters τ_m and τ_c using Eqs. (6.26) and (6.28) respectively
   (f) Solve for the updated field properties $\boldsymbol{v}_{\mathrm{h}}^{\mathrm{f,n}}$, $p_{\mathrm{h}}^{\mathrm{n}}$, and $\boldsymbol{v}_{\mathrm{h}}^{\mathrm{s,n}}$ at current time t^{n} using Eq. (6.29)

Similar to the quasi-monolithic formulation presented in Sect. 6.3, the fully-stabilized quasi-monolithic formulation also solves the combined fluid-structure system only once per time step. A matrix-free version of Krylov subspace iterative solvers are utilized to solve the equations system for the fluid-structure interaction. To scale the fluid-structure solver for large scale computations using distributed memory parallel cluster, we next present the parallel finite element implementation of three dimensional incompressible flow interacting with generic elastic structures for high Reynolds number flow.

6.6 Parallel Implementation

In this section, a parallel fluid-structure interaction strategy for supercomputers based on a hybrid approach is presented. Fluid-structure simulation codes pose serious challenges in terms of implementation and parallel efficiency for large scale problems due to non-local embedded interfaces. In this work, the fluid-structure code based on the stabilized formulation has been parallelized using two strategies: first, the Message Passing Interface (MPI) implementation [2] , where the parallelization is based on a standard domain decomposition master-slave strategy. Second, the OpenMP implementation [167]: to treat many-core processors and to exploit thread-level parallelism of multicore architecture. Since the communications between threads are done via shared memory, a reduction of the communication cost between processors is expected. This hybrid parallelization combines MPI tasks and OpenMP threads to exploit the different levels of parallelism. The present implementation of MPI/OpenMP is general and does not make any assumption regarding the geometry and mesh topologies of fluid-structure system.

The parallelization strategy involves implementation of the fluid and solid solvers to handle both distributed memory and shared memory using the message passing

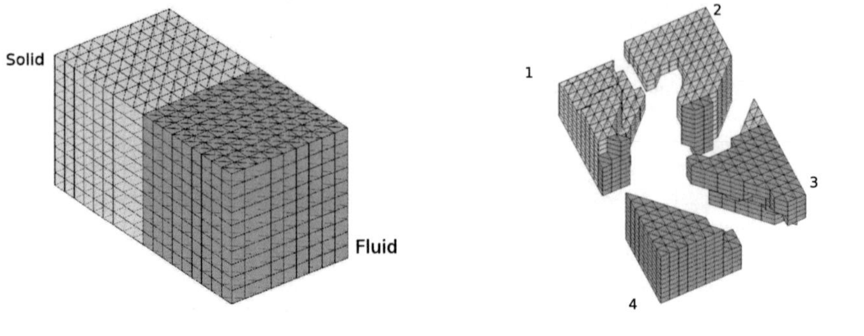

(a) initial fluid-structure mesh (b) decomposed fluid-structure mesh

Fig. 6.1 A graphical representation of the combined fluid (green)-structure (yellow) mesh into four subgrids labeled 1, 2, 3 and 4 respectively

interface [2] and open multi processing [167]. Since both the quasi-monolithic formulations solve the FSI variables $\boldsymbol{v}^{\mathrm{f,n}}$, $\boldsymbol{v}^{\mathrm{s,n}}$, $\boldsymbol{\varphi}^{\mathrm{s,n}}$ and p^{n} on the combined fluid-structure domain $\Omega_{\mathrm{h}}(t^{\mathrm{n}})$, at the domain decomposition level we do not distinguish between the fluid and solid domains. The combined fluid-solid mesh is decomposed via an automatic graph partitioner [118] into evenly sized subgrids which can be made-up of both fluid and solid or just fluid or just solid elements. While these subgrids communicate via MPI with a typial master-slave strategy, the internal loops are parallelized using OpenMP to take the advantage from the shared memory in multicore clusters. We next discuss specific details of the hybrid model, with emphasis on the parallel strategy and the linear solver.

Figure 6.1 demonstrates the domain decomposition strategy described above with the aid of graphical representation. One or more than one subgrid can be assigned to each processor. As mentioned earlier, the communication between the processors is performed with the aid of MPI protocol through a master processor and the element level matrix construction on multiple subgrids in a processor use the OpenMP. One of the key features of our implementation is that we never construct a global left-hand side (LHS) or right-hand side (RHS) matrices. Instead, we only construct element level matrices at subgrid level. The interface nodes on each subgrid communicate the nodal RHS matrix and GMRES(k)/CG AP vector product data through the master processor to include the contributions of the adjacent element present in a different subgrid. We have considered two types of linear solvers: firstly, we implemented conjugate gradient (CG) method for solving the ALE equation and secondly, restarted generalized minimal residual (GMRES-k) method for solving the combined fluid-structure equation. For both these methods, we have used diagonal preconditioners to improve the convergence. Apart from the parallel communication during the AP vector product, the implementation of the CG and GMRES(k) is largely standard, and more details on the linear solver algorithms can be found in [185] and [220]. We summarize the basic steps involved in the parallel quasi-monolithic formulation as follows:

Algorithm 3 Parallel implementation of second-order quasi-monolithic formulation for fluid-structure interactions

I. Perform the domain decomposition using auto graph partitioner into M subgrids
II. Distribute the M subgrids to N number of processor. For load balancing, we assume that M is divisible by N and each processor gets P number of subgrids
1. On each processor with known solutions of the field variables $\boldsymbol{v}_{\mathrm{h}}^{\mathrm{f,n-1}}$, $\boldsymbol{v}_{\mathrm{h}}^{\mathrm{f,n-2}}$, $\boldsymbol{v}_{\mathrm{h}}^{\mathrm{s,n-1}}$, $\boldsymbol{v}_{\mathrm{h}}^{\mathrm{s,n-2}}$, $\boldsymbol{\varphi}_{\mathrm{h}}^{\mathrm{s,n-1}}$ for the assigned subgrid at time $t^{\mathrm{n-1}}$
2. Advance from $t^{\mathrm{n-1}}$ to t^{n} on each subgrid
   (a) Define the structural position $\boldsymbol{\varphi}_{\mathrm{h}}^{\mathrm{s,n}}$ using Eq. (6.7)
   (b) Determine positions of vertices of triangles on the fluid domain $\Omega_{\mathrm{h}}^{\mathrm{f}}(t^{\mathrm{n}})$ by solving Eq. (3.74) on $\Omega_{\mathrm{h}}^{\mathrm{f}}(0)$
   (c) Use result from (b) to determine the updated mesh on each subgrid
   (d) Evaluate $\check{\boldsymbol{v}}_{\mathrm{h}}^{\mathrm{f}}$ and $\boldsymbol{w}_{\mathrm{h}}^{\mathrm{n}}$ by Eqs. (6.30) and (6.31)
   (e) Determine the element level stabilization parameters τ_m and τ_c using Eqs. (6.26) and (6.28) respectively
   (f) Solve for the updated field properties $\boldsymbol{v}_{\mathrm{h}}^{\mathrm{f,n}}$, $p_{\mathrm{h}}^{\mathrm{n}}$, and $\boldsymbol{v}_{\mathrm{h}}^{\mathrm{s,n}}$ at current time t^{n} using Eq. (6.29)

6.7 Convergence and Verification

Here, numerical experiments are performed to assess the numerical properties of the quasi-monolithic formulation presented in the previous section. Moreover, the accuracy of the two-dimensional and three-dimensional nonlinear coupled FSI framework based on the quasi-monolithic (Sect. 6.3) and the fully stabilized quasi-monolithic (Sect. 6.5) formulations, respectively are verified.

6.7.1 Numerical Assessment of Temporal Accuracy

We begin by assessing the temporal accuracy of the second-order formulation discussed in Sect. 6.3.2. The details of the implementation can be found in [43]. A two-dimensional laminar incompressible flow around a linearly elastic semi-circular cylinder of diameter D placed in a channel is considered. The schematic is shown in Fig. 6.2. The computational domain is rectangular with size $[0, 6.5] \times [-0.5, 1]$. The semi-circular cylinder of $D = 1$ centered at $z_0 = [1.5, -0.5]$ is placed on the floor of the channel. A parabolic velocity profile of $\boldsymbol{v}^{\mathrm{f}} = (U(y,t), 0) = (g(t)\,(1+2y)\,(1-y)\,, 0)$ is prescribed at the inlet boundary $\Gamma_{\mathrm{in}}^{\mathrm{f}}$, where $g(t)$ is a relationship between $U(y,t)$ and t:

$$g(t) = \begin{cases} 0 & t \leq 0 \\ 1 - \cos(\frac{\pi}{4}t) & t \in (0, 2] \\ 1 & t > 2 \end{cases} . \tag{6.33}$$

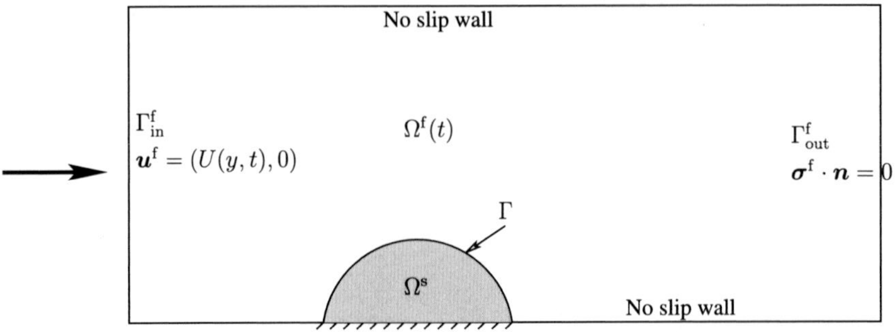

Fig. 6.2 Schematic and computational domain of elastic half-cylinder problem, and details of the boundary conditions

The top and bottom walls of the domain are considered rigid and a no-slip condition is imposed on those surfaces. A traction-free boundary condition is satisfied at the outlet boundary Γ^{f}_{out}.

The density and Lamé's constants for the elastic semi-circular cylinder are considered as $\rho^{s} = 1$, $\lambda^{s} = 500$ and $\mu^{s} = 50$ respectively. The fluid density and kinematic viscosity are $\rho^{f} = 1$ and $\nu^{f} = 1$ respectively. Time convergence study is performed by taking different time step sizes of $\Delta t = 1/40$, $1/80$, $1/120$, and $1/160$. For this problem, the spatial discretization of the domain consists of $\mathbb{P}_3/\mathbb{P}_2/\mathbb{P}_3$ higher-order iso-parametric finite elements for the fluid velocity, pressure and structural velocity, respectively. The fluid and structural mesh comprises of 1239 and 370 triangular elements respectively.

The temporal accuracy is determined by computing the error in the solutions of fluid velocity, pressure and structural displacement for the various values of Δt considered. The numerical solution at $\Delta t = 2 \times 10^{-5}$ is selected as the reference solution to compute the errors. The L_2 error of the interface position at $t = 1$ considering various Δt is computed as

$$||\text{error}\,(\theta)\,|| = \left(\left| \boldsymbol{\varphi}(\boldsymbol{X}, 1)_{\Delta t = 2\times 10^{-5}} - \boldsymbol{\varphi}_{\text{h}}(\boldsymbol{X}, 1) \right|_{\ell^2} \right) \quad \forall \ \boldsymbol{X} \in \Gamma, \tag{6.34}$$

where $\theta = \theta(z)$ is the angle measured for each of the interface nodes with respect to the positive X-axis. This error is shown in Fig. 6.3a. Moreover, Figs. 6.3b and 6.4 depict the errors in the structural velocity, fluid velocity and fluid pressure respectively for the different discrete norms, Lebesgue L_∞, L_2, Hilbert H^1, and Sobolev W^1_∞. It is evident that the slope of the plot is 2, which shows a second-order of temporal accuracy for the quasi-monolithic formulation presented in Sect. 6.3.2.

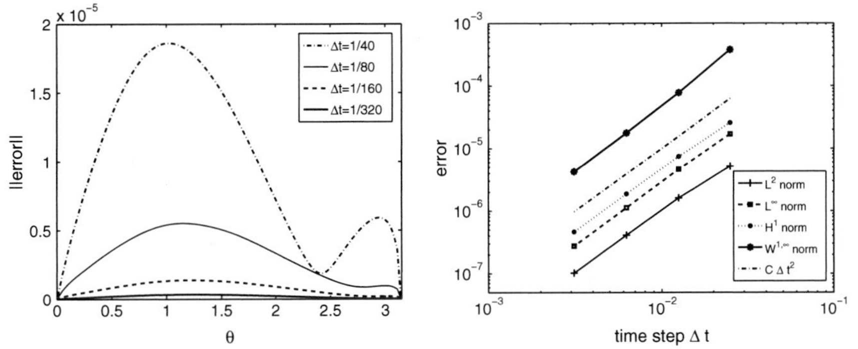

(a) (b)

Fig. 6.3 Elastic half-cylinder problem: **a** error in position along the interface at t = 1, where and $\theta = \theta(\boldsymbol{X})$ is the angle between $(\boldsymbol{X} - \boldsymbol{X}_0)$ and the positive X-axis. **b** 2nd order convergence of structural velocity in various norms

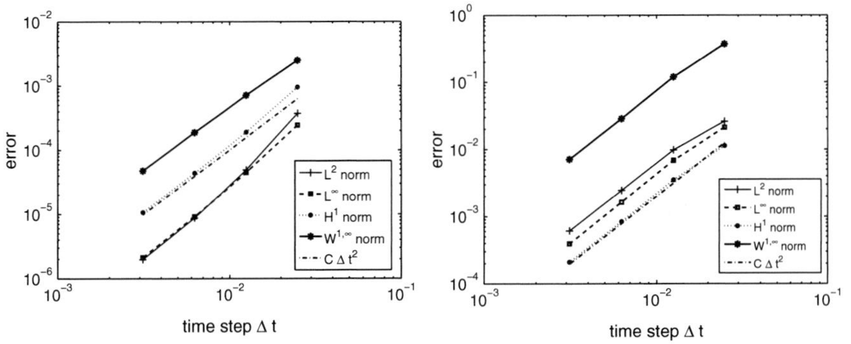

(a) (b)

Fig. 6.4 Second order convergence in various norms at time t=1 for fluid **a** velocity, **b** pressure

6.7.2 *Verification of 2D and 3D FSI Frameworks*

In the current section, a two-dimensional FSI benchmark given in [216] is performed to verify the 2D and 3D frameworks based on the formulations presented in Sects. 6.3 and 6.5 respectively. The benchmark consists of a stationary rigid cylinder with a flexible bar Ω^{s} attached to its end, placed in a channel flow $\Omega^{\mathrm{f}}(t)$, as shown in Fig. 6.5. The channel is modeled as a rectangular computational domain with length $L = 2.5$ and height $H = 0.41$. For the 3D framework, the width of the domain in the third dimension is considered as $W = 1$. The origin of the coordinate system lies at the bottom-left corner of the domain. The center of the cylinder is at $C = (0.2, 0.2)$ with $D = 0.1$ as its diameter. The length and thickness of the bar attached to the cylinder

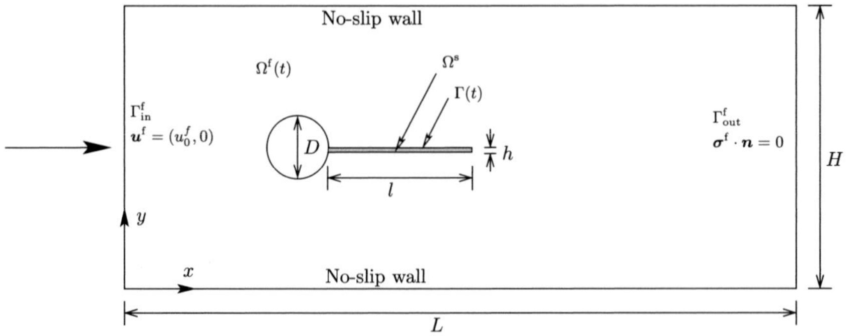

Fig. 6.5 Schematic and computational domain of cylinder-bar problem, and details of the boundary conditions and fluid-structure interface

are $l = 0.35$ and $h = 0.02$ respectively. The fluid-structure interface is denoted by Γ.

A parabolic velocity is prescribed at the inlet of the channel $\Gamma^{\mathrm{f}}_{\mathrm{in}}$ as

$$\boldsymbol{v}^{\mathrm{f}} = (U(y,t),0) = \frac{6.0}{0.1641} g(t)\bar{U} y\,(0.41 - y)\,, \tag{6.35}$$

where $\bar{U}$ represents the mean inlet velocity and $g(t)$ is defined as

$$g(t) = \begin{cases} 0 & t \leq 0 \\ 1 - \cos(\frac{\pi}{4}t) & t \in (0,2] \\ 1 & t > 2 \end{cases}\,. \tag{6.36}$$

A traction-free boundary condition is satisfied at the outlet boundary $\Gamma^{\mathrm{f}}_{\mathrm{out}}$. The stationary cylinder, the top and the bottom walls of the channel are assumed as rigid surfaces and a no-slip condition is imposed on these surfaces.

Two test cases are considered for the FSI benchmark of [216], consisting of a steady and unsteady FSI problem. For the steady FSI case, the non-dimensional parameters are given as $Re = 20$, $\bar{U} = 0.2$, $\rho^{\mathrm{s}}/\rho^{\mathrm{f}} = 1$, $\nu^{\mathrm{s}} = 0.4$ and $E/(\rho^{\mathrm{f}}\bar{U}^2) = 3.5 \times 10^4$. Similarly, the non-dimensional parameters for the unsteady FSI case are $Re = 100$, $\bar{U} = 1.0$, $\rho^{\mathrm{s}}/\rho^{\mathrm{f}} = 10$, $\nu^{\mathrm{s}} = 0.4$ and $E/(\rho^{\mathrm{f}}\bar{U}^2) = 1.4 \times 10^3$.

6.7.2.1 2D Quasi-Monolithic Formulation

We first consider the two-dimensional implementation of the quasi-monolithic formulation discussed in Sect. 6.3. The spatial discretization meshes considered for the two FSI cases are identical. The higher-order finite-element fluid and structural meshes consist of 9156 $\mathbb{P}_2/\mathbb{P}_1$ and 1064 $\mathbb{P}_2$ elements, respectively.

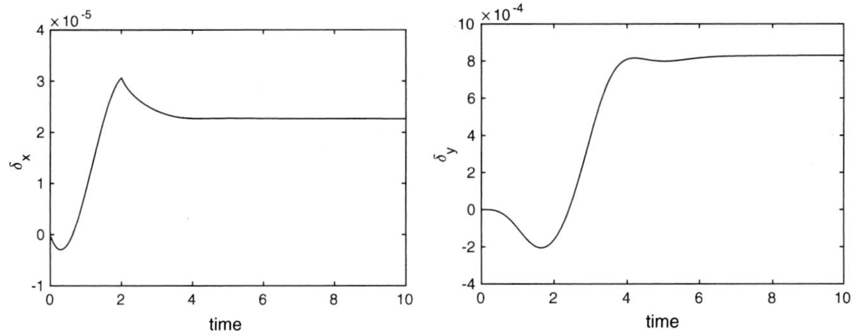

Fig. 6.6 Time history of dynamic forces drag (left) and lift (right) for cylinder bar problem simulated using 2D quasi-monolithic FSI solver with $\rho^{\mathrm{s}}/\rho^{\mathrm{f}} = 1$, $Re_D = 20$ and $E/\rho^{\mathrm{f}} U_0^2 = 3.5 \times 10^4$

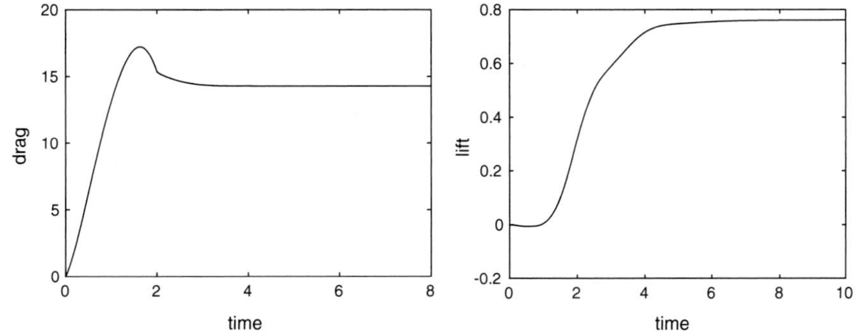

Fig. 6.7 Time history of dynamic forces drag (left) and lift (right) acting on a cylinder with flexible bar simulated using 2D quasi-monolithic FSI solver. Simulation parameter considered are $\rho^{\mathrm{s}}/\rho^{\mathrm{f}} = 1$, $Re_D = 20$ and $E/\rho^{\mathrm{f}} U_0^2 = 3.5 \times 10^4$

The first steady state FSI case is carried out at a constant $\Delta t = 0.005$. The evolution of the displacements of the tip of the trailing edge of the flexible bar in the streamwise and cross-stream directions are shown in Fig. 6.6. The steady state response of the FSI case can be clearly seen from the figure. The drag and lift forces on the cylinder-bar system have been plotted in Fig. 6.7. The results of the 2D computation for the steady state FSI problem are summarized in Table 6.1, where they are compared with the result from [216]. It is observed that the results agree very well with the benchmark data. The maximum difference between the computation results and the benchmark data is approximately 1%.

The unsteady FSI case at $Re = 100$ are performed considering a constant $\Delta t = 0.001$. The tip displacements of the trailing edge of the flexible bar and the fluid forces on the combined cylinder-bar system are shown in Figs. 6.8 and 6.9, respectively. The unsteady periodic results are summarized in Table 6.2 and compared with that of the benchmark in [216]. A good agreement with the results from the literature

Table 6.1 Summary of simulation results for the steady FSI problem simulated using the 2D quasi-monolithic formulation

Quantity	Present	Turek et al. [216]
$u_x \times 10^{-4}$	0.2265(0.22%)	0.2270
$u_y \times 10^{-3}$	0.830(1.09%)	0.821
Drag	14.28(0.07%)	14.27
Lift	0.761(0.26%)	0.763

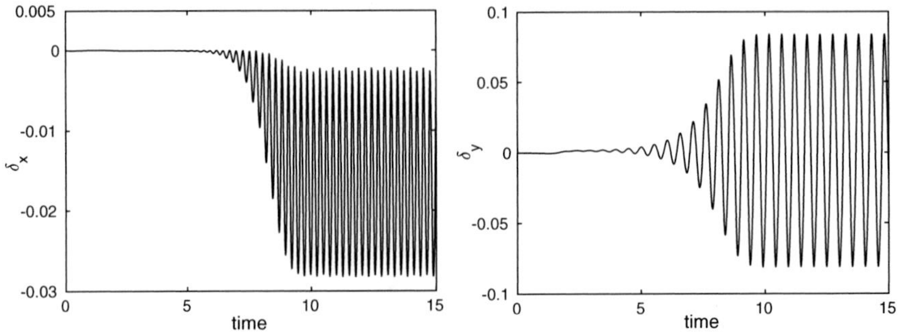

Fig. 6.8 Time history of trailing edge tip-displacements in lateral (left) and transverse (right) directions for cylinder bar problem simulated using 2D quasi-monolithic FSI solver with $\rho^s/\rho^f = 10$, $Re_D = 100$ and $E/\rho^f U_0^2 = 1.4 \times 10^3$

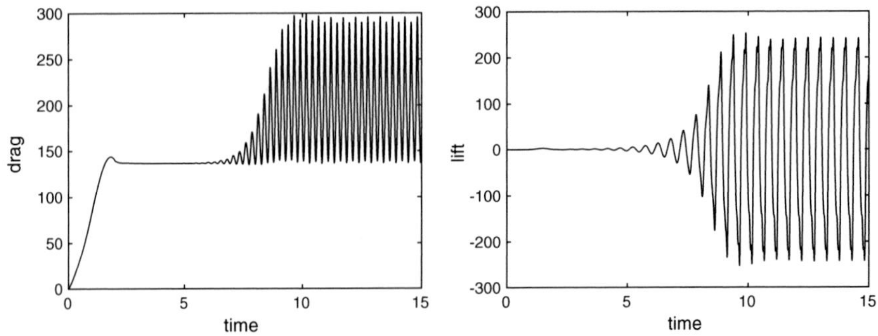

Fig. 6.9 Time history of dynamic forces drag (left) and lift (right) acting on a cylinder with flexible bar simulated using 2D quasi-monolithic FSI solver. Simulation parameter considered are $\rho^s/\rho^f = 10$, $Re_D = 100$ and $E/\rho^f U_0^2 = 1.4 \times 10^3$

is observed. The percentage difference in the unsteady component compared to the benchmark results is given in the parentheses.

Table 6.2 Summary of simulation results for the unsteady FSI problem simulated using the 2D quasi-monolithic formulation

Quantity	Present		Turek et al.[216]	
	Value	Frequency	Value	Frequency
$u_x \times 10^{-3}$	-15.25 ± 12.65 (1.68%)	3.8	-14.58 ± 12.44	3.8
$u_y \times 10^{-3}$	1.30 ± 82.43 (2.28%)	1.9	1.23 ± 80.60	2.0
Drag	215 ± 75.25 (2.03%)	3.8	208 ± 73.75	3.8
Lift	2.70 ± 239 (2.05%)	2.0	-0.88 ± 234.20	2.0

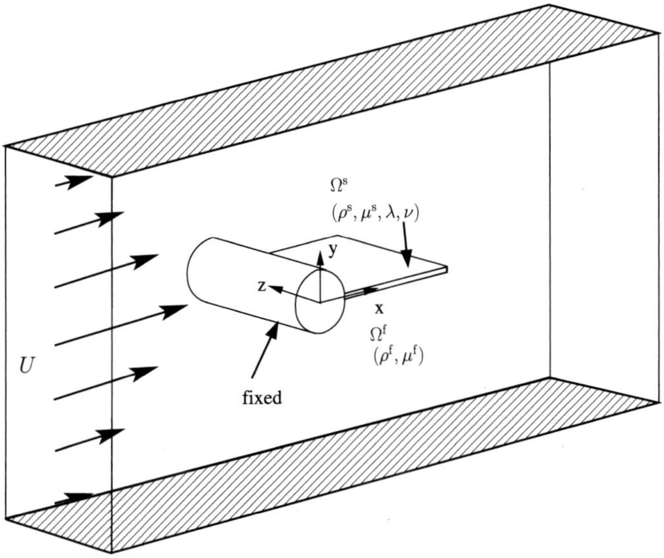

Fig. 6.10 Schematic of 3D rigid fixed cylinder attached with a flexible plate in a channel flow

6.7.2.2 3D Fully-Stabilized Quasi-Monolithic Formulation

Next, we consider the three-dimensional fully-stabilized quasi-monolithic formulation discussed in Sect. 6.5. Here, we perform pseudo-2D computations taking the three-dimensional mesh to be one element thick in the third dimension. A typical three-dimensional computational domain for the problem is shown in Fig. 6.10. The boundary conditions are similar to the set-up in 2D computational domain in Fig. 6.5, but also includes the periodic condition on the sides in the third dimension.

The fluid mesh comprises of 15348 six-node wedge elements and 16342 nodes. Identical order elements are used for both fluid velocity and pressure interpolations in

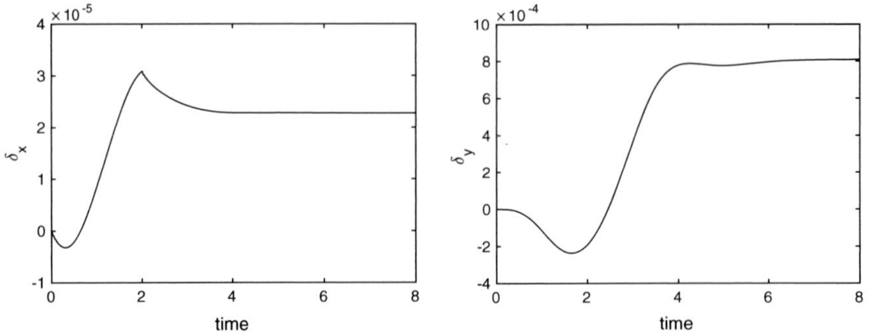

Fig. 6.11 Time history of trailing edge tip-displacements in lateral (left) and transverse (right) directions for cylinder bar problem simulated using fully stabilized 3D quasi-monolithic formulation with $\rho^{\mathrm{s}}/\rho^{\mathrm{f}} = 1$, $Re_D = 20$ and $E/\rho^{\mathrm{f}} U_0^2 = 3.5 \times 10^4$

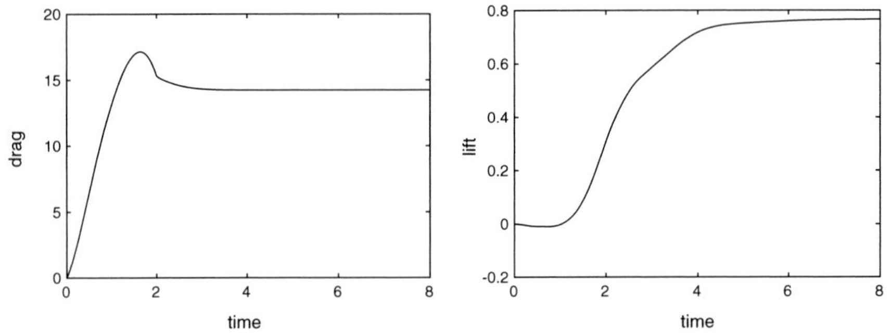

Fig. 6.12 Time history of dynamic forces drag (left) and lift (right) acting on a cylinder with flexible bar for the numerical simulations carried out using fully stabilized 3D quasi-monolithic formulation. Simulation parameter considered are $\rho^{\mathrm{s}}/\rho^{\mathrm{f}} = 1$, $Re_D = 20$ and $E/\rho^{\mathrm{f}} U_0^2 = 3.5 \times 10^4$

the fully-stabilized formulation. The computation results for the steady and unsteady cases are shown in Figs. 6.11, 6.12, 6.13and 6.14. The results are also summarized in Tables 6.3 and 6.4 and compared to the benchmark results for the cases $Re = 20$ and 100, respectively. The percentage error between the computation and the benchmark results is shown in the parentheses, where we observe a maximum error of less than 2%. Therefore, the 3D fully-stabilized quasi-monolithic formulation provides a good agreement with the benchmark solutions in [216].

6.7.3 Verification for Flapping Dynamics

After the verification of the quasi-monolithic FSI framework for the benchmark FSI problem, we demonstrate the framework for two-dimensional flapping dynamics

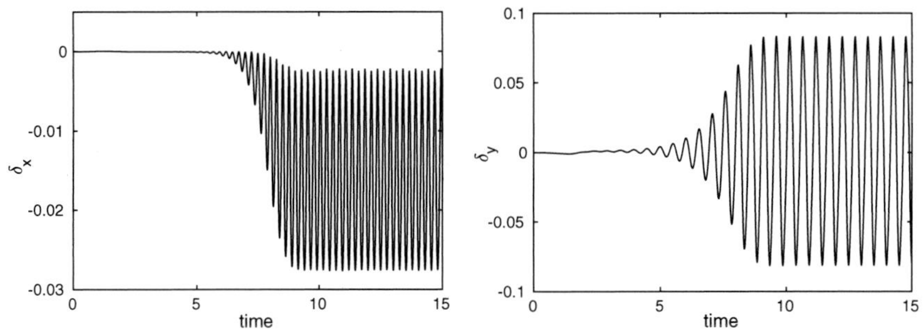

Fig. 6.13 Time history of trailing edge tip-displacements in lateral (left) and transverse (right) directions for cylinder bar problem with $\rho^s/\rho^f = 10$, $Re_D = 100$ and $E/\rho^f U_0^2 = 1.4 \times 10^3$

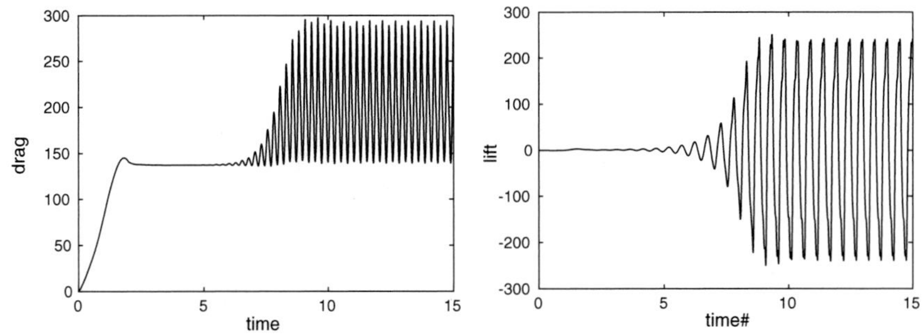

Fig. 6.14 Time history of dynamic forces drag (left) and lift (right) for cylinder bar problem with $\rho^s/\rho^f = 10$, $Re_D = 100$ and $E/\rho^f U_0^2 = 1.4 \times 10^3$

Table 6.3 Summary of simulation results for the steady FSI problem simulated using the fully stabilized 3D quasi-monolithic formulation

Quantity	Present	Turek et al. [216]
$u_x \times 10^{-3}$	0.0227(0%)	0.0227
$u_y \times 10^{-3}$	0.810(1.34%)	0.821
Drag	14.24(0.21%)	14.27
Lift	0.754(1.18%)	0.763

Table 6.4 Summary of simulation results for the unsteady FSI problem simulated using the fully stabilized 3D quasi-monolithic formulation

Quantity	Present		Turek et al.[216]	
	Value	Frequency	Value	Frequency
$u_x \times 10^{-3}$	$-14.90 \pm 12.60(1.26\%)$	3.8	-14.58 ± 12.44	3.8
$u_y \times 10^{-3}$	$0.80 \pm 81.06(0.82\%)$	1.9	1.23 ± 80.60	2.0
Drag	$216 \pm 75.0(1.69\%)$	3.8	208 ± 73.75	3.8
Lift	$0.50 \pm 239.05(1.77\%)$	2.0	-0.88 ± 234.20	2.0

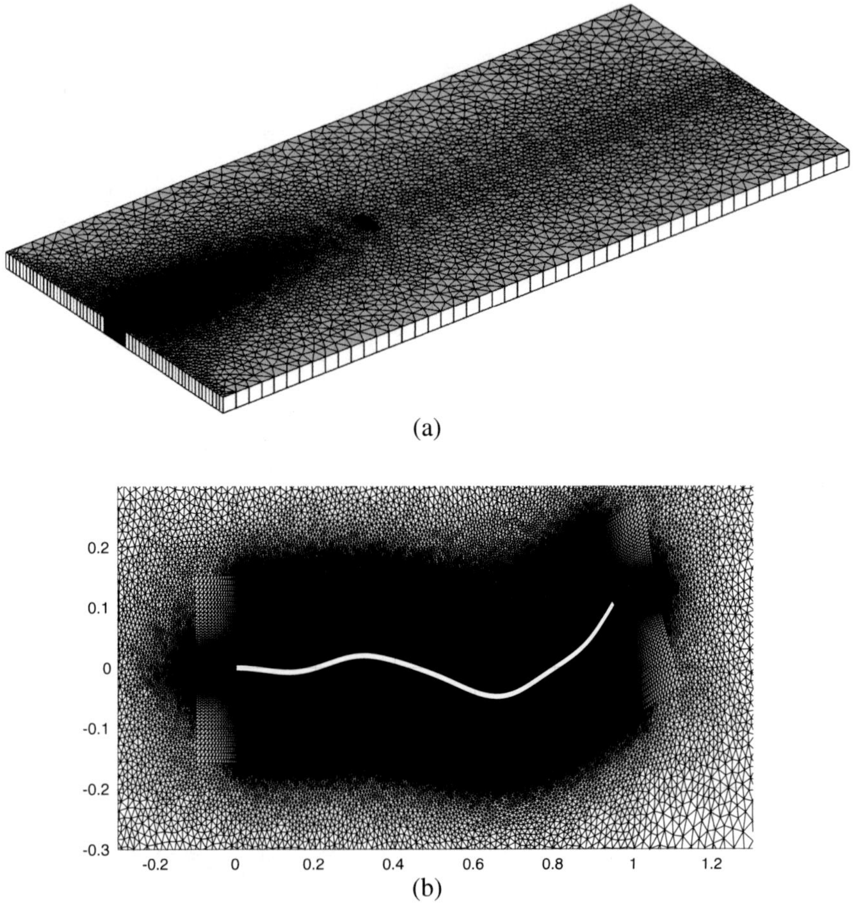

Fig. 6.15 3D finite-element fluid mesh considered for the validation of the 3D coupled fluid-structure solver for $m^* = 0.1,\ Re = 1000$ and $K_B = 0.0001$: **a** isometric view, **b** closeup view of the fluid boundary layer mesh around the foil while performing flapping motion

of a thin flexible foil of length L subjected to a uniform axial flow. The leading edge of the foil is clamped, while the trailing edge is free to undergo flapping. The non-dimensional parameters for the demonstration are mass-ratio $m^* = 0.1$, Reynolds numbers $Re = 1000$ and non-dimensional bending rigidity $K_B = 0.0001$. The results are compared with the computational data in [51].

The 2D formulation is verified by considering a rectangular computational domain of size $[-2L, 20L] \times [-5L, 5L]$. The origin is located at the leading edge center of the foil. The thickness h of the foil is assumed very less than its length, $h \ll L$ and $h/L = 0.01$. A uniform velocity $\boldsymbol{v}^{\mathrm{f}} = (U_0, 0)$ is applied at the inlet boundary $\Gamma^{\mathrm{f}}_{\mathrm{in}}$ and a traction-free condition is satisfied at the outlet $\Gamma^{\mathrm{f}}_{\mathrm{out}}$. The freestream velocity condition $\boldsymbol{v}^{\mathrm{f}} = (U_0, 0)$ is given at the top and bottom sides of the computational

Table 6.5 Numerical comparison of the fully-stabilized 3D quasi-monolithic solver against reference results of [51, 139] for $m^* = 0.1,\ Re = 1000$ and $K_B = 0.0001$

Quantity	Connell et al. [51] (2D simulations)	2D solver	3D solver
A/L	0.2540	0.2440	0.2420
$\delta_y^{\mathrm{rms}}/L$	–	0.0846	0.0826
fL/U_0	0.9152	0.9495	0.955

domain. The mesh consists of higher-order finite-element fluid and structural meshes comprising of 15699 and 792 triangular $\mathbb{P}_2/\mathbb{P}_1$ and $\mathbb{P}_2$ respectively.

The 3D formulation is verified by constructing a pseudo-2D domain, extruding the 2D mesh by a single element in the third dimension. Periodic conditions are imposed in the spanwise direction. An isometric view of the 3D fluid mesh consisting of 116461 six-node wedge elements and 117402 nodes is depicted in Fig. 6.15a. The close-up view of the fluid boundary layer mesh around the foil is also shown in Fig. 6.15b.

The results obtained from the 2D and 3D formulations for the non-dimensional parameters $m^* = 0.1,\ Re = 1000$ and $K_B = 0.0001$ are summarized in Table 6.5. The flapping amplitude, frequency and trailing edge root mean square displacement are compared with the results from the literature [51]. A reasonable agreement is found among the flapping properties from both the 2D and 3D formulations.

To summarize, we discussed monolithic strategy for dealing with coupled fluid-structure interaction problems in this chapter. Different quasi-monolithic formulations were presented and verified with standard FSI benchmarks available in the literature. Finally, flapping dynamics of a flexible foil was demonstrated with the help of the underlying presented formulations. Next, we deal with the second type of strategy for FSI problems, i.e., partitioned type of techniques.

Acknowledgements Some parts of this Chapter have been taken from the PhD thesis of Pardha S. Gurugubelli carried out at the National University of Singapore and supported by the Ministry of Education, Singapore.

Chapter 7
Partitioned Fluid-Structure Interaction Methods

7.1 Introduction

In the previous chapters, we formulated the finite element discretization of the fluid-structure interaction equations and looked into the monolithic type of coupling methods. An essential requirement of such a coupled system is the accurate description of the fluid–structure interface. In the monolithic technique, this feature is naturally taken care by the formulation as the dynamic equilibrium is easily satisfied. In this chapter, we look into the treatment of the fluid-structure interface and interface conditions for partitioned type of methods. While partitioning refers to the decomposition into the fluid and structural subdomains at a given time level, staggering is the sequence in which time integration is carried out along the interface between the subdomains, by satisfying the kinematic and dynamic equilibrium conditions at the interface.

For such problems under the partitioned staggered approach, one faces two key challenges. The first challenge is how to deal with the displacement and traction transfer across different mesh discretizations in the fluid and structural subdomains. The discretizations of the subdomains can be generally non-matching. The issue then arises into how to couple the subdomains in space while satisfying the interface equilibrium conditions in an accurate and conservative manner. These give rise to different spatial coupling techniques at the fluid-structure interface.

The second challenge is the stability, accuracy and convergence of the staggered time integration scheme at the fluid-structure interface which is used to couple the different subdomains in time. With regard to temporal coupling, the techniques can be categorized into (1) loosely coupled, and (2) strongly coupled schemes. Loosely coupled schemes involve explicit time integration where the fluid and structural subdomains are evolved in time once at each time step. These schemes are simple and have the advantage of low computational cost, but they often suffer from numerical instabilities and temporal inaccuracy. The latter involves predictor-corrector sub-iterations to ensure convergence of the properties at the interface at each time step.

R. K. Jaiman and V. Joshi, *Computational Mechanics of Fluid-Structure Interaction*,
https://doi.org/10.1007/978-981-16-5355-1_7

In this chapter, we discuss the various types of spatial and temporal coupling techniques helpful in partitioned methods for fluid-structure coupling. We carry out error analysis for some of the coupling methods and the verification with results from the literature. Finally, we demonstrate the partitioned staggered coupling technique for a problem of vortex-induced vibration of an offshore riser subjected to uniform flow current.

7.2 Spatial Coupling Techniques

This section deals with the different kinds of coupling techniques based on the spatial discretization of the fluid-structure interface. It discusses how the displacement and tractions can be transferred across the interface in a conservative and accurate manner. First, we define some notations to be used in the upcoming subsections and then dive into the details of the techniques.

Let N_i^{f} and N_j^{s} denote the finite element shape functions associated with node i and j of the fluid and structural mesh respectively at the fluid-structure interface Γ. The approximate tractions at the corresponding nodes of the discrete mesh are given by $\tilde{\boldsymbol{t}}_i^{\mathrm{f}} \in L^2(\Omega^{\mathrm{f}}(t))$ and $\tilde{\boldsymbol{t}}_j^{\mathrm{s}} \in L^2(\Omega^{\mathrm{s}})$. The continuum fields for the traction are thus given by $\boldsymbol{t}^{\mathrm{f}}$ and $\boldsymbol{t}^{\mathrm{s}}$ as

$$\boldsymbol{t}^{\mathrm{f}} \approx \sum_{i=1}^{m^{\mathrm{f}}} N_i^{\mathrm{f}} \tilde{\boldsymbol{t}}_i^{\mathrm{f}}, \tag{7.1}$$

$$\boldsymbol{t}^{\mathrm{s}} \approx \sum_{j=1}^{m^{\mathrm{s}}} N_j^{\mathrm{s}} \tilde{\boldsymbol{t}}_j^{\mathrm{s}}, \tag{7.2}$$

where m^{f} and m^{s} denote the number of fluid and structural nodes on the interface respectively and $\tilde{\boldsymbol{t}}_i^{\mathrm{f}}$ and $\tilde{\boldsymbol{t}}_j^{\mathrm{s}}$ denote the discrete values of the traction for the fluid and structural interfaces respectively.

Based on the type of spatial coupling, the techniques can be categorized into: (1) point-to-point mapping, (2) point-to-element projection, and (3) common-refinement projection.

7.2.1 *Point-to-Point Mapping*

In the case of interaction of fluid and multiple components of a flexible multibody system, it is essential that there is accurate and conservative data transfer across the fluid-structure interface. For problems considering large deformations of the structure, global conservation methods such as point-to-point mapping (e.g., radial basis function with compact support) are preferred which involve interpolation with the help of scattered data points. The benefit with such mapping is that it does not require connectivity of the mesh while transferring the data across the fluid-structure interface.

Remark 7.1 The global conservation property is satisfied by equating the virtual work done by the forces in the structural domain (δW^{s}) by that carried out by the fluid loads (δW^{f}) at the fluid-structure interface. The interpolation with the help of radial basis function is constructed in such a manner that the global conservation is satisfied across the interface. Let the interpolation of the fluid displacement at the interface be given by $\boldsymbol{u}^{\text{f}} = \boldsymbol{H}\boldsymbol{u}^{\text{s}}$, where $\boldsymbol{u}^{\text{s}}$ is the structural displacement to be transferred and $\boldsymbol{H}$ is some interpolation matrix. The force transfer from the fluid nodes to the structural nodes is represented as $\boldsymbol{f}^{\text{s}} = \boldsymbol{H}^T \boldsymbol{f}^{\text{f}}$, where $\boldsymbol{f}^{\text{s}}$ and $\boldsymbol{f}^{\text{f}}$ denote the contributions of the forces from the structural and fluid domains, respectively. The virtual work done at the fluid-structure interface can be expressed as

$$\delta W^{\text{f}} = \delta \boldsymbol{u}^{\text{f}} \cdot \boldsymbol{f}^{\text{f}} = (\delta \boldsymbol{u}^{\text{f}})^T \boldsymbol{f}^{\text{f}}, \tag{7.3}$$

$$= (\delta \boldsymbol{u}^{\text{s}})^T \boldsymbol{H}^T \boldsymbol{f}^{\text{f}}, \tag{7.4}$$

$$= (\delta \boldsymbol{u}^{\text{s}})^T \boldsymbol{f}^{\text{s}}, \tag{7.5}$$

$$= \delta \boldsymbol{u}^{\text{s}} \cdot \boldsymbol{f}^{\text{s}} = \delta W^{\text{s}}, \tag{7.6}$$

Thus, if an approximation matrix $\boldsymbol{H}$ exists, it can be utilized for both displacement and force interpolations across the interface [24, 89]. Here, we discuss the construction of the interpolation matrix $\boldsymbol{H}$ with the help of radial basis functions.

7.2.1.1 Review of Radial Basis Functions

We next review the concept of radial basis functions. A radial basis function (RBF) is radially symmetric and forms the basis of the interpolation matrix. A d-variate interpolation function $g(\boldsymbol{x})$ from a given scattered data $\{g_1, g_2, \ldots, g_N\}$ at the arbitrary distinct source locations $X^{\text{c}} = \{\boldsymbol{x}_1^{\text{c}}, \boldsymbol{x}_2^{\text{c}}, \ldots, \boldsymbol{x}_N^{\text{c}}\} \subseteq \mathbb{R}^d$ is given as

$$g(\boldsymbol{x}) = \sum_{j=1}^{N} \alpha_j \phi(||\boldsymbol{x} - \boldsymbol{x}_j^{\text{c}}||), \tag{7.7}$$

where $||\cdot||$ is the Euclidean norm and α_j are the weights for the basis functions. If the weights are known, the interpolated values at the requested target data points $X^{\text{t}} = \{\boldsymbol{x}_1^{\text{t}}, \boldsymbol{x}_2^{\text{t}}, \ldots, \boldsymbol{x}_M^{\text{t}}\}$ are thus given by

$$g(\boldsymbol{x}_i^{\text{t}}) = \sum_{j=1}^{N} \alpha_j \phi(||\boldsymbol{x}_i^{\text{t}} - \boldsymbol{x}_j^{\text{c}}||), \ 1 \leq i \leq M, \tag{7.8}$$

where the matrix $\phi(||\boldsymbol{x}_i^{\text{t}} - \boldsymbol{x}_j^{\text{c}}||)$ is also known as kernel matrix for the interpolation.

The coefficients α_j are determined by the condition, $g(\boldsymbol{x}_j^{\mathrm{c}}) = g_j,\ 1 \leq j \leq N$. Therefore, the interpolation recovers to the exact values at the scattered control points, i.e., for $\boldsymbol{x}^{\mathrm{t}} = \boldsymbol{x}^{\mathrm{c}}$,

$$g_j = g(\boldsymbol{x}_j^{\mathrm{c}}) = \sum_{j=1}^{N} \alpha_j \phi(||\boldsymbol{x}_i^{\mathrm{c}} - \boldsymbol{x}_j^{\mathrm{c}}||),\ 1 \leq i \leq N, \tag{7.9}$$

which is a system of linear equations with α_j as unknowns and let $(\boldsymbol{A}_{\mathrm{cc}})_{ij} = \phi(||\boldsymbol{x}_i^{\mathrm{c}} - \boldsymbol{x}_j^{\mathrm{c}}||)$. To have a guaranteed solvability for these coefficients, the symmetric matrix $\boldsymbol{A}_{\mathrm{cc}}$ should be positive definite which depends on the positive-definiteness of the radial basis function $\phi(||\cdot||)$.

Remark 7.2 The positive definite property can be imparted to a radial basis function by adding polynomials to the interpolant as

$$g(\boldsymbol{x}^{\mathrm{t}}) = \sum_{j=1}^{N} \alpha_j \phi(||\boldsymbol{x}^{\mathrm{t}} - \boldsymbol{x}_j^{\mathrm{c}}||) + q(\boldsymbol{x}^{\mathrm{t}}), \tag{7.10}$$

where $q(\boldsymbol{x}^{\mathrm{t}})$ is a polynomial. Now, with additional degrees of freedom of the coefficients of the polynomial, the condition $\sum_{j=1}^{N} \alpha_j p(\boldsymbol{x}_j^{\mathrm{t}}) = 0$, has to be satisfied for unique solvability of the coefficients, for any polynomial $p(\boldsymbol{x})$ with degree less than or equal to the degree of $q(\boldsymbol{x})$. A minimal degree of the polynomial $q(\boldsymbol{x}^{\mathrm{t}})$ depends on the radial basis function.

The radial basis function can be selected among many alternatives, namely Gaussian, multiquadric, polyharmonic, etc., satisfying certain conditions for the positive definite property. Besides the positive definiteness property, the property of localization or compact support is also beneficial for a radial basis function. Such localization ensures that the system matrix is sparse, which is helpful in its inversion. A good example of functions having compact support is the class of Wendland's function, which are always positive definite up to a maximal space dimension and have smoothness C^{2k}. They are of the form

$$\phi(r) = \begin{cases} p(r), & 0 \leq r \leq 1, \\ 0, & r > 1, \end{cases} \tag{7.11}$$

where $p(r)$ is a univariate polynomial. Considering a truncated power function $\phi_l(r) = (1-r)^l$ which satisfies the positive definiteness property for $l \geq \lfloor d/2 \rfloor + 1$, d being the spatial dimensions and $\lfloor x \rfloor$ represents an integer n such that $n \leq x < n+1$. Using this definition, a new class of functions is constructed $\phi_{l,k}(r)$ which are positive definite in the dimension d and are C^{2k} with degree $\lfloor d/2 \rfloor + 3k + 1$ [237]. If one defines an operator $I(f)(r) = \int_r^{\infty} f(t)t dt$, these new class of functions can be constructed via integration as $\phi_{l,k}(r) = I^k \phi_l(r)$, which can be represented by

$$\phi_{l,k}(r) = \sum_{n=0}^{k} \beta_{n,k} r^n \phi_{l+2k-n}(r), \tag{7.12}$$

where $\beta_{0,0} = 1$ and

$$\beta_{j,k+1} = \sum_{n=j-1}^{k} \beta_{n,k} \frac{[n+1]_{n-j+1}}{(l+2k-n+1)_{n-j+2}}, \tag{7.13}$$

where $[q]_{-1} = 1/(q+1)$, $[q]_0 = 1$, $[q]_l = q(q-1)\dots(q-l+1)$ and $(q)_0 = 1$, $(q)_l = q(q+1)\dots(q+l-1)$. It has also been shown in [237] that a function $\Phi_{l,k}(x) = \phi_{l,k}(||x||)$ has strictly positive Fourier transform and produces a positive definite radial basis function, except when $d = 1$ and $k = 0$. Furthermore, it has been proven that the function is $2k$ times continuously differentiable and their polynomial degree is minimal for a given smoothness, and they are related to certain Sobolev spaces [237, 238, 241]. Considering Wendland's C^2 function for the three-dimensional study ($d = 3$) with $k = 1$ and choosing $l = \lfloor d/2 \rfloor + k + 1$ as $\phi_{3,1}(r)$, the function can be written as

$$\phi(||\boldsymbol{x}||) = \begin{cases} (1-||\boldsymbol{x}||)^4(1+4||\boldsymbol{x}||), & 0 \le ||\boldsymbol{x}|| \le 1, \\ 0, & ||\boldsymbol{x}|| > 1. \end{cases} \tag{7.14}$$

7.2.1.2 Application to Fluid-Structure Interaction Framework

For the three-dimensional fluid-structure interaction problems and the use of Wendland's C^2 function, we can employ a linear polynomial $q(\boldsymbol{x}) = \lambda_0 + \lambda_1 x + \lambda_2 y + \lambda_3 z$, which is exactly reproduced and recovers any rigid body translation and rotation in the mapping. Note that the radial basis function can be scaled by a compact support radius r which gives the advantage of covering enough number of interpolation points depending on the application. This scaling, however, does not have any effect on the positive definiteness and the compact support properties of the function. The scaled Wendland's C^2 function is given by

$$\phi(||\boldsymbol{x}||/r) = \begin{cases} (1-||\boldsymbol{x}||/r)^4(1+4||\boldsymbol{x}||/r), & 0 \le ||\boldsymbol{x}||/r \le 1, \\ 0, & ||\boldsymbol{x}||/r > 1. \end{cases} \tag{7.15}$$

Next, we discuss how the interpolation matrix obtained from radial basis functions can be used to obtain the displacements and fluid tractions across the fluid-structure interface. Suppose the structural displacement field is known at the fluid-structure interface as $\boldsymbol{u}_I^{\mathrm{s}}$ and we want to interpolate the fluid displacement at the interface $\boldsymbol{u}_I^{\mathrm{f}}$. Based on the RBF interpolation, the coefficients β_j can be found by the fact that

$\boldsymbol{u}_I^{\mathrm{s}} = \boldsymbol{C}_{\mathrm{ss}}\boldsymbol{\beta}$, where $\boldsymbol{C}_{\mathrm{ss}}$ is given as

$$\boldsymbol{C}_{\mathrm{ss}} = \begin{bmatrix} 0 & 0 & 0 & 0 & 1 & 1 & \dots & 1 \\ 0 & 0 & 0 & 0 & x_1^{\mathrm{c}} & x_2^{\mathrm{c}} & \dots & x_N^{\mathrm{c}} \\ 0 & 0 & 0 & 0 & y_1^{\mathrm{c}} & y_2^{\mathrm{c}} & \dots & y_N^{\mathrm{c}} \\ 0 & 0 & 0 & 0 & z_1^{\mathrm{c}} & z_2^{\mathrm{c}} & \dots & z_N^{\mathrm{c}} \\ 1 & x_1^{\mathrm{c}} & y_1^{\mathrm{c}} & z_1^{\mathrm{c}} & \phi_{1,1}^{\mathrm{c,c}} & \phi_{1,2}^{\mathrm{c,c}} & \dots & \phi_{1,N}^{\mathrm{c,c}} \\ 1 & x_2^{\mathrm{c}} & y_2^{\mathrm{c}} & z_2^{\mathrm{c}} & \phi_{2,1}^{\mathrm{c,c}} & \phi_{2,2}^{\mathrm{c,c}} & \dots & \phi_{2,N}^{\mathrm{c,c}} \\ . & . & . & . & . & . & . & . \\ . & . & . & . & . & . & . & . \\ . & . & . & . & . & . & . & . \\ 1 & x_N^{\mathrm{c}} & y_N^{\mathrm{c}} & z_N^{\mathrm{c}} & \phi_{N,1}^{\mathrm{c,c}} & \phi_{N,2}^{\mathrm{c,c}} & \dots & \phi_{N,N}^{\mathrm{c,c}} \end{bmatrix}, \tag{7.16}$$

where $\phi_{i,j}^{\mathrm{c,c}} = \phi(||\boldsymbol{x}_i^{\mathrm{c}} - \boldsymbol{x}_j^{\mathrm{c}}||)$, where $\boldsymbol{x}_i^{\mathrm{c}}$ are the coordinates of the points on the structural interface at Γ. The displacement field of the fluid nodes at the interface can be interpolated by $\boldsymbol{u}_I^{\mathrm{f}} = \boldsymbol{A}_{\mathrm{fs}}\boldsymbol{C}_{\mathrm{ss}}^{-1}\boldsymbol{u}_I^{\mathrm{s}}$, where $\boldsymbol{A}_{\mathrm{fs}}$ is given by

$$\boldsymbol{A}_{\mathrm{fs}} = \begin{bmatrix} 1 & x_1^{\mathrm{t}} & y_1^{\mathrm{t}} & z_1^{\mathrm{t}} & \phi_{1,1}^{\mathrm{t,c}} & \phi_{1,2}^{\mathrm{t,c}} & \dots & \phi_{1,N}^{\mathrm{t,c}} \\ 1 & x_2^{\mathrm{t}} & y_2^{\mathrm{t}} & z_2^{\mathrm{t}} & \phi_{2,1}^{\mathrm{t,c}} & \phi_{2,2}^{\mathrm{t,c}} & \dots & \phi_{2,N}^{\mathrm{t,c}} \\ . & . & . & . & . & . & . & . \\ . & . & . & . & . & . & . & . \\ . & . & . & . & . & . & . & . \\ 1 & x_M^{\mathrm{t}} & y_M^{\mathrm{t}} & z_M^{\mathrm{t}} & \phi_{M,1}^{\mathrm{t,c}} & \phi_{M,2}^{\mathrm{t,c}} & \dots & \phi_{M,N}^{\mathrm{t,c}} \end{bmatrix}, \tag{7.17}$$

where $\phi_{i,j}^{\mathrm{t,c}} = \phi(||\boldsymbol{x}_i^{\mathrm{t}} - \boldsymbol{x}_j^{\mathrm{c}}||)$, $\boldsymbol{x}_i^{\mathrm{t}}$ being the coordinates of the fluid points on the interface. Therefore, the mapping of the structural displacements to the fluid displacements can be described as $\boldsymbol{u}_I^{\mathrm{f}} = \boldsymbol{A}_{\mathrm{fs}}\boldsymbol{C}_{\mathrm{ss}}^{-1}\boldsymbol{u}^{\mathrm{s}} = \boldsymbol{H}\boldsymbol{u}_I^{\mathrm{s}}$.

The same technique can be applied for the transfer of tractions from fluid points to the structural points at the interface by considering the control points to be that of the fluid side of the interface and target points as that of the structural side. Similar to the finite element method, the local support of RBF makes the system matrices to be sparse along the interface. Moreover, the Wendland's function provides positive definite property for the matrix computation.

Apart from the transfer of data along the fluid-structure interface, the fluid mesh nodes inside the fluid domain can also be displaced based on the radial basis function mapping where the matrix $\boldsymbol{A}_{\mathrm{fs}}$ is constructed based on the volumetric nodes of the fluid domain rather than just the fluid-structure interface carried out previously, i.e., the target nodes now contain the volumetric data of the fluid domain.

7.2.1.3 Convergence of RBF Mapping: Static Load Data Transfer

In this section, we carry out a systematic convergence analysis of the radial basis function mapping approach described in the previous section to transfer data across

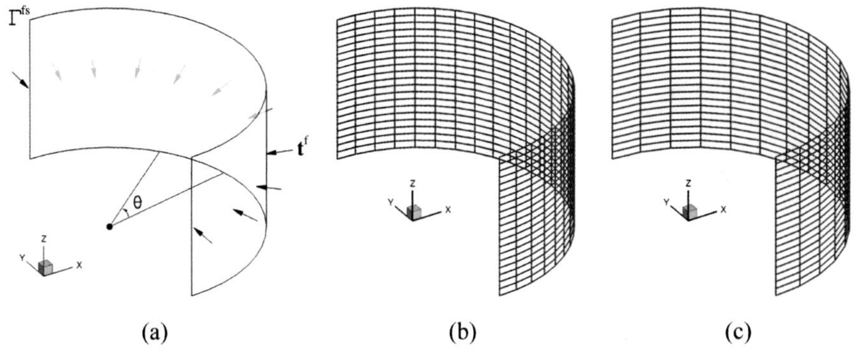

Fig. 7.1 Convergence test for radial basis function mapping: **a** schematic of the traction transfer from the fluid to the structural meshes along Γ^{fs}, **b** fluid and **c** structural meshes for the case $A^f/A^s = 0.67$ and $n_c^s = 16$

the fluid-structure interface. To accomplish this, we consider a semi-circular surface as the fluid-structure interface with varying discretization on the fluid and the structural sides of the interface as shown in Fig. 7.1. In the schematic, the outer domain consists of fluid (Ω^f) and the inner domain is the structure (Ω^s). A fluid traction acts on the fluid side of the fluid-structure interface Γ^{fs} as

$$\mathbf{t}^f = \mathbf{t}^f(\theta, z) = -\left(\frac{1}{2}\rho^f U_\infty^2 (1 - 4\sin^2\theta) + \rho^f g z\right) \begin{pmatrix} 0.5 \cos\theta \\ 0.5 \sin\theta \\ 0 \end{pmatrix}, \tag{7.18}$$

where $\rho^f = 1000$ kg/m^3, $U_\infty = 1.0$ m/s, $g = 9.81$ m/s^2, z is the Z-coordinate and $\theta \in [-\pi/2, \pi/2]$ is the angle shown in Fig. 7.1a. This prescribed load represents the static pressure along the Z-direction as a result of potential flow around a cylinder. We only consider the downstream half of the cylinder for the convergence study. The support radius r for the radial basis function is selected as 2.

We employ two ratios of mismatch between the fluid and the structural meshes. The discretization in the Z-direction is kept constant at the number of elements of $n_z^f = n_z^s = 20$. The number of elements along the circumference of the semi-circle is varied on the fluid (n_c^f) as well as structural (n_c^s) meshes. The element area can thus be written as $A^f = \pi R z/(n_c^f n_z^f)$ and $A^s = \pi R z/(n_c^s n_z^s)$, where $R = 1$ and $z = 1$ are the radius and height of the semi-circular cylinder, respectively. The refinement is carried out such that the area mismatch is $A^f/A^s \in [0.67, 2]$. The refinement degree of the structural mesh is $n_c^s \in [16, 32, 64, 128, 256]$ which is kept equivalent for the two cases of mismatch considered. The different meshes on the fluid and structural sides at Γ^{fs} are shown in Figs. 7.1b, c for $A^f/A^s = 0.67$ and $n_c^s = 16$.

The interpolated traction values at the structural nodes are shown for the representative case of $A^f/A^s = 0.67$ and $n_c^s = 16$ in Fig. 7.2. The relative error in the

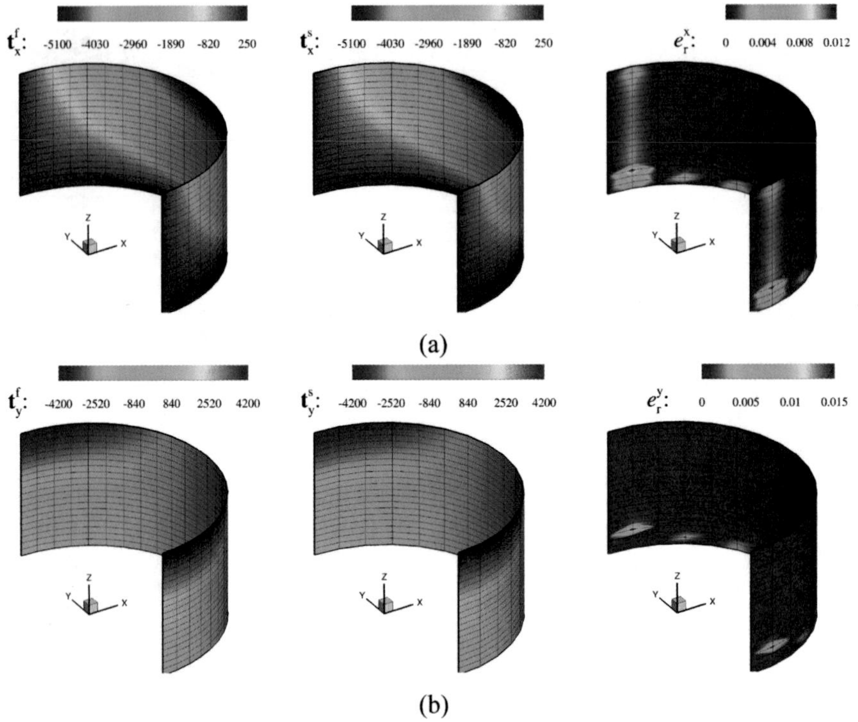

Fig. 7.2 Comparison of interpolation via radial basis function mapping for $A^f/A^s = 0.67$ and $n_c^s = 16$ for the fluid (left) and structural (middle) meshes for traction in **a** X-direction, and **b** Y-direction. The relative error of the traction values are also shown for the structural mesh (right)

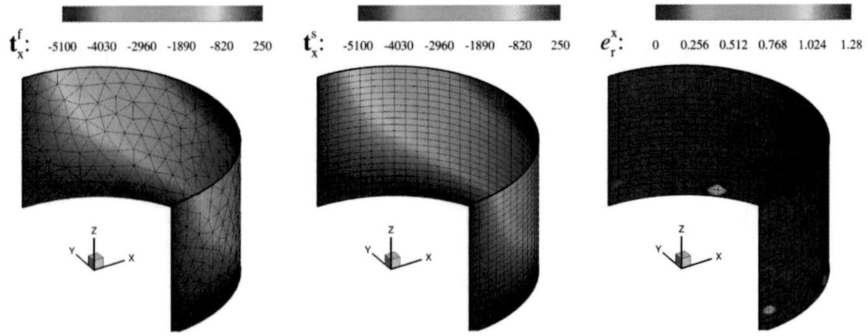

Fig. 7.3 Comparison of interpolation via radial basis function mapping for $A^f/A^s = 2$ with unstructured triangular mesh for the fluid (left) and structured quadrilateral mesh for the solid (middle) for traction in X-direction when $n_z^s = 20$ and $n_c^s = 40$. The relative error of the traction values are also shown for the structural mesh (right)

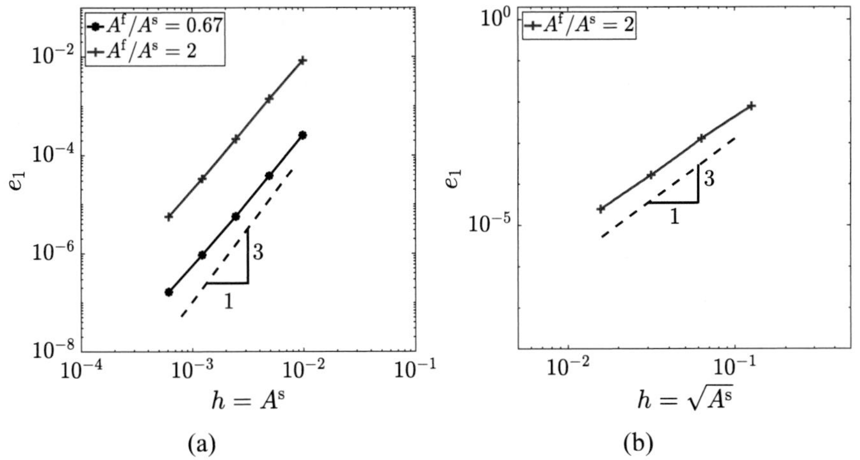

Fig. 7.4 Error convergence for the radial basis function interpolation method for varying mesh mismatch on the fluid and structural sides of the fluid-structural interface for **a** structured grid, and **b** unstructured grid with different shapes

interpolation is quantified as

$$e_r = \frac{|\mathbf{t}^s - \mathbf{t}^s(\theta, z)|}{|\mathbf{t}^s(\theta, z)|}, \tag{7.19}$$

where $\mathbf{t}^s$ is the interpolated traction values at the structural nodes and $\mathbf{t}^s(\theta, z)$ is the exact value at the corresponding nodes based on Eq. (7.18). The contour of the relative error is shown in Fig. 7.2 (right) for the structural mesh. It is observed that the error is less than 1.5% for the cases considered. We also quantify the convergence of the RBF interpolation by evaluating the error in the transfer of the traction as

$$e_1 = \frac{||\mathbf{t}^s - \mathbf{t}^s(\theta, z)||_2}{||\mathbf{t}^s(\theta, z)||_2}, \tag{7.20}$$

The behavior of the error with mesh refinement has been plotted in Fig. 7.4a where convergence of order close to 3 is observed which is consistent with the Wendland's C^2 function interpolation. Note that in this case, $h = A^s$ as the number of elements in the Z-direction is constant.

A further analysis is carried out to quantify the error for unstructured non-matching meshes across the fluid and structural domains at the interface consisting of different shapes such as triangular mesh on the fluid and quadrilateral on the structural side. In this case, the number of elements in the structural mesh $n_z^s \in [10, 20, 40, 80]$ and $n_c^s \in [20, 40, 80, 160]$ such that $A^f/A^s = 2$. As the element size is varying in both the directions, the element size is defined as $h = \sqrt{A^s}$. The convergence for the unstructured meshes across the interface is shown in Fig. 7.4b where a third order of convergence is observed, as expected. The interpolation data is shown in Fig. 7.3 for the representative mesh with $n_z^s = 20$ and $n_c^s = 40$.

The error convergence study for the RBF interpolation technique shows a higher order of convergence and independence of the mesh connectivity, emphasizing its generality for scattered data interpolation and efficiency for fluid-structure interaction problems.

7.2.2 Point-to-Element Projection

We next discuss the second category of spatial coupling techniques, which is point-to-element projection. Two types of point-to-element projection schemes are commonly used for FSI problems, viz., (1) Nodal projection, and (2) Quadrature projection.

7.2.2.1 Nodal Projection

At the fluid structure interface Γ, the kinematic equilibrium conditions impose that $\boldsymbol{u}^{\mathrm{f}} = \boldsymbol{u}^{\mathrm{s}}$. Based on the property of the shape functions for the fluid and structural subdomains, the displacement of the fluid nodes at Γ can be expressed as

$$\tilde{\boldsymbol{u}}_j^{\mathrm{f}} = \sum_{i=1}^{m^{\mathrm{s}}} c_{ji} \tilde{\boldsymbol{u}}_i^{\mathrm{s}}, \tag{7.21}$$

where $\tilde{\boldsymbol{u}}_j^{\mathrm{f}}$ is the discrete value of the displacement at the fluid node j, c_{ji} is an approximation matrix based on the shape functions of the discretization and m^{s} is the number of structural nodes on the interface. Let the virtual displacement of the fluid nodes be given by

$$\tilde{\boldsymbol{u}}^{\mathrm{f}} = \sum_{j=1}^{m^{\mathrm{f}}} D_j \tilde{\boldsymbol{u}}_j^{\mathrm{f}}, \tag{7.22}$$

where m^{f} is the number of fluid nodes and D_j is some interpolatory function of the discrete values $\tilde{\boldsymbol{u}}_j^{\mathrm{f}}$. Therefore, the virtual work for the fluid tractions can be written as

$$\delta W^{\mathrm{f}} = \int_{\Gamma^{\mathrm{f}}} \boldsymbol{t}^{\mathrm{f}} \tilde{\boldsymbol{u}}^{\mathrm{f}} d\Gamma = \sum_{j=1}^{m^{\mathrm{f}}} \int_{\Gamma^{\mathrm{f}}} \boldsymbol{t}^{\mathrm{f}} (D_j \tilde{\boldsymbol{u}}_j^{\mathrm{f}}) d\Gamma = \sum_{j=1}^{m^{\mathrm{f}}} \boldsymbol{F}_j^{\mathrm{f}} \tilde{\boldsymbol{u}}_j^{\mathrm{f}}, \tag{7.23}$$

where $\boldsymbol{F}_j^{\mathrm{f}} = \int_{\Gamma^{\mathrm{f}}} \boldsymbol{t}^{\mathrm{f}} D_j d\Gamma$. From Eq. (7.21),

$$\delta W^{\mathrm{f}} = \sum_{j=1}^{m^{\mathrm{f}}} \boldsymbol{F}_j^{\mathrm{f}} \sum_{i=1}^{m^{\mathrm{s}}} c_{ji} \tilde{\boldsymbol{u}}_i^{\mathrm{s}} = \sum_{i=1}^{m^{\mathrm{s}}} \left(\sum_{j=1}^{m^{\mathrm{f}}} \boldsymbol{F}_j^{\mathrm{f}} c_{ji} \right) \tilde{\boldsymbol{u}}_i^{\mathrm{s}} = \sum_{i=1}^{m^{\mathrm{s}}} \boldsymbol{F}_i^{\mathrm{s}} \tilde{\boldsymbol{u}}_i^{\mathrm{s}} = \delta W^{\mathrm{s}}, \tag{7.24}$$

where $\boldsymbol{F}_i^{\mathrm{s}} = \sum_{j=1}^{m^{\mathrm{f}}} \boldsymbol{F}_j^{\mathrm{f}} c_{ji}$ is the structural force.

The nodal projection scheme consists of the following steps [62]:

1. Pair each fluid mesh point S_j on Γ^{f} with the closest structural element Ω_e^{s} on Γ^{s}
2. Find the natural coordinates of the projection $\boldsymbol{\chi}_j$ on the structural element of the fluid point S_j
3. Interpolate the structural displacement at the projected point to obtain the fluid displacement using the shape functions of the structural discretization, i.e.,

$$\tilde{\boldsymbol{u}}_j^{\mathrm{f}} = \tilde{\boldsymbol{u}}^{\mathrm{f}}(S_j) = \tilde{\boldsymbol{u}}^{\mathrm{s}}(\boldsymbol{\chi}_j) = \sum_{i=1}^{i_e} N_i^{\mathrm{s}}(\boldsymbol{\chi}_j) \tilde{\boldsymbol{u}}_{S_j}^{\mathrm{s}}, \tag{7.25}$$

i_e being the number of element nodes in the structural element.

Comparing Eqs. (7.21) and (7.25), $c_{ji} = N_i^{\mathrm{s}}(\boldsymbol{\chi}_j)$. Therefore,

$$\boldsymbol{F}_i^{\mathrm{s}} = \sum_{j=1}^{m^{\mathrm{f}}} \boldsymbol{F}_j^{\mathrm{f}} N_i^{\mathrm{s}}(\boldsymbol{\chi}_j). \tag{7.26}$$

Hence, with the knowledge of the load vectors at the node points of the fluid interface, the loads on the structural points can be interpolated by projecting the fluid node to the structural interface (Fig. 7.5).

7.2.2.2 Quadrature Projection

For obtaining an accurate estimate of the traction on the structural subdomain, our aim is to calculate $\tilde{\boldsymbol{t}}_j^{\mathrm{s}}$ for given $\tilde{\boldsymbol{t}}_i^{\mathrm{f}}$, N_i^{f} and N_j^{s}. We use the Galerkin weighted residual method for minimizing the residual $\boldsymbol{t}^{\mathrm{s}} - \boldsymbol{t}^{\mathrm{f}}$ to get the required approximation for load vector on the structural interface. Multiplying by N_i^{s} on both sides of the residual and integrating over the interface boundary, we get

$$\int_{\Gamma} N_i^{\mathrm{s}} \boldsymbol{t}^{\mathrm{s}} \mathrm{d}\Gamma = \int_{\Gamma} N_i^{\mathrm{s}} \boldsymbol{t}^{\mathrm{f}} \mathrm{d}\Gamma. \tag{7.27}$$

Using Eq. (7.1),

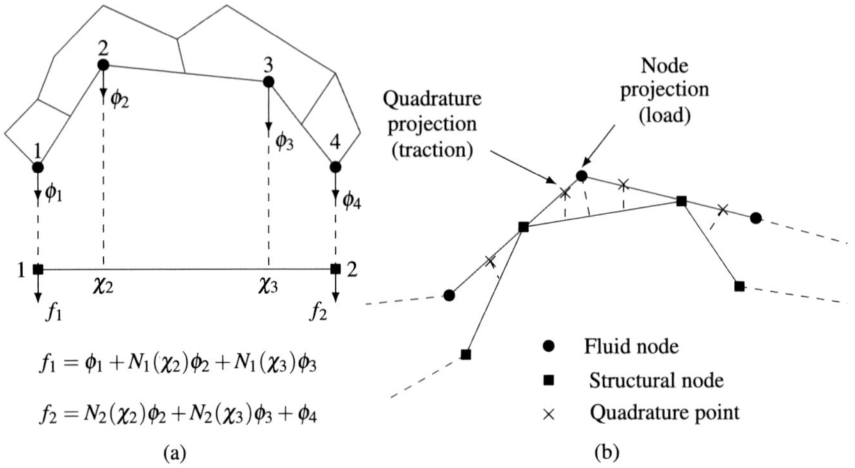

Fig. 7.5 **a** Nodal projection method, **b** comparison of nodal and quadrature projection techniques

$$\int_{\Gamma} N_i^{\mathrm{s}} N_j^{\mathrm{s}} \tilde{\boldsymbol{t}}_j^{\mathrm{s}} \mathrm{d}\Gamma = \int_{\Gamma} N_i^{\mathrm{s}} N_j^{\mathrm{f}} \tilde{\boldsymbol{t}}_j^{\mathrm{f}} \mathrm{d}\Gamma \tag{7.28}$$

The tractions on the structural mesh are thus given as

$$\tilde{\boldsymbol{t}}_k^{\mathrm{s}} = \left[M_{ki}^{\mathrm{s}} \right]^{-1} \left\{ f_i^{\mathrm{s}} \right\}, \tag{7.29}$$

where

$$\left[M_{ki}^{\mathrm{s}} \right] = \int_{\Gamma} N_i^{\mathrm{s}} N_k^{\mathrm{s}} d\Gamma, \qquad \left\{ f_i^{\mathrm{s}} \right\} = \int_{\Gamma} N_i^{\mathrm{s}} N_j^{\mathrm{f}} \tilde{\boldsymbol{t}}_j^{\mathrm{f}} d\Gamma. \tag{7.30}$$

Therefore, for evaluating the traction on the fluid-structure interface for the structural equation, the above mass matrix and the force vector need to be evaluated which consists of the shape functions of both the structural and fluid subdomains. For matching meshes across Γ, this operation is straightforward. However, for non-matching meshes, the inconsistency in the shape functions can lead to integrations across discontinuities.

The challenge here is to evaluate f_i^{s} across non-matching meshes at Γ. There are two ways of obtaining the expression via quadrature projection from the fluid to structure interface [40]:

1. Loop over structural elements and introduce quadrature points in each element. The vector is then given by

$$\left\{f_i^{\mathrm{s}}\right\} = \sum_{i=1}^{m_{\mathrm{el}}^{\mathrm{s}}} \sum_{g=1}^{ngp} N_i^{\mathrm{s}}(\boldsymbol{\chi}_g^{\mathrm{s}}) \det(\boldsymbol{J}) w(g) N_j^{\mathrm{f}} \tilde{\boldsymbol{t}}_j^{\mathrm{f}}(\boldsymbol{\chi}_g^{\mathrm{s}}), \tag{7.31}$$

where $m_{\mathrm{el}}^{\mathrm{s}}$ is the number of structural elements, ngp is the number of Gauss quadrature points on an element, $\boldsymbol{J}$ is the Jacobian of the transformation from global to natural coordinates, $w(g)$ is the Gauss weights at point $\boldsymbol{\chi}_g^{\mathrm{s}}$. Here, the structural shape functions are known at the quadrature points and the fluid traction is interpolated at these points. This scheme is not conservative because there can be some fluid elements without any quadrature points.

2. Loop over fluid elements and introduce quadrature points in each element. In this case, the vector is

$$\left\{f_i^{\mathrm{s}}\right\} = \sum_{i=1}^{m_{\mathrm{el}}^{\mathrm{f}}} \sum_{g=1}^{ngp} N_i^{\mathrm{s}}(\boldsymbol{\chi}_g^{\mathrm{f}}) \det(\boldsymbol{J}) w(g) N_j^{\mathrm{f}} \tilde{\boldsymbol{t}}_j^{\mathrm{f}}(\boldsymbol{\chi}_g^{\mathrm{f}}), \tag{7.32}$$

where $m_{\mathrm{el}}^{\mathrm{f}}$ is the number of fluid elements at the fluid-structure interface. In this technique, the shape functions on the structural interface are evaluated by using optimized search algorithms to find out the corresponding structural element of the fluid traction Gauss quadrature points $\boldsymbol{\chi}_g^{\mathrm{f}}$. The looping over all the fluid elements ensures that the fluid traction is transferred completely to the structural interface, thus maintaining the conservation property. However, some structural nodes may not receive traction in doing so. Therefore, sometimes it is essential to either take higher order Gauss integration or recursively divide the fluid element so that all the structural nodes receive traction.

7.2.3 Common-Refinement Projection

Both the nodal and quadrature projection schemes are conservative but do not satisfy the traction continuity due to inaccurate transfer of loads from the fluid to the structural subdomain. Therefore, such schemes lead to local errors for non-matching meshes [109].

The common-refinement projection is a special data structure for transferring data between non-matching meshes with varying degree of mismatch. We construct a common-refinement surface between the fluid and structural boundaries at Γ (Fig. 7.6). It consists of polygons that subdivide the input boundary meshes of the structure and fluid subdomains simultaneously. Each sub-element of a common-refinement mesh has two geometrical realizations, in general, which are different but must be close to each other to obtain a physically consistent data transfer. The topology of the common-refinement sub-elements are defined by the intersection of the surface elements of the input meshes. In this technique, the load vector defined in Eq. (7.30) is given by

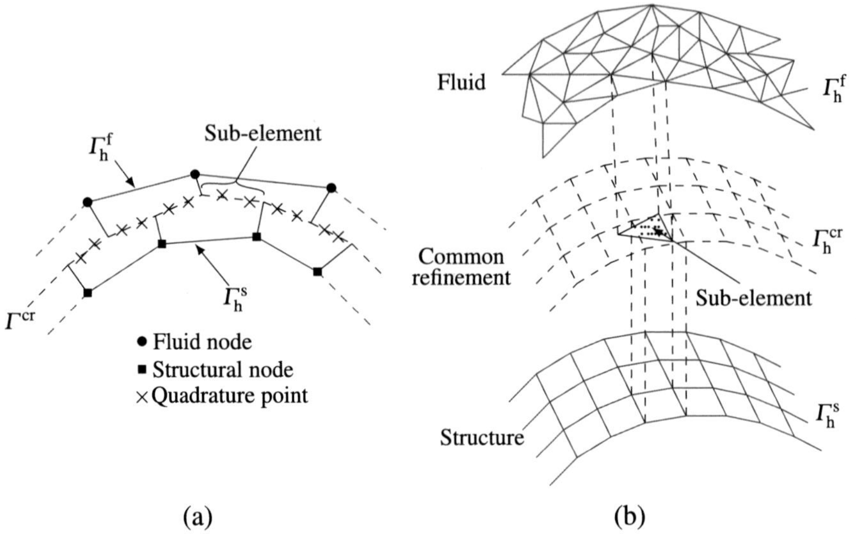

Fig. 7.6 Common-refinement method for load projection in **a** 2D and **b** 3D

$$\left\{f_i^{\mathrm{s}}\right\} = \sum_{j=1}^{n_{\mathrm{el}}^{\mathrm{cr}}} \sum_{g=1}^{ngp} N_i^{\mathrm{s}}(\boldsymbol{\chi}_g^{\mathrm{cr}}) \det(\boldsymbol{J}) w(g) N_j^{\mathrm{f}} \tilde{\boldsymbol{t}}_j^{\mathrm{f}}(\boldsymbol{\chi}_g^{\mathrm{cr}}), \tag{7.33}$$

where $n_{\mathrm{el}}^{\mathrm{cr}}$ is the number of sub-elements in the common-refinement interface and $\boldsymbol{\chi}_g^{\mathrm{cr}}$ is the quadrature projection of the common-refinement points on the structural interface. More details about this projection can be found in [106]. The error analyses for load transfer across a static circular arc, conducted in [106] shows an optimal convergence for source-based common-refinement and exact transfer for target-based common-refinement scheme (Fig. 7.7a). Furthermore, the grid mismatch study showed that the common-refinement scheme performed within the interpolant error for all the cases considered (Fig. 7.7b), while the errors depended strongly on the grid mismatch for the nodal and quadrature projection schemes.

7.3 Temporal Coupling Techniques

In the previous section, we discussed the various types of spatial coupling techniques at the fluid-structure interface for accurate and stable transfer of forces and displacements. Next, we focus on the temporal coupling where the sequence of the data transfer at the interface across the partitioned subdomains is discussed. These can be classified into (1) loosely coupled, and (2) strongly coupled schemes. They are discussed in the following subsections.

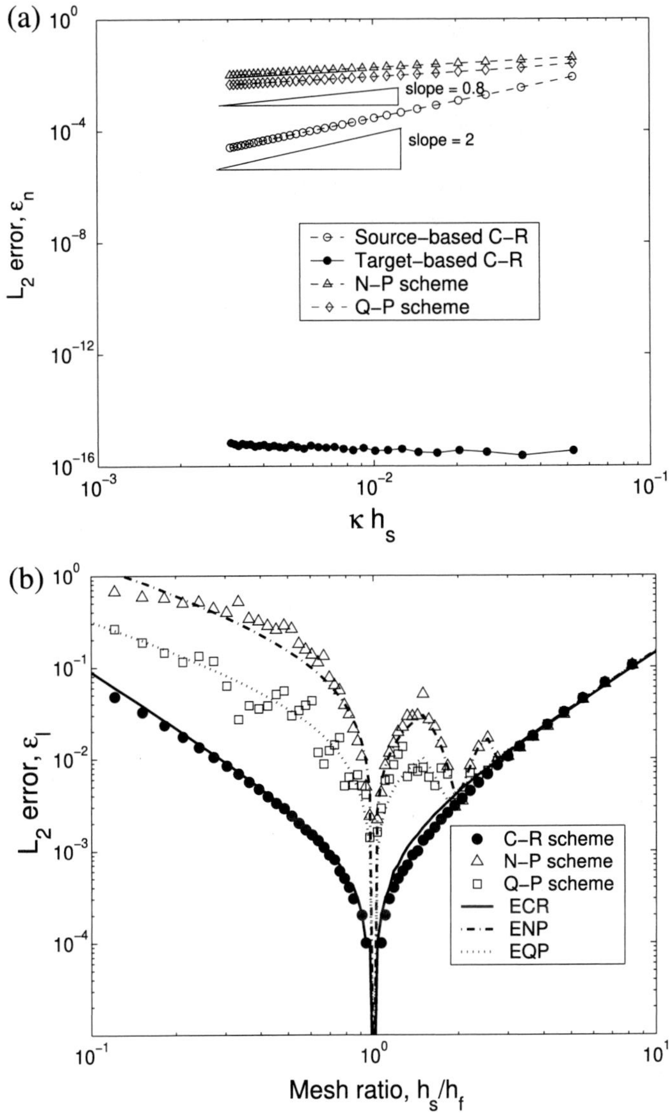

Fig. 7.7 Error assessment for the common-refinement scheme: **a** accuracy of load transfer schemes in L_2 norm error ε_n from analytical solution, and **b** dependence of load vector error on relative mesh ratio h_s/h_f for the load transfer schemes

7.3.1 Staggered Loosely Coupled Techniques

Loosely coupled techniques consist of explicit time integration of the fluid-structure interface data transfer. They are straightforward and easy to implement and are very computationally efficient. Here, we discuss conventional sequential staggered, generalized serial staggered and combined interface boundary condition methods.

7.3.1.1 Conventional Sequential Staggered Method

The most popular temporal coupling scheme is called the conventional sequential staggered (CSS) scheme shown in Fig. 7.8. The algorithm consists of the following steps: For each $t \in [t^{\mathrm{n}}, t^{\mathrm{n}+1}]$,

1. Solve structural equation using the known traction $\boldsymbol{t}^{\mathrm{s,n}}$ to obtain the structural velocity $\boldsymbol{v}^{\mathrm{s,n+1}}$
2. Apply kinematic equilibrium condition at the interface by imposing structural motion

$$\boldsymbol{v}^{\mathrm{f,n+1}} = (1-\alpha)\boldsymbol{v}^{\mathrm{s,n+1}} + \alpha \boldsymbol{v}^{\mathrm{s,n}}, \text{ on } \Gamma, \text{ where } \alpha \in [0, 1] \tag{7.34}$$

3. Set $\dot{\boldsymbol{u}}^{\mathrm{f,n+1}} = \boldsymbol{v}^{\mathrm{f,n+1}}$ and update the fluid mesh displacements and solve the ALE fluid equations for updated values of velocity and pressure
4. Extract new interface traction $\boldsymbol{t}^{\mathrm{f,n+1}}$ and satisfy dynamic equilibrium at the interface

$$\boldsymbol{t}^{\mathrm{s,n+1}} = (1-\beta)\boldsymbol{t}^{\mathrm{f,n+1}} + \beta \boldsymbol{t}^{\mathrm{f,n}}, \text{ on } \Gamma, \text{ where } \beta \in [0, 1] \tag{7.35}$$

In the algorithm, α and β are the weighting parameters to interpolate the structural velocity and the fluid tractions respectively between the time interval t^{n} and $t^{\mathrm{n}+1}$. When $\alpha = \beta = 1$, the solution at the previous time step is transferred to the respective subdomains, giving a method that is called parallel conventional staggered method [64] as the fluid and structural subdomains can start their computations at the same time level and perform their inner subdomain integration in a parallel way. Figure 7.8 refers to the case when $\alpha = \beta = 0$.

The CSS scheme may suffer significantly from destabilizing effects introduced through the interface discretization. High nonlinearity in the problem can make the one-sided approximation inaccurate. More precisely, the numerical treatment of boundary conditions create inconsistencies in omitting and retaining some terms on the neighboring discrete space and time slabs along the interface. Depending on the direction of the interface acceleration, high under- or over-prediction of pressure may occur due to lack of energy equilibrium across the fluid-structure interface. In other words, any small error in the interface displacements imposed onto the fluid by the structure may result in large errors in the fluid pressure. In general, due to a

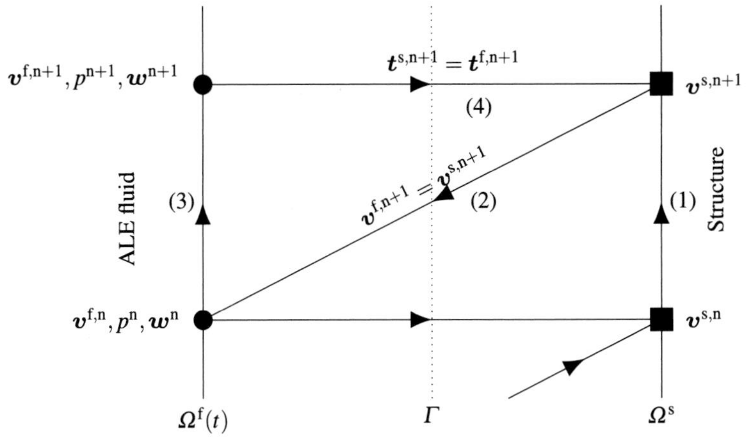

Fig. 7.8 Conventional sequential staggered (CSS) scheme for $\alpha = \beta = 0$

time lag between the structural and fluid subdomains, the CSS technique is at most first-order energy accurate due to $\mathcal{O}(\Delta t)$ in time on Γ.

7.3.1.2 Generalized Serial Staggered Method

The temporal accuracy of the scheme can be improved by applying a structural prediction step based on the higher-order interface velocity extrapolation and momentum averaging. It consists of the following steps: For each $t \in [t^n, t^{n+1}]$,

1. Solve structural equation using the known traction at interface $\boldsymbol{t}^{I,n}$ to obtain the structural velocity $\boldsymbol{v}^{s,n+1}$
2. Apply kinematic equilibrium condition at the interface by predicting interface velocity

$$\boldsymbol{v}^{I,n+1} = \boldsymbol{v}^{s,n} + \Delta t(\alpha_0 \dot{\boldsymbol{v}}^{s,n} - \alpha_1 \dot{\boldsymbol{v}}^{s,n-1}), \text{ on } \Gamma \tag{7.36}$$

3. Set $\dot{\boldsymbol{u}}^{f,n+1} = \boldsymbol{v}^{I,n+1}$ and update the fluid mesh displacements and solve the ALE fluid equations for updated values of velocity and pressure
4. Extract new interface traction $\boldsymbol{t}^{f,n+1}$ and satisfy dynamic equilibrium at the interface

$$\boldsymbol{t}^{I,n+1} = (1-\beta)\boldsymbol{t}^{f,n+1} + \beta \boldsymbol{t}^{f,n}, \text{ on } \Gamma, \text{ where } \beta \in [0, 1] \tag{7.37}$$

In the GSS algorithm, $\alpha_0 = 3/2$ and $\alpha_1 = 1/2$ are required for second-order accuracy in the satisfaction of the kinematic condition and $\beta = 1/2$ for the improved momentum or load averaging.

7.3.1.3 Combined Interface Boundary Condition Method

In the temporal coupling techniques discussed in the previous subsections, the governing equations of each subdomain have no direct influence along the equilibrium conditions satisfied at the fluid-structure interface, which are enforced sequentially with a time lag. In the combined interface boundary condition (CIBC) technique, the interface solution is influenced explicitly by the neighboring subdomains. The approach achieves improved precision and numerical stability by solving the transformed conditions for the interface quantities based on their spatial and temporal derivatives [181].

The combined residual operators for the kinematic and dynamic equilibrium conditions can be constructed as

$$\mathscr{R}^{D,N}\left(\rho^{\mathrm{f}}\frac{\partial \boldsymbol{v}^{\mathrm{s}}}{\partial t}, \frac{\partial \boldsymbol{t}^{\mathrm{s}}}{\partial t}, \frac{\partial \boldsymbol{t}^{\mathrm{f}}}{\partial t}, \frac{\partial \boldsymbol{t}^{\mathrm{f}}}{\partial n}\right) = \boldsymbol{0}, \text{ along } \Gamma, \tag{7.38}$$

where ρ^{f} is the density of the fluid, $\partial/\partial t$ and $\partial/\partial n$ denote the time derivative and normal spatial derivative of the interface quantities respectively. The idea relies on constructing a local discrete energy-preserving property for the staggered stencil between the pair of differential equations. Based on the current solutions on both sides of the interface, these operators calculate successive corrections to the staggered solutions on Γ.

The conventional kinematic and dynamic equilibrium conditions can be transformed into

$$\frac{\partial \boldsymbol{t}^{\mathrm{f}}}{\partial n^{\mathrm{f}}} = \rho^{\mathrm{f}}\frac{\partial \boldsymbol{v}^{\mathrm{s}}}{\partial t}, \quad \text{on } \Gamma, \tag{7.39}$$

$$\frac{\partial \boldsymbol{t}^{\mathrm{s}}}{\partial t}\cdot \boldsymbol{n}^{\mathrm{s}} = \frac{\partial \boldsymbol{t}^{\mathrm{f}}}{\partial t}\cdot \boldsymbol{n}^{\mathrm{f}}, \quad \text{on } \Gamma. \tag{7.40}$$

Combining the above equations and using the fact that $\boldsymbol{n}^{\mathrm{f}} = -\boldsymbol{n}^{\mathrm{s}}$, the relation for the structural velocity at the interface and the fluid traction are obtained as

$$\rho^{\mathrm{f}}\frac{\partial \boldsymbol{v}^{\mathrm{s}}}{\partial t} + \omega\frac{\partial \boldsymbol{t}^{\mathrm{s}}}{\partial t} = \frac{\partial \boldsymbol{t}^{\mathrm{f}}}{\partial n^{\mathrm{f}}} - \omega\frac{\partial \boldsymbol{t}^{\mathrm{f}}}{\partial t}, \quad \text{on } \Gamma^{\mathrm{s}}, \tag{7.41}$$

$$\frac{\partial \boldsymbol{t}^{\mathrm{f}}}{\partial n^{\mathrm{f}}} + \omega\frac{\partial \boldsymbol{t}^{\mathrm{f}}}{\partial t} = \rho^{\mathrm{f}}\frac{\partial \boldsymbol{v}^{\mathrm{s}}}{\partial t} - \omega\frac{\partial \boldsymbol{t}^{\mathrm{s}}}{\partial t}, \quad \text{on } \Gamma^{\mathrm{f}}, \tag{7.42}$$

where ω is a positive dimensional parameter and small enough to ensure that the interface energy is always stable [106]. It provides appropriate combination of the applied traction forces with the appropriate corrections into the interfacial acceleration.

The above relationships can be written in the explicit staggered form as

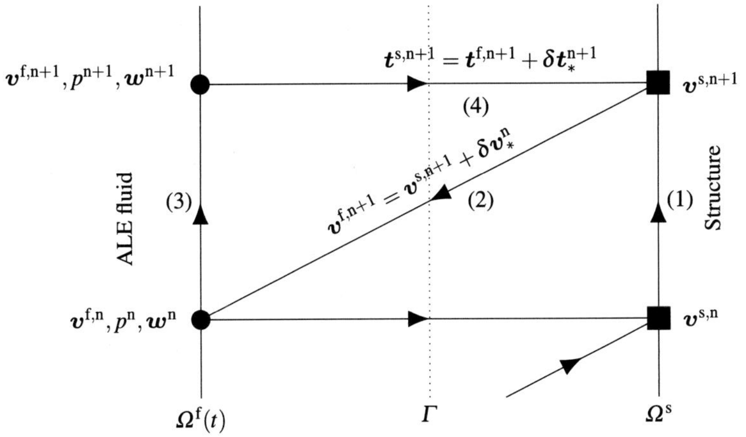

Fig. 7.9 Combined interface boundary condition (CIBC) scheme

$$\rho^{\mathrm{f}}\left(\frac{\partial \boldsymbol{v}^{\mathrm{s}}}{\partial t}\right)^{\mathrm{n}} = \left(\frac{\partial \boldsymbol{t}^{\mathrm{f}}}{\partial n^{\mathrm{f}}}\right)^{\mathrm{n}} - \omega\left[\left(\frac{\partial \boldsymbol{t}^{\mathrm{f}}}{\partial t}\right)^{\mathrm{n}} - \left(\frac{\partial \boldsymbol{t}^{\mathrm{s}}}{\partial t}\right)^{\mathrm{n}}\right], \quad \text{on } \Gamma^{\mathrm{s}}, \tag{7.43}$$

$$\left(\frac{\partial \boldsymbol{t}^{\mathrm{f}}}{\partial t}\right)^{\mathrm{n+1}} = -\left(\frac{\partial \boldsymbol{t}^{\mathrm{s}}}{\partial t}\right)^{\mathrm{n+1}} + \frac{1}{\omega}\left[\left(\rho^{\mathrm{f}}\frac{\partial \boldsymbol{v}^{\mathrm{s}}}{\partial t}\right)^{\mathrm{n+1}} - \left(\frac{\partial \boldsymbol{t}^{\mathrm{f}}}{\partial n^{\mathrm{f}}}\right)^{\mathrm{n+1}}\right], \quad \text{on } \Gamma^{\mathrm{f}}. \tag{7.44}$$

The above equations provide the corrections for the velocity and tractions at the fluid-structure interface. Instead of applying the equilibrium conditions directly, we apply a corrected velocity and traction at Γ as

$$\boldsymbol{v}^{\mathrm{f,n+1}} = \boldsymbol{v}^{\mathrm{s,n+1}} + \delta \boldsymbol{v}_*^{\mathrm{n}}, \quad \text{on } \Gamma^{\mathrm{s}}, \tag{7.45}$$

$$\boldsymbol{t}^{\mathrm{s,n+1}} = \boldsymbol{t}^{\mathrm{f,n+1}} + \delta \boldsymbol{t}_*^{\mathrm{n+1}}, \quad \text{on } \Gamma^{\mathrm{f}}, \tag{7.46}$$

where the corrections are given by

$$\delta \boldsymbol{v}_*^{\mathrm{n}} = \frac{\Delta t}{\rho^{\mathrm{f}}}\left[\left(\frac{\partial \boldsymbol{t}^{\mathrm{f}}}{\partial n^{\mathrm{f}}}\right)^{\mathrm{n}} - \omega\left\{\left(\frac{\partial \boldsymbol{t}^{\mathrm{f}}}{\partial t}\right)^{\mathrm{n}} - \left(\frac{\partial \boldsymbol{t}^{\mathrm{s}}}{\partial t}\right)^{\mathrm{n}}\right\}\right], \quad \text{on } \Gamma^{\mathrm{s}}, \tag{7.47}$$

$$\delta \boldsymbol{t}_*^{\mathrm{n+1}} = \Delta t\left[-\left(\frac{\partial \boldsymbol{t}^{\mathrm{s}}}{\partial t}\right)^{\mathrm{n+1}} + \frac{1}{\omega}\left\{\left(\rho^{\mathrm{f}}\frac{\partial \boldsymbol{v}^{\mathrm{s}}}{\partial t}\right)^{\mathrm{n+1}} - \left(\frac{\partial \boldsymbol{t}^{\mathrm{f}}}{\partial n^{\mathrm{f}}}\right)^{\mathrm{n+1}}\right\}\right], \quad \text{on } \Gamma^{\mathrm{f}}. \tag{7.48}$$

At each time step, the prediction of the displacement and the correction of tractions are formed in a sequential manner. The algorithm for CIBC consists of the following steps: For each $t \in [t^{\mathrm{n}}, t^{\mathrm{n+1}}]$ (Fig. 7.9),

1. Solve structural equation using the known traction at interface $\boldsymbol{t}^{\mathrm{s,n}}$ to obtain the structural velocity $\boldsymbol{v}^{\mathrm{s,n+1}}$

2. Solve for the velocity prediction $\delta \boldsymbol{v}_*^{\mathrm{n}}$ given in Eq. (7.47) and apply kinematic equilibrium condition at the interface

$$\boldsymbol{v}^{\mathrm{f,n+1}} = \boldsymbol{v}_*^{\mathrm{n+1}} = \boldsymbol{v}^{\mathrm{s,n+1}} + \delta \boldsymbol{v}_*^{\mathrm{n}} \tag{7.49}$$

3. Set $\dot{\boldsymbol{u}}^{\mathrm{f,n+1}} = \boldsymbol{v}^{\mathrm{f,n+1}}$ and update the fluid mesh displacements and solve the ALE fluid equations for updated values of velocity and pressure
4. Solve for the traction correction $\delta \boldsymbol{t}_*^{\mathrm{n+1}}$ in Eq. (7.48) and satisfy dynamic equilibrium at the interface

$$\boldsymbol{t}^{\mathrm{s,n+1}} = \boldsymbol{t}_*^{\mathrm{n+1}} = \boldsymbol{t}^{\mathrm{f,n+1}} + \delta \boldsymbol{t}_*^{\mathrm{n+1}} \tag{7.50}$$

Note that we only considered explicit corrections in the staggered algorithm. It can be extended to a semi-implicit predictor-corrector procedure by forming the structural velocity as predictor and traction as corrector. This forms the basis of the strongly coupled techniques which are discussed in the next subsection.

7.3.2 Strongly Coupled Techniques

In this section, we discuss the partitioned iterative coupling for the FSI system, which is considered to be a strongly coupled technique. We begin with the coupled linearized matrix form and discuss the quasi-Newton updates that are carried out to correct the fluid forces on the structure. Then, the algorithm for the coupling is presented for the block-type partitioned system.

Consider the decomposition of the set of degrees of freedom (DOFs) of the fluid-structure system into interior DOFs and the DOFs at the fluid-structure interface. With the help of Newton-Raphson linearization, the coupled fluid-structure system in a partitioned format can be expressed as

$$\begin{bmatrix} \boldsymbol{A}^{\mathrm{ss}} & \boldsymbol{0} & \boldsymbol{0} & \boldsymbol{A}^{\mathrm{Is}} \\ \boldsymbol{A}^{\mathrm{sI}} & \boldsymbol{I} & \boldsymbol{0} & \boldsymbol{0} \\ \boldsymbol{0} & \boldsymbol{A}^{\mathrm{If}} & \boldsymbol{A}^{\mathrm{ff}} & \boldsymbol{0} \\ \boldsymbol{0} & \boldsymbol{0} & \boldsymbol{A}^{\mathrm{fI}} & \boldsymbol{I} \end{bmatrix} \begin{Bmatrix} \Delta \boldsymbol{u}^{\mathrm{s}} \\ \Delta \boldsymbol{u}^{\mathrm{I}} \\ \Delta \boldsymbol{q}^{\mathrm{f}} \\ \Delta \boldsymbol{f}^{\mathrm{I}} \end{Bmatrix} = \begin{Bmatrix} \mathscr{R}^{\mathrm{s}} \\ \mathscr{R}_{\mathrm{D}}^{\mathrm{I}} \\ \mathscr{R}^{\mathrm{f}} \\ \mathscr{R}_{\mathrm{N}}^{\mathrm{I}} \end{Bmatrix}, \tag{7.51}$$

where $\Delta \boldsymbol{u}^{\mathrm{s}}$ denotes the increment in the structural displacement, $\Delta \boldsymbol{u}^{\mathrm{I}}$ and $\Delta \boldsymbol{f}^{\mathrm{I}}$ represent the increments in the displacement and the forces along the fluid-structure interface. The increment in the unknowns associated with the fluid domain is denoted by $\Delta \boldsymbol{q}^{\mathrm{f}} = (\Delta \boldsymbol{v}^{\mathrm{f}}, \Delta p)$. On the right hand side, $\mathscr{R}^{\mathrm{s}}$ and $\mathscr{R}^{\mathrm{f}}$ represent the weighted residuals of the structural and stabilized flow equations respectively, whereas $\mathscr{R}_{\mathrm{D}}^{\mathrm{I}}$ and $\mathscr{R}_{\mathrm{N}}^{\mathrm{I}}$ denote the residuals corresponding to the imbalances during the enforcement of the kinematic (Dirichlet) condition [Eq. (3.66)] and the dynamic (Neumann) condition [Eq. (3.67)] at the fluid-structure interface respectively.

The block matrices on the left-hand side are described as follows. The matrix comprising of the mass, damping and stiffness matrices of the structural system for the non-interface structural DOFs is given by $\boldsymbol{A}^{\mathrm{ss}}$. The transformation of the fluid forces at the interface to obtain the force vector for the structural system is denoted by $\boldsymbol{A}^{\mathrm{Is}}$. The mapping of the structural displacements to the fluid-structure interface satisfying the kinematic condition is represented by $\boldsymbol{A}^{\mathrm{sI}}$, and $\boldsymbol{I}$ is an identity matrix. $\boldsymbol{A}^{\mathrm{fI}}$ denotes the transfer of the fluid forces to the fluid-structure interface satisfying the dynamic equilibrium condition and $\boldsymbol{A}^{\mathrm{If}}$ is associated with the ALE mapping of the fluid spatial points. Finally, $\boldsymbol{A}^{\mathrm{ff}}$ comprises of the transient, convection, diffusion and stabilization terms for the Navier-Stokes fluid system.

The idea behind the strong coupling procedure is to construct the cross-coupling between the fluid and the structure without forming the off-diagonal Jacobian term [$\boldsymbol{A}^{\mathrm{Is}}$ in Eq. (7.51)] with the help of nonlinear iterations. This cross-coupling effect gives rise to force corrections at the interface resulting in what we call here nonlinear iterative force correction (NIFC) procedure. This correction relies on an input-output relationship between the structural displacement and the force transfer at each nonlinear iteration. This feedback process can be considered as a nonlinear generalization of the steepest descent method which transforms a divergent fixed-point iteration to a stable and convergent update of the approximate forces at the interface degrees of freedom [110]. Unlike the brute-force iterations in the strongly coupled FSI which lead to severe numerical instabilities for low structure-to-fluid mass ratios, the NIFC procedure provides a desired stability to the partitioned fluid-structure coupling, without the explicit evaluation of the off-diagonal Jacobian term. As derived in [110], the idea of partitioning is to eliminate the off-diagonal term $\boldsymbol{A}^{\mathrm{Is}}$ to facilitate the staggered sequential updates for strongly coupled fluid-structure system. Through static condensation, Eq. (7.51) can be re-written as

$$\begin{bmatrix} \boldsymbol{A}^{\mathrm{ss}} & \mathbf{0} & \mathbf{0} & \mathbf{0} \\ \boldsymbol{A}^{\mathrm{sI}} & \boldsymbol{I} & \mathbf{0} & \mathbf{0} \\ \mathbf{0} & \boldsymbol{A}^{\mathrm{If}} & \boldsymbol{A}^{\mathrm{ff}} & \mathbf{0} \\ \mathbf{0} & \mathbf{0} & \mathbf{0} & \boldsymbol{A}^{\mathrm{II}} \end{bmatrix} \begin{Bmatrix} \Delta \boldsymbol{u}^{\mathrm{s}} \\ \Delta \boldsymbol{u}^{\mathrm{I}} \\ \Delta \boldsymbol{q}^{\mathrm{f}} \\ \Delta \boldsymbol{f}^{\mathrm{I}} \end{Bmatrix} = \begin{Bmatrix} \mathscr{R}^{\mathrm{s}} \\ \mathscr{R}^{\mathrm{I}}_{\mathrm{D}} \\ \mathscr{R}^{\mathrm{f}} \\ \widetilde{\mathscr{R}}^{\mathrm{I}}_{\mathrm{N}} \end{Bmatrix}. \tag{7.52}$$

In the nonlinear interface force correction, we form the iterative scheme of the following matrix-vector product form

$$\Delta \boldsymbol{f}^{\mathrm{I}} = \left(\boldsymbol{A}^{\mathrm{II}}\right)^{-1} \widetilde{\mathscr{R}}^{\mathrm{I}}_{\mathrm{N}}, \tag{7.53}$$

where $\left(\boldsymbol{A}^{\mathrm{II}}\right)^{-1}$ is not constructed explicitly. Instead, the force correction vector $\Delta \boldsymbol{f}^{\mathrm{I}}$ at the nonlinear iteration (subiteration) k can be constructed by successive matrix-vector products. This process essentially provides the control for the interface fluid force $\boldsymbol{f}^{\mathrm{I}} = \int_{\Gamma} \boldsymbol{\sigma}^{\mathrm{f}} \cdot \boldsymbol{n}^{\mathrm{f}} \mathrm{d}\Gamma$ to stabilize strong fluid-structure interaction at low structure-to-fluid mass ratio. The scheme proceeds in a similar fashion as the predictor-corrector schemes by constructing the iterative interface force correction at each iteration. Let the error in the interface fluid force between the initial and first nonlinear iteration be $\Delta \boldsymbol{E}^{\mathrm{I}}_{(0)} = \boldsymbol{f}^{\mathrm{I}}_{(1)} - \boldsymbol{f}^{\mathrm{I}}_{(0)}$. Similarly, the force at the iteration k + 1 is given by

$$\boldsymbol{f}^{\mathrm{I}}_{(\mathrm{k}+1)} = \boldsymbol{f}^{\mathrm{I}}_{(\mathrm{k})} + \Delta \boldsymbol{f}^{\mathrm{I}}_{(\mathrm{k})} = \boldsymbol{f}^{\mathrm{I}}_{(\mathrm{k})} + \left[\left(\boldsymbol{A}^{\mathrm{II}}\right)^{-1} \widetilde{\mathscr{R}}^{\mathrm{I}}_{\mathrm{N}}\right]_{(\mathrm{k})}. \tag{7.54}$$

For the iterative correction of the fluid forces, a power method is considered for the above matrix problem. We assume an iteration matrix $\boldsymbol{M}$ which is diagonalizable in such a way that $\boldsymbol{M}\boldsymbol{v}_{(\mathrm{k})} = \lambda_{(\mathrm{k})}\boldsymbol{v}_{(\mathrm{k})}$ for each k and eigenvalues $\lambda_{(\mathrm{k})}$ are distinct and nonzero with the corresponding eigenvectors $\boldsymbol{v}_{(\mathrm{k})}$. The correction to the forces is then constructed with the aid of the error vector $\Delta \boldsymbol{E}^{\mathrm{I}}_{(0)}$ as

$$\boldsymbol{f}^{\mathrm{I}}_{(\mathrm{k}+1)} = \boldsymbol{f}^{\mathrm{I}}_{(\mathrm{k})} + \boldsymbol{M}^{\mathrm{k}} \Delta \boldsymbol{E}^{\mathrm{I}}_{(0)}, \tag{7.55}$$

which can be written in terms of successive estimates as

$$\boldsymbol{f}^{\mathrm{I}}_{(\mathrm{k}+1)} = \boldsymbol{f}^{\mathrm{I}}_{(0)} + \sum_{i=0}^{\mathrm{k}} \boldsymbol{M}^{i} \Delta \boldsymbol{E}^{\mathrm{I}}_{(0)}, \quad \text{for } \mathrm{k} = 1, 2, \ldots \tag{7.56}$$

The error vectors $\Delta \boldsymbol{E}^{\mathrm{I}}_{(\mathrm{k})}$ can then be expressed in terms of the eigenvalues and eigenvectors λ and $\boldsymbol{v}$ respectively to obtain a sequence of transformation for the force vector $\boldsymbol{f}^{\mathrm{I}}_{(\mathrm{k}+1)}$ similar to the Aitken's iterated Δ^2 process [110].

This interface force correction can also be interpreted as a quasi-Newton update

$$\Delta \boldsymbol{f}^{\mathrm{I}}_{(\mathrm{k}+1)} = \Delta \boldsymbol{f}^{\mathrm{I}}_{(\mathrm{k})} + \boldsymbol{\Lambda}_{(\mathrm{k})} \Delta \boldsymbol{E}^{\mathrm{I}}_{(0)}, \tag{7.57}$$

where $\Delta \boldsymbol{f}^{\mathrm{I}}_{(k+1)} = \boldsymbol{f}^{\mathrm{I}}_{(\mathrm{k}+1)} - \boldsymbol{f}^{\mathrm{I}}_{(\mathrm{k})}$, $\Delta \boldsymbol{f}^{\mathrm{I}}_{(\mathrm{k})} = \boldsymbol{f}^{\mathrm{I}}_{(\mathrm{k})} - \boldsymbol{f}^{\mathrm{I}}_{(\mathrm{k}-1)}$ and $\boldsymbol{\Lambda}_{(\mathrm{k})} = (\boldsymbol{M}^{\mathrm{k}} - \boldsymbol{M}^{\mathrm{k}-1})$ is an $n \times n$ matrix. There are three possible alternatives for the matrix $\boldsymbol{\Lambda}_{(\mathrm{k})}$, namely, scalar, diagonal and full matrix. We consider $\boldsymbol{\Lambda}_{(\mathrm{k})} = \alpha_{(\mathrm{k})}\mathbf{I}$ for the iterative quasi-Newton update, which can be considered as a minimal residual iteration method when $(\boldsymbol{y}, \Delta \boldsymbol{f}^{\mathrm{I}}_{(\mathrm{k}+1)}) = 0$ for some $\boldsymbol{y}$, where $(\cdot, \cdot)$ denotes the standard inner product. Thus, we have

$$(\boldsymbol{y}, \Delta \boldsymbol{f}^{\mathrm{I}}_{(\mathrm{k}+1)}) = (\boldsymbol{y}, \Delta \boldsymbol{f}^{\mathrm{I}}_{(\mathrm{k})}) + \alpha_{(\mathrm{k})}(\boldsymbol{y}, \Delta \boldsymbol{E}^{\mathrm{I}}_{(0)}) = 0, \tag{7.58}$$

$$\implies \alpha_{(\mathrm{k})} = -\frac{(\boldsymbol{y}, \Delta \boldsymbol{f}^{\mathrm{I}}_{(\mathrm{k})})}{(\boldsymbol{y}, \Delta \boldsymbol{E}^{\mathrm{I}}_{(0)})}. \tag{7.59}$$

It can be observed that the choice of $\boldsymbol{y} = \Delta \boldsymbol{E}^{\mathrm{I}}_{(0)}$ minimizes $||\Delta \boldsymbol{f}^{\mathrm{I}}_{(\mathrm{k}+1)}||$ and this type of iterative procedure is similar to the minimal residual method [31, 32].

The algorithm for the NIFC scheme, shown in Fig. 7.10 can be summarized as follows: In a typical nonlinear iteration k,

1. Solve structural equation using the known traction at interface $\boldsymbol{f}^{\mathrm{s,n}}_{(\mathrm{k})}$ to obtain the structural velocity $\boldsymbol{v}^{\mathrm{s,n+1}}_{(\mathrm{k}+1)}$

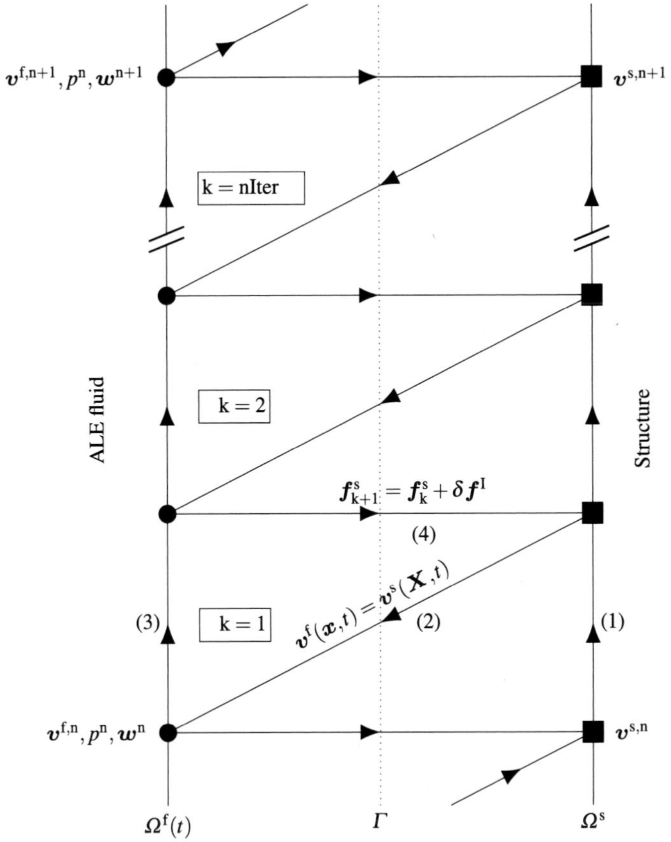

Fig. 7.10 Strong coupling with nonlinear iterative force correction for the fluid-structure system

2. Apply kinematic equilibrium condition and the ALE compatibility condition at the interface. This is accomplished as follows:
   - Mesh displacement $\boldsymbol{u}^{\mathrm{m,n+1}}_{(\mathrm{k+1})}$ is equated to structural displacement as

$$\boldsymbol{u}^{\mathrm{m,n+1}}_{(\mathrm{k+1})} = \boldsymbol{u}^{\mathrm{s,n+1}}_{(\mathrm{k+1})} \text{ on } \Gamma \tag{7.60}$$

   - Conservation property between moving elements in the fluid domain is satisfied by equating fluid velocity to mesh velocity as

$$\boldsymbol{v}^{\mathrm{f,n+}\alpha^{\mathrm{f}}}_{(\mathrm{k+1})} = \boldsymbol{w}^{\mathrm{n+}\alpha^{\mathrm{f}}}_{(\mathrm{k+1})} \text{ on } \Gamma, \tag{7.61}$$

where

$$\boldsymbol{w}_{(k+1)}^{n+\alpha^f} = \frac{\boldsymbol{u}_{(k+1)}^{m,n+1} - \boldsymbol{u}_{(k+1)}^{m,n}}{\Delta t} = \boldsymbol{v}_{(k+1)}^{s,n+\alpha^s} \text{ on } \Gamma \tag{7.62}$$

3. Evaluate the mesh velocity of spatial points and solve for ALE fluid equation to get updated velocity $\boldsymbol{v}_{(k+1)}^{f,n+1}$ and pressure $p_{(k+1)}^{n+1}$
4. Compute the updated hydrodynamic forces on the interface Γ and apply the NIFC force correction, thus satisfying the dynamic equilibrium as

$$\boldsymbol{f}_{(k+1)}^{s,n+\alpha^s} = \boldsymbol{f}_{(k+1)}^{I} = \boldsymbol{f}_{(k)}^{I} + \Delta \boldsymbol{f}_{(k)}^{I} \tag{7.63}$$

This concludes the description about the strongly coupled partitioned iterative techniques, where we have discussed one type of method known as nonlinear iterative force correction (NIFC). Next, we consider the common-refinement projection for the spatial coupling and the NIFC algorithm for the strong temporal coupling and assess the scheme by performing numerical tests on benchmark problems and then demonstrate the partitioned framework for a practical application.

7.4 Common-Refinement Projection with Nonlinear Iterative Force Correction

In this section, we investigate the effectiveness and accuracy of the common-refinement projection scheme along with the nonlinear iterative force correction scheme in the partitioned coupling between the fluid and the structural domains. The scheme is then applied to validate a problem of three-dimensional FSI consisting of non-matching meshes across the two fluid and structural domains. Finally, the partitioned coupling is demonstrated for a flow across an offshore riser.

7.4.1 Error Analysis and Convergence Study

A typical FSI simulation deals with an intact interface between the fluid and the structural domains without any gaps or overlaps. At the fluid-structure interface, the fluid forces are transferred to the structural surface and the structural displacement is projected from the structure to the fluid surface. Such data transfer across the interface is repeated multiple times over a single time step, and predominantly leads to two types of errors in the data transfer: (1) error during a single transfer from one surface to another across a non-matching mesh, and (2) error resulting due to repeated transfers. These errors are analyzed and assessed for the common-refinement projection scheme in this section.

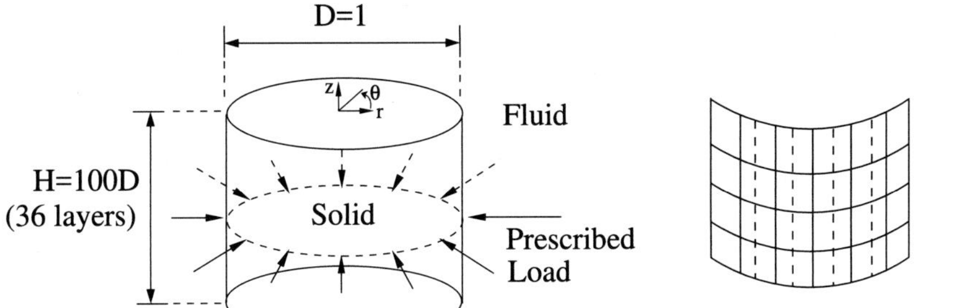

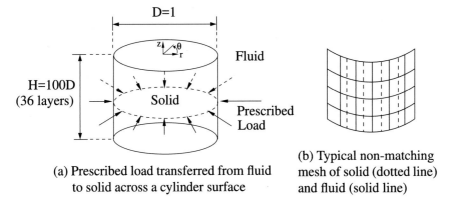

(a) Prescribed load transferred from fluid to solid across a cylinder surface

(b) Typical non-matching mesh of solid (dotted line) and fluid (solid line)

Fig. 7.11 3D non-matching meshes for static error analysis of common-refinement method

7.4.1.1 Static Data Transfer

The first type of error resulting from a single transfer of data across the interface consisting of non-matching meshes is investigated here. Two intact surfaces with different mesh sizes are employed for quantification of this error, one of which is termed as fluid boundary and the other surface represents the deformable structure/solid. The geometry of both the surfaces resembles a circular cylinder with diameter $D = 1$ and height $H = 100D$, as shown in Fig. 7.11a. The surfaces are discretized into $w_z = 36$ uniform elements in the Z-direction. Along the circumference of the cylinder, the surface is discretized into N_s and N_f uniform elements. Thus, the total number of elements on the fluid and solid surfaces are $w_z N_f$ and $w_z N_s$, respectively. A typical patch of the non-matching fluid and solid surface meshes is shown in Fig. 7.11b.

The size of the element on each surface is defined by the area of each element as

$$A_s = \frac{\pi DH}{w_z N_s}, \quad A_f = \frac{\pi DH}{w_z N_f}. \tag{7.64}$$

Furthermore, the degree of non-matching meshes along the fluid-structure interface is given by the mesh ratio between the two surfaces, A_s/A_f. Various mesh refinements are generated by $N_s, N_f \in \{36, 54, 108, 162, 216\}$ leading to the mesh ratio of $A_s/A_f \in [1/6, 6]$. For the static load data transfer, a prescribed load is applied on the nodes of the fluid surface mesh at position $\boldsymbol{x}^{\mathrm{f}}$, which is then transferred to the structural surface mesh by the common-refinement technique. This is carried out for each set of the generated meshes. The prescribed load is given by the estimation of the static pressure along the Z-direction generated by a potential flow around the cylinder as

Table 7.1 Dependence of load vector error ε_1 on different mesh ratios A_s/A_f

N_s	N_f	A_s/A_f	Error, ε_1
36	216	6.0	2.94896×10^{-3}
36	162	4.5	2.88308×10^{-3}
36	108	3.0	2.69495×10^{-3}
36	54	1.5	1.6836×10^{-6}
36	36	1.0	2.51×10^{-16}
54	36	0.6667	1.76919×10^{-6}
108	36	0.3333	3.30083×10^{-3}
162	36	0.2222	3.29819×10^{-3}
216	36	0.1667	3.32599×10^{-3}

$$\boldsymbol{t}^{\mathrm{s}} = \boldsymbol{t}^{\mathrm{s}}(\theta, z) = -\left(\frac{1}{2}\rho^{\mathrm{f}} U_{\infty}^{2}(1 - 4\sin^2\theta) + \rho^{\mathrm{f}} g z\right)\begin{pmatrix} 0.5\cos\theta \\ 0.5\sin\theta \\ 0 \end{pmatrix} \tag{7.65}$$

where $(\theta, z) \in \Gamma^{\mathrm{fs}}$ is the cylindrical position vector on the surface of the cylinder and $U_\infty = 1$. The origin of the cylindrical coordinates lies at the center of the top surface of the cylinder. Let $(\theta_j^{\mathrm{s}}, z_j^{\mathrm{s}})$ be the position vector of node j on the solid surface mesh and $\boldsymbol{T}_j^{\mathrm{s}}$ be the corresponding load transferred to that node, then the relative error ε_1 is quantified as

$$\varepsilon_1 = \frac{\sum_j ||\boldsymbol{T}_j^{\mathrm{s}} - \boldsymbol{t}^{\mathrm{s}}(\theta_j^{\mathrm{s}}, z_j^{\mathrm{s}})||_2}{\sum_j ||\boldsymbol{t}^{\mathrm{s}}(\theta_j^{\mathrm{s}}, z_j^{\mathrm{s}})||_2}, \tag{7.66}$$

where $||\cdot||_2$ is the ℓ_2 norm. This error is computed for each mesh ratio ranging from 0.1667 to 6.0 and is summarized in Table 7.1. It can be observed that the common-refinement projection performs well within the interpolation error for all the mesh ratios. It is worth to note that the error is consistent for both $A_s/A_f > 1$ and $A_s/A_f < 1$ as the overlay mesh surface constructed in the common-refinement method involves both the fluid and the structural meshes.

A spatial convergence study is also carried out to further analyze the common-refinement technique by considering $A_s/A_f = 1.5$ and $A_s/A_f = 0.6667$ as the reference mesh ratios. Both N_s and N_f are increased simultaneously while keeping the mesh ratio fixed to reduce the error introduced by the spatial discretization. The relative error ε_1 computed for the mesh convergence is shown in Fig. 7.12. The slope of the convergence plot shows a second-order convergence rate implying that the common-refinement method is optimally accurate up to the geometric interpolation.

Capability of the common-refinement projection to handle different mesh shapes is assessed by considering a transfer of data from a triangular (fluid) to a quadrilateral

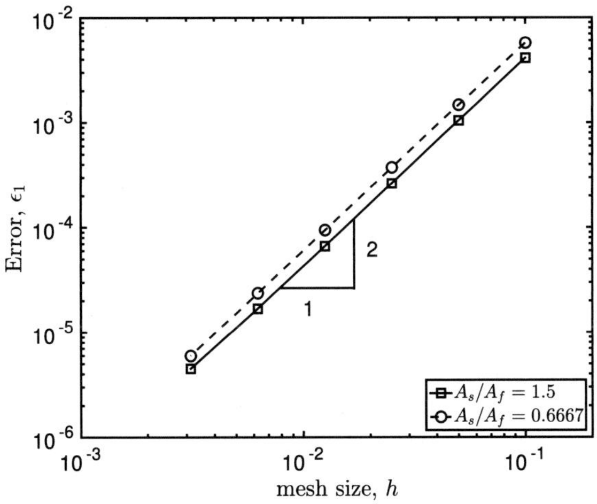

Fig. 7.12 Spatial mesh convergence study of common-refinement method for non-matching meshes

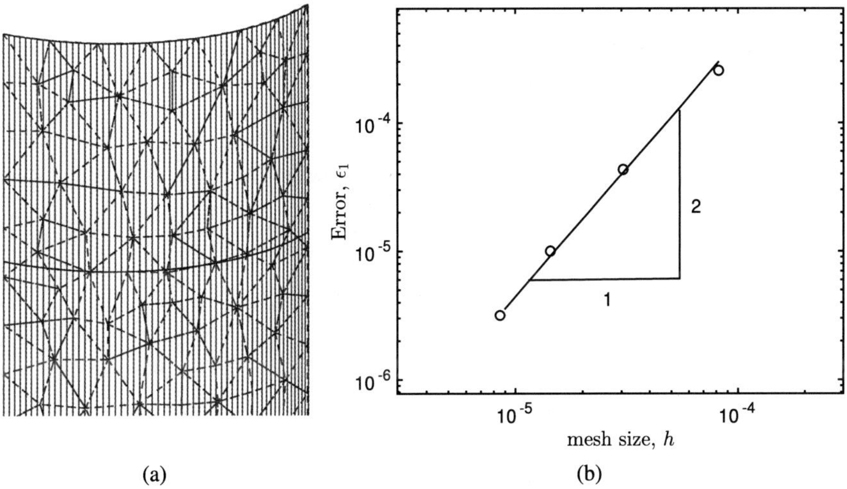

Fig. 7.13 Demonstration of common-refinement method for triangular-to-quadrilateral data transfer: **a** the representative surface meshes, and **b** the spatial error ε_1 versus mesh size h

(solid) mesh. The traction on the fluid mesh is prescribed as given by Eq. (7.65) and the error is computed by Eq. (7.66). Different mesh sizes (h ranging from 0.04 to 0.2) are employed for the study and the error is shown in Fig. 7.13, where a second-order of accuracy is obtained for load transfer across meshes of different shapes, thus demonstrating the generality of the technique.

7.4.1.2 Transient Data Transfer

This section analyses the second error resulting from the repeated transfer of data across the non-matching meshes. To accomplish this, a long deformable cylinder in a flow channel is considered with varying degrees of mismatches along the fluid-structure interface. The schematic of the problem is shown in Fig. 7.14a. The cylinder has a diameter of D and is modeled as an elastic tube of length $50D$. The inlet and outlet boundaries are at a distance of $10D$ and $30D$ from the center of the cylinder, respectively. The ends of the cylinder are fixed at Γ_{top} and Γ_{bottom} boundaries. A freestream velocity of $u^{\mathrm{f}} = U$ is given at the inlet boundary Γ_{in}, where u^{f} is the X-component of the fluid velocity $\boldsymbol{v}^{\mathrm{f}} = (u^{\mathrm{f}}, v^{\mathrm{f}}, w^{\mathrm{f}})$. A stress-free boundary condition is satisfied at the outlet boundary Γ_{out}. A slip boundary condition is imposed on Γ_{top} and Γ_{bottom}, while a no-slip condition is satisfied on the surface of the cylinder.

Similar to the static load data transfer, the fluid and the solid surfaces are decomposed into $w_\theta = 32$ and $w_z = 25$ elements along the circumference and the spanwise directions, respectively, giving a mesh ratio of $A_s/A_f = 1$, where the areas of the fluid and the solid elements are identical. Then, a mismatch across the meshes is generated by fixing the fluid mesh and rotating the solid mesh along the spanwise axis of the cylinder. This kind of mismatch is depicted in Fig. 7.14b. Corresponding to each fluid element k and the solid element j, a sub-element on the common-refinement interface is generated which is the intersection of the projected elements from both the subdomains. Consider $A_{k,j}$ to be the area of the sub-element corresponding to the fluid element k and the solid element j. Thus, the degree of mismatch with respect to the fluid element k ($\delta_k^{F\to S}$), can be defined as

$$\delta_k^{F\to S} = \left(1 - \frac{\max_j A_{k,j}}{\sum_j A_{k,j}}\right). \tag{7.67}$$

Furthermore, the degree of mismatch with respect to the fluid surface ($\delta^{F\to S}$) can be written as the mean of the degree of mismatch with respect to each fluid element, i.e.,

$$\delta^{F\to S} = \frac{1}{N}\sum_{k=1}^{N} \delta_k^{F\to S}. \tag{7.68}$$

It is found that, with this definition, $\delta_{F\to S} \in [0, 0.5]$ for the case. A $\delta_{F\to S} = 0$ represents a matching mesh and $\delta_{F\to S} = 0.5$ depicts a staggered configuration between the fluid and the solid meshes (Fig. 7.14c). Now, to assess the common-refinement technique for the transient data transfer, five sets of meshes with different degrees of mismatches are constructed for the analysis. The matching mesh case is considered as reference to compute the error associated with the degree of mismatch.

For the various mesh configurations described above, the characteristic response of the deformable cylinder, including the in-line displacement, cross-flow displacement and the force coefficients (C_d and C_l) are quantified and compared. The error in these quantities is computed as

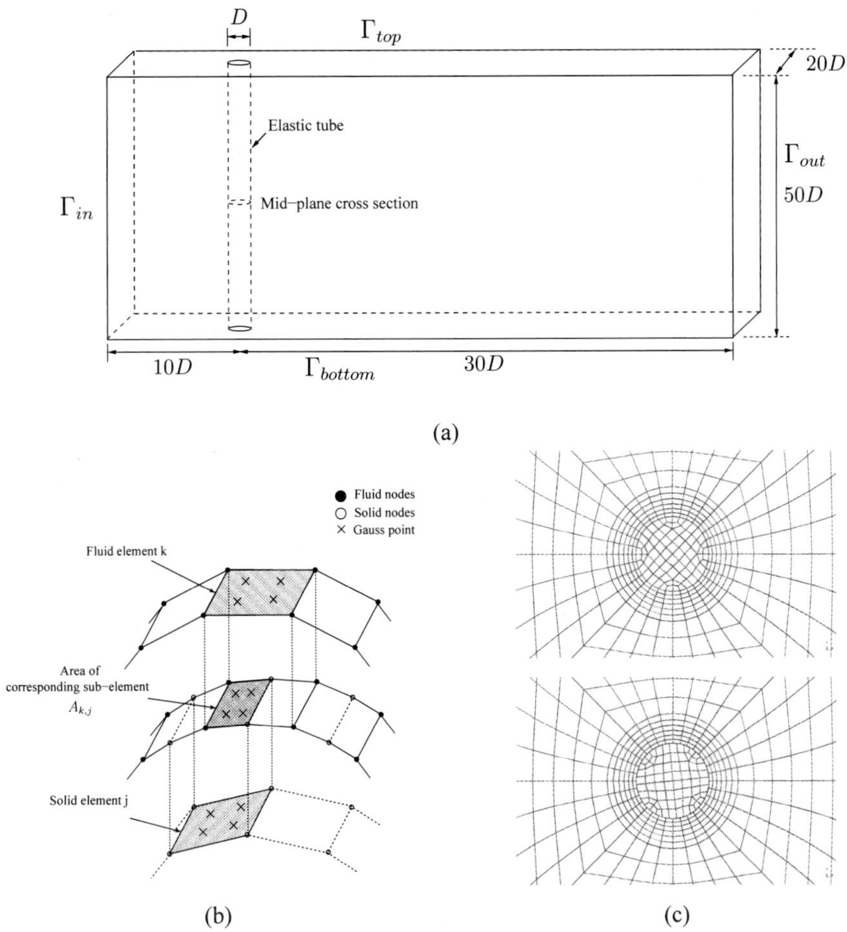

Fig. 7.14 Computational setup for the transient analysis of common-refinement discretization: **a** schematic diagram of physical setup for the deformable tube problem in a uniform flow; **b** sketch of non-matching meshes and their projected area on the common-refinement surface; and **c** the meshes on the mid-plane cross section for matching ($\delta_{F\to S} = 0$) and non-matching ($\delta_{F\to S} = 0.5$) scenarios

$$\varepsilon_2 = \frac{||\boldsymbol{R} - \boldsymbol{R}_{\delta_{F\to S}=0}||_\infty}{||\boldsymbol{R}_{\delta_{F\to S}=0}||_\infty} \tag{7.69}$$

where $\boldsymbol{R}$ is a vector consisting of the temporal response of the non-matching mesh, $\boldsymbol{R}_{\delta_{F\to S}=0}$ is the corresponding response of the matching mesh and $||\cdot||_\infty$ is the infinity norm. The computed error is tabulated in Table 7.2. A very small difference in the responses across the non-matching cases is observed. This establishes the reliability and accuracy of the common-refinement method for the transfer of data across the non-matching meshes at the fluid-structure interface.

Table 7.2 The relative error of characteristic response between the matching and non-matching meshes

$\delta_{F\to S}$	$x/D(\times 10^{-3})$	$y/D(\times 10^{-3})$	$C_l(\times 10^{-3})$	$C_d(\times 10^{-3})$
0.1	1.2512	2.1213	0.0791	1.2382
0.2	1.6620	3.0710	0.1102	1.5931
0.3	3.6998	3.9819	0.1301	1.6318
0.4	5.6904	4.8199	0.1691	1.8075
0.5	7.5399	5.6292	0.2005	2.1933

Table 7.3 Average elapsed time for each component in an iteration collected from 500 steps of simulation

Component	Average elapsed time (s)	Percentage (%)
Fluid	17.82	89.30
Solid	2.14	10.69
Common refinement	1.77×10^{-3}	0.01

The percentage is computed with respect to the average elapsed time for each iteration. The common refinement method implemented has insignificant contribution to the total time spent

7.4.1.3 Performance

Finally, the CPU performance of the implementation of the common-refinement technique along with the nonlinear iterative force correction (NIFC) is discussed here. As a result of NIFC, the coupled FSI formulation undergoes several nonlinear iterations at each time step. For each iteration, the elapsed time consists of solving the fluid and structural equations along with the common-refinement data transfer. The time taken for each component is clocked to assess the performance at each iteration, considering the reference case in Sect. 7.4.1.2. The fluid domain consists of 33450 eight-node hexahedron elements with 35,984 nodes and the solid domain has 2314 nodes comprising of 1800 eight-node hexahedron elements. The interface surface consists of 800 quadrilateral elements with 832 nodes. The computation is carried out in parallel with 2 cores in Intel(R) Xeon(R) CPU E5-2630L v2 @ 2.40 GHz. Table 7.3 summarizes the average time elapsed for each component in each iteration, considering 500 time steps. Generally, the solver takes about four iterations in each time step to satisfy the conference tolerance of 10^{-5}.

It can be seen that the common-refinement data transfer does not take significant amount of time, while most of the computational time is taken for solving the fluid and the solid equations. This is consistent with the observations in [109].

The parallel performance of the implementation is assessed by collecting the elapsed time for data transfer using different number of processors. The fluid and the solid surfaces are 100 units long in length and width and are discretized into 5040 $\times$ 1000 four-node quadrilateral elements. An arbitrary load data is specified at each node on one of the surface and is projected to the other surface by the common-

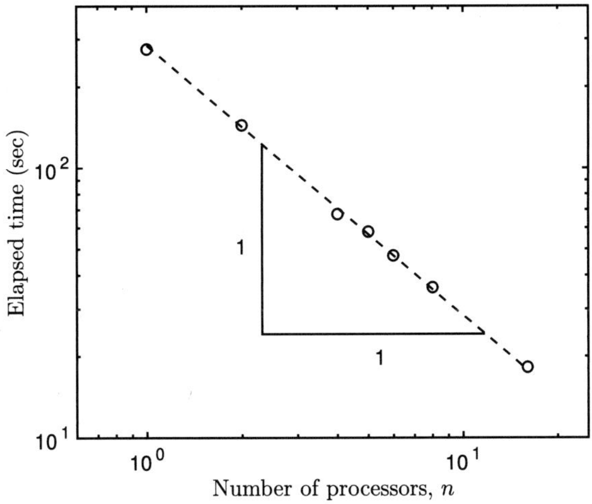

Fig. 7.15 Parallel performance of the common-refinement method. The number of processors tested ranges from 2 to 16. The dotted line is the best fitted line for data collected, which has a negative gradient of one

refinement technique. The elapsed time for the data transfer across the surfaces by different number of processors ranging from 2 to 16 is shown in Fig. 7.15. The slope of the plot indicates an inverse proportionality between the elapsed time and the number of processors, confirming the scalability of the common-refinement method.

7.4.2 Three-Dimensional FSI with Non-Matching Meshes

After assessing the common-refinement projection technique for non-matching meshes by error analysis, the three-dimensional partitioned fluid-structure interaction formulation is verified and validated for the benchmark problem of elastic bar attached to a circular cylinder, FSI-III case in [216]. The problem consists of low structure-to-fluid density ratio and $Re = 200$ based on the diameter of the cylinder. A thin flexible bar is clamped behind a fixed rigid non-rotating circular cylinder, as shown in Fig. 7.16a. The non-dimensional parameters for benchmark problem are given in Table 7.4.

The boundary conditions applied on the computational domain are identical to that in [216]. A no-slip condition is imposed on the surface of the cylinder, the flexible bar, the top and the bottom boundaries. The fluid-structure interface is thus the boundary between the flexible bar and the fluid domain. At the inlet boundary, a parabolic velocity is specified as

$$u^{\mathrm{f}}(0, y) = 1.5\bar{U}\frac{y(H-y)}{\left(\frac{H}{2}\right)^2} = 1.5\bar{U}\frac{4.0}{0.1681}y(0.41-y), \tag{7.70}$$

Table 7.4 FSI parameters for unsteady cylinder-bar problem at $Re = 200$ and $\rho^s/\rho^f = 1.0$

Parameters	Benchmark
Cylinder diameter, D	0.1 m
Mean inlet velocity, $\bar{U}$	1.0 m/s
Fluid density, ρ^f	1000 kg/m^3
Bar thickness w	0.02 m
Bar length, L	0.35 m
Structure density, ρ^s	1000 kg/m^3
Young modulus, E	5.6×10^5 Pa
Reynolds number, Re	200
Density ratio, $\rho^r = \frac{\rho^s}{\rho^f}$	1.0
Poisson's ratio, ν^s	0.4

Table 7.5 Mesh convergence and validation of FSI-III case

Mesh	Fluid elements	Solid elements	$A_{x,max}$	$A_{y,max}$	f_y
M1	7171	136	2.68 (9.39%)	32.94 (0.15%)	5.15 (4.90%)
M2	11770	544	2.55 (4.08%)	31.89 (3.04%)	5.34 (1.34%)
M3	25882	1280	2.45	32.89	5.42
Benchmark	9216	1280	2.68	35.34	5.3

The percentage differences are calculated by using M3 result as the reference

where u^f is the X-component of the fluid velocity $\boldsymbol{v}^f = (u^f, v^f, w^f)$, $\bar{U}$ denotes the mean inlet velocity and $H = 4.1D$ is the height of the computational domain between Γ_{top} and Γ_{bottom}. A stress-free condition is satisfied at the outlet boundary.

The fluid and the structural domains are discretized using finite elements. In the fluid domain, a boundary layer region is constructed around the cylinder and the bar, while a triangular mesh is formed outside the boundary layer region. To begin with, mesh convergence study is performed to ensure sufficient mesh resolution for both the fluid as well as structural domains. Matching meshes along the fluid-structure interface are employed to verify the coupling, thus eliminating the effect of the non-matching discretization. A typical mesh for the study is shown in Fig. 7.16b. Three sets of matching meshes with increasing refinement with details about the mesh and the response characteristics are shown in Table 7.5. Thus, M3 is observed to achieve convergence and is selected as the reference for further study.

The error in the frequency response is defined as

$$\varepsilon_1 = \frac{|f_y - f_{y,M3}|}{f_{y,M3}}. \tag{7.71}$$

where f_y is the frequency of the cross-flow response for 5 consecutive time periods and $f_{y,M3}$ is the value obtained for the reference mesh M3. The variation of this error with mesh size is shown in Fig. 7.17. A slight deviation from the theoretical

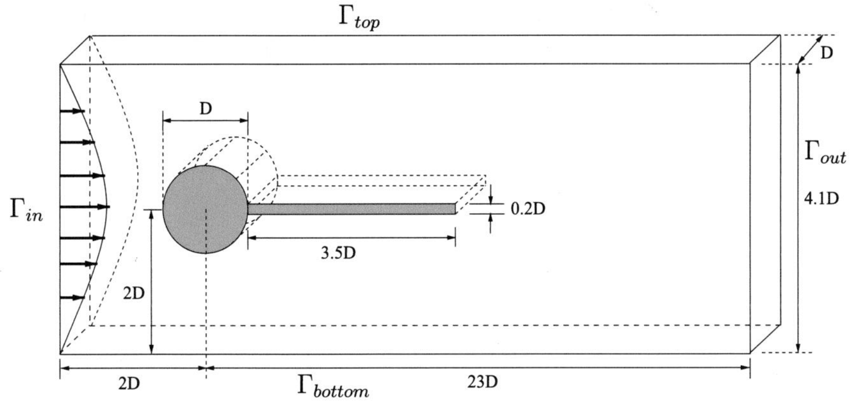

(a) Schematic of unsteady cylinder-bar problem

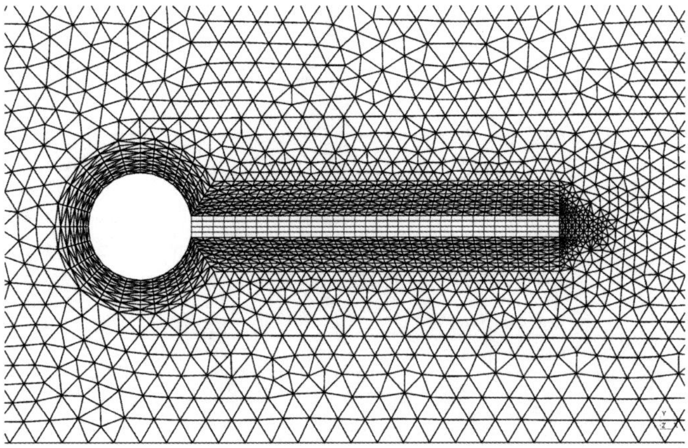

(b) Representative matching mesh M1

Fig. 7.16 Unsteady cylinder-bar problem for the verification and convergence study. Details of meshing parameters are listed in Table 7.5

second-order convergence is observed which may be due to other spatial and temporal discretization errors of the coupled system.

The maximum tip displacements ($A_{x,max}$ and $A_{y,max}$) and the transverse frequency f_y are observed to have a good agreement with the benchmark results. The temporal variation of the tip displacement of the bar for mesh M3 is compared with the benchmark result in Fig. 7.18, where both the components of the displacements agree well with the results from the literature.

Next, FSI computations are carried out for the non-matching meshes across the fluid-structure interface employing different mesh ratios. Figure 7.19 depicts a typical mesh for such non-matching mesh configuration. Mesh ratios varying from 0.25 to 4.0 are employed so that situations involving load transfer from coarse to fine mesh

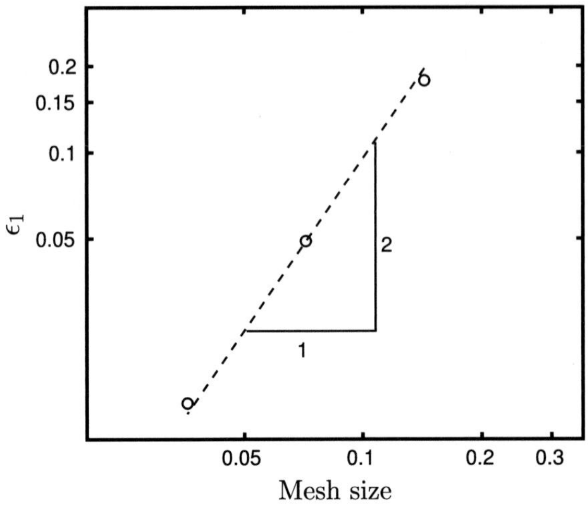

Fig. 7.17 Mesh convergence of FSI-III case for non-matching meshes using hybrid CR-NIFC formulation. The dotted line indicates a theoretical second-order accuracy

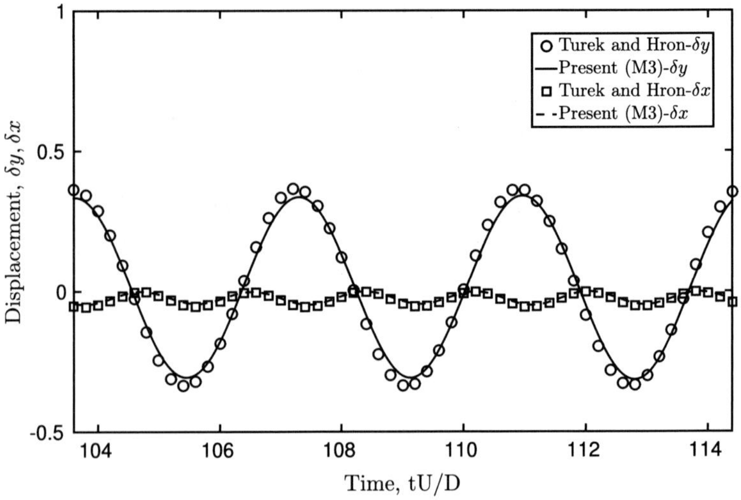

Fig. 7.18 Comparisons of tip displacements between the present study using M3 mesh and the reference data

Table 7.6 Sensitivity and assessment of FSI results for varying mesh ratios h_s/h_f

h_s/h_f	Fluid elements	Solid elements	$A_{x,max}$	$A_{y,max}$	f_y
4.0	25882	136	2.45	32.73	5.467
2.0	25882	544	2.47	32.97	5.455
1.0	25882	1280	2.45	32.89	5.468
0.5	25882	5120	2.46	32.95	5.455
0.25	25882	20480	2.44	32.98	5.442

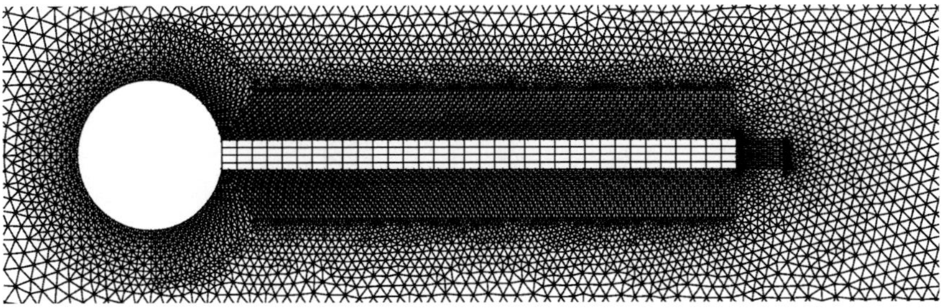

(a) X-Y view

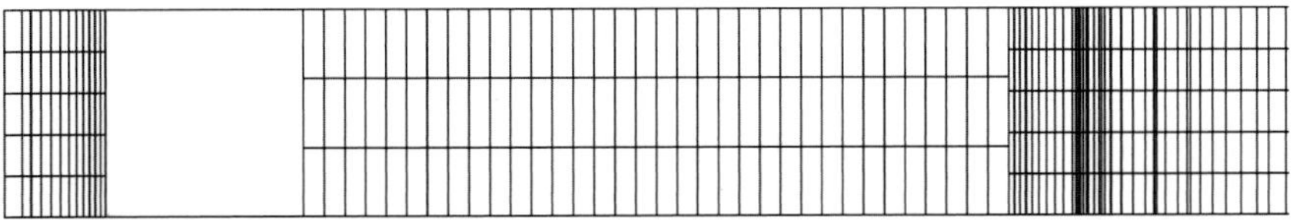

(b) Y-Z view

Fig. 7.19 A representative non-matching mesh configuration for the cylinder-bar system

and vice-versa are considered. The response characteristics of the tip displacement and the frequency for each mesh ratio h_s/h_f are shown in Table 7.6.

The results for different mesh ratios are observed to be close to each other. The flow field is compared by plotting the instantaneous Z-vorticity contours for representative mesh ratios, viz., $h_s/h_f = 0.5, 1.0, 2.0$ in Fig. 7.20 at the instant where the tip of the bar is at the maximum displacement. The visualization shows very similar flow patterns indicating qualitative accuracy of the three-dimensional common-refinement projection technique employing the nonlinear iterative force correction for the fluid-structure interaction problems.

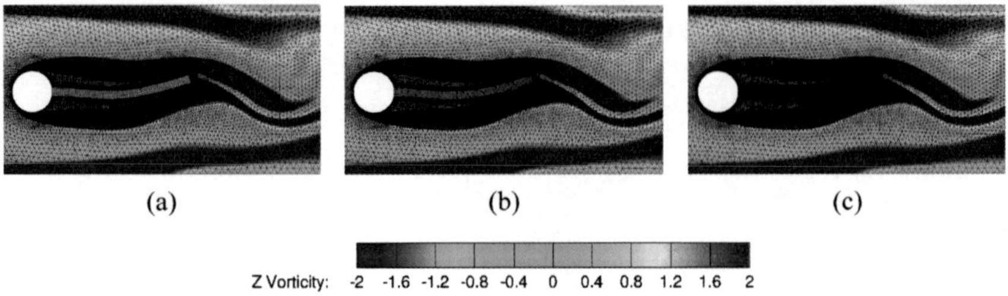

Fig. 7.20 Instantaneous Z-vorticity contours and meshes for representative mesh ratio: **a** fine to coarse mesh, $h_s/h_f = 2.0$, **b** matching mesh, $h_s/h_f = 1.0$, (c) coarse to fine mesh, $h_s/h_f = 0.5$

7.4.3 Application to Offshore Riser VIV

Vortex-induced vibrations (VIV) of a riser can be detrimental to the offshore operations due to fatigue failure in complex ocean environment. Therefore, its prediction and control is imperative in the riser design strategies. In this section, the partitioned fluid-structure formulation based on common-refinement technique for non-matching meshes at the fluid-structure interface is demonstrated for such a riser subjected to ocean current and strong added-mass effects. We consider a uniform flow across the riser and compare our results with the experiment conducted in [1].

The computational domain along with the boundary conditions applied on the riser are shown in Fig. 7.21. The inlet (Γ_{in}) and the outlet (Γ_{out}) boundaries are at a distance of $10D$ and $30D$ from the center of the riser, respectively, D being the diameter of the riser. The side boundaries are $10D$ from the center of the riser. The length of the riser spans $481.5D$ in the Z-direction. At the inlet boundary, a freestream velocity of $u^{\mathrm{f}} = U$ along the X-axis is applied, where u^{f} is the X-component of the fluid velocity $\boldsymbol{v}^{\mathrm{f}} = (u^{\mathrm{f}}, v^{\mathrm{f}}, w^{\mathrm{f}})$. Slip boundary condition is satisfied at the top and bottom boundaries and a traction-free condition is imposed at the outlet boundary. Both ends of the riser are pinned with a tension T applied at the top of the riser, while no-slip condition is applied on the riser surface.

The computational domain can be divided into fluid and structural subdomains for spatial discretization. The fluid and structural meshes are shown in Fig. 7.22. The fluid mesh consists of 9×10^5 nodes with 1.2×10^6 unstructured hexahedral elements and the structural mesh of the riser contains 8.7×10^4 nodes with 9.7×10^4 hexahedral elements. The circumference of the riser cross-section is divided into 96 and 120 elements on the fluid and solid meshes, respectively. In the fluid mesh, a boundary layer region considering $y^+ < 1$ in the wall-normal direction is constructed. In the spanwise direction, both the fluid and the structural subdomains are discretized into 100 layers. The time step size for the computation is selected as $\Delta t U/D = 0.1$. The non-dimensional parameters for the problem can be summarized as

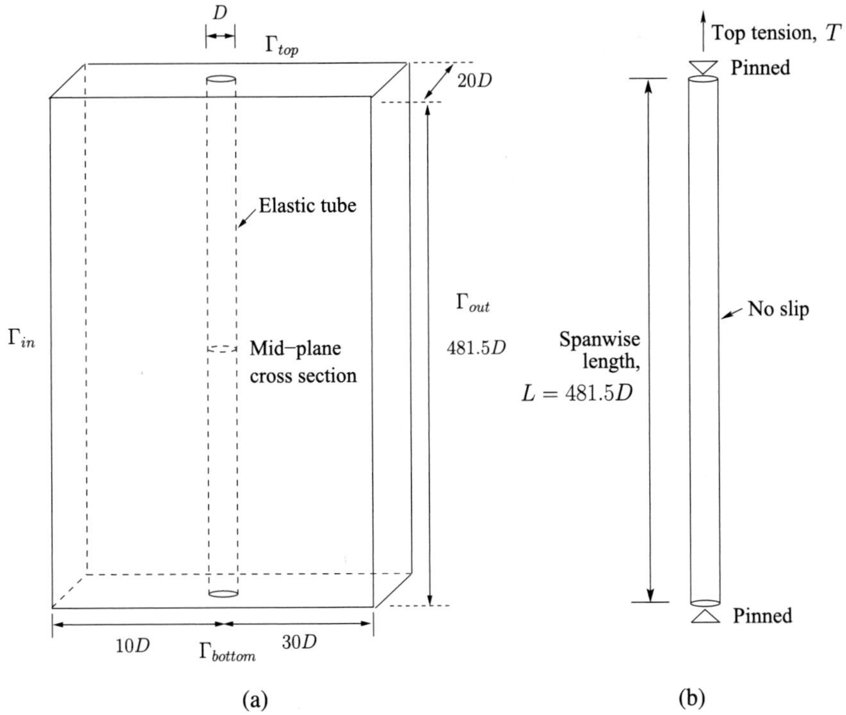

Fig. 7.21 Flow past a flexible offshore riser: **a** schematic of computational setup, **b** boundary conditions applied on the pinned-pinned tensioned riser

$$Re = \frac{\rho^{\mathrm{f}} U D}{\mu^{\mathrm{f}}} = 4000, \tag{7.72}$$

$$\frac{EI}{\rho^{\mathrm{f}} U^2 D^4} = 2.1158 \times 10^7, \tag{7.73}$$

$$\frac{T}{\rho^{\mathrm{f}} U^2 D^2} = 5.10625 \times 10^4, \tag{7.74}$$

$$m^* = \frac{m^{\mathrm{s}}}{\frac{\pi}{4} D^2 L \rho^{\mathrm{f}}} = 2.23, \tag{7.75}$$

where I is the second moment of area of the cross-section of the riser and m^{s} is the mass of the riser.

7.4.3.1 Response Characteristics

The amplitude response of the riser is compared to that of the experiment in Fig. 7.23 where the temporal evolution of the cross-flow displacement at the posi-

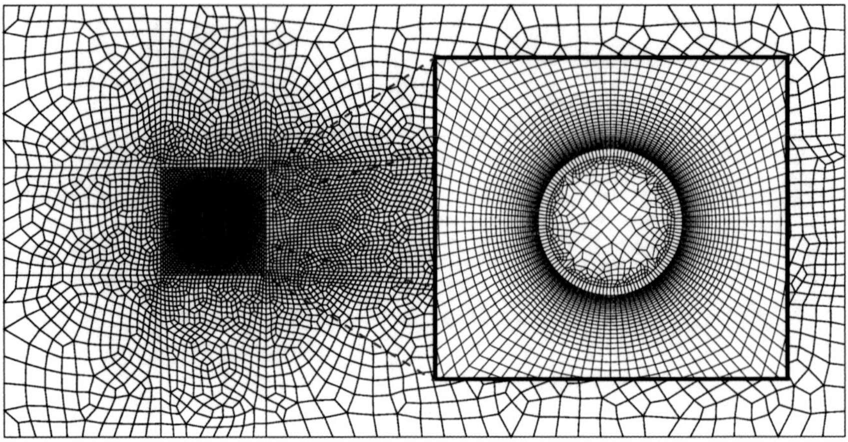

Fig. 7.22 A long riser under uniform flow: two-dimensional layer of the unstructured non-matching computational mesh. The inset shows the magnified view of the non-matching fluid-structure interface. The mesh is extruded in the third-dimension while maintaining a non-matching spanwise mesh

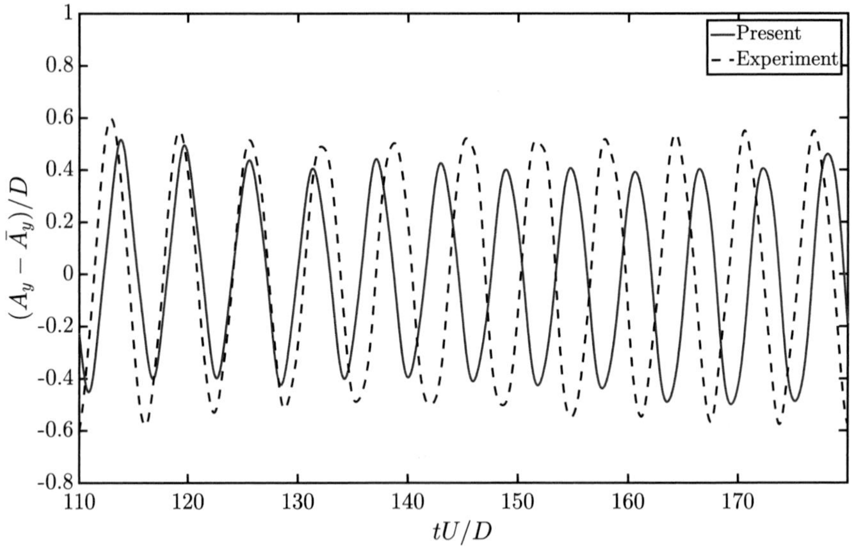

Fig. 7.23 Riser response under uniform current flow: time history of the cross-flow displacement at $z/L = 0.55$

tion $z/L = 0.55$ is shown. A multi-frequency response is observed from the spectral analysis of the amplitude response in Fig. 7.24a. Moreover, the in-line response frequency ($fD/U = 0.3516$) is observed to be twice that of the cross-flow frequency ($fD/U = 0.1758$).

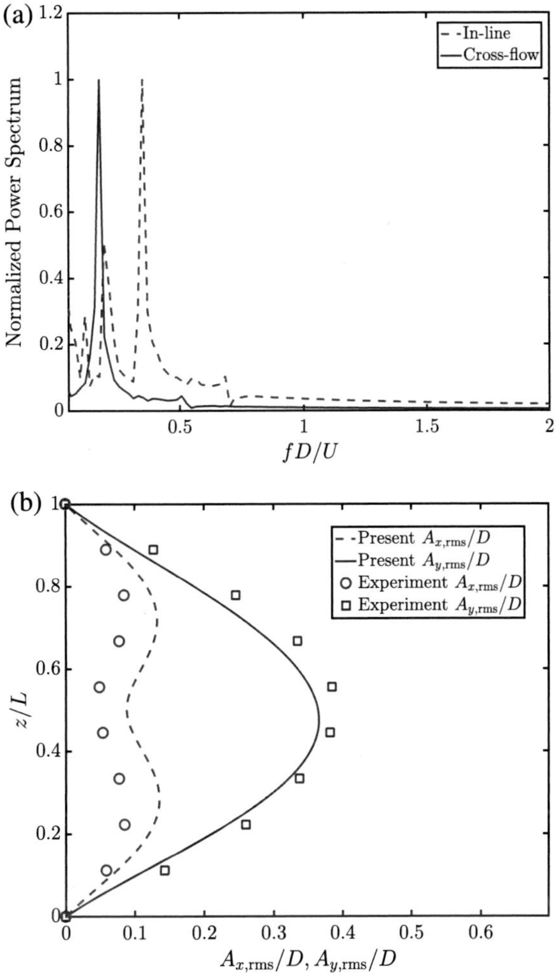

Fig. 7.24 Riser response under uniform current flow at $Re = 4000$: **a** power spectrum of the in-line and the cross-flow amplitudes along the riser, **b** comparison of the root mean square values of the in-line and cross-flow displacements along the riser with that of the experiment

The root mean square (rms) values of the amplitude response along the riser span is also compared with the experiment in Fig. 7.24b. The rms values are computed as follows: Let A_x and A_y be the amplitude response of a point along the riser in the in-line and cross-flow directions, respectively and $\overline{A_x}$ and $\overline{A_y}$ be the corresponding temporal average of the responses in that location. Thus, the rms values are calculated as

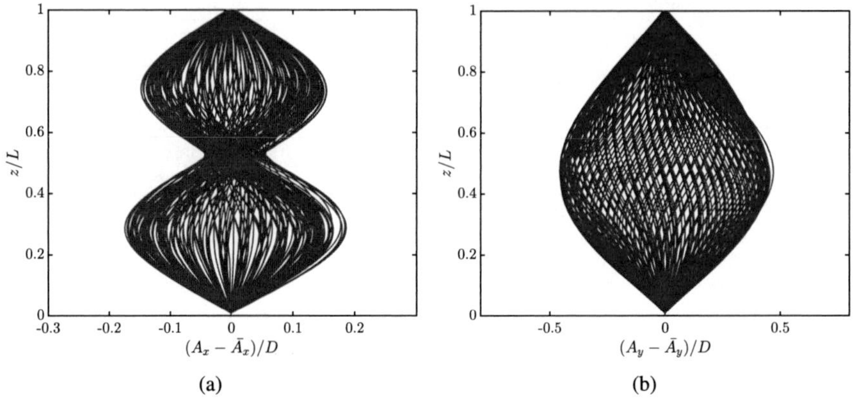

Fig. 7.25 Riser response envelope under uniform current flow: **a** in-line and **b** cross-flow directions. The riser is vibrating in the fundamental mode for the cross-flow, and the second mode for the in-line directions

$$A_{x,\mathrm{rms}} = \sqrt{\frac{1}{N}\sum_{i=1}^{N}(A_{x,i} - \overline{A_x})^2}, \tag{7.76}$$

$$A_{y,\mathrm{rms}} = \sqrt{\frac{1}{N}\sum_{i=1}^{N}(A_{y,i} - \overline{A_y})^2}, \tag{7.77}$$

where N denotes the number of samples collected in time. A good agreement of the cross-flow amplitude is observed in Fig. 7.24b, while some over-prediction is seen for the in-line vibrations. The riser oscillates in a dominant second mode response and first mode response in the in-line and cross-flow directions, respectively, as can be observed from the response envelope plotted in Fig. 7.25. These observations corroborate the phenomenon of dual resonance from the spectral study. Furthermore, a standing wave-like pattern is observed in the response in Fig. 7.26, where the riser response along its span is plotted with time.

7.4.3.2 Vortex Patterns

The vortex patterns in the wake of the riser are visualized by plotting the Z-vorticity contours at different locations $z/L \in [0.11, 0.88]$ along the span of the riser in Fig. 7.27a. The vibration amplitude of the riser is also displayed along the riser surface. The vortex shedding pattern is quite complex near the locations having larger amplitude of vibration. While we observe a 2S shedding mode at most of the locations, a 2P mode is also observed at some locations. The three-dimensional vortical structures in the wake of the riser are visualized by iso-surfaces of Q-criterion colored by the Z-vorticity in Fig. 7.27b. More intense structures are observed at

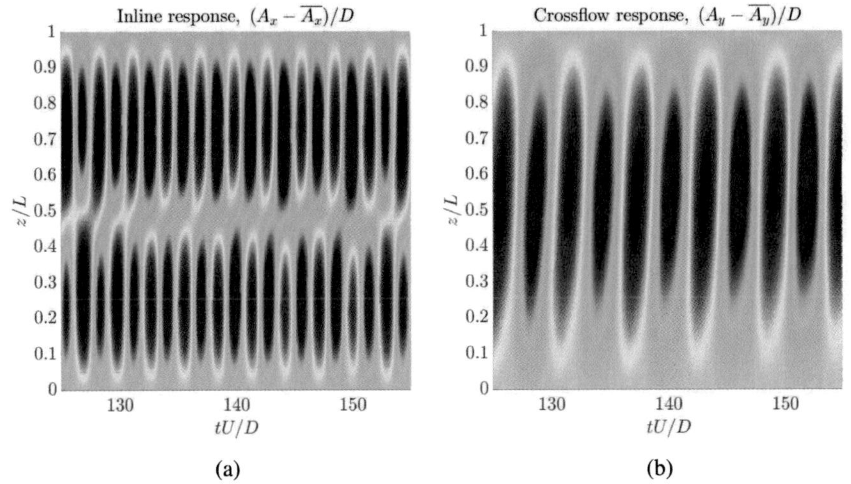

Fig. 7.26 Standing wave riser response under uniform current flow: **a** in-line; **b** cross-flow

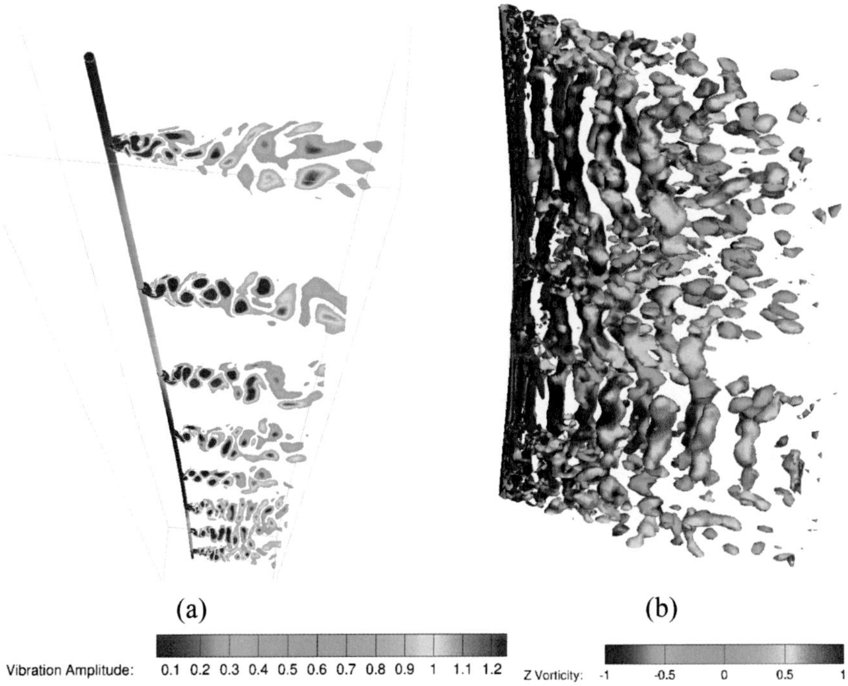

Fig. 7.27 **a** Vibration amplitude (surface of riser) and Z-vorticity contours (cross section) at different positions along the riser; **b** Instantaneous isosurfaces of Q-criterion $(-\frac{1}{2}\frac{\partial u_i}{\partial x_j}\frac{\partial u_j}{\partial x_i})$ at $Q = 0.1$, colored by Z-vorticity.

the location having large amplitude of vibration. A more detailed study on the flow physics of VIV of a riser has been carried out in [113].

To summarize, this chapter discussed various partitioned techniques with respective to spatial as well as temporal coupling along the fluid-structure interface. A more robust common-refinement projection spatial coupling for non-matching meshes and stable nonlinear iterative force correction (NIFC) are discussed. Error and convergence analysis are performed for the partitioned coupled algorithm leading to validation with the benchmark results of strongly coupled FSI problem. Finally, the scheme is demonstrated for vortex-induced vibration of an offshore riser subjected to uniform ocean current, where reasonable results are obtained considering non-matching meshes across the fluid-structure interface.

Acknowledgements The first author wishes to acknowledge support from the Center for the Simulation of Advanced Rockets (CSAR) funded by the US Department of Energy and from the Computational Science and Engineering (CSE) program at University of Illinois. Both authors would like to acknowledge the National Research Foundation, the Singapore Maritime Institute and A*STAR-SERC for their financial support.

Chapter 8
Two-Phase Fluid-Structure Interaction

8.1 Introduction

Two-phase fluid-structure interaction is omnipresent. It has applications from offshore pipelines carrying oil or gas [42, 195], marine vessels subjected to free-surface ocean waves, blood flow through arteries and veins, to multiphase flow inside heat exchangers [229]. Here, we focus on the application to two-phase internal flow in offshore pipelines subjected to turbulent ocean currents. These pipelines can undergo fluid-elastic instabilities and self-excited vibrations [25, 168], which may lead to the failure of the structure and operational delay due to highly nonlinear effects of fluid-structure interaction.

Two-phase FSI consists of complex nonlinear interfacial interactions and challenges associated with the boundary conditions, for example, the satisfaction of the no-slip condition at the structure near highly-deformable two-phase fluid interface, the evolution of the two-phase fluid-fluid interface, and the accurate movement of the fluid-structure interface while satisfying the equilibrium conditions. The no-slip condition at the fluid-structure interface can be satisfied by either considering a body-conforming mesh with the structure as Lagrangian and solving the two-phase fluid equations in arbitrary Lagrangian-Eulerian (ALE) description, or by representing the boundary conditions by a fictitious force field via immersed boundary approach [150, 196]. The ALE framework is beneficial in accurate modeling of the boundary-layer vorticity flux and near-wall turbulence at the fluid-structure interface.

Techniques which employ the boundary-conforming mesh for the fluid-structure interaction can be classified into monolithic and partitioned schemes. The fluid and structural equations are solved simultaneously by assembling into a single block matrix in the monolithic approach [23, 72, 90]. This approach is quite expensive computationally, albeit being robust and stable for low structure-to-fluid mass ratios. They lack the flexibility and modularity in utilizing existing fluid and structural solvers [23, 72, 90, 140]. In contrast, the partitioned approach solves the fluid and structural equations in a sequential manner, giving flexibility in the coupling procedure with minimal changes to the existing blocks of the code. For solving such a large

R. K. Jaiman and V. Joshi, *Computational Mechanics of Fluid-Structure Interaction*,
https://doi.org/10.1007/978-981-16-5355-1_8

scale computation of offshore structures interacting with the two-phase fluid flow, the structures can consist of multiple bodies such as rigid bodies, beams, shells, cables, etc. Using a partitioned approach for such complex coupling gives some attractive advantages with regard to iterative solvers, preconditioning strategies, scalability and parallel processing. Moreover, the discretizations in space and time can be chosen independently for the fluid and structural solvers. Thus, partitioned coupling of the two-phase fluid-structure equations has been considered in this chapter.

Numerical treatment of the fluid-fluid two-phase interface involving immiscible fluids poses certain challenges owing to the complexity in the representation and evolution of the interface [174]. This representation of the continuum interface between the two phases can be carried out by either interface-tracking or interface-capturing methods. In the former, the boundary separating the two fluid sub-domains is tracked explicitly, for example, front-tracking [217], particle tracking [228] and arbitrary Lagrangian-Eulerian [55] techniques. Although the interface location can be accurately predicted by tracking the moving boundary or markers at the interface by these techniques, large topological changes in the interface motion may lead to numerical difficulties, requiring remeshing which can be computationally expensive in three-dimensions. On the contrary, interface-capturing methods do not explicitly track the interface. The fluid-fluid interface is represented implicitly using a field function on the computational domain on a fixed Eulerian mesh. Level-set, volume-of-fluid and phase-field approaches fall under interface-capturing techniques. The implicit evolution of the field function gives the advantage of capturing even highly topologically changing interface such as breaking and merging.

Apart from the accurate capturing of the fluid-fluid interface, the discontinuity of the physical quantities such as density, viscosity and pressure across the interface poses a numerical challenge. In particular, the discontinuities in these properties can lead to spurious unphysical oscillations in the solution resulting in an unbounded behavior of the two-phase flow system. Thus, the background fixed mesh is required to be sufficiently resolved to capture these discontinuities demanding high cost of computation. Accurate modeling of surface tension or capillary effects along with mass conservation property are crucial for physically consistent solution of the two-phase flow system.

Among the types of interface-capturing techniques, the volume-of-fluid (VOF) and level-set are widely used methods. The VOF utilizes a volume-of-fluid function to extract the volume fraction of each fluid phase in every computational cell. It conserves mass accurately for incompressible flow, but the calculation of the curvature of the interface and normal can be quite tedious due to interface reconstruction [177, 189, 217]. The smearing of the interface due to numerical diffusion adds to the complexity. The level-set approach, on the other hand, constructs a signed-distance function (level-set function) of a discretization point to the interface and can provide a non-smeared interface [191]. The zero-level-set of the distance function gives a sharp-interface description and the curvature of the interface can be accurately approximated. However, mass conservation is a challenge for the level-set method due to the numerical dissipation as a result of discretization, and the re-initialization process which keeps the level-set function as a signed distance function [248]. Techniques consisting of high-order discretization [203], improved re-initialization [170, 184, 203], a coupled particle tracking/level-set [60, 61, 127], and a coupled level-set/volume-of-fluid [204, 235, 242] have been proposed to circumvent the mass conservation issue, but they tend to make the overall scheme more complex and expensive computationally. Another attempt for imparting the mass conservation was made as conservative level-set method in [160, 161] in which the signed distance function was replaced with a hyperbolic tangent function. However, re-initialization was necessary to maintain the width of the hyperbolic tangent profile. The extended finite element method (XFEM) is a recent improvement in the level-set technique. It utilizes the enrichment of shape functions of the elements along the interface region [188]. Therefore, it can be quite tedious to implement the level-set and volume-of-fluid methods in three-dimensions for unstructured grids over complex geometries owing to the geometric reconstruction and manipulation required at the fluid-fluid interface.

The interface-capturing techniques can be further classified into sharp-interface and diffuse-interface descriptions. In the former, the interface is treated as infinitely thin with the physical properties such as density and viscosity having a jump discontinuity at the interface between the two phases. The latter description assumes a gradual and smooth variation of these physical properties across the interface of a finite thickness. The physical properties are a function of a conserved order parameter which is solved by minimization of a free energy functional derived from thermodynamic arguments [9]. The interface location is defined by the contour levels of the phase-field order parameter in this case. In contrast to the sharp-interface description, the jump conditions at the two-phase interface are replaced by a rapid, but smooth variation of the order parameter and other physical quantities in the thin interfacial region. It has been shown to approach the sharp-interface limit as the interface thickness approaches to zero asymptotically [8].

The diffuse-interface description based phase-field models originate from the thermodynamically consistent theories of phase transitions by minimization of gradient energy across the two-phase interface. The phase-indicator order parameter has a value of 1 in one phase and -1 in the other, and is solved on a fixed Eulerian mesh to evolve the interface. Due to the diffused interface, jump conditions are not required to be satisfied explicitly at the interface. Furthermore, the surface tension effects are modeled as a function of the order parameter. As a result, these models do not require any re-initialization or reconstruction at the interface. The mass conservation property can also be imposed in a relatively simple manner. Given that the mesh is refined sufficiently, the phase-field models can handle any topological changes in the interface easily. Therefore, they offer attractive physical properties for the modeling of two-phase systems for a broad range of conditions with interface topological changes while maintaining the mass conservation.

In the phase-field methods, the evolution of the two-phase interface is carried out by solving a transport equation in the form of a gradient flow of the energy functional, either in the $L^2(\Omega)$ norm (Allen-Cahn (AC) equation [6]) or in the $H^{-1}(\Omega)$ norm (Cahn-Hilliard (CH) equation [35]). The AC equation is a second-order differential equation and avoids the requirement of equation splitting, as is the case with

CH equation (fourth-order) while using lower-order polynomial discretizations. For the phase-field models, the energy stability and mass conservation properties are strongly influenced by the treatment of the nonlinearity of the double-well potential function and the Lagrange multiplier, respectively. The nonlinearities can be handled by various techniques reviewed in [212] to obtain an energy-stable scheme. An unconditional energy-stable scheme is obtained using the mid-point approximation of the derivative of the double-well potential [57]. Furthermore, mass conservation property can be imposed easily by a Lagrange multiplier technique in the conventional AC equation [30, 33, 115, 123, 183]. Therefore, owing to its simplicity, we consider the Allen-Cahn phase-field equation for the two-phase order parameter computations.

In this chapter, we develop the mathematical background involving the phase-field methods followed by the variational discretization of the Allen-Cahn equation under the positivity preserving framework. We present the coupling algorithm between the Navier-Stokes and the Allen-Cahn equations using partitioned approach and verify as well as validate the developed two-phase fluid solver. We then focus on the two-phase fluid-structure coupling to form the two-phase fluid-structure interaction framework.

8.2 Governing Equations for Two-Phase Flow

We begin by briefly discussing the governing equations for the two-phase flow framework consisting of incompressible Navier-Stokes and Allen-Cahn equations.

8.2.1 *The Navier-Stokes Equations*

Consider a d-dimensional spatial fluid domain $\Omega^{\mathrm{f}}(t) \subset \mathbb{R}^d$ with a piecewise smooth boundary $\Gamma^{\mathrm{f}}(t)$. Let the boundary be decomposed into three components, the Dirichlet boundary $\Gamma^{\mathrm{f}}_D(t)$, the Neumann boundary $\Gamma^{\mathrm{f}}_N(t)$ and the fluid-structure boundary $\Gamma^{\mathrm{fs}}(t)$ at time t. The domain $\Omega^{\mathrm{f}}(t)$ consists of two immiscible, incompressible and Newtonian fluid phases occupying the sub-domains $\Omega^{\mathrm{f}}_1(t)$ and $\Omega^{\mathrm{f}}_2(t)$ with a boundary $\Gamma^{\mathrm{ff}}(t)$ between them as shown in Fig. 8.1.

The governing equations for the one-fluid formulation for a viscous, incompressible and immiscible two-phase system in the ALE framework are given as

$$\rho^{\mathrm{f}} \left.\frac{\partial \boldsymbol{v}^{\mathrm{f}}}{\partial t}\right|_{\chi} + \rho^{\mathrm{f}} (\boldsymbol{v}^{\mathrm{f}} - \boldsymbol{w}) \cdot \nabla \boldsymbol{v}^{\mathrm{f}} = \nabla \cdot \boldsymbol{\sigma}^{\mathrm{f}} + \mathbf{sf} + \rho^{\mathrm{f}} \boldsymbol{b}^{\mathrm{f}}, \qquad \text{on } \Omega^{\mathrm{f}}(t) \times [0, T], \tag{8.1}$$

$$\nabla \cdot \boldsymbol{v}^{\mathrm{f}} = 0, \qquad \text{on } \Omega^{\mathrm{f}}(t) \times [0, T] \tag{8.2}$$

where $\boldsymbol{v}^{\mathrm{f}}$ represents the fluid velocity defined for each spatial point $\boldsymbol{x}^{\mathrm{f}}$ in $\Omega^{\mathrm{f}}(t)$, ρ^{f} is the fluid density, **sf** denotes the surface tension singular force replaced by the continuum surface force in the diffuse-interface description, $\boldsymbol{b}^{\mathrm{f}} = \boldsymbol{g}$ is the body force

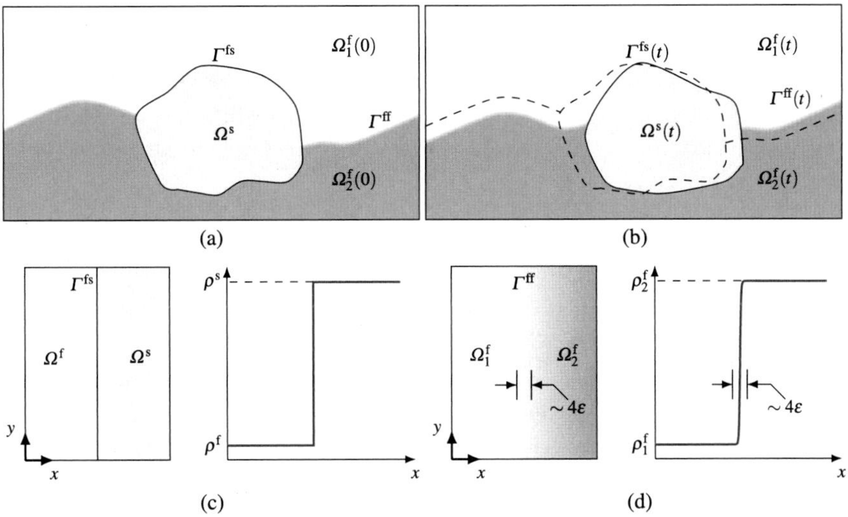

Fig. 8.1 Schematic of two-phase fluid-structure interaction at **a** the initial configuration $t = 0$, and **b** some deformed configuration of the structure at time $t > 0$. $\Omega^{\text{f}}(0)$, Ω^{s} and $\Omega^{\text{f}}(t)$, $\Omega^{\text{s}}(t)$ are the fluid and the structural domains at $t = 0$ and some time $t > 0$ respectively with **c** sharp fluid-structure interface, and **d** diffused fluid-fluid interface, smeared using the internal length scale parameter ε. Γ^{fs} and Γ^{ff} denote the fluid-structure and fluid-fluid interfaces, respectively

applied on the fluid such as the gravitational force, with $\boldsymbol{g}$ being the acceleration due to gravity. The density and viscosity are dependent on the order parameter ϕ which evolves the fluid-fluid interface as

$$\rho^{\text{f}}(\phi) = \frac{1+\phi}{2}\rho_1^{\text{f}} + \frac{1-\phi}{2}\rho_2^{\text{f}}, \tag{8.3}$$

$$\mu^{\text{f}}(\phi) = \frac{1+\phi}{2}\mu_1^{\text{f}} + \frac{1-\phi}{2}\mu_2^{\text{f}}, \tag{8.4}$$

where ρ_i^{f} and μ_i^{f} are the density and dynamic viscosity of the ith phase of the fluid respectively.

8.2.2 *The Fluid-Fluid Interface*

In the sharp fluid-fluid interface description, the velocity continuity and the pressure-jump condition are required to be satisfied at the interface,

$$\boldsymbol{v}^{\text{f}}_{\Omega_1^{\text{f}}} = \boldsymbol{v}^{\text{f}}_{\Omega_2^{\text{f}}}, \qquad \text{on } \Gamma^{\text{ff}}(t) \times [0, T], \tag{8.5}$$

$$(\boldsymbol{\sigma}^{\text{f}}_{\Omega_1^{\text{f}}} - \boldsymbol{\sigma}^{\text{f}}_{\Omega_2^{\text{f}}}) \cdot \boldsymbol{n}_{\Gamma^{\text{ff}}} = \sigma\kappa\boldsymbol{n}_{\Gamma^{\text{ff}}}, \qquad \text{on } \Gamma^{\text{ff}}(t) \times [0, T], \tag{8.6}$$

where $(\cdot)_{\Omega_i^f}$ denotes the argument in the fluid phase i, $\boldsymbol{n}_{\Gamma^{ff}}$ is the normal to the fluid-fluid interface, σ is the surface tension coefficient between the two fluid phases and κ is the curvature of the interface denoted by $\kappa = -\nabla \cdot \boldsymbol{n}_{\Gamma^{ff}}$. The surface tension singular force in the Navier-Stokes equations [Eq. (8.1)] which models the surface tension is thus written as $\mathbf{sf} = \sigma\kappa\delta_{\Gamma^{ff}}\boldsymbol{n}_{\Gamma^{ff}}$, where $\delta_{\Gamma^{ff}}$ is the one-dimensional Dirac delta function given as

$$\delta_{\Gamma^{ff}} = \begin{cases} 1, & \text{for } \boldsymbol{x}^f \in \Gamma^{ff}(t), \\ 0, & \text{otherwise}. \end{cases} \tag{8.7}$$

As mentioned before, the sharp-interface description based on the moving mesh ALE framework is not trivial for complex three-dimensional fluid-fluid interfaces. Therefore, here we consider the diffuse fluid-fluid interface description in which the interface is assumed to have a finite thickness, $\mathcal{O}(\varepsilon)$, on which the physical properties of the two phases vary gradually based on the phase indicator field ϕ. The diffuse-interface description of the fluid-fluid interface recovers to the classical jump discontinuity conditions [Eqs. (8.5–8.6)] for the sharp-interface description asymptotically as $\varepsilon \to 0$ [8]. The singular force in the diffuse-interface description is replaced by a continuum surface force (CSF) [29], which depends on the order parameter (ϕ). Several forms of $\mathbf{sf}(\phi)$ used in the literature are reviewed in [120, 121]. We consider the following definition:

$$\mathbf{sf}(\phi) = \sigma\varepsilon\alpha_{sf}\nabla \cdot (|\nabla\phi|^2\boldsymbol{I} - \nabla\phi \otimes \nabla\phi), \tag{8.8}$$

where ε is the interface thickness parameter defined in the Allen-Cahn phase-field equation and $\alpha_{sf} = 3\sqrt{2}/4$ is a constant.

8.2.3 The Allen-Cahn Equation

Next, we review some of the concepts of free energy functional and look into its minimization which leads to the Allen-Cahn equation. In the diffuse-interface description, the interface has a finite thickness (represented by ε) where the phase-indicator order parameter $\phi(\boldsymbol{x}^f, t)$ varies gradually. The evolution of this interface and its dynamics are governed by the Ginzburg-Landau energy functional $\mathcal{E}(\phi)$, which can be expressed as

$$\mathcal{E}(\phi) = \int_{\Omega^f(t)} \left(\frac{\varepsilon^2}{2}|\nabla\phi|^2 + F(\phi)\right) d\Omega. \tag{8.9}$$

The functional consists of two components: the interfacial energy which depends on the gradient of the order parameter, and the bulk or free energy of mixing which is expressed as a double-well potential function depending on the local value of $\phi(\boldsymbol{x}^f, t)$. There exist two minima of the potential function corresponding to the two stable phases of the fluid. The system minimizes the functional by searching for the stable phases and thus evolving the interface.

The expression for the evolution of the interface can be derived by the gradient flow of the energy functional which can lead to either Cahn-Hilliard or Allen-Cahn equations. Owing to its simplicity in numerical implementation, the Allen-Cahn equation is chosen and can be derived as

$$\partial_t \phi = -\left(\frac{\delta \mathscr{E}(\phi)}{\delta \phi}\right). \tag{8.10}$$

Here, $\partial_t \phi$ is the partial temporal derivative of the order parameter $\phi(\boldsymbol{x}^{\mathrm{f}}, t)$. The double-well potential function is chosen as $F(\phi) = \frac{1}{4}(\phi^2 - 1)^2$ which gives the two stable phases as $\phi = 1$ and $\phi = -1$ with an interface of finite thickness between them over which ϕ varies from -1 to 1. To determine the equilibrium profile of the interface, we minimize the functional by taking its variational derivative and equating it to zero. The variational derivative of the energy functional with respect to ϕ is called the chemical potential, which is given by

$$\frac{\delta \mathscr{E}}{\delta \phi} = -\varepsilon^2 \nabla^2 \phi + F'(\phi). \tag{8.11}$$

The minimization of the energy functional involves the competing effects of the diffusive term $\nabla^2 \phi$ and the reactive term $F'(\phi)$. The equilibrium interface profile is obtained by finding the solution of Eq. (8.11) as:

$$\phi(z) = \tanh\left(\frac{z}{\sqrt{2}\varepsilon}\right), \tag{8.12}$$

where z is the coordinate normal to the interface. The equilibrium interface thickness denoted by $\varepsilon_{\mathrm{eqm}}$ is defined as the distance over which ϕ varies from -0.9 to 0.9 which can be evaluated as $2\sqrt{2}\varepsilon \tanh^{-1}(0.9) = 4.164\varepsilon$.

Based on the equilibrium conditions and the sharp-interface limit, the constant α_{sf} in the surface tension modeling term (Eq. 8.8) must satisfy the condition [120]:

$$\varepsilon \alpha_{\mathrm{sf}} \int_{-\infty}^{\infty} \left(\frac{d\phi}{dz}\right)^2 dz = 1, \tag{8.13}$$

which gives $\alpha_{\mathrm{sf}} = 3\sqrt{2}/4$ after we substitute the expression of the equilibrium interface profile $\phi(z)$ given in Eq. (8.12). This gives rise to the continuum surface force expression for surface tension and capillary force modeling.

Although the conventional Allen-Cahn equation does not conserve mass, we consider its convective form with a Lagrange multiplier for mass conservation. We discuss the strong and variational weak forms of the equation along with its finite element discretization and energy-stable schemes in the following sections.

8.2.3.1 Strong Differential Form

The modified AC equation with both local and global Lagrange multiplier terms to conserve mass [115, 123], in the ALE reference coordinates is given as

$$\left.\frac{\partial \phi}{\partial t}\right|_{\chi} + (\boldsymbol{v}^{\mathrm{f}} - \boldsymbol{w}) \cdot \nabla \phi$$
$$-\gamma\left(\varepsilon^2 \nabla^2 \phi - F'(\phi) + \beta(t)\sqrt{F(\phi)}\right) = 0, \qquad \text{on } \Omega^{\mathrm{f}}(t) \times [0, T], \tag{8.14}$$
$$\phi = \phi_D, \qquad \text{on } \Gamma_D^{\phi}(t) \times [0, T], \tag{8.15}$$
$$\nabla \phi \cdot \boldsymbol{n}^{\phi} = 0, \qquad \text{on } \Gamma_N^{\phi}(t) \times [0, T], \tag{8.16}$$
$$\phi = \phi_0, \qquad \text{on } \Omega^{\mathrm{f}}(0), \tag{8.17}$$

where $\boldsymbol{v}^{\mathrm{f}}$ and $\boldsymbol{w}$ are the fluid and the mesh velocities respectively and γ is a mobility parameter with units of $[T^{-1}]$ which is selected as 1 for simplicity. The Dirichlet condition on the order parameter is given by ϕ_D satisfied at the Dirichlet boundary $\Gamma_D^{\phi}(t)$. $\boldsymbol{n}^{\phi}$ is the unit outward normal to the Neumann boundary $\Gamma_N^{\phi}(t)$, and ϕ_0 denotes the initial condition on the order parameter. The term $F'(\phi)$ denotes the derivative of $F(\phi)$ with respect to ϕ and $\beta(t)$ is the time-dependent part of the Lagrange multiplier which can be derived with the help of incompressibility condition and the given boundary conditions as

$$\beta(t) = \frac{\displaystyle\int_{\Omega^{\mathrm{f}}(t)} F'(\phi) d\Omega}{\displaystyle\int_{\Omega^{\mathrm{f}}(t)} \sqrt{F(\phi)} d\Omega}. \tag{8.18}$$

The Lagrange multiplier is written in such a way that for $K(\phi) = 0.5(\phi^3/3 - \phi)$ so that $K'(\phi) = \sqrt{F(\phi)}$,

$$\int_{\Omega^{\mathrm{f}}(t)} K(\phi) d\Omega = \text{constant}. \tag{8.19}$$

8.2.3.2 Semi-Discrete Variational Form

The Allen-Cahn equation mathematically can be seen as a nonlinear convection-diffusion-reaction (CDR) equation, for which detailed variational formulation was presented in Chap. 4 to stabilize the solution in convection- and reaction-dominated regimes.

The order parameter ϕ solved by the Allen-Cahn equation is used to interpolate the physical properties of the two fluid phases, viz., density and viscosity, which should always be positive. Unwanted oscillations in the numerical solution of ϕ in the convection- and reaction-dominated regimes can lead to negative values of these properties, thus producing unstable and unphysical results. Therefore, we extend the positivity-preserving variational (PPV) technique discussed in Chapter 4 to the phase-field equation after variational temporal discretization of the equation.

It is known that variational time integrators have the energy conserving property in comparison to Runge-Kutta time integration [136]. To impart this property and for consistency in the formulation, we consider the generalized-α time integration technique [44, 111]. The expressions for the temporal discretization can be written as:

$$\phi^{\mathrm{n+1}} = \phi^{\mathrm{n}} + \Delta t \partial_t \phi^{\mathrm{n}} + \gamma^{\mathrm{f}} \Delta t (\partial_t \phi^{\mathrm{n+1}} - \partial_t \phi^{\mathrm{n}}), \tag{8.20}$$

$$\partial_t \phi^{\mathrm{n}+\alpha_{\mathrm{m}}^{\mathrm{f}}} = \partial_t \phi^{\mathrm{n}} + \alpha_{\mathrm{m}}^{\mathrm{f}} (\partial_t \phi^{\mathrm{n+1}} - \partial_t \phi^{\mathrm{n}}), \tag{8.21}$$

$$\phi^{\mathrm{n}+\alpha^{\mathrm{f}}} = \phi^{\mathrm{n}} + \alpha^{\mathrm{f}} (\phi^{\mathrm{n+1}} - \phi^{\mathrm{n}}), \tag{8.22}$$

where Δt is the time step size, $\alpha_{\mathrm{m}}^{\mathrm{f}}$, α^{f} and γ^{f} are the generalized-α parameters defined in Eq. (5.23). The time discretized Allen-Cahn equation is therefore,

$$\partial_t \phi^{\mathrm{n}+\alpha_{\mathrm{m}}^{\mathrm{f}}} + (\boldsymbol{v}^{\mathrm{f}} - \boldsymbol{w}) \cdot \nabla \phi^{\mathrm{n}+\alpha^{\mathrm{f}}} - \varepsilon^2 \nabla^2 \phi^{\mathrm{n}+\alpha^{\mathrm{f}}} + F'(\phi^{\mathrm{n}+\alpha^{\mathrm{f}}}) - \beta(t^{\mathrm{n}+\alpha^{\mathrm{f}}}) \sqrt{F(\phi^{\mathrm{n}+\alpha^{\mathrm{f}}})} = 0. \tag{8.23}$$

For energy stability, $F'(\phi^{\mathrm{n}+\alpha^{\mathrm{f}}})$ is written as [57]

$$F'(\phi^{\mathrm{n}+\alpha^{\mathrm{f}}}) = \frac{F(\phi^{\mathrm{n+1}}) - F(\phi^{\mathrm{n}})}{\phi^{\mathrm{n+1}} - \phi^{\mathrm{n}}}. \tag{8.24}$$

Let $K'(\phi) = \sqrt{F(\phi)} = 0.5(\phi^2 - 1)$ be written in the discretized form as

$$\boxed{K'(\phi^{\mathrm{n}+\alpha^{\mathrm{f}}}) = \sqrt{F(\phi^{\mathrm{n}+\alpha^{\mathrm{f}}})} = \frac{K(\phi^{\mathrm{n+1}}) - K(\phi^{\mathrm{n}})}{\phi^{\mathrm{n+1}} - \phi^{\mathrm{n}}}} \tag{8.25}$$

The above expressions in Eqs. (8.24) and (8.25) are designed to provide the energy stability property to the variational formulation. A detailed derivation of the discrete energy law showing energy stability is provided in [115].

Using Eq. (8.22) to replace the expression for $\phi^{\mathrm{n+1}}$ in $F'(\phi^{\mathrm{n}+\alpha^{\mathrm{f}}})$ and $\sqrt{F(\phi^{\mathrm{n}+\alpha^{\mathrm{f}}})}$ and rearranging, the Allen-Cahn equation can be written in the form of convection-diffusion-reaction equation as follows:

$$\partial_t \phi^{\mathrm{n}+\alpha_{\mathrm{m}}^{\mathrm{f}}} + \hat{\boldsymbol{v}} \cdot \nabla \phi^{\mathrm{n}+\alpha^{\mathrm{f}}} - \hat{k} \nabla^2 \phi^{\mathrm{n}+\alpha^{\mathrm{f}}} + \hat{s} \phi^{\mathrm{n}+\alpha^{\mathrm{f}}} - \hat{f}(t^{\mathrm{n}+\alpha^{\mathrm{f}}}) = 0, \quad \text{on } \Omega^{\mathrm{f}}(t), \tag{8.26}$$

where $\hat{\boldsymbol{v}}$, $\hat{k}$, $\hat{s}$ and $\hat{f}$ are the modified convection velocity, diffusion coefficient, reaction coefficient and the source terms respectively, given as

$$\hat{\boldsymbol{v}} = \boldsymbol{v}^{\mathrm{f}} - \boldsymbol{w}, \tag{8.27}$$

$$\hat{k} = \varepsilon^2, \tag{8.28}$$

$$\begin{aligned}\hat{s} = &\frac{1}{4}\bigg[\frac{(\phi^{\mathrm{n}+\alpha^{\mathrm{f}}})^2}{(\alpha^{\mathrm{f}})^3} - \bigg(\frac{3}{(\alpha^{\mathrm{f}})^3} - \frac{4}{(\alpha^{\mathrm{f}})^2}\bigg)\phi^{\mathrm{n}+\alpha^{\mathrm{f}}}\phi^{\mathrm{n}} \\ &+ \bigg(\frac{3}{(\alpha^{\mathrm{f}})^3} - \frac{8}{(\alpha^{\mathrm{f}})^2} + \frac{6}{\alpha^{\mathrm{f}}}\bigg)(\phi^{\mathrm{n}})^2 - \frac{2}{\alpha^{\mathrm{f}}}\bigg] \\ &- \frac{\beta(t^{\mathrm{n}+\alpha^{\mathrm{f}}})}{2}\bigg[\frac{\phi^{\mathrm{n}+\alpha^{\mathrm{f}}}}{3(\alpha^{\mathrm{f}})^2} + \frac{1}{3}\bigg(-\frac{2}{(\alpha^{\mathrm{f}})^2} + \frac{3}{\alpha^{\mathrm{f}}}\bigg)\phi^{\mathrm{n}}\bigg],\end{aligned} \tag{8.29}$$

$$\begin{aligned}\hat{f} = &-\frac{1}{4}\bigg[\bigg(-\frac{1}{(\alpha^{\mathrm{f}})^3} + \frac{4}{(\alpha^{\mathrm{f}})^2} - \frac{6}{\alpha^{\mathrm{f}}} + 4\bigg)(\phi^{\mathrm{n}})^3 + \bigg(\frac{2}{\alpha^{\mathrm{f}}} - 4\bigg)\phi^{\mathrm{n}}\bigg] \\ &+ \frac{\beta(t^{\mathrm{n}+\alpha^{\mathrm{f}}})}{2}\bigg[\frac{1}{3}\bigg(\frac{1}{(\alpha^{\mathrm{f}})^2} - \frac{3}{\alpha^{\mathrm{f}}} + 3\bigg)(\phi^{\mathrm{n}})^2 - 1\bigg].\end{aligned} \tag{8.30}$$

Next, for the variational formulation, we define the space of trial solution as $\mathcal{S}_{\phi}^{\mathrm{h}}$ and that of the test function as $\mathcal{V}_{\phi}^{\mathrm{h}}$ such that

$$\mathcal{S}_{\phi}^{\mathrm{h}} = \big\{\phi_{\mathrm{h}} \mid \phi_{\mathrm{h}} \in H^1(\Omega^{\mathrm{f}}(t)), \phi_{\mathrm{h}} = \phi_D \text{ on } \Gamma_D^{\phi}(t)\big\}, \tag{8.31}$$

$$\mathcal{V}_{\phi}^{\mathrm{h}} = \big\{\hat{w}_{\mathrm{h}} \mid \hat{w}_{\mathrm{h}} \in H^1(\Omega^{\mathrm{f}}(t)), \hat{w}_{\mathrm{h}} = 0 \text{ on } \Gamma_D^{\phi}(t)\big\}. \tag{8.32}$$

The variational finite element statement for the Allen-Cahn equation using the PPV technique is thus written as: find $\phi_{\mathrm{h}}(\boldsymbol{x}^{\mathrm{f}}, t^{\mathrm{n}+\alpha^{\mathrm{f}}}) \in \mathcal{S}_{\phi}^{\mathrm{h}}$ such that $\forall \hat{w}_{\mathrm{h}} \in \mathcal{V}_{\phi}^{\mathrm{h}}$,

$$\begin{aligned}&\int_{\Omega^{\mathrm{f}}(t)} \Big(\hat{w}_{\mathrm{h}}\partial_t\phi_{\mathrm{h}} + \hat{w}_{\mathrm{h}}(\hat{\boldsymbol{v}} \cdot \nabla\phi_{\mathrm{h}}) + \nabla\hat{w}_{\mathrm{h}} \cdot (\hat{k}\nabla\phi_{\mathrm{h}}) + \hat{w}_{\mathrm{h}}\hat{s}\phi_{\mathrm{h}} - \hat{w}_{\mathrm{h}}\hat{f}\Big)d\Omega \\ &+ \sum_{\mathrm{e}=1}^{\mathrm{n_{el}}}\int_{\Omega^{\mathrm{e}}} \Big(\big(\hat{\boldsymbol{v}} \cdot \nabla\hat{w}_{\mathrm{h}}\big)\tau\big(\partial_t\phi_{\mathrm{h}} + \hat{\boldsymbol{v}} \cdot \nabla\phi_{\mathrm{h}} - \nabla \cdot (\hat{k}\nabla\phi_{\mathrm{h}}) + \hat{s}\phi_{\mathrm{h}} - \hat{f}\big)\Big)d\Omega^{\mathrm{e}} \\ &+ \sum_{\mathrm{e}=1}^{\mathrm{n_{el}}}\int_{\Omega^{\mathrm{e}}} \chi\frac{|\mathcal{R}(\phi_{\mathrm{h}})|}{|\nabla\phi_{\mathrm{h}}|}k_s^{\mathrm{add}}\nabla\hat{w}_{\mathrm{h}} \cdot \bigg(\frac{\hat{\boldsymbol{v}} \otimes \hat{\boldsymbol{v}}}{|\hat{\boldsymbol{v}}|^2}\bigg) \cdot \nabla\phi_{\mathrm{h}} d\Omega^{\mathrm{e}} \\ &+ \sum_{\mathrm{e}=1}^{\mathrm{n_{el}}}\int_{\Omega^{\mathrm{e}}} \chi\frac{|\mathcal{R}(\phi_{\mathrm{h}})|}{|\nabla\phi_{\mathrm{h}}|}k_c^{\mathrm{add}}\nabla\hat{w}_{\mathrm{h}} \cdot \bigg(\boldsymbol{I} - \frac{\hat{\boldsymbol{v}} \otimes \hat{\boldsymbol{v}}}{|\hat{\boldsymbol{v}}|^2}\bigg) \cdot \nabla\phi_{\mathrm{h}} d\Omega^{\mathrm{e}} = 0,\end{aligned} \tag{8.33}$$

where the first and the second lines represent the Galerkin and SUPG stabilization terms respectively. Note that the SUPG formulation rather than the combined GLS-SGS methodology in the linear stabilization of PPV helps to maintain the mass conservation property [115]. The third and the fourth lines denote the positivity preserving nonlinear stabilization terms which impose the positivity property at the element matrix level. The residual of the Allen-Cahn equation is given by

$$\mathscr{R}(\phi_h) = \partial_t \phi_h + \hat{\boldsymbol{v}} \cdot \nabla \phi_h - \nabla \cdot (\hat{k} \nabla \phi_h) + \hat{s}\phi_h - \hat{f}, \tag{8.34}$$

and the stabilization parameter τ is written as

$$\tau = \left[\left(\frac{2}{\Delta t} \right)^2 + \hat{\boldsymbol{v}} \cdot \boldsymbol{G} \hat{\boldsymbol{v}} + 9\hat{k}^2 \boldsymbol{G} : \boldsymbol{G} + \hat{s}^2 \right]^{-1/2}, \tag{8.35}$$

where $\boldsymbol{G}$ is the contravariant metric tensor defined in Eq. (5.40). For the current context of the AC equation, the PPV parameters are defined as:

$$\chi = \frac{2}{|\hat{s}|h + 2|\hat{\boldsymbol{v}}|}, \tag{8.36}$$

$$k_s^{\mathrm{add}} = \max\left\{ \frac{||\hat{\boldsymbol{v}}| - \tau|\hat{\boldsymbol{v}}|\hat{s}|h}{2} - (\hat{k} + \tau|\hat{\boldsymbol{v}}|^2) + \frac{\hat{s}h^2}{6}, 0 \right\}, \tag{8.37}$$

$$k_c^{\mathrm{add}} = \max\left\{ \frac{|\hat{\boldsymbol{v}}|h}{2} - \hat{k} + \frac{\hat{s}h^2}{6}, 0 \right\}, \tag{8.38}$$

where $|\hat{\boldsymbol{v}}|$ is the magnitude of the convection velocity and h is the characteristic element length which is selected as the streamline element length.

8.3 Partitioned Iterative Coupling

In this section, we present the nonlinear partitioned iterative coupling between the different fields of equations. First, we discuss the two-phase flow coupling between the Navier-Stokes and the Allen-Cahn equations, i.e., without any moving ALE meshes for the fluid-structure interface ($\boldsymbol{w} = \boldsymbol{0}$). Then, we extend this coupling with the structure to get the two-phase fluid-structure framework.

8.3.1 Two-Phase Flow System

The fully-coupled linearized matrix form of the two-phase flow system after Newton-Raphson linearization can be written as

$$\left[\boldsymbol{A}^{\mathrm{ff}}\right]\left\{\Delta \boldsymbol{q}^{\mathrm{f}}\right\} = \left\{\mathscr{R}^{\mathrm{f}}\right\}, \tag{8.39}$$

where $\Delta \boldsymbol{q}^{\mathrm{f}} = (\Delta \boldsymbol{v}^{\mathrm{f}}, \Delta p, \Delta \phi)$ is the vector of the unknowns consisting of fluid velocity, pressure and the order parameter, $\mathscr{R}^{\mathrm{f}}$ is the weighted residual of the stabilized two-phase flow equations, and $\boldsymbol{A}^{\mathrm{ff}}$ is

$$\boldsymbol{A}^{\mathrm{ff}} = \begin{bmatrix} \boldsymbol{K}_{\Omega^{\mathrm{f}}} & -\boldsymbol{G}_{\Omega^{\mathrm{f}}} & \boldsymbol{D}_{\Omega^{\mathrm{f}}} \\ \boldsymbol{G}^T_{\Omega^{\mathrm{f}}} & \boldsymbol{C}_{\Omega^{\mathrm{f}}} & \boldsymbol{0} \\ \boldsymbol{G}_{AC} & \boldsymbol{0} & \boldsymbol{K}_{AC} \end{bmatrix}, \tag{8.40}$$

where $\boldsymbol{K}_{\Omega^{\mathrm{f}}}$ is the stiffness matrix of the momentum equation consisting of inertia, convection, viscous and stabilization terms, $\boldsymbol{G}_{\Omega^{\mathrm{f}}}$ is the gradient operator, $\boldsymbol{G}^T_{\Omega^{\mathrm{f}}}$ is the divergence operator for the continuity equation and $\boldsymbol{C}_{\Omega^{\mathrm{f}}}$ is the pressure-pressure stabilization term. $\boldsymbol{D}_{\Omega^{\mathrm{f}}}$ contains the terms in the momentum equation which depend on the order parameter, $\boldsymbol{G}_{AC}$ consists of the term in the Allen-Cahn equation depending on the fluid velocity and $\boldsymbol{K}_{AC}$ is the stiffness matrix for the Allen-Cahn equation consisting of inertia, convection, diffusion, reaction and stabilization terms.

The two-phase flow system in Eq. (8.40) is now decoupled for the nonlinear partitioned coupling into two subsystems: Navier-Stokes and Allen-Cahn solves, for which the linear system of equations can be summarized as

$$\begin{bmatrix} \boldsymbol{K}_{\Omega^{\mathrm{f}}} & -\boldsymbol{G}_{\Omega^{\mathrm{f}}} \\ \boldsymbol{G}^T_{\Omega^{\mathrm{f}}} & \boldsymbol{C}_{\Omega^{\mathrm{f}}} \end{bmatrix} \begin{Bmatrix} \Delta \boldsymbol{v}^{\mathrm{f}} \\ \Delta p \end{Bmatrix} = \begin{Bmatrix} \widetilde{\mathscr{R}}_{\mathrm{m}} \\ \widetilde{\mathscr{R}}_{\mathrm{c}} \end{Bmatrix}, \tag{8.41}$$

$$\left[\boldsymbol{K}_{AC}\right]\left\{\Delta \phi\right\} = \left\{\widetilde{\mathscr{R}}(\phi)\right\}, \tag{8.42}$$

where $\widetilde{\mathscr{R}}_{\mathrm{m}}$, $\widetilde{\mathscr{R}}_{\mathrm{c}}$ and $\widetilde{\mathscr{R}}(\phi)$ denote the weighted residuals of the stabilized momentum, continuity and the Allen-Cahn equations respectively. Note that the cross-coupling terms in the matrices $\boldsymbol{D}_{\Omega^{\mathrm{f}}}$ and $\boldsymbol{G}_{AC}$ have been neglected as we decouple the equations in a partitioned iterative manner described below.

The algorithm of the iterative coupling between the Navier-Stokes and the Allen-Cahn equations is shown in Algorithm 1. The Navier-Stokes equations provide a predictor fluid velocity and the Allen-Cahn equation is then solved to update the order parameter to interpolate density, viscosity and capillary forces in the Navier-Stokes equations. Consider the known quantities, velocity $\boldsymbol{v}^{\mathrm{f}}(\boldsymbol{x}^{\mathrm{f}}, t^{\mathrm{n}})$, pressure $p(\boldsymbol{x}^{\mathrm{f}}, t^{\mathrm{n}})$ and the order parameter $\phi(\boldsymbol{x}^{\mathrm{f}}, t^{\mathrm{n}})$ at time t^{n}. In the first step of a nonlinear iteration k, the Navier-Stokes equations are solved for fluid velocity and pressure. The computed fluid velocity is then transferred to the Allen-Cahn solve in the second step. The third step involves solving the Allen-Cahn equation to obtain the updated order parameter $\phi_{(\mathrm{k}+1)}$. Finally, the density $\rho^{\mathrm{f}}(\phi)$, viscosity $\mu^{\mathrm{f}}(\phi)$ and surface tension force $\mathbf{sf}(\phi)$ are interpolated based on the updated order parameter value, and transferred to the

Navier-Stokes solve for the next iteration. The variables for the next time step $t^{\mathrm{n}+1}$ are updated at the end of the convergence of the nonlinear iteration and the solver is advanced in time.

Algorithm 1 Partitioned coupling of implicit Navier-Stokes and Allen-Cahn solvers

Given $\boldsymbol{v}^{\mathrm{f},0}$, p^0, ϕ^0

Loop over time steps, $\mathrm{n} = 0, 1, \cdots$

Start from known variables $\boldsymbol{v}^{\mathrm{f,n}}$, p^{n}, ϕ^{n}

Predict the solution:

$$\boldsymbol{v}^{\mathrm{f,n+1}}_{(0)} = \boldsymbol{v}^{\mathrm{f,n}}$$
$$p^{\mathrm{n+1}}_{(0)} = p^{\mathrm{n}}$$
$$\phi^{\mathrm{n+1}}_{(0)} = \phi^{\mathrm{n}}$$

Loop over the nonlinear iterations, $\mathrm{k} = 0, 1, \cdots$ until convergence

[1] **Navier-Stokes Implicit Solve**

(a) Interpolate solution:
$$\boldsymbol{v}^{\mathrm{f,n}+\alpha^{\mathrm{f}}}_{(\mathrm{k}+1)} = \boldsymbol{v}^{\mathrm{f,n}} + \alpha^{\mathrm{f}}(\boldsymbol{v}^{\mathrm{f,n+1}}_{(\mathrm{k})} - \boldsymbol{v}^{\mathrm{f,n}})$$
$$p^{\mathrm{n+1}}_{(\mathrm{k}+1)} = p^{\mathrm{n+1}}_{(\mathrm{k})}$$

(b) Solve for $\Delta\boldsymbol{v}^{\mathrm{f,n}+\alpha^{\mathrm{f}}}$ and $\Delta p^{\mathrm{n+1}}$ in Eq. (8.41)

(c) Correct solution:
$$\boldsymbol{v}^{\mathrm{f,n}+\alpha^{\mathrm{f}}}_{(\mathrm{k}+1)} = \boldsymbol{v}^{\mathrm{f,n}+\alpha^{\mathrm{f}}}_{(\mathrm{k}+1)} + \Delta\boldsymbol{v}^{\mathrm{f,n}+\alpha^{\mathrm{f}}}$$
$$p^{\mathrm{n+1}}_{(\mathrm{k}+1)} = p^{\mathrm{n+1}}_{(\mathrm{k}+1)} + \Delta p^{\mathrm{n+1}}$$

(d) Update solution:
$$\boldsymbol{v}^{\mathrm{f,n+1}}_{(\mathrm{k}+1)} = \boldsymbol{v}^{\mathrm{f,n}} + \frac{1}{\alpha^{\mathrm{f}}}(\boldsymbol{v}^{\mathrm{f,n}+\alpha_{\mathrm{f}}}_{(\mathrm{k}+1)} - \boldsymbol{v}^{\mathrm{f,n}})$$
$$p^{\mathrm{n+1}}_{(\mathrm{k}+1)} = p^{\mathrm{n+1}}_{(\mathrm{k}+1)}$$

[2] $\boldsymbol{v}^{\mathrm{f,n+1}}_{(\mathrm{k}+1)}$ →

[3] **Allen-Cahn Implicit Solve**

(a) Interpolate solution:
$$\phi^{\mathrm{n}+\alpha^{\mathrm{f}}}_{(\mathrm{k}+1)} = \phi^{\mathrm{n}} + \alpha^{\mathrm{f}}(\phi^{\mathrm{n+1}}_{(\mathrm{k})} - \phi^{\mathrm{n}})$$

(b) Solve for $\Delta\phi^{\mathrm{n}+\alpha^{\mathrm{f}}}$ in Eq. (8.42)

(c) Correct solution:
$$\phi^{\mathrm{n}+\alpha^{\mathrm{f}}}_{(\mathrm{k}+1)} = \phi^{\mathrm{n}+\alpha^{\mathrm{f}}}_{(\mathrm{k}+1)} + \Delta\phi^{\mathrm{n}+\alpha^{\mathrm{f}}}$$

(d) Update solution:
$$\phi^{\mathrm{n+1}}_{(\mathrm{k}+1)} = \phi^{\mathrm{n}} + \frac{1}{\alpha^{\mathrm{f}}}(\phi^{\mathrm{n}+\alpha^{\mathrm{f}}}_{(\mathrm{k}+1)} - \phi^{\mathrm{n}})$$

← [4] $\phi^{\mathrm{n+1}}_{(\mathrm{k}+1)}$ to interpolate $\rho^{\mathrm{f}}(\phi)$, $\mu^{\mathrm{f}}(\phi)$, $\mathbf{sf}(\phi)$, $\boldsymbol{b}^{\mathrm{f}}(\phi)$

8.3.2 *Two-Phase Fluid-Structure Interaction System*

For extending the coupling of the two-phase flow system to the structure, we similarly couple the structural equation in a partitioned manner using the nonlinear iterative force correction (NIFC) technique discussed in Chap. 7. The structural equation can be that of a rigid, flexible, or multi-body system. The algorithmic steps for the partitioned iterative coupling of the implicit two-phase fluid-structure interaction solver are shown in Algorithm 2.

Algorithm 2 Partitioned coupling of implicit two-phase fluid-structure interaction solver

Given $\boldsymbol{v}^{\mathrm{f},0}$, p^0, ϕ^0, $\boldsymbol{u}^{\mathrm{s},0}$

Loop over time steps, $\mathrm{n} = 0, 1, \cdots$

Start from known variables $\boldsymbol{v}^{\mathrm{f,n}}$, p^{n}, ϕ^{n}, $\boldsymbol{u}^{\mathrm{s,n}}$

Predict the solution:

$$\boldsymbol{v}^{\mathrm{f,n+1}}_{(0)} = \boldsymbol{v}^{\mathrm{f,n}}; \quad p^{\mathrm{n+1}}_{(0)} = p^{\mathrm{n}}; \quad \phi^{\mathrm{n+1}}_{(0)} = \phi^{\mathrm{n}}; \quad \boldsymbol{u}^{\mathrm{s,n+1}}_{(0)} = \boldsymbol{u}^{\mathrm{s,n}}$$

Loop over the nonlinear iterations, $\mathrm{k} = 0, 1, \cdots$ until convergence

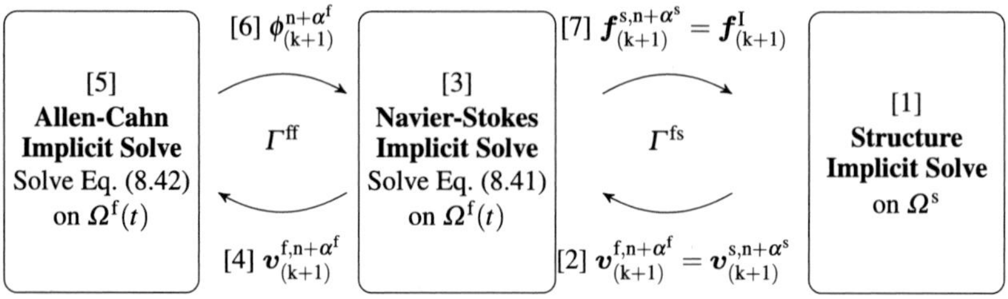

8.3.3 General Remarks

The stabilization methods help in considering equal-order interpolations for all the quantities ($\boldsymbol{v}^{\mathrm{f}}$, p, ϕ) for the finite element discretization. The nonlinear errors of the implicit systems of the Navier-Stokes and Allen-Cahn equations are minimized by Newton-Raphson technique at each time step. We found that around 3–4 nonlinear iterations are sufficient to obtain a reasonably converged solution of the two-phase fluid-structure interaction system. Furthermore, each nonlinear iteration consists of a single pass through the various sub-systems of the coupling (Navier-Stokes, Allen-Cahn and structure). This feature, along with the NIFC technique, helps to reduce the computational time without compromising with the accuracy and stability of the solution. The partitioned-block type feature of the solver leads to flexibility and ease in its implementation to the existing variational solvers.

The linearized matrix systems are formed and stored using the Harwell-Boeing sparse matrix format. The linear system of equations is solved by the generalized minimal residual (GMRES) algorithm proposed in [186], which uses the modified Gram-Schmidt orthogonalization and relies on preconditioned Krylov subspace iterations.

The coupled solver implementation relies on a hybrid parallelism for parallel computing. A standard master-slave strategy is employed for the distributed memory clusters by message passing interface (MPI) which is based on domain decomposition strategy [2]. This parallel implementation takes the advantage of the state-of-the-art hierarchical memory and parallel architectures.

8.4 Numerical Tests

In this section, we present some benchmark numerical tests to assess the two-phase flow coupling and the two-phase fluid-structure interaction solver under the partitioned framework. The location of the two-phase fluid interface is calculated by linearly interpolating the order parameter ϕ such that $\phi = 0$ represents the interface.

8.4.1 Verification of the Allen-Cahn Implementation

We consider the volume-conserved motion by curvature in two-dimensions [134] to verify the standalone Allen-Cahn solver. The equation is solved on a square computational domain $[0, 1] \times [0, 1]$ with different mesh resolution and satisfying periodic boundary condition on all the sides.

The initial condition is selected as:

$$\begin{aligned}\phi(x, y, 0) =& 1 + \tanh\left(\frac{R_1 - \sqrt{(x-0.25)^2 + (y-0.25)^2}}{\sqrt{2}\varepsilon}\right)\\ &+ \tanh\left(\frac{R_2 - \sqrt{(x-0.57)^2 + (y-0.57)^2}}{\sqrt{2}\varepsilon}\right),\end{aligned} \tag{8.43}$$

where $R_1 = 0.1$ and $R_2 = 0.15$ are the radii of the two circles centered at $(0.25, 0.25)$ and $(0.57, 0.57)$, respectively. The evolution of the radii of the two circles is recorded with time and compared with the results obtained in [134] for $\varepsilon = 0.01$. Figure 8.2 shows the problem set-up and the evolution of the radii till $t = 100$ with a time step size of 0.1. The results show good agreement with the reference data.

We further analyze the number of elements required to sufficiently resolve the interface. Let the number of elements across the equilibrium interface thickness $\varepsilon_{\text{eqm}} = 4.164\varepsilon$ be denoted by N_ε. We perform the test for $N_\varepsilon \in [3, 10]$ and quantify the percentage error (e_1) with the radii at the final time $t = 100$ which is defined as

$$e_1 = \frac{|R_i - R_{\text{ref}}|}{R_{\text{ref}}} \times 100, \tag{8.44}$$

with R_i being the radius corresponding to different resolutions at $t = 100$ and R_{ref} is the radius obtained for the finest resolution $N_\varepsilon = 10$ at $t = 100$. The results have been summarized in Table 8.1. It is observed that $N_\varepsilon > 3$ and $N_\varepsilon > 6$ are required to get accurate solution within 1% error for the large circle (circle 2) and the small circle (circle 1) respectively. This is due to the different curvatures of the two circles. Decrease in the radius increases the curvature, thus leading to more number of elements required to maintain sufficient resolution of the interface.

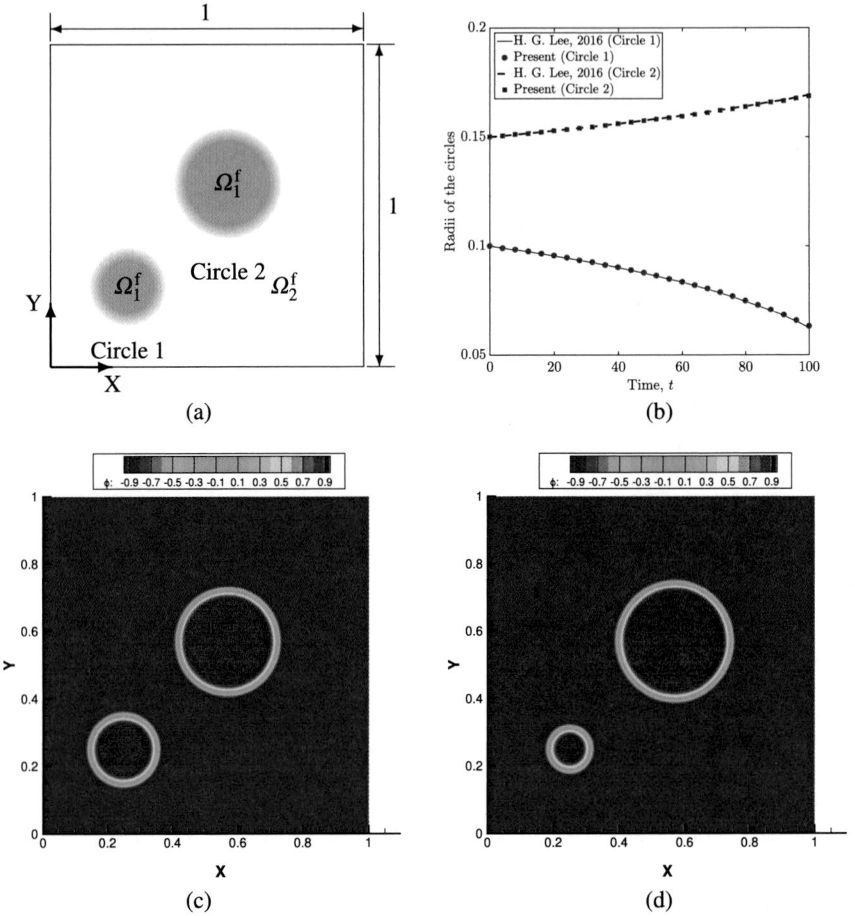

Fig. 8.2 Evolution of the radii of two-dimensional circles: **a** schematic diagram showing the computational domain, **b** validation of the evolution of the radii of the two circles with the literature [134], and the contour plots of the order parameter ϕ at **c** $t = 0$, and **d** $t = 100$. In **a**, Ω_1^f and Ω_2^f are the two phases with periodic boundary conditions imposed on all the sides

Table 8.1 Percentage error in the final radii of the two circles for different mesh resolutions

$N_\varepsilon \rightarrow$	3	4	5	6	7	8
Circle 1	10.39	5.29	3.26	1.54	0.91	0.58
Circle 2	1.28	0.60	0.36	0.20	0.15	0.02

Furthermore, we study the behavior of the solver under spatial and temporal refinements. The L^2 error of the solution over the whole domain at $t = 50$ is quantified under the mesh and temporal refinements. The L^2 error is calculated as:

$$e_2 = \frac{||\Phi - \Phi_{\text{ref}}||_2}{||\Phi_{\text{ref}}||_2}, \tag{8.45}$$

where the vector consisting of the order parameter values of the whole domain at $t = 50$ for the respective refinement is denoted by Φ, Φ_{ref} represents the same vector for the finest resolution, and $||\cdot||_2$ is the L^2 norm. Uniform spatial refinement is carried out by varying the element size from $h = 1/72$ until $h = 1/2304$. Time step size from $\Delta t = 0.4$ to $\Delta t = 0.025$ by a uniform factor of 2 is selected for the temporal refinement. $h = 1/2304$ and $\Delta t = 3.90625 \times 10^{-4}$ are selected as the reference solution for evaluating the error norm.

The spatial and temporal convergence plots (shown in Fig. 8.3) depict a second-order accuracy of the solver. The nonlinear energy-stability in the discretization helps to maintain the second-order temporal accuracy, which has been analytically proven in [115].

Finally, the conservation of mass or the order parameter is quantified by the percentage change in ϕ over the time with respect to its initial value. The mass loss is quantified as approximately 5.4577×10^{-5} % for this example. Notice that the mass conservation is dependent on the solver tolerances for the nonlinear iterations as well as linear GMRES iterations. For this case, the linear and nonlinear tolerance are chosen as 10^{-15} and 10^{-4} respectively. It can be deduced that the solver conserves mass reasonably for the variational finite element framework within the tolerance limit.

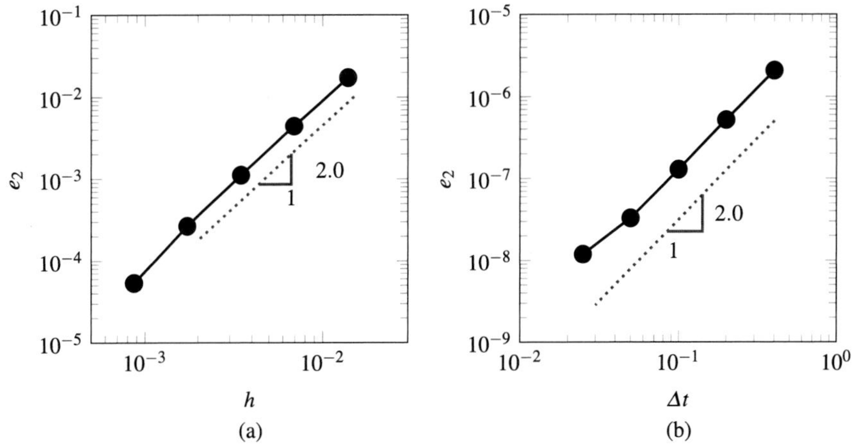

Fig. 8.3 Convergence study for the present method through dependence of non-dimensionalized L^2 error (e_2) as a function of: **a** uniform mesh refinement h, and **b** uniform temporal refinement Δt

8.4.2 Two-Phase Flow Coupling

Here, we perform numerical tests dealing with the coupling between the Navier-Stokes and Allen-Cahn equations for the two-phase flow framework. Tests such as Laplace-Young law, sloshing tank problem and three-dimensional dam break are solved by the two-phase flow solver.

8.4.2.1 Laplace-Young Law

Laplace-Young law states that for a static bubble in a two-phase fluid system, the pressure difference (Δp) across the interface is equal to the ratio of the surface tension coefficient (σ) to the radius of curvature (R) of the bubble,

$$\Delta p = p_{\text{in}} - p_{\text{out}} = \frac{\sigma}{R}. \tag{8.46}$$

At equilibrium, the fluid velocity $\boldsymbol{v}^{\text{f}} = \mathbf{0}$ and the pressure gradient balances the surface tension force in the momentum equation, i.e.,

$$\nabla p = \sigma \varepsilon \frac{3\sqrt{2}}{4} \nabla \cdot (|\nabla \phi|^2 \boldsymbol{I} - \nabla \phi \otimes \nabla \phi). \tag{8.47}$$

We verify this law for our implementation of the two-phase flow solver at high density ratio.

For verifying the law, we consider a computational domain of size $\Omega^{\text{f}} = [0, 4] \times [0, 4]$ with a uniform structured mesh of grid size $1/200$. The bubble radius is varied as 0.2, 0.25, 0.3, 0.35 and 0.4 units, along with the surface tension coefficient as 0.5, 0.25 and 0.05 units. The density and dynamic viscosity of the two phases are selected as $\rho_1^{\text{f}} = 1000$, $\rho_2^{\text{f}} = 1$, $\mu_1^{\text{f}} = 10$ and $\mu_2^{\text{f}} = 0.1$ so that the density and viscosity ratios are 1000 and 100 respectively. Periodic boundary conditions are imposed on all the boundaries and the initial condition of the bubble is given as:

$$\phi(x, y, 0) = -\tanh\left(\frac{R - \sqrt{(x - x_c)^2 + (y - y_c)^2}}{\sqrt{2}\varepsilon}\right), \tag{8.48}$$

where R is the radius of the bubble with its centre at $(x_c, y_c) = (2, 2)$. The interface thickness parameter is $\varepsilon = 0.01$. The pressure difference is measured after 5000 time steps with time step size of $\Delta t = 0.01$ s. The computational domain and the results are shown in Fig. 8.4. It is observed that the pressure difference in Fig. 8.4(b) shows a good agreement with the Laplace-Young law, thus demonstrating the coupling between the Navier-Stokes and the Allen-Cahn equations for high-density ratio.

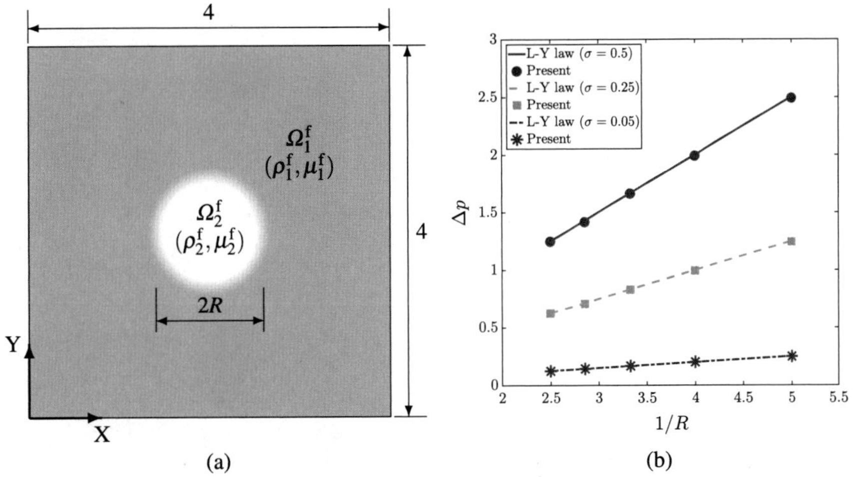

Fig. 8.4 Laplace-Young law: **a** schematic diagram showing the computational domain, **b** comparison of the pressure difference across the interface in a static bubble obtained from the simulation with the Laplace-Young law. In **a**, Ω_1^f and Ω_2^f are the two fluid phases with densities $\rho_1^f = 1000$ and $\rho_2^f = 1$, viscosities $\mu_1^f = 10$, $\mu_2^f = 0.1$ and acceleration due to gravity $\boldsymbol{g} = (0, 0, 0)$ and all the boundaries have periodic boundary condition

8.4.2.2 Sloshing Tank Problem

We further conduct studies on a sloshing tank problem. The domain consists of a rectangular tank $\Omega^f \in [0, 1] \times [0, 1.5]$ discretized in space by varying resolutions characterized by the number of elements in the equilibrium interface thickness, N_ε. High density and viscosity ratios are considered for the two fluid phases: $\rho_1^f = 1000$, $\rho_2^f = 1$, $\mu_1^f = 1$ and $\mu_2^f = 0.01$ with gravity $\boldsymbol{g} = (0, -1, 0)$. The capillary effects due to the surface tension have been neglected in this case. The initial condition of the order parameter is:

$$\phi(x, y, 0) = -\tanh\left(\frac{y - (1.01 + 0.1\sin((x - 0.5)\pi))}{\sqrt{2}\varepsilon}\right). \tag{8.49}$$

With this initial condition, we let the fluid slosh under gravity inside the closed tank and monitor the interface evolution. A Dirichlet boundary condition of $p = 0$ is prescribed at the top boundary and all the remaining walls are regarded as slip walls. The problem set-up is shown in Fig. 8.5a with the contour of the initial condition of the order parameter depicted in Fig. 8.5b.

A series of experiments are performed to assess (1) the effectiveness of the positivity preserving variational (PPV) technique, (2) required resolution at the equilibrium interface thickness (N_ε), (3) temporal convergence for sufficient accuracy, and (4) the effect of the interface thickness parameter ε on the solution. To accomplish this, an L^2 error (e_3) is quantified which is defined as

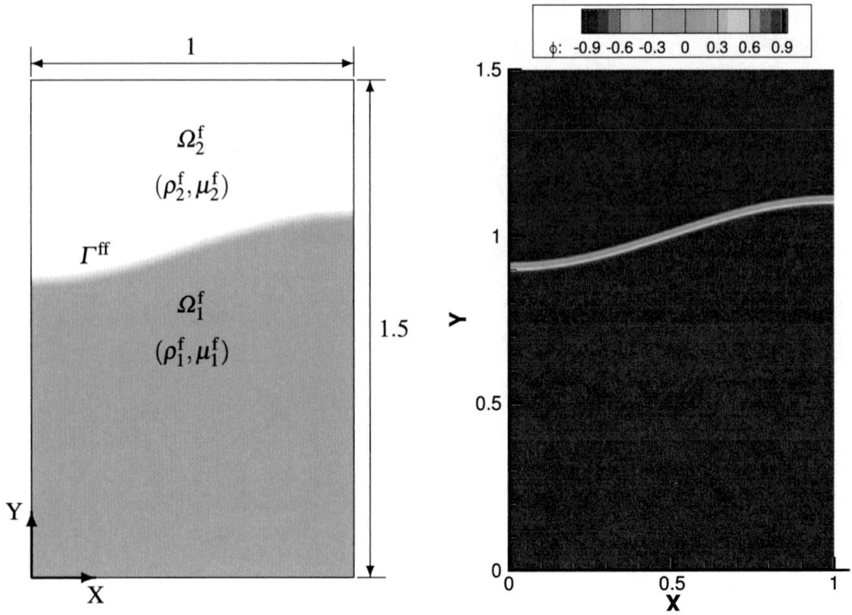

(a) (b)

Fig. 8.5 Sloshing in a rectangular tank: **a** schematic diagram showing the computational domain, **b** contour plot of the order parameter ϕ at $t = 0$. In **a**, Ω_1^f and Ω_2^f are the two fluid phases with densities $\rho_1^f = 1000$ and $\rho_2^f = 1$, viscosities $\mu_1^f = 1$, $\mu_2^f = 0.01$ and acceleration due to gravity $\boldsymbol{g} = (0, -1, 0)$ and the walls of the tank have slip boundary condition with $p = 0$ at the upper boundary

$$e_3 = \frac{||\Phi - \Phi_{\text{ref}}||_2}{||\Phi_{\text{ref}}||_2}, \tag{8.50}$$

where Φ is the temporal solution of the interface elevation at the left boundary, Φ_{ref} is the solution of the finest resolution associated with the respective experiment and $|| \cdot ||_2$ is the L^2 norm.

Effectiveness of the PPV Technique: We start by emphasizing the benefits of the PPV technique in the phase-field formulation. As mentioned before, the order parameter ϕ needs to be bounded to ensure positive values of density and viscosity. Oscillations in the solution due to unresolved mesh can lead to unphysical results. The PPV method adds diffusion to the regions of high gradients in the solution, thus reducing these oscillations and maintaining positivity.

To demonstrate this, we select two test cases: (a) with the nonlinear PPV stabilization terms and (b) without nonlinear PPV terms. The other crucial parameters are $\Delta t = 0.001$, $\varepsilon = 0.01$ with $N_\varepsilon = 3$ and $N_\varepsilon = 4$. The minimum and maximum values of the order parameter solution at the left boundary are quantified as the bounds of the numerical solution. The results are summarized in Table 8.2.

Table 8.2 Bounds in the solution of interface evolution at the left boundary with and without PPV consideration

N_ε	$\min(\phi)$		$\max(\phi)$	
	non-PPV	PPV	non-PPV	PPV
3	−1.0544	−1.0	1.0154	0.9999
4	−1.0065	−1.0	1.0023	0.9999

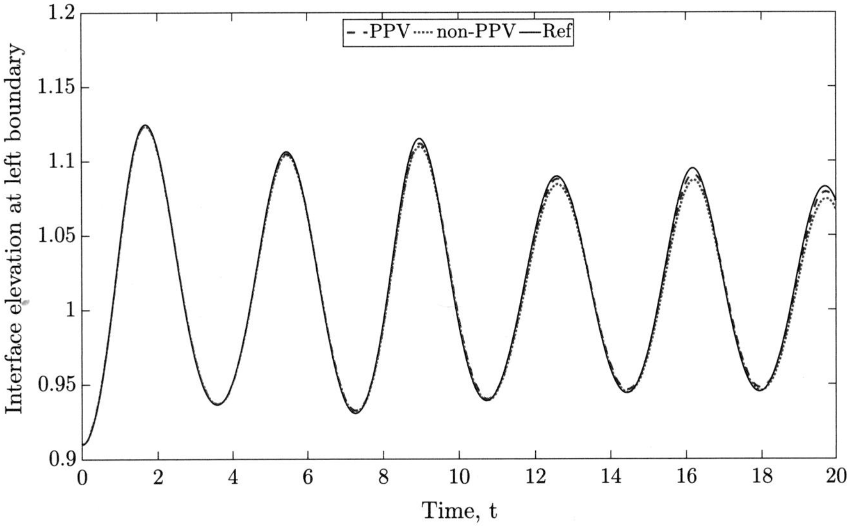

Fig. 8.6 Sloshing tank problem: Effect of using the PPV technique on the evolution of the interface. $\varepsilon = 0.01$, $\Delta t = 0.001$ and $N_\varepsilon = 3$. The reference solution is obtained at $N_\varepsilon = 8$

It can be observed that positivity is preserved and boundedness is maintained in the PPV technique. The solution is unbounded for the case when the nonlinear PPV terms are absent. The evolution of the interface at the left boundary is plotted for the cases considered in Fig. 8.6 for $N_\varepsilon = 3$. The reference solution in the plot is for a refined case $N_\varepsilon = 8$, $\Delta t = 0.001$ and $\varepsilon = 0.01$. The error percentage based on the reference solution can be evaluated from the plot and the error is 0.3548% and 0.1905% for non-PPV and PPV based solution respectively, reflecting lesser error for the PPV method. However, the interpolated density and viscosity values are chopped off when $\phi > 1$ and $\phi < -1$ to avoid any negative values. This chopping can lead to less accurate solution for non-PPV method as seen in Fig. 8.6 compared to the PPV-based method.

Effect of the Number of Elements in the Equilibrium Interfacial Thickness (N_ε): For the coupled Navier-Stokes and Allen-Cahn system, we quantify the required resolution of the mesh to capture the interface dynamics accurately. This mesh resolution is characterized by N_ε which represents the number of elements in the equilibrium

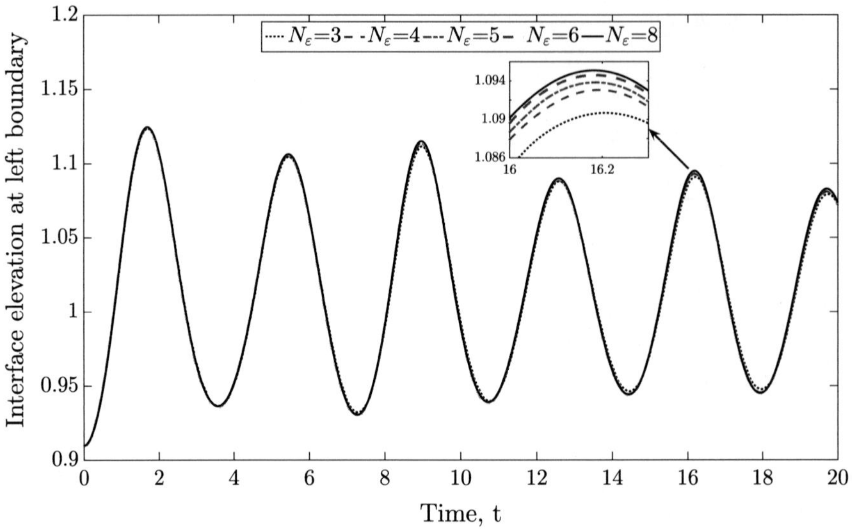

Fig. 8.7 Sloshing tank problem: Effect of N_ε on the evolution of the interface. $\varepsilon = 0.01$ and $\Delta t = 0.001$ are kept constant

Table 8.3 Error (e_3) in the solution for different N_ε

$N_\varepsilon \rightarrow$	3	4	5	6
$e_3(\times 10^{-3})$	2.491	1.094	0.564	0.208

interface thickness. The evolution of the interface with varying N_ε has been shown in Fig. 8.7 with fixed $\varepsilon = 0.01$ and $\Delta t = 0.001$. Consequently, the error is quantified in Table 8.3 where the reference solution is selected as the case when $N_\varepsilon = 8$.

It can be concluded from the plots that N_ε of 4 or 5 is sufficient to accurately capture the interface dynamics. It is obvious that an increase in N_ε will lead to more accurate solution (although the difference is very small), but it will lead to over-resolution of the mesh, thus increasing the cost of computation. Therefore, a compromise needs to be reached between accuracy and computational cost.

Effect of the Time Step Size (Δt): Next, we conduct time convergence study by decreasing the time step size Δt while keeping $N_\varepsilon = 4$ and $\varepsilon = 0.01$. The evolution of the interface is compared in Fig. 8.8. The difference in the solution is noticeable till $\Delta t = 0.01$. With solution at $\Delta t = 0.001$ chosen as the reference solution, the error is tabulated in Table 8.4.

Effect of the Interfacial Thickness Parameter (ε): We further study the effect of the interface thickness parameter ε on the solution. It is known that $\varepsilon \rightarrow 0$ gives the sharp-interface limit. A decrease in ε will indeed give more accurate representation of the interface, but it will lead to high computational cost as the number of elements required to capture the sharp interface increases. Different ε values are chosen with

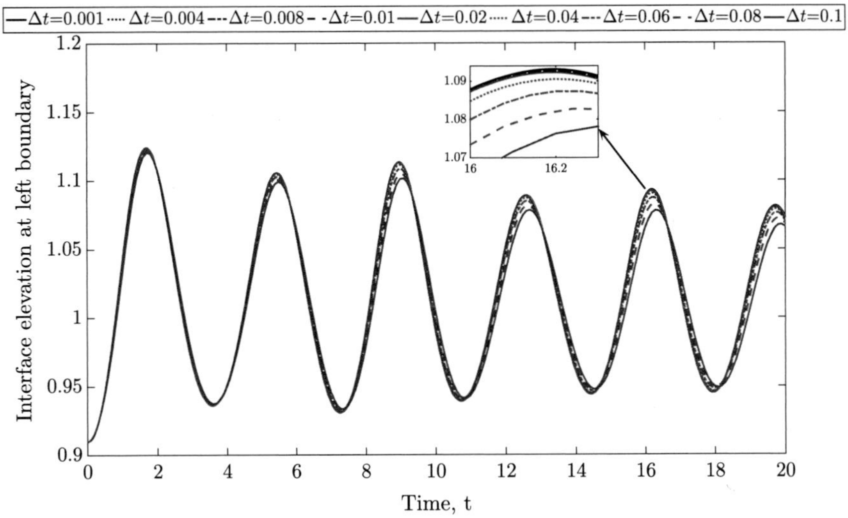

Fig. 8.8 Sloshing tank problem: Effect of Δt on the evolution of the interface. $\varepsilon = 0.01$ and $N_\varepsilon = 4$ are kept constant

Table 8.4 Error (e_3) in the solution for different Δt

$\Delta t \rightarrow$	0.1	0.08	0.06	0.04	0.02	0.01	0.008	0.004
$e_3(\times 10^{-3})$	10.839	6.959	3.759	1.418	0.457	0.478	0.474	0.324

fixed $\Delta t = 0.01$ and $N_\varepsilon = 4$ and the results are shown in Fig. 8.9. Taking the $\varepsilon = 0.005$ case as the reference solution, the error is quantified in Table 8.5. In this particular case, a selection of $\varepsilon = 0.01$ is a good compromise between accuracy and computational cost.

Finally, we select the converged parameters from the tests carried out, viz., $N_\varepsilon = 4$, $\Delta t = 0.01$ and $\varepsilon = 0.01$ and compare our results with that obtained using extended finite element method (XFEM) in [188] in Fig. 8.10.

XFEM involves a local enrichment of the pressure interpolation shape functions and solves the level-set equation to capture the interface dynamics along with the reinitialization procedure. The PPV implementation is much simpler and does not involve any geometric manipulations or reinitialization. We can observe from the figure that the diffuse-interface approach in the present scenario gives sufficiently accurate results.

8.4.2.3 Three-Dimensional Dam-Break with Obstacle

Next, we proceed to demonstrate the capability of the solver to capture the topological changes in the interface as it breaks or merges. Two-dimensional dam-break

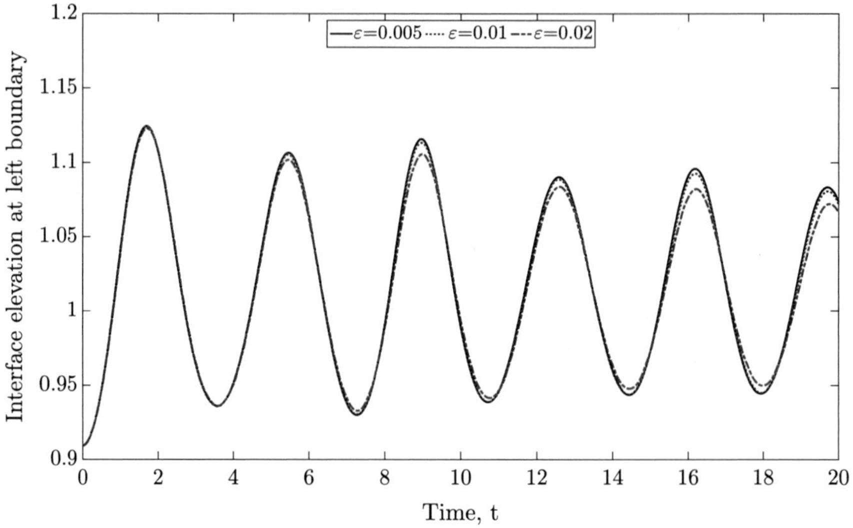

Fig. 8.9 Sloshing tank problem: Effect of ε on the evolution of the interface. $\Delta t = 0.01$ and $N_\varepsilon = 4$ are kept constant

Table 8.5 Error (e_3) in the solution for different ε

$\varepsilon \rightarrow$	0.02	0.01
$e_3(\times 10^{-3})$	5.469	1.269

problem implementing the PPV algorithm is studied in [115] and we present results for three-dimensional dam-break problem here. The computational domain consists of a cuboid of size $L \times B \times H$, where $L = 3.22$, $B = 1$ and $H = 1$ with a water column of size $L_w \times B_w \times H_w$, with $L_w = 1.228$, $B_w = 1$ and $H_w = 0.55$ as shown in Fig. 8.11. A cuboid obstacle of size $l \times b \times h$, where $l = 0.161$, $b = 0.403$ and $h = 0.161$ is placed in the closed tank. We let the water column break under gravity and interact with the obstacle to produce breaking and merging of the two-phase interface. We quantify the pressure values to compare with the experimental results at the two locations (with respect to the axis orientation given in Fig. 8.11):

- $(2.3955, 0.021, -0.5255)$: Pressure value (P1)
- $(2.3955, 0.061, -0.5255)$: Pressure value (P2).

The notation in the parentheses is the name of the probe used in the experimental study performed at Maritime Research Institute Netherlands (MARIN) [125, 126]. The variation in the water level (air-water interface) is recorded at two locations: $(X, Z) = (2.724, -0.5)$ and $(X, Z) = (2.228, -0.5)$ corresponding to H1 and H2 probes of the experiment, respectively.

The three-dimensional domain is discretized spatially into 430,000 nodes consisting of 2.5 million tetrahedral elements. The interface thickness parameter is selected

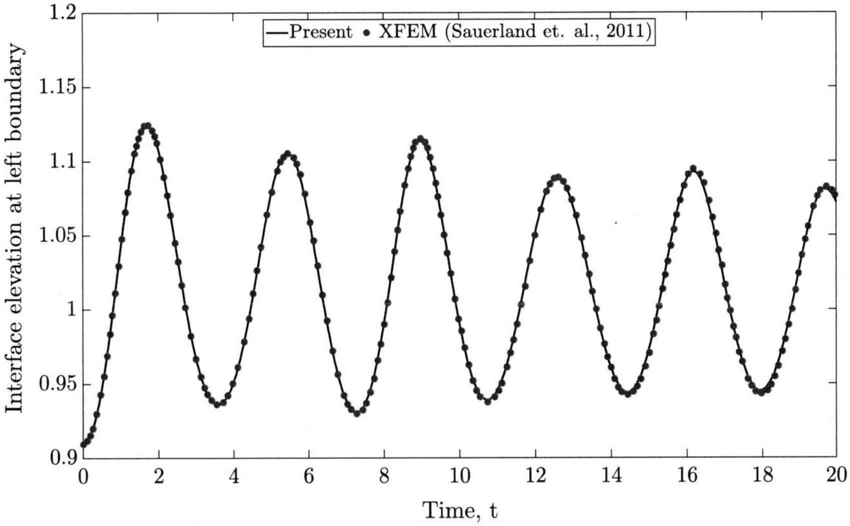

Fig. 8.10 Comparison of the solution considering $N_\varepsilon = 4$, $\Delta t = 0.01$ and $\varepsilon = 0.01$ with XFEM-based level set approach in [188]

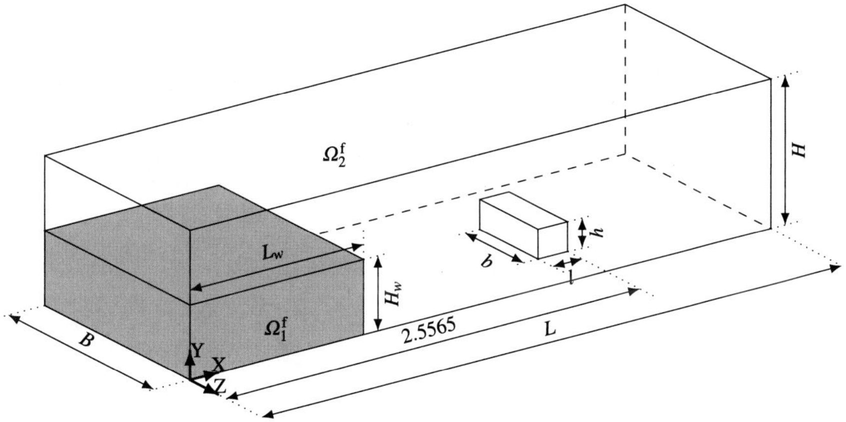

Fig. 8.11 Schematic diagram showing the computational domain for three-dimensional dam-break with obstacle. Ω_1^f and Ω_2^f are the two phases with $\rho_1^f = 1000$, $\rho_2^f = 1.225$, $\mu_1^f = 1.002 \times 10^{-3}$ and $\mu_2^f = 1.983 \times 10^{-5}$, acceleration due to gravity is taken as $\boldsymbol{g} = (0, -9.81, 0)$ and slip boundary condition is imposed on all the boundaries

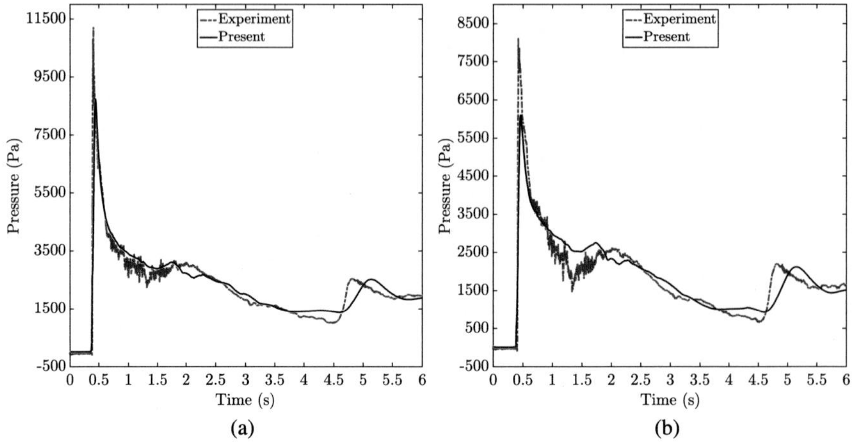

Fig. 8.12 Three-dimensional dam-break problem: temporal evolution of the pressure at the probe points: **a** P1 and **b** P2

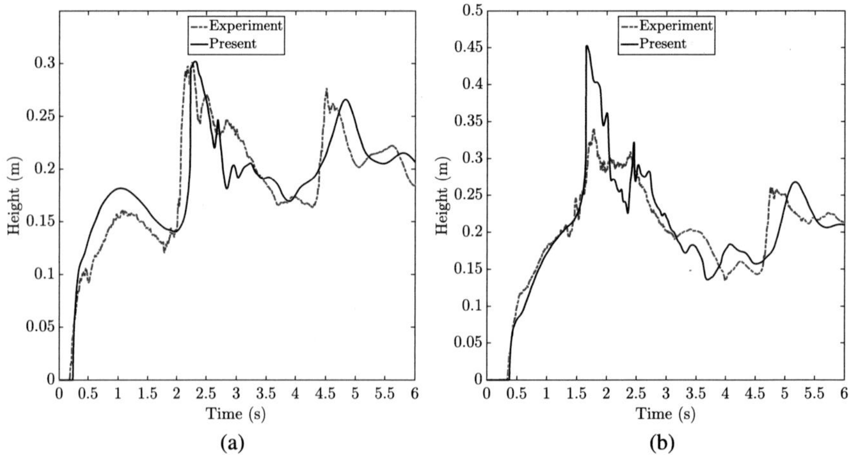

Fig. 8.13 Three-dimensional dam-break problem: temporal evolution of the height at the probe points: **a** H1 and **b** H2

as $\varepsilon = 0.02$ and the time step size $\Delta t = 0.005$. The computation was simulated for 1200 time steps and took 2.2 hrs on 24 CPUs. Figures 8.12 and 8.13 depict the comparison of the results obtained from the present simulation with that of the experiment. The data of the experiment for comparison is taken from [103]. The iso-contours of water ($\phi > 0$) are shown at different time instances in Fig. 8.14 and show a good qualitative agreement with the results from the literature [5, 59, 125].

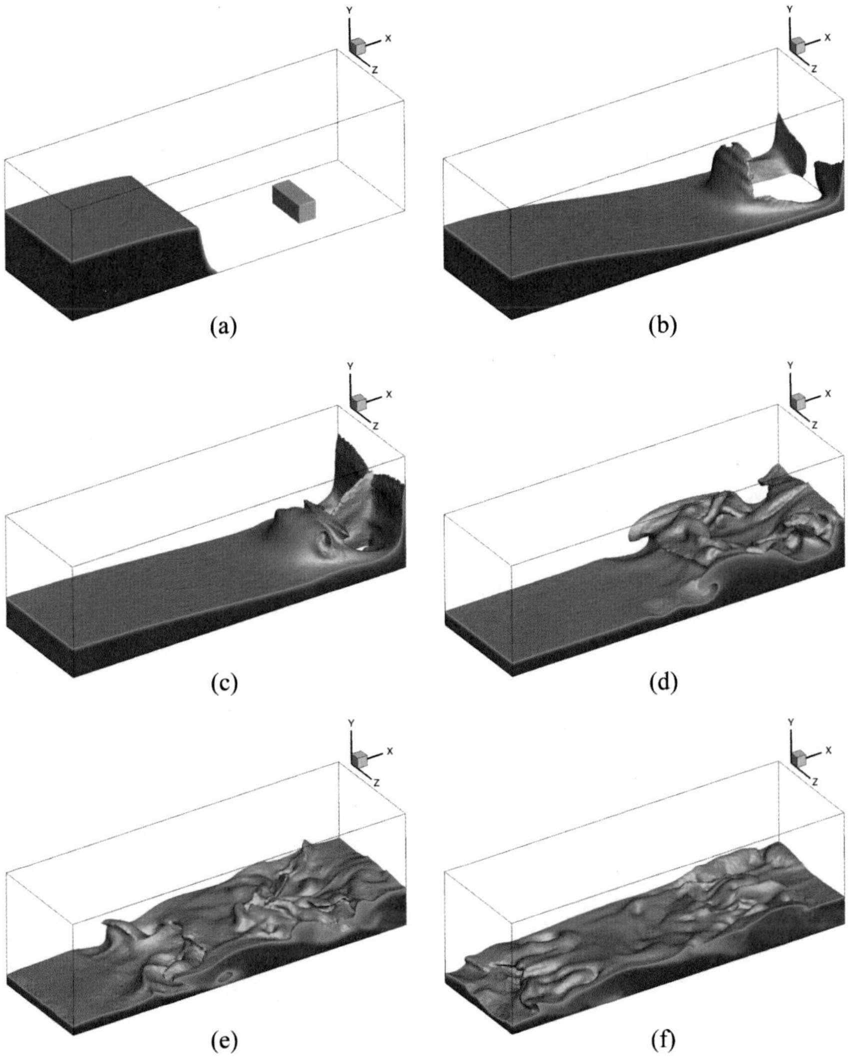

Fig. 8.14 Three-dimensional dam-break problem: temporal evolution of the iso-contours of water ($\phi > 0$) at time: **a** 0.125 s, **b** 0.75 s, **c** 1.125 s, **d** 2 s, **e** 2.5 s and **f** 3 s

8.4.3 Two-Phase Fluid-Structure Interaction Coupling

After the coupled two-phase flow equations (Navier-Stokes and Allen-Cahn) have been assessed, we next move on to the structural coupling with the two-phase flow system. This will form the two-phase fluid-structure interaction framework in the partitioned coupling format. To accomplish this, we perform decay tests involving interaction of the free-surface with a rigid circular cylinder under pure translation and a rectangular barge under pure rotation.

8.4.3.1 Heave Decay Test Under Translation

We start the assessment of the two-phase FSI system by considering free heave motion of a circular cylinder at the free-surface. The computational domain, $\Omega \in [0, 90D] \times [0, 14.6D] \times [0, 17D]$ consists of a circular cylinder of diameter $D = 0.1524$m placed at an offset of $0.167D$ from the free-surface of water, shown in Fig. 8.15. We let the cylinder heave under gravity, $\boldsymbol{g} = (0, -9.81, 0)$ until the heave decays to its equilibrium position.

The density and viscosity of the two phases of the fluid are $\rho_1^{\mathrm{f}} = 1000$, $\rho_2^{\mathrm{f}} = 1.2$, $\mu_1^{\mathrm{f}} = 10^{-3}$ and $\mu_2^{\mathrm{f}} = 1.8 \times 10^{-5}$. The structural density is half that of the denser fluid, i.e., $\rho^{\mathrm{s}} = 500$. The high density ratio of the two phases, $\rho^* = \rho_1^{\mathrm{f}}/\rho_2^{\mathrm{f}} = 833.3$ along with the low structure-to-fluid density ratio, $(\rho^{\mathrm{s}}/\rho_1^{\mathrm{f}} = 0.5)$ makes this problem more challenging for FSI solvers. The initial condition for the order parameter is given as

$$\phi(x, y, 0) = -\tanh\left(\frac{y}{\sqrt{2}\varepsilon}\right). \tag{8.51}$$

The Reynolds number is evaluated as $Re = \rho_1^{\mathrm{f}} U_{cyl} D/\mu_1^{\mathrm{f}} \approx 30,000$ based on the maximum velocity achieved by the cylinder, its diameter and viscosity of the denser fluid. To model the turbulent effects, we employ a hybrid RANS/LES model discussed in Chap. 10 in detail.

The computational mesh for this cylinder decay test is shown in Fig. 8.16. The mesh is constructed consisting of a boundary layer grid enveloping the cylinder with the first layer at a distance from the cylindrical surface such that $y^+ \sim 1$ (Fig. 8.16b).

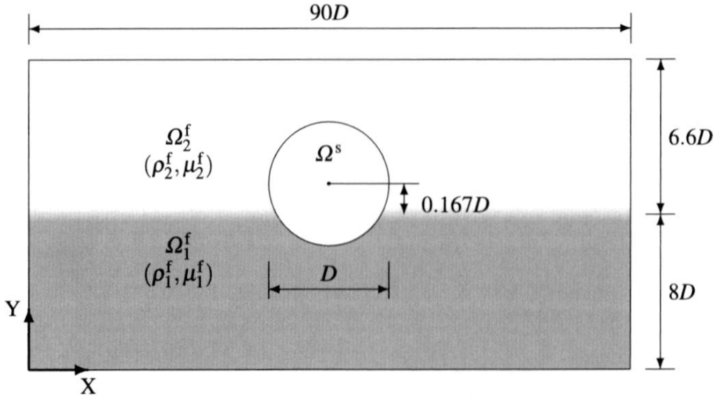

Fig. 8.15 Schematic of the decay response of a cylinder of diameter D under gravity in the X-Y cross-section. The computational domain extends a distance of $17D$ in the Z-direction

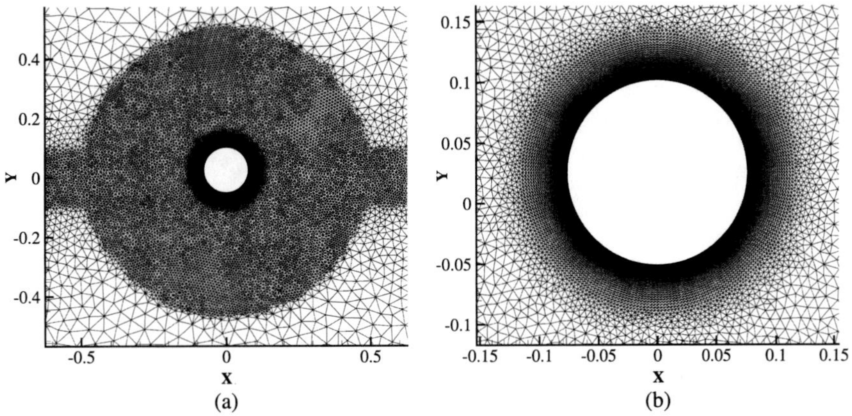

Fig. 8.16 Mesh employed for the decay response of a cylinder interacting with the free-surface: **a** the refined mesh around the cylinder to capture the flow vortices near the structure, and **b** the boundary layer mesh around the cylinder with $y^+ \sim 1$ for hybrid RANS/LES modeling

Furthermore, the mesh is refined around the cylinder till a radius of $3.3D$ to capture the vortices produced due to the heaving motion at the free-surface (Fig. 8.16a), and the interfacial region is also refined to accurately capture the air-water interface in accordance with the suggestions in Sect. 8.4.2.2 so that at least 4 elements exist in the equilibrium interface region. The mesh consists of 72,604 number of nodes with 144,566 triangular elements in the two-dimensional plane perpendicular to the axis of the cylinder. Three-dimensional domain is created by extruding the two-dimensional mesh in the Z-direction by 7 layers. The no-slip condition is satisfied at the structural surface and a slip condition is imposed on all the other boundaries.

We first conduct temporal convergence on the two-dimensional mesh by fixing the interface thickness parameter as $\varepsilon = 0.01$ and starting with a time step size of $\Delta t = 2 \times 10^{-2}$ and decreasing it by a factor of 2 till $\Delta t = 6.25 \times 10^{-4}$. The decaying heave of the cylinder for varying time steps are shown in Fig. 8.17a. We quantify the error as

$$e_1 = \frac{||\boldsymbol{\eta} - \boldsymbol{\eta}_{\text{ref}}||_2}{||\boldsymbol{\eta}_{\text{ref}}||_2}, \tag{8.52}$$

where $\boldsymbol{\eta}$ represents the temporal evolution of the heave motion of the cylinder for corresponding time step, $\boldsymbol{\eta}_{\text{ref}}$ is the same for the finest time step ($\Delta t = 6.25 \times 10^{-4}$) and $|| \cdot ||_2$ represents the standard Euclidean L^2 norm. The temporal convergence based on the error evaluated is plotted in Fig. 8.17b, giving a convergence rate of 1.6.

We next conduct a spatial convergence study on the two-dimensional mesh by varying the thickness parameter ε such that suggested 4 number of elements are maintained in the interfacial region. The mesh consists of 35,954 (71,300 elements), 72,604 (144,566 elements), and 205,094 (409,494 elements) nodes for $\varepsilon = 0.02$,

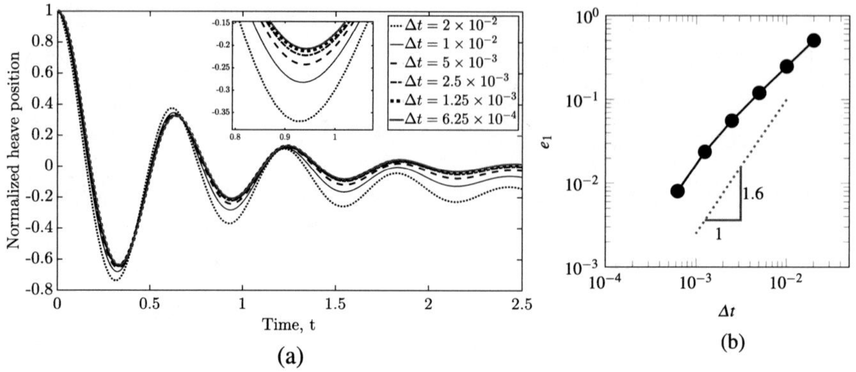

Fig. 8.17 Temporal convergence study for the decay test of a circular cylinder: **a** heave motion employing temporal refinement, and **b** the dependence of non-dimensionalized L^2 error (e_1) as a function of uniform temporal refinement Δt

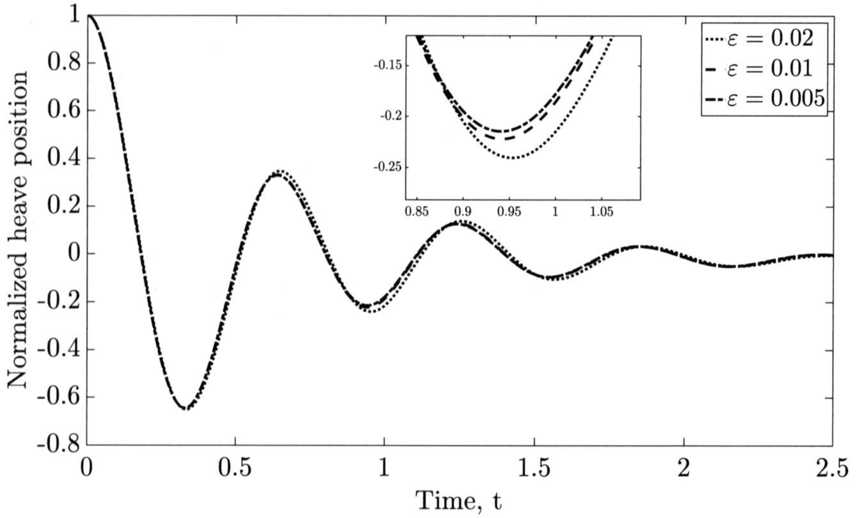

Fig. 8.18 Dependence of spatial grid convergence on the interfacial thickness parameter ε for the decay test of a circular cylinder at free-surface

0.01 and 0.005 respectively. The results of the heave motion are depicted in Fig. 8.18 for various values of ε. We observe a very minor difference in the results for different ε. Therefore, we select the interface thickness parameter of $\varepsilon = 0.01$ and a time step of $\Delta t = 0.0025$ for the validation test, based on the convergence studies.

The validation of the heave of the circular cylinder is carried out on a three-dimensional mesh with the selected convergence parameters. The mesh here consists of 580,000 nodes with 1.01 million six-node wedge elements. The computation was carried out by 48 CPUs and took 11.63 hrs. The nonlinear convergence criterion

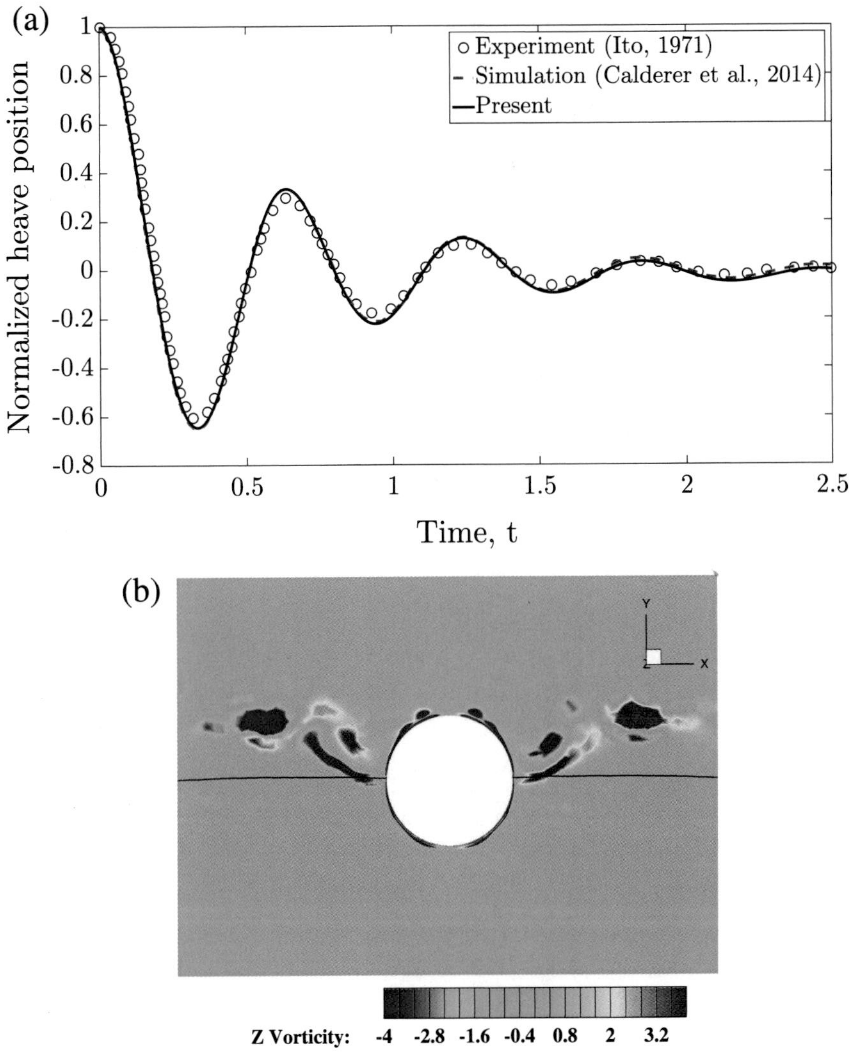

Fig. 8.19 Decay test of a circular cylinder under translation along the free-surface: **a** validation of the heave motion of the cylinder at the free-surface with the experimental [104] and simulation [36] studies, and **b** Z-vorticity contours around the cylinder at $t = 1.5$ s with the free-surface indicated at $\phi = 0$

with tolerance of 5×10^{-4} is reached in around 4 nonlinear iterations in a particular time step by the two-phase FSI solver. The results are compared with that of the experiment [104] and the simulation [36] in Fig. 8.19a where a good agreement is found. The Z-vorticity contours along with the interface at $\phi = 0$ are shown in Fig. 8.19b. The resolution of the vortices around the cylinder can be improved by refining the mesh surrounding it.

8.4.3.2 Decay Test Under Rotation Along Free-Surface

We next perform the decay test of a rectangular barge under pure rotation at the free-surface. The computational domain is $\Omega \in [0, 58.3L] \times [0, 6.3L] \times [0, 3L]$ with a barge inclined at an angle of $\theta = 15^\circ$ from the free-surface level as depicted in Fig. 8.20a. We let the barge rotate under gravity and the rotational amplitude decays with time. The barge has a length, height and width of $L = 0.3$, $H = 0.2$ and $W = 3L$ respectively. The center of gravity of the barge levels with the free-surface and its mass moment of inertia and rotational damping matrices respectively are given as

$$\boldsymbol{I}^{\mathrm{s}} = \begin{bmatrix} 0 & 0 & 0 \\ 0 & 0 & 0 \\ 0 & 0 & 0.236 \end{bmatrix} \mathrm{kg} \cdot \mathrm{m}^2, \qquad \boldsymbol{C}_\theta = \begin{bmatrix} 0 & 0 & 0 \\ 0 & 0 & 0 \\ 0 & 0 & 0.275 \end{bmatrix} \mathrm{kg} \cdot \frac{\mathrm{m}^2}{\mathrm{s}}. \tag{8.53}$$

The physical properties of the two phases of the fluid are $\rho_1^{\mathrm{f}} = 1000$, $\rho_2^{\mathrm{f}} = 1.2$, $\mu_1^{\mathrm{f}} = 10^{-3}$, $\mu_2^{\mathrm{f}} = 1.8 \times 10^{-5}$ with acceleration due to gravity $\boldsymbol{g} = (0, -9.81, 0)$. The initial condition for the order parameter is set as

$$\phi(x, y, 0) = -\tanh\left(\frac{y}{\sqrt{2}\varepsilon}\right). \tag{8.54}$$

The computational mesh, constructed with similar characteristics as that for the circular cylinder in Sect. 8.4.3.1, is shown in Fig. 8.20b. A rectangular boundary layer mesh around the barge and a refined grid along the free-surface interface is constructed to capture the boundary layer physics and the free-surface respectively. The two-dimensional mesh is extruded by 10 layers in the Z-direction. The Reynolds number is based on the maximum velocity achieved by the upper corner of the barge, its length, and the density of the denser fluid, i.e., $Re = \rho_1^{\mathrm{f}} U_{corner} L/\mu_1^{\mathrm{f}} \approx 99{,}500$. The mesh consists of 1,025,684 nodes with 1,856,930 six-node wedge elements. The computation is carried out by 72 CPUs with a computational time of 22.2 hrs.

The rotational heave motion of the barge is plotted by considering the converged parameters in Fig. 8.21a where a good agreement is found with the experimental [117] and the numerical [36] results in the literature. The Z-vorticity contours are also shown in Fig. 8.21b at $t = 1.4$ s.

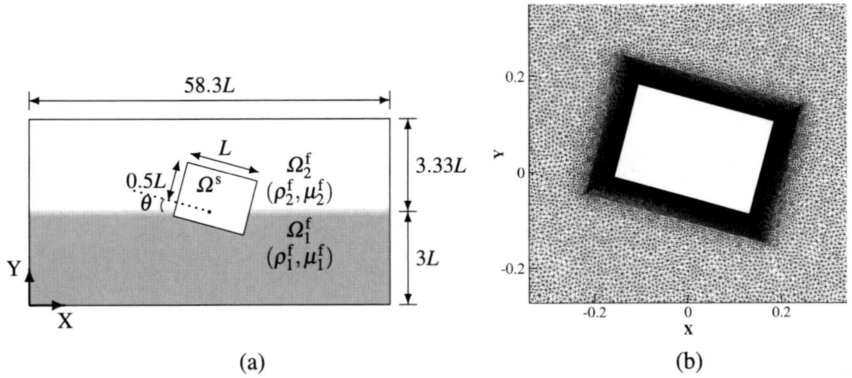

Fig. 8.20 Rotation of a rectangular barge under gravity: **a** schematic of the computational setup of a barge of length $L = 0.3$, height $H = 0.2$ and width $W = 3L$ in the X-Y cross-section, and **b** zoom-in view of the computational mesh near the barge with boundary layer. The computational domain extends a distance of $3L$ in the Z direction

This concludes the convergence and validation studies of the coupled two-phase FSI solver in a partitioned format. We next demonstrate the solver for a practical problem of wave-structure interaction and uniform flow across a flexible pipeline with internal two-phase flow.

8.5 Application to Wave-Structure Interaction

In this section, we demonstrate the two-phase flow partitioned framework based on phase-field method on an application of wave-structure interaction of free-surface waves interacting with a fixed truncated cylinder. We first briefly present a theoretical background on ocean waves.

Propagation of ocean waves is a very complex phenomenon owing to their irregular nature and nonlinear effects. Numerical modeling of this phenomenon adds to the complexity and is very challenging. The simplest model is linear or first-order wave theory based on the assumptions of incompressible, inviscid and irrotational fluid with uniform density, planar and monochromatic waves. Boundary conditions at the free-surface which are nonlinear in nature are further simplified in the model [42, 195]. Second-order Stokes wave theory developed in [201] gives a more accurate model and involves the nonlinearities of the waves. The notations used in the description of the theory are depicted in Fig. 8.22.

Let T, $\omega = 2\pi/T$ and $k = 2\pi/\lambda$ be the time period, angular frequency and the wave number of the wave respectively. According to the second-order Stokes wave theory, the free-surface profile of the wave can be written as

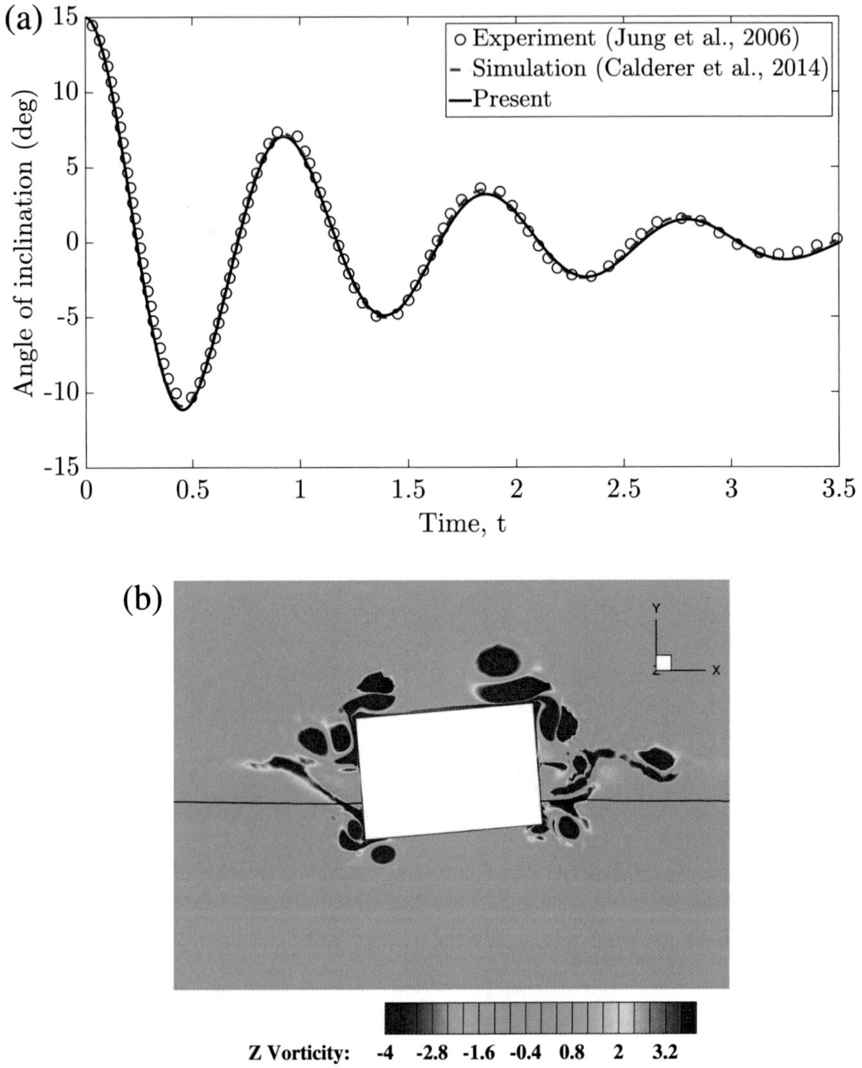

Fig. 8.21 Decay test under rotation: **a** validation of the rotational motion of the rectangular barge at the free-surface with the experimental [117] and simulation [36] studies, and **b** Z-vorticity contours around the barge at $t = 1.4$ s with the free-surface indicated at $\phi = 0$

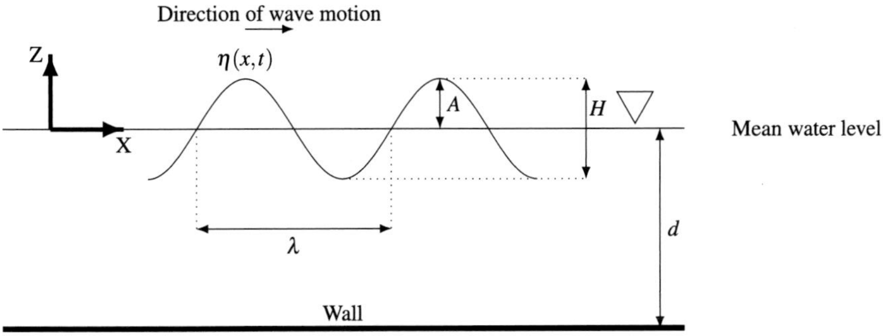

Fig. 8.22 Notations and coordinate system in the description of the second-order Stokes wave theory: $\eta(x, t)$ is the profile of the free-surface, d is the water depth, A is the amplitude of the wave, $H = 2A$ is the height of the wave and λ is the wavelength

$$\eta(x,t) = \frac{H}{2}\cos(kx - \omega t) + \frac{\pi H^2}{8\lambda}\frac{(\cosh(kd))(2 + \cosh(2kd))}{\sinh^3(kd)}\cos[2(kx - \omega t)]. \tag{8.55}$$

The above profile can be generated by the following horizontal (u) and vertical (w) components of the fluid velocity:

$$u(x,z,t) = \frac{Hgk}{2\omega}\frac{\cosh[k(d+z)]}{\cosh(kd)}\cos(kx - \omega t) + \frac{3H^2\omega k}{16}\frac{\cosh[2k(d+z)]}{\sinh^4(kd)}\cos[2(kx - \omega t)], \tag{8.56}$$

$$w(x,z,t) = \frac{Hgk}{2\omega}\frac{\sinh[k(d+z)]}{\cosh(kd)}\sin(kx - \omega t) + \frac{3H^2\omega k}{16}\frac{\sinh[2k(d+z)]}{\sinh^4(kd)}\sin[2(kx - \omega t)]. \tag{8.57}$$

The wave run-up on a submerged structure is characterized by the run-up ratio, R/A which is defined as the ratio of the highest elevation of the free-surface of the fluid at the front-face of the structure to the amplitude of the incident incoming ocean wave. This quantity depends on two non-dimensional parameters, viz., the wave steepness kA and the wave scattering parameter ka, given as

$$kA = \frac{2\pi A}{\lambda}, \tag{8.58}$$

$$ka = \frac{2\pi a}{\lambda}, \tag{8.59}$$

with A and λ denoting the amplitude and wavelength of the incoming wave respectively and a is the radius of the cylinder. An increase in the steepness of the incoming wave (kA) or an increase in the cross-sectional width of the structure (ka) leads to an increase in resistance to the flow, thus increasing the run-up ratio. The quantification of the run-up ratio with kA and ka is of a particular interest in ocean engineering. Some of the works in the literature dealing with the run-up ratio across structures of varying cross-sectional area can be found in [154, 211].

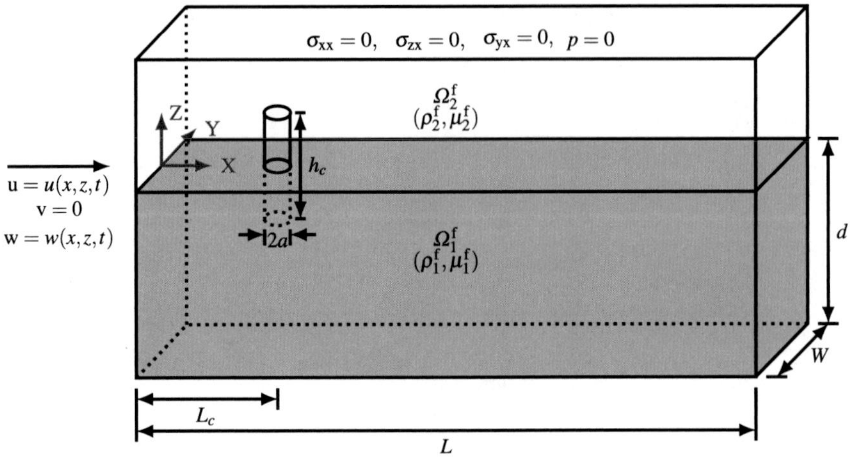

Fig. 8.23 A schematic of the wave-structure interaction problem. The computational setup and boundary conditions are shown for the Navier-Stokes equations. Here, $\boldsymbol{v}^{\mathrm{f}} = (\mathrm{u}, \mathrm{v}, \mathrm{w})$ denotes the components of the fluid velocity which are given by Eqs. (8.56) and (8.57) with $\mathrm{v} = 0$. Stress-free boundary condition with zero pressure is given at the top surface of the wave tank. Slip boundary condition is imposed at the X-Z plane at $y = -1$ and $y = 1$. All other boundaries have the no-slip condition. Moreover, zero Neumann boundary condition is imposed for the order parameter ϕ on all the boundaries

In the current demonstration, we set up a numerical wave tank to simulate a wave run-up problem across a vertical truncated cylinder, which is one of the most commonly used structural members in offshore industry, for example, gravity-based structures, semi-submersibles and tension-leg platforms. We quantify the run-up on the circular cylinder by varying kA and keeping ka constant. The computational set-up is similar to that in [211] and the schematic of the computational domain is shown in Fig. 8.23.

The numerical wave tank of 24 m × 2 m × 2 m, i.e., $L = 24$ m and $W = 2$ m is set up. The water depth is $d = 1.2$ m with the draft of the submerged truncated cylinder (diameter of $2a = 0.2$ m and total height $h_c = 0.8$ m) as 0.4 m. The center of the cylinder is placed at a distance of $L_c = 3.6$ m from the left boundary of the computational domain. The inlet velocity profiles are prescribed by Eqs. (8.56) and (8.57) to model the incoming second-order Stokes waves and stress-free boundary condition with zero pressure is imposed at the top boundary. Slip condition is satisfied at the two sides corresponding to $y = -1$ m and $y = 1$ m. No-slip condition is imposed at the bottom and the right boundaries. The initial condition of the order parameter is such that the interface lies at $z = 0$ m with Ω_1^{f} phase representing water ($\rho_1^{\mathrm{f}} = 1000$, $\mu_1^{\mathrm{f}} = 1.002 \times 10^{-3}$) and Ω_2^{f} depicting air ($\rho_2^{\mathrm{f}} = 1.225$, $\mu_2^{\mathrm{f}} = 1.983 \times 10^{-5}$). The phase-field interface thickness parameter is chosen as $\varepsilon = 0.02$ as initial numerical tests suggested negligible change in the solution with a much sharper interface. The time step size is selected as $\Delta t = 0.0025$ with the total number of time

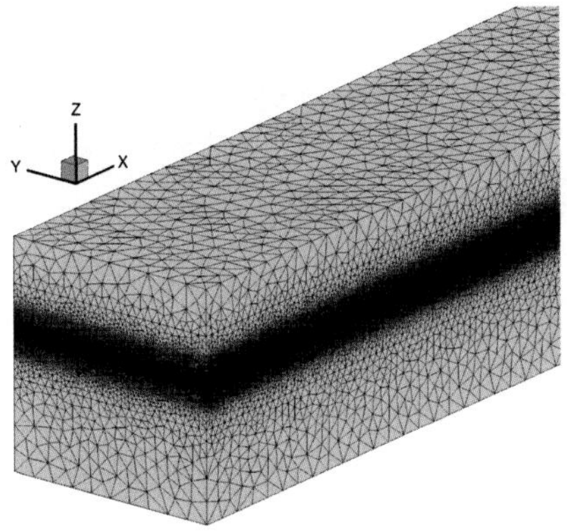

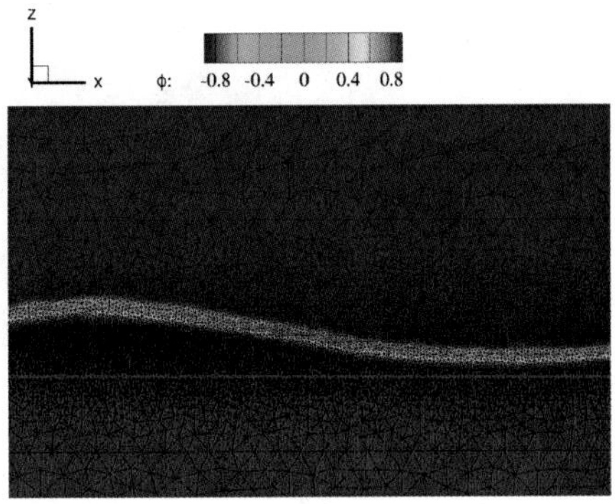

Fig. 8.24 Three-dimensional computational mesh and contours of order parameter for the run-up problem: **a** refined interfacial region with unstructured finite element mesh, **b** contour plot of order parameter ϕ superimposed with the unstructured mesh

steps as 5000. Again, the capillary effects due to the surface tension are neglected for the modeling of the free-surface waves.

Based on the experimental study in [154], $ka = 0.208$ is selected giving a wavelength of $\lambda = 3.0208$ m. For Stokes waves, the time period can be evaluated as a function of wavelength as $\lambda = \frac{gT^2}{2\pi}\tanh\left(\frac{2\pi d}{\lambda}\right)$, which gives $T = 1.40045$ s. Therefore, kA values are varied by changing the amplitude of the incoming wave and the run-up ratio R/A is quantified for the truncated cylinder.

A three-dimensional unstructured mesh is constructed by open-source mesh generator [77]. A refined region to capture the free-surface is formed from $z = -0.2$ to $z = 0.2$ where a gradual coarsening of the mesh is observed from mesh size $\delta = 0.01$ at $z = 0$ to $\delta = 0.02$. This resolution ensures that at least 4 number of elements lie in the equilibrium interface region. Moreover, the mesh enveloping the cylinder is refined to a value of $\delta = 0.01$, thus giving the total grid node count of around 5.6 million with approximately 34 million tetrahedral unstructured elements (Fig. 8.24). The computation is carried out in 600 CPUs and took approximately 6.5 hrs to reach 5000 time steps.

The observations are recorded at two locations- $x = 2$ m and $x = 3.5$ m (front-face of the cylinder) to evaluate the free-surface wave elevation which is determined by linearly interpolating the order parameter and finding the free-surface at $\phi = 0$. The amplitude of first harmonic of the free-surface elevation of the incident wave at $x = 2$ m gives the wave amplitude by taking the Fourier transform of the elevation η as $A = A(\omega_1)$, where ω_1 is the first harmonic frequency of the incident wave. The wave run-up on the front-face of the cylinder at $x = 3.5$ m is evaluated by taking the mean of the maximum amplitude from each wave cycle as $R = \frac{1}{M}\sum_{n=1}^{M}\eta_n$, where η_n is the peak amplitude of the free-surface elevation η at $x = 3.5$ m of the nth wave

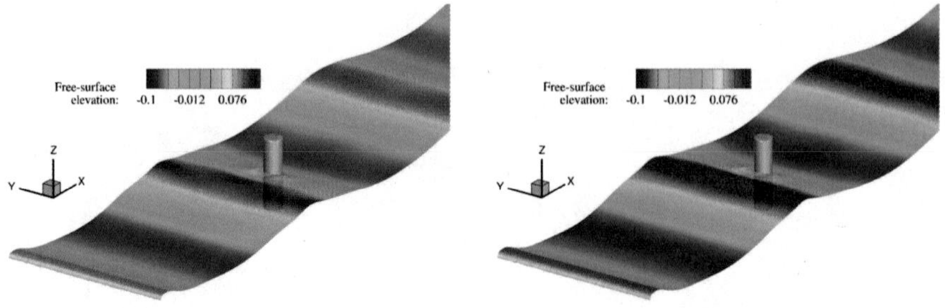

(a) (b)

Fig. 8.25 Water wave run-up on a truncated cylinder: snapshots of iso-contours $\phi = 0$ colored by the free-surface elevation for two representative wave steepness parameters at $t/T = 8.926$: **a** $kA = 0.1837$, and **b** $kA = 0.2146$

cycle and M is the total number of wave cycles which is equal to 5 in this case. Thus, the wave run-up ratio is calculated as R/A. To exclude any transient behavior, the post-processing is performed in the time interval $t/T \in [4, 9]$.

The results are analyzed in a similar manner as done in the experiment [154]. The run-up on the cylinder is quantified for a range of $kA \in [0.04, 0.2282]$. The iso-contours of the free-surface colored by the free-surface elevation are shown in Fig. 8.25 for two different kA values at $t/T = 8.926$. The variation of the elevation of the incident wave and the run-up for various kA is depicted in Fig. 8.26a–c. A secondary kink in the run-up at high kA is observed similar to the findings in the literature. Finally, the run-up ratio is compared with the theories and results from the literature in Fig. 8.26d where we observe a good agreement with the experimental observations.

8.6 Application to Flexible Riser FSI with Internal Two-Phase Flow

In this section, we extend the demonstration to two-phase fluid-structure interaction problem. We consider a complex problem of a riser with internal two-phase flow, subjected to an external uniform current flow. The riser is modeled by Euler-Bernoulli beam equation using linear modal analysis. A schematic for the problem is shown in Fig. 8.27.

The computational domain is described as follows. Let the outer diameter of the riser be D with a span of $L = 20D$. The inflow and outflow boundaries are at a distance of $10D$ and $30D$ from the center of the riser respectively. The side walls are equidistant at a distance of $15D$ from the center of the riser. A uniform current of $\boldsymbol{v}^{\mathrm{f}} = (U_\infty, 0, 0)$ is prescribed at the inflow boundary with a no-slip condition at the outer surface of the riser. Stress-free condition is satisfied at the outflow boundary and

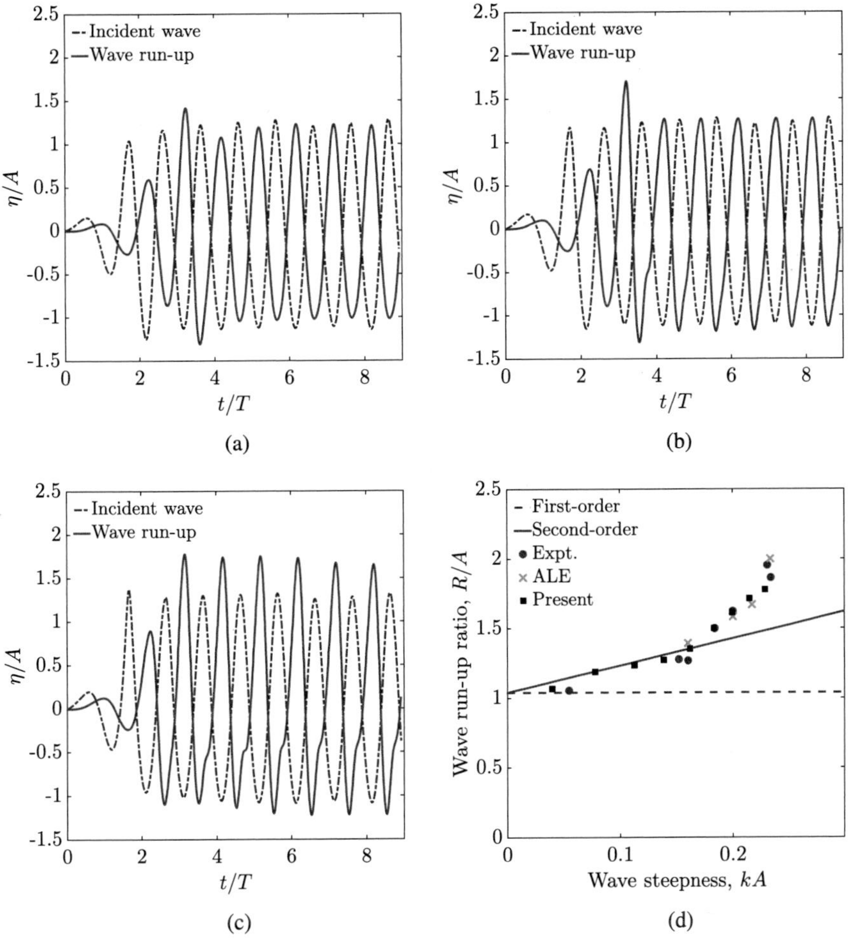

Fig. 8.26 Water wave run-up on a truncated cylinder: Time histories of the free-surface elevations at the probe locations for three representative wave steepness parameters: **a** $kA = 0.0779$, **b** $kA = 0.1381$, **c** $kA = 0.2146$, and **d** comparison of the wave run-up ratio with the results from the literature at $ka = 0.208$: The first- and second-order results are from frequency domain potential theory, the experimental results are from the campaign in [154] and the ALE results are from the free-surface ALE solver in [122]. Time and free-surface elevation are non-dimensionalized by time period T and wave amplitude A respectively

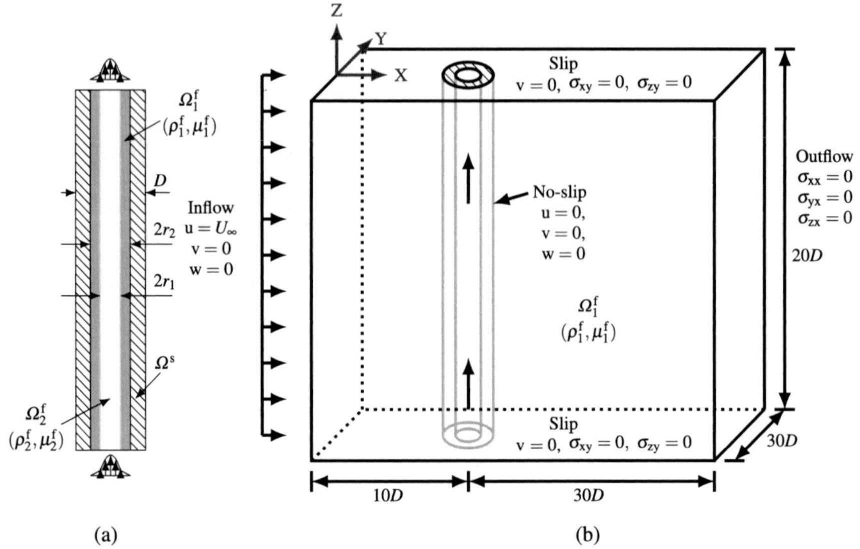

Fig. 8.27 Schematic of the uniform flow past a flexible riser with an internal two-phase flow: **a** the $X - Z$ cross-section of the flexible riser with the internal flow velocity profile at the inlet and outlet of the pipe, and **b** the computational setup and boundary conditions employed for the demonstration. Here, $\boldsymbol{v}^{\mathrm{f}} = (\mathrm{u}, \mathrm{v}, \mathrm{w})$ denotes the components of the fluid velocity and the hatched area shows the flexible structure (riser)

slip condition on all the other boundaries is imposed. Let the riser be subjected to the fluid phase Ω_1^{f} externally. The interior of the riser has internal diameter $2r_2$ with an initial concentric profile for the two phases with the interface at a radius of r_1 from the riser axis separating the two phases Ω_2^{f} and Ω_1^{f} (see Fig. 8.27a). A Z-velocity profile (co-annular, laminar and fully developed flow consisting of immiscible Newtonian fluids) is prescribed at the inlet of the riser for the internal flow and a stress-free condition is imposed at its outlet. The internal velocity profile is such that no-slip condition is satisfied along the internal surface of the riser. The prescribed internal velocity profile is given by [157]

$$\boldsymbol{v}^{\mathrm{f}} = (0, 0, \mathrm{w}), \tag{8.60}$$

$$\mathrm{w}(R) = \begin{cases} \frac{C[1-(r^*)^2+\mu^*((r^*)^2-R^2)]}{(r^*)^{n+3}(\mu^*-1)+1}, & 0 \leq R \leq r^*, \\ \frac{C[1-R^2]}{(r^*)^{n+3}(\mu^*-1)+1}, & r^* \leq R \leq 1, \end{cases} \tag{8.61}$$

where $R = \sqrt{x^2 + y^2}/r_2$, $r^* = r_1/r_2$, $\mu^* = \mu_1^{\mathrm{f}}/\mu_2^{\mathrm{f}}$ and $C = (n+3)/2$, where $n = 1$ for a circular tube, $r_1 = 0.2$, $r_2 = 0.4$ and $\boldsymbol{g} = (0, 0, 0)$. The initial condition for the order parameter is written as

$$\phi(x, y, 0) = \begin{cases} \tanh\left(\frac{\sqrt{x^2+y^2}-r_1}{\sqrt{2}\varepsilon}\right), & \sqrt{x^2+y^2} \leq r_2, \\ 1, & \text{elsewhere}. \end{cases} \tag{8.62}$$

A linear flexible structural solver is employed for modeling the riser. The riser is modeled by the Euler-Bernoulli beam equation written in the eigenspace where the structural displacement can be represented as a linear combination of the eigenmodes. The details about the linear modal analysis can be found in [116]. The non-dimensional parameters for the fluid-structure interaction of the riser with the internal flow are defined as:

$$Re = \frac{\rho_1^f U_\infty D}{\mu_1^f} = \frac{\rho_2^f U_\infty D}{\mu_2^f}, \quad m^* = \frac{m^s}{\pi D^2 L \rho_1^f/4}, \quad \rho^* = \frac{\rho_1^f}{\rho_2^f}, \quad \mu^* = \frac{\mu_1^f}{\mu_2^f},$$
$$U_r = \frac{U_\infty}{f_1 D}, \quad P^* = \frac{P}{\rho_1^f U_\infty^2 D^2}, \quad EI^* = \frac{EI}{\rho_1^f U_\infty^2 D^4}, \tag{8.63}$$

where Re, m^*, ρ^*, μ^*, U_r, P^* and EI^* denote the non-dimensional quantities, viz., Reynolds number, mass ratio, density ratio, viscosity ratio, reduced velocity, axial tension and flexural rigidity of the riser modeled as a beam. f_1 denotes the frequency of the first eigenmode. Two cases with different Reynolds number are considered. The non-dimensional numbers related to the cases are given in Table 8.6.

The computational mesh for the demonstration is depicted in Fig. 8.28. The mesh is divided into two parts: external and internal. The external mesh contains the fluid domain external to the riser surface. The mesh consists of a refined wake region to capture the flow structures as the current interacts with the riser and produced vortex-induced vibrations (VIV). A two-dimensional cross-section of the mesh is shown in Fig. 8.28b. This two-dimensional layer is extruded in the third dimension by 130 layers such that the element size in the spanwise direction is $\sim 0.15D$. Some of the layers are shown in Fig. 8.28a for clarity. On the other hand, the internal mesh to capture the internal two-phase flow is much more refined to accurately represent the fluid-fluid interface. It is extruded 500 layers in the third dimension. This leads to the overall three-dimensional mesh consisting of 5.14 million grid points with 10.08 million six-node wedge elements.

A brief physical insight is provided on the response of the riser for the two cases and the effect of the external flow VIV on the internal two-phase flow regimes. Pertaining to the reduced velocity lying in the "lock-in" range of VIV, amplitude

Table 8.6 Non-dimensional parameters for the two cases considered for the present study

Cases	Re	m^*	ρ^*	μ^*	U_r	P^*	EI^*	ρ^s/ρ_1^f
Case 1	100	2.89	100	100	5	0.34	5872.8	6.68
Case 2	1000	2.89	100	100	5	0.34	5872.8	6.68

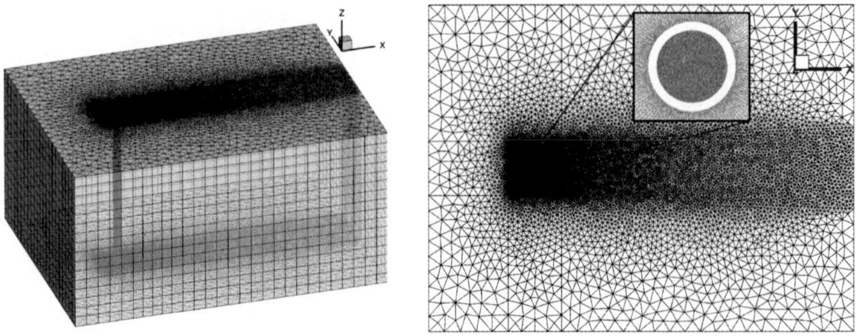

(a) (b)

Fig. 8.28 Computational mesh for the VIV of a riser with an internal two-phase flow: **a** three-dimensional view of the mesh, and **b** two-dimensional cross-section of the mesh with refined wake region behind the riser and refined internal region of the riser to capture the interface between the two phases in the internal flow accurately

of the order of the diameter of the riser is expected. Further parametric studies are required to be conducted to thoroughly understand the physics of this coupled problem, which is beyond the scope of this book.

8.6.1 Amplitude Response and Flow Patterns

The time history of the amplitude of the riser for the two cases has been plotted in Fig. 8.29 at the mid-point along its span where it is found to be maximum. The two cases are simulated till the time to obtain at least 4 cycles of VIV in the response. The temporal variation of the response amplitude along the riser in the cross-flow and in-line directions are shown in Figs. 8.30 and 8.31 for Cases 1 and 2 respectively. A standing wave-like pattern is observed from the plots. A spectral analysis conducted on the amplitude response at $z/L = 0.5$ shows a single dominating non-dimensional frequency (fU_∞/D) of 0.1709 in the cross-flow and 0.3662 in the in-line directions for Case 1, indicating the lock-in phenomenon. For Case 2, the in-line oscillation depicts a multi-modal response consisting of fU_∞/D as 0.1953 and 0.3662 and the cross-flow vibration indicates a single frequency response with $fU_\infty/D = 0.1709$.

The contours of the Z-vorticity along the span of the riser with the internal flow visualization are shown in Figs. 8.32 and 8.33 for Case 1 and Case 2, respectively. Case 2 shows the onset of turbulent wake pertaining to the irregularity in the vortex patterns. The fluid-fluid interface in the two-phase internal flow seems to be captured qualitatively. The change of the internal flow pattern from co-annular regime to bubble/slug flow regime can also be observed due to the effect of the external VIV of the riser.

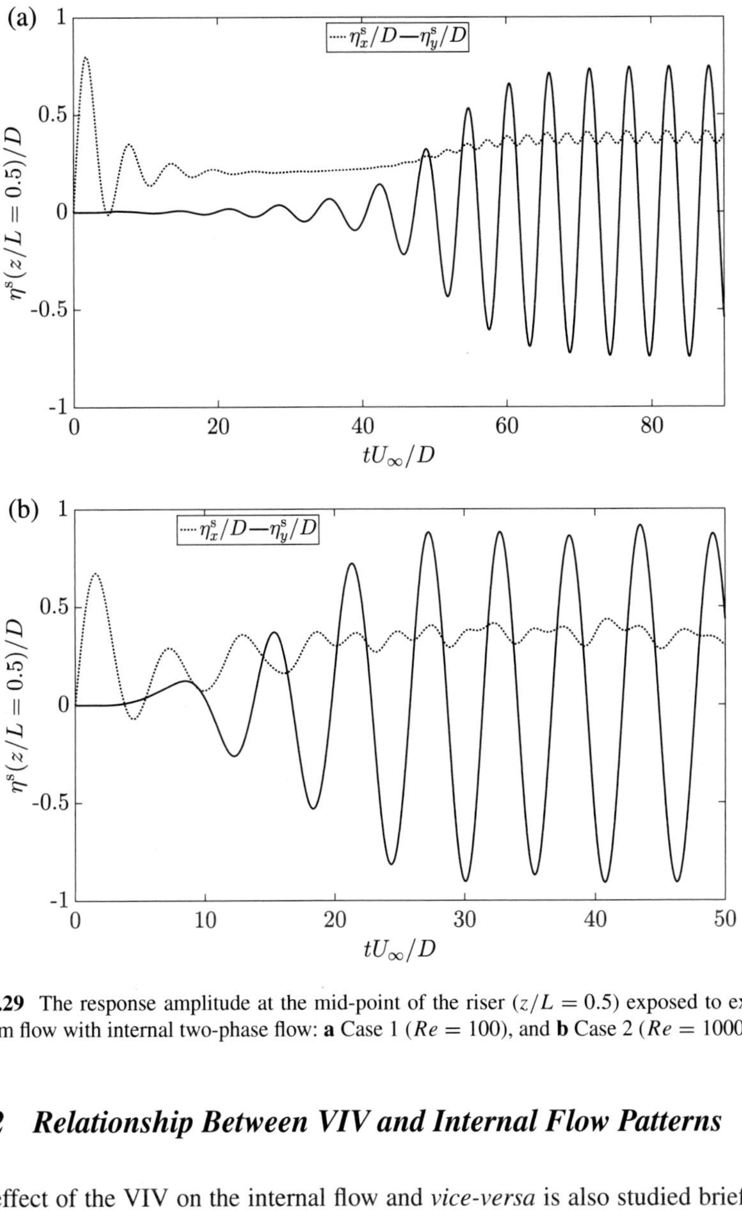

Fig. 8.29 The response amplitude at the mid-point of the riser ($z/L = 0.5$) exposed to external uniform flow with internal two-phase flow: **a** Case 1 ($Re = 100$), and **b** Case 2 ($Re = 1000$)

8.6.2 Relationship Between VIV and Internal Flow Patterns

The effect of the VIV on the internal flow and *vice-versa* is also studied briefly by conducting the same experiment without the internal flow, i.e., a solid riser with the same non-dimensional parameters considered for Case 1. The results do not show any significant change in the amplitude response along the riser, meaning that for the parameters considered in the current scenario, the internal flow has no effect on the external VIV of the riser. This may be due to the dominant effect of the VIV compared to the inertia of the internal two-phase fluid. The effect of VIV on the internal two-phase flow, on the other hand, seems to be significant as shown from the flow contours

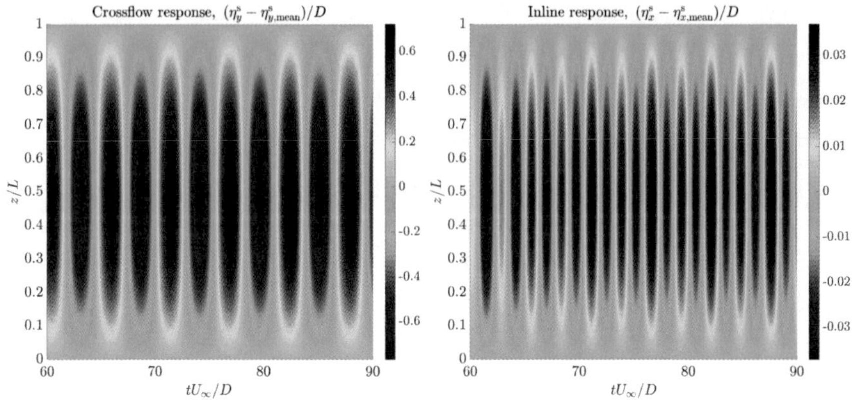

(a) (b)

Fig. 8.30 Response amplitudes of the riser along the span with non-dimensional time tU_∞/D for Case 1: **a** cross-flow, and **b** in-line

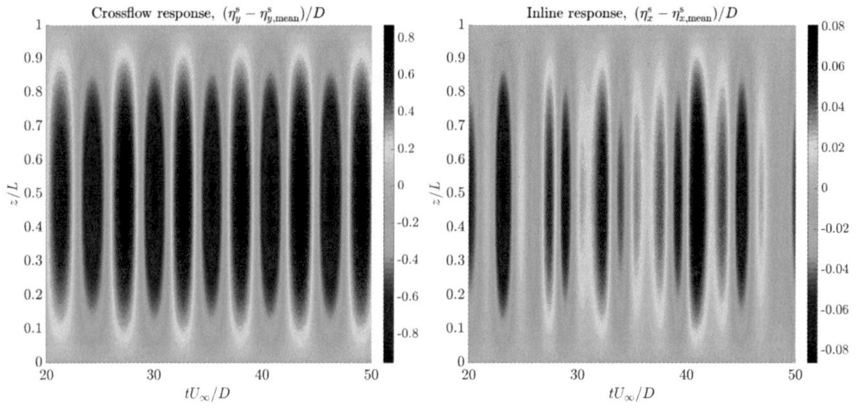

(a) (b)

Fig. 8.31 Response amplitudes of the riser along the span with non-dimensional time tU_∞/D for Case 2: **a** cross-flow, and **b** in-line

in Figs. 8.32c and 8.33c. A transition from the co-annular initial two-phase flow regime to an elongated bubble/slug flow pattern is observed. This prediction through coupled two-phase FSI simulation can be beneficial to improve multiphase flow assurance. However, a rigorous parametric study is needed to quantify the effects of structural parameters, reduced velocity and mass ratio on the internal flow regimes.

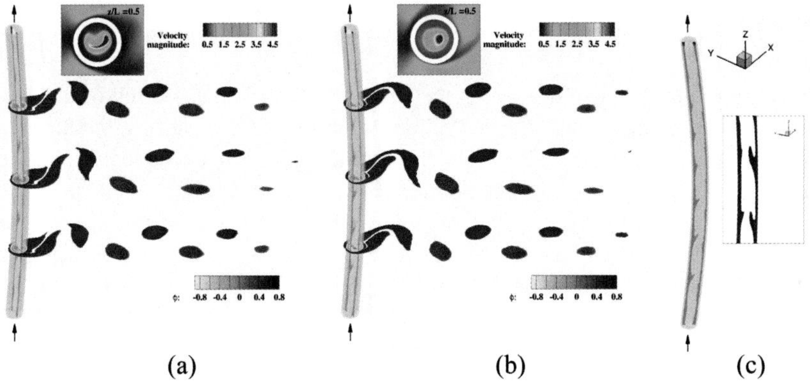

Fig. 8.32 Contour plots for the VIV of a riser at $Re = 100$ with internal two-phase flow at tU_∞/D: **a** 80, **b** 90 and **c** the internal flow along the riser. The Z-vorticity contours are shown at three cross-sections along the riser, viz., $z/L \in [0.25, 0.5, 0.75]$ and are colored with red for positive vorticity and blue for negative vorticity. The inset figure provides the velocity magnitude at the mid-section of the riser. The interior two-phase flow of the riser is visualized by the contours of order parameter $\phi > 0$ at an arbitrary plane passing through the axis of the deformed riser

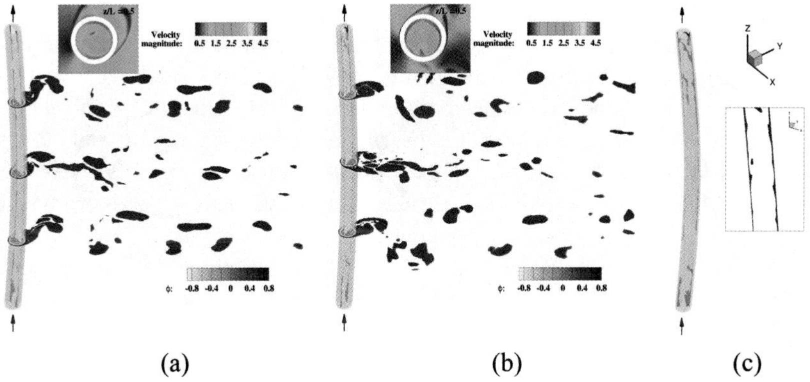

Fig. 8.33 Contour plots for the VIV of a riser at $Re = 1000$ with internal two-phase flow at tU_∞/D: **a** 40, **b** 50 and **c** the internal flow along the riser. The Z-vorticity contours are shown at three cross-sections along the riser, viz., $z/L \in [0.25, 0.5, 0.75]$ and are colored with red for positive vorticity and blue for negative vorticity. The inset figure provides the velocity magnitude at the mid-section of the riser. The interior two-phase flow of the riser is visualized by the contours of order parameter $\phi > 0$ at an arbitrary plane passing through the axis of the deformed riser

8.7 Interface-Capturing via Mesh Adaptivity

In the current chapter, we focused on the fixed grid interface-capturing of the fluid-fluid interface for two-phase flows. Although the interface-capturing methods can capture the breaking and merging of the fluid-fluid interface in a simple manner compared to interface-tracking, the computational grid may need to be refined throughout the domain as the location of the evolving interface is not known *a-priori*. This, in turn increases the cost of computation for a desired accuracy of the problem. A promising way, in this scenario, is to refine the mesh adaptively during the simulation run-time so that the focus area of refinement is confined and the accuracy and efficiency of the interface-capturing methods can be increased.

Some of the refinement strategies dealing with the mesh adaptivity are mesh movement or repositioning (r-methods), mesh enrichment (h/p-methods) and adaptive remeshing (m-methods) [141]. Among these, h-refinement is attractive as it deals with adding extra degrees of freedom near the region of interest in the domain. On the other hand, the p-refinement elevates the order of approximation by utilizing higher-order polynomials. The benefits of h-refinement also include conservation of the physical quantity and easy vectorization and parallelization. This refinement can be carried out either by subdivision of the elements into equal parts or by recursive subdivision of the largest edge side of the element. To optimize such refinement process, error indicators based on *a posteriori* residual estimates are formed which rely on gradients of the concerned quantities or the residuals of the equations being solved. Commonly used estimators are residual-based, hierarchical-based [14] and ZZ-type [250] error estimators. Theory behind these error estimates and their derivation for different equations can be found in [223–225]. Basic strategies and algorithms for adaptive methods from a computational perspective are discussed in [190].

Error estimates for the non-conservative Allen-Cahn equation have been discussed in [15, 16, 65, 73, 119, 245]. Mesh refinement of the conservative Allen-Cahn equation coupled with Navier-Stokes equations has been studied in [206, 222, 246]. An open-source library for adaptive mesh refinement [124] was utilized in [222], where a tree data structure was used to store the refinement hierarchy, while a mesh distribution strategy was used in [206, 246]. A disadvantage of such tree-type data structure for the hierarchical mesh is that it can take a good amount of the computational storage. On the other hand, the algorithm of [43] helps to avoid such data structures, thus easing the implementation and decreasing the load on the computational memory. A recent work utilized the algorithm in [43] to adaptively refine the grid based on error estimates on the Allen-Cahn equation coupled with the Navier-Stokes flow system in [114].

Chapter 9
Flexible Multibody Fluid-Structure Interaction

9.1 Introduction

Interaction between interconnected multiple rigid or flexible bodies with the fluid can be found in engineering applications such as underwater robotics, helicopter rotor dynamics, offshore wind turbines and oil/gas platforms, bio-inspired flying vehicles, among others. Such complex fluid-structure interactions comprise of large rigid body displacement and rotation followed by local deformation of the flexible structures as a result of the fluid forces acting on them. Due to the strong coupling between the fluid and the structure, such large displacements can alter the flow field around the multibody system and in turn affect the fluid loading on the structures. Such interactions affect the performance and stability of the multibody system. Of particular interest in the current chapter are the offshore floaters connecting with marine risers and mooring lines and the bio-inspired wing design of bat flight.

In offshore engineering, long slender flexible structures such as marine risers and mooring lines are subjected to turbulent ocean currents and the interaction can lead to vortex-induced vibrations (VIV). These flexible pipelines are generally connected to a rigid body drill-ship or a floating platform, which interacts with the free-surface waves and currents. The prediction and control of these flow-induced oscillations is very crucial in the offshore industry. On the other hand, study of flexible multibody system relating to biological wings of birds and bats can be imperative in developing highly efficient bio-inspired wing designs helpful for micro-air vehicles (MAVs) and drones with applications to search and rescue, payload transportation to remote locations and surveillance. The musculoskeletal system of these animals consists of multiple components of bodies such as bones (humerus, radius, metacarpals and phalanges) connected to each other with the help of joints. Further, highly flexible membranes attach these bones together to form the anatomical structure of the wing. Investigation of these multibody system can be highly beneficial in improving the flight performance and enhancing flight agility and control of the bio-inspired flying vehicles.

Numerical methods for solving such fluid-structure interaction problems can be classified into monolithic and partitioned approaches. As discussed in the previous

R. K. Jaiman and V. Joshi, *Computational Mechanics of Fluid-Structure Interaction*,
https://doi.org/10.1007/978-981-16-5355-1_9

chapters, the benefits of partitioned coupling compared to monolithic approaches lie in the flexibility and modularity of using the existing stable and well established fluid and/or structural solvers [23, 72, 90, 107, 108]. Therefore, for such highly complex flexible multibody fluid-structure system, partitioned coupling approaches are preferred.

Aside from the numerical instabilities due to the relative inertia of the fluid and the structure, another challenge for the multibody system is the relative inertia between the components of the multibody system. The physical properties of the different components may vary which could lead to instability in the structural system [17]. Similar to the fluid-structure system, these multiple structural components can be coupled in a monolithic or partitioned manner. Significant work has been carried out for such monolithic coupling employing energy preserving (EP) and energy decaying (ED) schemes for multibody structural interactions. The performance of the ED schemes is better compared to EP schemes for a physically stiff nonlinear system [19, 192]. Algebraic constraints in the multibody system can be responsible for oscillations in the solution, whereby the nonlinearities provide a mechanism to transfer the energy from low to high frequency modes. Therefore, presence of high frequency numerical dissipation is helpful for such multibody system. Various ED schemes have been proposed to deal with such systems which are physically stiff [12, 20, 21].

Among the partitioned coupling techniques, a loosely coupled staggered technique applied to the problem of multibody system has been proposed in [243] with improved stability and accuracy. In the technique, the constraints are written in a mathematically equivalent partial differential equation with a coupling parameter for the dimensional consistency of velocity. But, the method is simplified by a force decomposition technique to estimate the fluid forces, which reduces the complexity with the fluid-structure coupling. The method was extended for conjugate heat transfer problems in [148]. It is worth noting that although many methods have been proposed to deal with such multibody systems, none have been proven to be stable for all possible scenarios. Therefore, avoiding any complexity in the coupling, an unconditionally stable and accurate monolithic scheme for the multibody system with constraints [18] is presented in the current chapter. A discontinuous Galerkin time integration algorithm based on energy decay inequality is employed for the formulation and the constraints are enforced using a Lagrange multiplier technique.

In this chapter, the description and variational formulation of the co-rotational multibody structural framework for solving such multibody problems along with joints or constraints is presented. Thereafter, the framework is utilized to validate the problem of VIV of an offshore riser where a line-to-surface and surface-to-line coupling algorithm is described, followed by the demonstration of the multibody system of floater-mooring-riser system. As a second application of the flexible multibody framework, we consider the radial basis function based node-to-node coupling at the fluid-structure interface with validation of flow across a pitching plate. Finally, three-dimensional flapping dynamics of a bat is demonstrated by the flexible multibody framework.

9.2 Three-Dimensional Flexible Multibody Structural Framework

We begin by presenting the structural equations written in the co-rotational form under small strain assumption followed by the description of the constraint equations for the connections or joints between the multiple components of the multibody system.

9.2.1 Co-Rotational Multibody Structural Framework for Small Strain Problems

Consider a d-dimensional structural domain of a multibody component i in a multibody system, $\Omega_i^s(0) \subset \mathbb{R}^d$ with material coordinates $\boldsymbol{X}^s$ at time $t = 0$ consisting of a piecewise smooth boundary Γ^s. Let $\boldsymbol{\varphi}^s(\boldsymbol{X}^s, t) : \Omega_i^s(0) \to \Omega_i^s(t)$ be a one-to-one mapping between the reference coordinates $\boldsymbol{X}^s$ at $t = 0$ and the corresponding position in the deformed configuration $\Omega_i^s(t)$. If $\boldsymbol{u}^s(\boldsymbol{X}^s, t)$ represents the structural displacement due to the deformation, $\boldsymbol{\varphi}^s(\boldsymbol{X}^s, t) = \boldsymbol{X}^s + \boldsymbol{u}^s$.

The structural equation for the structural component i of the multibody system can be written as

$$\rho^s \frac{\partial^2 \boldsymbol{\varphi}^s}{\partial t^2} = \nabla \cdot \boldsymbol{\sigma}^s(\boldsymbol{E}(\boldsymbol{u}^s)) + \rho^s \boldsymbol{b}^s, \quad \text{on } \Omega_i^s(0) \times [0, T], \tag{9.1}$$

$$\boldsymbol{v}^s = \boldsymbol{v}_D^s, \quad \text{on } \Gamma_D^s \times [0, T], \tag{9.2}$$

$$\boldsymbol{\sigma}^s \cdot \boldsymbol{n}^s = \boldsymbol{\sigma}_N^s, \quad \text{on } \Gamma_N^s \times [0, T], \tag{9.3}$$

$$\boldsymbol{\varphi}^s = \boldsymbol{\varphi}_0^s, \quad \text{on } \Omega_i^s(0), \tag{9.4}$$

$$\boldsymbol{v}^s = \boldsymbol{v}_0^s, \quad \text{on } \Omega_i^s(0), \tag{9.5}$$

where ρ^s is the structural density, $\boldsymbol{b}^s$ is the body force per unit density acting on Ω_i^s and $\boldsymbol{\sigma}^s$ is the first Piola-Kirchhoff stress tensor which is a function of the Cauchy-Green Lagrangian strain tensor $\boldsymbol{E}(\boldsymbol{u}^s) = (1/2)[(\boldsymbol{I} + \nabla \boldsymbol{u}^s)^T(\boldsymbol{I} + \nabla \boldsymbol{u}^s) - \boldsymbol{I}]$, $\boldsymbol{I}$ being the identity tensor. The Dirichlet condition on the structural velocity $\boldsymbol{v}^s = \partial \boldsymbol{\varphi}^s / \partial t = \partial \boldsymbol{u}^s / \partial t$ on the Dirichlet boundary Γ_D^s is given by $\boldsymbol{v}_D^s$. The Neumann condition on the stress tensor is denoted by $\boldsymbol{\sigma}_N^s$ on Γ_N^s with $\boldsymbol{n}^s$ being the outward normal to Γ_N^s. The initial conditions on the position vector and the structural velocity are denoted as $\boldsymbol{\varphi}_0^s$ and $\boldsymbol{v}_0^s$ respectively.

For small strain problems, the above equation can be simplified further. Consider a fixed inertial frame of reference $Ox_1x_2x_3$ with a component i of the multibody system. The initial reference or undeformed configuration of the body at time $t = 0$ is denoted by $\Omega_i^s(0)$ having an orthonormal material coordinate system $AX_1X_2X_3$ as shown in Fig. 9.1a. The body then undergoes deformation under the action of internal or/and external forces to the deformed configuration $\Omega_i^s(t)$ at time t. The

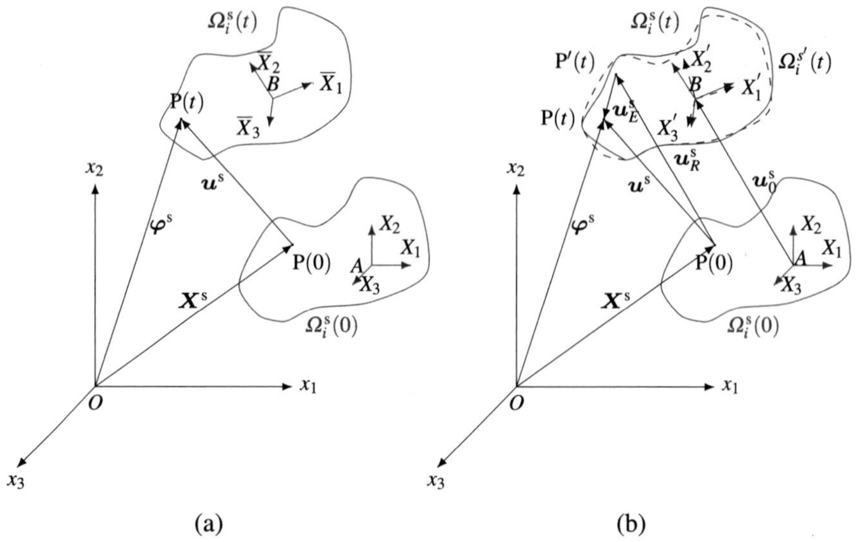

Fig. 9.1 The reference and deformed configurations ($\Omega_i^s(0)$ and $\Omega_i^s(t)$ respectively) of a component i of the multibody system in the inertial coordinate system $Ox_1x_2x_3$

deformed coordinate system $B\overline{X}_1\overline{X}_2\overline{X}_3$ in $\Omega_i^s(t)$ consists of basis vectors which are neither orthogonal nor unit vectors since the material lines are deformed along with the structure. Now, consider a point P which displaces by $\boldsymbol{u}^s$ from the reference (P(0)) to the deformed configuration (P(t)). In the deformed state, the position of P is written as $\boldsymbol{\varphi}^s = \boldsymbol{X}^s + \boldsymbol{u}^s$. Thus, the deformation gradient tensor from $AX_1X_2X_3$ to $B\overline{X}_1\overline{X}_2\overline{X}_3$ can be expressed as $\boldsymbol{F} = \boldsymbol{I} + \nabla \boldsymbol{u}^s$, where $\nabla(\cdot)$ is the gradient with respect to the material coordinates $\boldsymbol{X}^s$.

Noting that the basis vectors in $B\overline{X}_1\overline{X}_2\overline{X}_3$ are not orthonormal, we define a new orthogonal coordinate system $BX_1^{'}X_2^{'}X_3^{'}$ in the deformed configuration $\Omega_i^{s'}(t)$ (Fig. 9.1b). Considering $\hat{\boldsymbol{F}}$ as the deformation mapping from $BX_1^{'}X_2^{'}X_3^{'}$ to $B\overline{X}_1\overline{X}_2\overline{X}_3$ and $\boldsymbol{R}$ as the rotation matrix mapping the orthonormal coordinates $AX_1X_2X_3$ to $BX_1^{'}X_2^{'}X_3^{'}$, the deformation gradient tensor can now be written as $\boldsymbol{F} = \boldsymbol{R}\hat{\boldsymbol{F}}$. As can be seen, the deformation gradient tensor is decomposed as a product of rotation tensor (representing rigid body rotation) and a deformation gradient tensor $\hat{\boldsymbol{F}}$, depicting the deformation of the body after rigid body motion.

Individual component of a multibody system can be assumed to undergo small deformation although there are large relative rotations at the connections or joints. Consequently, the displacement at any point can be decomposed into a rigid body motion and an elastic counterpart [227]. The total displacement can therefore be written as a combination of a rigid body motion $\boldsymbol{u}_R^s$ and an elastic deformation relative to the configuration $\Omega_i^{s'}(t)$, $\boldsymbol{u}_E^s$. The total displacement is thus, $\boldsymbol{u}^s = \boldsymbol{u}_R^s + \boldsymbol{u}_E^s$. The rigid body displacement can be expressed as $\boldsymbol{u}_R^s = \boldsymbol{u}_0^s + \boldsymbol{R}\boldsymbol{X}^s - \boldsymbol{X}^s$. Thus, the gradient of the displacement with respect to material coordinates $\boldsymbol{X}^s$ is

$$\nabla \boldsymbol{u}^{\mathrm{s}} = \nabla \boldsymbol{u}_R^{\mathrm{s}} + \nabla \boldsymbol{u}_E^{\mathrm{s}} \tag{9.6}$$
$$= \boldsymbol{R} - \boldsymbol{I} + \nabla \boldsymbol{u}_E^{\mathrm{s}} \tag{9.7}$$
$$\boldsymbol{I} + \nabla \boldsymbol{u}^{\mathrm{s}} = \boldsymbol{R} + \nabla \boldsymbol{u}_E^{\mathrm{s}} \tag{9.8}$$

Next, the Cauchy-Green Lagrangian strain tensor can be written as,

$$\boldsymbol{E} = \frac{1}{2}[(\boldsymbol{R} + \nabla \boldsymbol{u}_E^{\mathrm{s}})^T(\boldsymbol{R} + \nabla \boldsymbol{u}_E^{\mathrm{s}}) - \boldsymbol{I}], \tag{9.9}$$
$$= \frac{1}{2}[\boldsymbol{R}^T\boldsymbol{R} + \boldsymbol{R}^T\nabla \boldsymbol{u}_E^{\mathrm{s}} + (\nabla \boldsymbol{u}_E^{\mathrm{s}})^T\boldsymbol{R} + (\nabla \boldsymbol{u}_E^{\mathrm{s}})^T\nabla \boldsymbol{u}_E^{\mathrm{s}} - \boldsymbol{I}], \tag{9.10}$$
$$= \frac{1}{2}[\boldsymbol{R}^T\nabla \boldsymbol{u}_E^{\mathrm{s}} + (\nabla \boldsymbol{u}_E^{\mathrm{s}})^T\boldsymbol{R} + (\nabla \boldsymbol{u}_E^{\mathrm{s}})^T\nabla \boldsymbol{u}_E^{\mathrm{s}}], \tag{9.11}$$

which can be simplified by neglecting higher order terms in $\nabla \boldsymbol{u}_E^{\mathrm{s}}$ considering small strain problems as

$$\boldsymbol{E} \approx \frac{1}{2}[\boldsymbol{R}^T\nabla \boldsymbol{u}_E^{\mathrm{s}} + (\nabla \boldsymbol{u}_E^{\mathrm{s}})^T\boldsymbol{R}]. \tag{9.12}$$

The small elastic deformations $\boldsymbol{u}_E^{\mathrm{s}}$ can be evaluated in the deformed coordinate system by modeling the multibody components such as cable, beam, membrane, shell, etc.

Remark 9.1 The rotation tensor $\boldsymbol{R}$ can be parameterized using either scalar or vector parameterization. Here, we discuss the vector parameterization using the conformal rotation vector. Consider the vector parameterization of rotation of an angle ϕ about the normal $\boldsymbol{n}$ as $\boldsymbol{p} = p(\phi)\widehat{\boldsymbol{n}}$, where $\widehat{\boldsymbol{n}}$ is the unit vector along $\boldsymbol{n}$. Therefore, the rotation tensor is given by the Rodrigues' rotation formula, $\boldsymbol{R} = \boldsymbol{I} + \xi_1(\phi)\tilde{\boldsymbol{p}} + \xi_2(\phi)\tilde{\boldsymbol{p}}\tilde{\boldsymbol{p}}$, where $\tilde{\boldsymbol{a}}$ is a skew-symmetric tensor formed of the components of a vector $\boldsymbol{a} = [a_1 \;\; a_2 \;\; a_3]^T$ as,

$$\tilde{\boldsymbol{a}} = \begin{bmatrix} 0 & -a_3 & a_2 \\ a_3 & 0 & -a_1 \\ -a_2 & a_1 & 0 \end{bmatrix}, \tag{9.13}$$

and $\xi_1(\phi) = \sin\phi/p(\phi)$ and $\xi_2(\phi) = (1 - \cos\phi)/(p(\phi))^2$ are even functions of the rotation angle. Different choices of the generating function $p(\phi)$ leads to the different types of vector parameterization such as linear, Cayley-Gibbs-Rodrigues, Cartesian rotation vector, etc. A technique based on conformal transformation on Euler parameters (known as conformal rotation vector (CRV)) leads to the rotation matrix $\boldsymbol{R}$ parameterized by Wiener-Milenković parameters [149, 239]. Here, the generating function is chosen as $p(\phi) = 4\tan(\phi/4)$. Thus, the rotation tensor can be re-written as [75],

$$\boldsymbol{R} = \frac{1}{(4 - c_0)^2}[(c_0^2 + 8c_0 - 16)\boldsymbol{I} + 2\boldsymbol{c}\boldsymbol{c}^T + 2c_0\tilde{\boldsymbol{c}}], \tag{9.14}$$

where $\boldsymbol{c} = 4\hat{\boldsymbol{n}}\tan(\phi/4)$ and $c_0 = (1/8)(16 - ||\boldsymbol{c}||^2)$. This choice of parameters avoids any singularities for particular values of rotation angle. Redundancy in the description of the rotation tensor is also prevented by selecting a minimal set of three parameters in this case [19, 74].

9.2.2 Constraints for Joints Connecting Multiple Bodies

The constraint equations at the connections or joints between the components of the multibody system is described in this section. Consider a constraint joint between the bodies Ω_1^{s} and Ω_2^{s} at points A and B respectively. The coordinate systems at the reference configurations $\Omega_1^{\mathrm{s}}(0)$ and $\Omega_2^{\mathrm{s}}(0)$ are denoted by $AX_1X_2X_3$ and $BY_1Y_2Y_3$ respectively. Similarly, the co-rotated coordinate systems in the deformed configuration are represented as $CX_1'X_2'X_3'$ and $DY_1'Y_2'Y_3'$ respectively, as shown in Fig. 9.2. The unit vectors of the deformed coordinate systems are $\boldsymbol{e}_k^A = \boldsymbol{R}^A\boldsymbol{R}_0^A\boldsymbol{i}_k$ and $\boldsymbol{e}_l^B = \boldsymbol{R}^B\boldsymbol{R}_0^B\boldsymbol{i}_l$, where $\boldsymbol{R}^A$ and $\boldsymbol{R}_0^A$ denote the rotation matrix mapping $AX_1X_2X_3$ to $CX_1'X_2'X_3'$ and the inertial coordinate system $Ox_1x_2x_3$ to $AX_1X_2X_3$ respectively for the point A. Here, $\boldsymbol{i}_k$ denotes the unit vectors in the inertial coordinate system.

Let the scalar and the vector product between the unit vectors be given as $s_{kl} = (\boldsymbol{e}_k^A)^T\boldsymbol{e}_l^B$ and $\boldsymbol{v}_{kl} = \tilde{\boldsymbol{e}}_k^A\boldsymbol{e}_l^B$ respectively. Note that the position vector of the

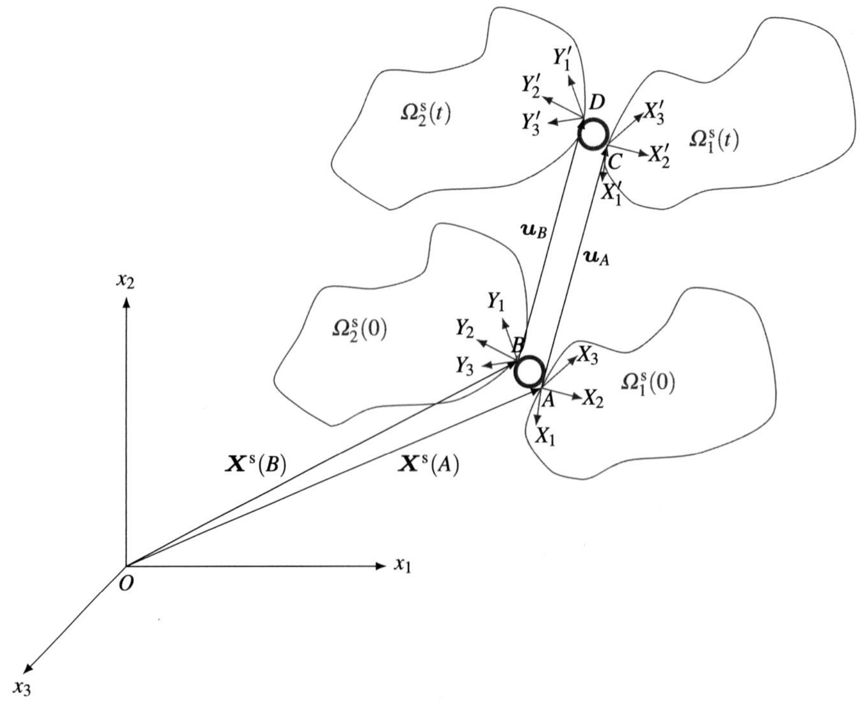

Fig. 9.2 The reference and deformed configurations of a constraint joint connecting two bodies Ω_1^{s} and Ω_2^{s} of a multibody system in the inertial coordinate system $Ox_1x_2x_3$

two locations A and B at the joint in the reference configuration is the same, i.e., $\boldsymbol{X}^{\mathrm{s}}(A) = \boldsymbol{X}^{\mathrm{s}}(B)$ and $\boldsymbol{R}_0^A = \boldsymbol{R}_0^B$. The relative displacement of B with respect to A is given by $\boldsymbol{u}_{B/A} = \boldsymbol{u}_B - \boldsymbol{u}_A$. Assuming the prescribed relative displacement and relative rotation between the two bodies at the joint in the deformed configuration to be $\boldsymbol{u}_m$ along the unit vector $\boldsymbol{e}_m^A$ and ϕ_m about the same unit vector, respectively, the relative displacement and rotation constraint equations can be written as

$$(\boldsymbol{e}_m^A)^T \boldsymbol{u}_{B/A} - \boldsymbol{u}_m = 0, \tag{9.15}$$

$$s_{kk}\sin(\phi_m) + s_{kl}\cos(\phi_m) = 0. \tag{9.16}$$

It can be observed that when there is no relative displacement at the joint ($\boldsymbol{u}_m = 0$), $(\boldsymbol{e}_k^A)^T \boldsymbol{u}_{B/A} = 0$, whereas when there is no relative rotation ($\phi_m = 0$), $s_{kl} = 0$. Further details can be found in [19].

9.3 Variational Formulation for the Flexible Multibody Framework

With the continuum equations being described in the previous section for the flexible multibody framework for small strain problems, we write the weak variational form and matrix form for the equation as well as constraints in this section.

Let us define a set of finite weighting function space $\mathscr{V}_{\boldsymbol{\psi}^{\mathrm{s}}}^{\mathrm{h}}$ and a set of solution space $\mathscr{S}_{\boldsymbol{u}^{\mathrm{s}}}^{\mathrm{h}}$. Thus, the variational statement of the structural equation reads: find $\boldsymbol{u}_{\mathrm{h}}^{\mathrm{s}} \in \mathscr{S}_{\boldsymbol{u}^{\mathrm{s}}}^{\mathrm{h}}$ such that $\forall \boldsymbol{\psi}_{\mathrm{h}}^{\mathrm{s}} \in \mathscr{V}_{\boldsymbol{\psi}^{\mathrm{s}}}^{\mathrm{h}}$,

$$\int_{t^{\mathrm{n}}}^{t^{\mathrm{n+1}}} \left(\int_{\Omega_i^{\mathrm{s}}} \rho^{\mathrm{s}} \frac{\partial^2 \boldsymbol{u}_{\mathrm{h}}^{\mathrm{s}}}{\partial t^2} \cdot \boldsymbol{\psi}_{\mathrm{h}}^{\mathrm{s}} d\Omega + \int_{\Omega_i^{\mathrm{s}}} \boldsymbol{\sigma}^{\mathrm{s}} : \nabla \boldsymbol{\psi}_{\mathrm{h}}^{\mathrm{s}} d\Omega \right) dt$$

$$= \int_{t^{\mathrm{n}}}^{t^{\mathrm{n+1}}} \left(\int_{\Omega_i^{\mathrm{s}}} \rho^{\mathrm{s}} \boldsymbol{b}^{\mathrm{s}} \cdot \boldsymbol{\psi}_{\mathrm{h}}^{\mathrm{s}} d\Omega + \int_{\Gamma_i^{\mathrm{s}}} \boldsymbol{\sigma}_N^{\mathrm{s}} \cdot \boldsymbol{\psi}_{\mathrm{h}}^{\mathrm{s}} d\Gamma \right) dt, \tag{9.17}$$

where $\boldsymbol{\sigma}_N^{\mathrm{s}}$ is the Neumann boundary condition on the boundary Γ_N^{s}.

The kinematic constraints at the joints connecting the multibody components are represented by the equation, $\boldsymbol{c}_J(\boldsymbol{u}^{\mathrm{s}}) = 0$ which is further discussed later. The constraints are imposed by a Lagrange multiplier technique. The variational statement including the constraints can be written as: find $\boldsymbol{u}_{\mathrm{h}}^{\mathrm{s}} \in \mathscr{S}_{\boldsymbol{u}^{\mathrm{s}}}^{\mathrm{h}}$ such that $\forall \boldsymbol{\psi}_{\mathrm{h}}^{\mathrm{s}} \in \mathscr{V}_{\boldsymbol{\psi}^{\mathrm{s}}}^{\mathrm{h}}$,

$$\int_{t^{\mathrm{n}}}^{t^{\mathrm{n+1}}} \left(\int_{\Omega_i^{\mathrm{s}}} \rho^{\mathrm{s}} \frac{\partial^2 \boldsymbol{u}_{\mathrm{h}}^{\mathrm{s}}}{\partial t^2} \cdot \boldsymbol{\psi}_{\mathrm{h}}^{\mathrm{s}} d\Omega + \int_{\Omega_i^{\mathrm{s}}} \boldsymbol{\sigma}^{\mathrm{s}} : \nabla \boldsymbol{\psi}_{\mathrm{h}}^{\mathrm{s}} d\Omega + \int_{\Gamma_i^{\mathrm{s}}} \boldsymbol{c}_J'(\boldsymbol{u}_{\mathrm{h}}^{\mathrm{s}})^T \lambda_{\mathrm{h}} \cdot \boldsymbol{\psi}_{\mathrm{h}}^{\mathrm{s}} d\Gamma \right) dt$$

$$= \int_{t^{\mathrm{n}}}^{t^{\mathrm{n+1}}} \left(\int_{\Omega_i^{\mathrm{s}}} \rho^{\mathrm{s}} \boldsymbol{b}^{\mathrm{s}} \cdot \boldsymbol{\psi}_{\mathrm{h}}^{\mathrm{s}} d\Omega + \int_{\Gamma_i^{\mathrm{s}}} \boldsymbol{\sigma}_N^{\mathrm{s}} \cdot \boldsymbol{\psi}_{\mathrm{h}}^{\mathrm{s}} d\Gamma \right) dt, \tag{9.18}$$

$$\boldsymbol{c}_J(\boldsymbol{u}_{\mathrm{h}}^{\mathrm{s}}) = 0, \tag{9.19}$$

where λ_h is the Lagrange multiplier corresponding to the constraints and $\boldsymbol{c}'_J$ represents the Jacobian of $\boldsymbol{c}_J$. The above variational form can be written in the matrix form as

$$\int_{t^n}^{t^{n+1}} \Big(\boldsymbol{M}(\Delta\ddot{\boldsymbol{u}}^s(t)) + \boldsymbol{K}(\Delta\boldsymbol{u}^s(t)) + \boldsymbol{C}(\Delta\boldsymbol{u}^s(t))\Big)dt = \int_{t^n}^{t^{n+1}} \boldsymbol{F}^s(t)dt, \tag{9.20}$$

where $\boldsymbol{M}$, $\boldsymbol{K}$ and $\boldsymbol{C}$ are the mass, stiffness and constraint matrices of the multibody system respectively after the linearization using Newton-Raphson technique. The increment in the unknowns are denoted by Δ and $\boldsymbol{F}^s$ comprises of the body and external fluid forces acting on the multibody system.

Remark 9.2 Depending on the type of component of the structure, e.g., beam, cable, shell, membrane, etc., the construction of these matrices will vary. Detailed derivation for each component under co-rotational finite element framework can be found in [17].

Temporal discretization of the above equations is carried out with the help of an unconditionally stable energy decaying scheme based on discontinuous Galerkin linear in time approximation [17, 19–21] of Eq. (9.20) between the initial (t^n) and the final time (t^{n+1}). A linear approximation of the Lagrange multiplier (λ^h) is carried out over the time step, $t \in [t^n, t^{n+1}]$. The resulting discretized equations can be written as:

$$\boldsymbol{M}\frac{\dot{\boldsymbol{u}}^{s,-}_{n+1} - \dot{\boldsymbol{u}}^{s,-}_{n}}{\Delta t^s} + \boldsymbol{K}\frac{\boldsymbol{u}^{s,-}_{n+1} + \boldsymbol{u}^{s,+}_{n}}{2} + \frac{\lambda^{h,-}_{n+1} + \lambda^{h,+}_{n}}{2}\frac{C\left(\boldsymbol{u}^{s,-}_{n+1}\right) + C\left(\boldsymbol{u}^{s,-}_{n}\right)}{2} = \frac{\boldsymbol{F}^{s,+}_{n} + \boldsymbol{F}^{s,-}_{n+1}}{2}, \tag{9.21}$$

$$\boldsymbol{M}\frac{\dot{\boldsymbol{u}}^{s,+}_{n} - \dot{\boldsymbol{u}}^{s,-}_{n}}{\Delta t^s} - \boldsymbol{K}\frac{\boldsymbol{u}^{s,-}_{n+1} - \boldsymbol{u}^{s,+}_{n}}{6} + \frac{\lambda^{h,-}_{n+1} - \lambda^{h,+}_{n}}{6}\frac{C\left(\boldsymbol{u}^{s,+}_{n}\right) + C\left(\boldsymbol{u}^{s,-}_{n}\right)}{2} = \frac{\boldsymbol{F}^{s,+}_{n} - \boldsymbol{F}^{s,-}_{n+1}}{6}, \tag{9.22}$$

$$\frac{\boldsymbol{u}^{s,-}_{n+1} - \boldsymbol{u}^{s,-}_{n}}{\Delta t^s} = \frac{\dot{\boldsymbol{u}}^{s,-}_{n+1} + \dot{\boldsymbol{u}}^{s,+}_{n}}{2}, \tag{9.23}$$

$$\frac{\boldsymbol{u}^{s,+}_{n} - \boldsymbol{u}^{s,-}_{n}}{\Delta t^s} = -\frac{1}{6}\left[\dot{\boldsymbol{u}}^{s,-}_{n+1} - \dot{\boldsymbol{u}}^{s,-}_{n} - \alpha\left(\dot{\boldsymbol{u}}^{s,+}_{n} - \dot{\boldsymbol{u}}^{s,-}_{n}\right)\right], \tag{9.24}$$

where Δt^s and α are the time step size and tuning parameter for controlling numerical dissipation of the scheme. Taking $\alpha = 1$ ensures asymptotic annihilation and is selected in the current formulation. The notations $()^-_n$, $()^+_n$ and $()^-_{n+1}$ used in Eqs. (9.21-9.24) indicate the quantities at t^-_n, t^+_n and t^-_{n+1}, respectively. The scheme can be proven to be unconditionally stable based on energy decay inequality using the theory of time discontinuous Galerkin method applied to hyperbolic conservation laws [17, 20, 21].

Remark 9.3 It is to be noted that the generalized-α time integration method has been employed for the fluid equations which is unconditionally stable and second-order accurate for linear problems. On the other hand, an energy decaying scheme has been utilized for the multibody structural equations. As the energy conservation is not sufficient to yield a robust time integration scheme, high frequency numerical dissipation is necessary for nonlinear flexible multibody system.

Next, we discuss the details about the variational form of the constraint equations at the joints. Continuing with the constraint equations written for the joints in Sect. 9.2.2, we write the variational and matrix forms for the same joint. Let us define the virtual rotation vector for the relative motion at the joints as

$$\delta\boldsymbol{\psi}^A = \text{axial}(\delta\boldsymbol{R}^A(\boldsymbol{R}^A)^T), \tag{9.25}$$

$$\delta\boldsymbol{\psi}^B = \text{axial}(\delta\boldsymbol{R}^B(\boldsymbol{R}^B)^T), \tag{9.26}$$

where $\delta\boldsymbol{R}\boldsymbol{R}^T$ is the skew-symmetric tensor related to the variation and axial$(\cdot)$ is the vector associated with it (similar to Eq. (9.13)). Similar variation in the unit vectors in the deformed coordinate system can be written as $\delta\boldsymbol{e}_k^A = (\tilde{\boldsymbol{e}}_k^A)^T\delta\boldsymbol{\psi}^A$ and $\delta\boldsymbol{e}_l^B = (\tilde{\boldsymbol{e}}_l^B)^T\delta\boldsymbol{\psi}^B$. Moreover, the variation in the scalar and vector products defined in Sect. 9.2.2 are expressed as

$$\delta s_{kl} = (\delta\boldsymbol{\psi}^A - \delta\boldsymbol{\psi}^B)^T\boldsymbol{v}_{kl}, \tag{9.27}$$

$$\delta\boldsymbol{v}_{kl} = (\delta\boldsymbol{\psi}^A)^T\boldsymbol{D}_{kl}^{AB} - (\delta\boldsymbol{\psi}^B)^T\boldsymbol{D}_{lk}^{BA}, \tag{9.28}$$

where $\boldsymbol{D}_{kl}^{AB} = \tilde{\boldsymbol{e}}_k^A\tilde{\boldsymbol{e}}_l^B$ and $\boldsymbol{D}_{lk}^{BA} = \tilde{\boldsymbol{e}}_l^B\tilde{\boldsymbol{e}}_k^A$. A potential function for the constraint $V = \lambda C_{AB}$ and its variation $\delta V = \delta\lambda C_{AB} + \lambda\delta C_{AB}$, where λ is the Lagrange multiplier, can be defined for a given constraint equation $C_{AB} = 0$ (which could be related to any one of the constraints given in Eqs. (9.15) and (9.16)). Therefore, the virtual work done is $\delta W = \lambda\delta C_{AB} = (\delta\boldsymbol{q})^T\boldsymbol{F}_{AB}$, where $\delta\boldsymbol{q}$ and $\boldsymbol{F}_{AB}$ are the variation of generalized coordinates and the constraint forces respectively. Based on the type of constraint and the variations in Eqs. (9.27) and (9.28) and $\delta C_{AB} = \boldsymbol{B}\delta\boldsymbol{q}$, the constraint forces can be written as $\boldsymbol{F}_{AB} = \lambda\boldsymbol{B}^T$. The matrix $\boldsymbol{B}^T$ is the Jacobian of the constraint denoted by $\boldsymbol{c}'(\boldsymbol{u}_{\text{h}}^{\text{s}})^T$ in Eq. (9.18).

The increment of constraint forces after Newton-Raphson linearization can be expressed as

$$\Delta\boldsymbol{F}_{AB} = \frac{\partial\boldsymbol{F}_{AB}}{\partial\boldsymbol{q}}\Delta\boldsymbol{q} = \boldsymbol{K}_{AB}\Delta\boldsymbol{q}, \tag{9.29}$$

where $\boldsymbol{K}_{AB}$ is the stiffness matrix for the constraint. Now, we consider two cases.

Case 1: Constraint for Relative Displacement

When there is a prescribed relative displacement at the joint, the constraint is given by $C_{AB} = (\boldsymbol{e}_m^A)^T\boldsymbol{u}_{B/A} - u_m = 0$. Thus, the incremental vector of unknowns and the stiffness matrix are written as

$$\Delta \boldsymbol{q} = \begin{bmatrix} \Delta \boldsymbol{u}_A \\ \Delta \boldsymbol{\psi}_A \\ \Delta \boldsymbol{u}_B \\ \Delta \boldsymbol{u}_m \end{bmatrix}, \qquad \boldsymbol{K}_{AB} = \lambda \begin{bmatrix} \boldsymbol{0} & \tilde{\boldsymbol{e}}_m^A & \boldsymbol{0} & \boldsymbol{0} \\ -\tilde{\boldsymbol{e}}_m^A & \tilde{\boldsymbol{u}}_{A/B} \tilde{\boldsymbol{e}}_m^A & \tilde{\boldsymbol{e}}_m^A & \boldsymbol{0} \\ \boldsymbol{0} & -\tilde{\boldsymbol{e}}_m^A & \boldsymbol{0} & \boldsymbol{0} \\ \boldsymbol{0} & \boldsymbol{0} & \boldsymbol{0} & \boldsymbol{0} \end{bmatrix} \tag{9.30}$$

Case 2: Constraint for Relative Rotation

In the case of prescribed relative rotation, the constraint equation is given as $C_{AB} = s_{kk} \sin(\phi_m) + s_{kl} \cos(\phi_m) = 0$. The increment vector of the unknowns and the stiffness matrix are written as

$$\Delta \boldsymbol{q} = \begin{bmatrix} \Delta \boldsymbol{\psi}_A \\ \Delta \boldsymbol{u}_B \\ \Delta \boldsymbol{\phi}_m \end{bmatrix}, \qquad \boldsymbol{K}_{AB} = \lambda \begin{bmatrix} \boldsymbol{E}^T & \boldsymbol{E} & z \\ -\boldsymbol{E}^T & \boldsymbol{E} & -z \\ z^T & -z^T & -C_{AB} \end{bmatrix} \tag{9.31}$$

where $z = \boldsymbol{v}_{kk} \cos(\phi_m) - \boldsymbol{v}_{kl} \sin(\phi_m)$ and $\boldsymbol{E} = \boldsymbol{D}_{kk}^{AB} \sin(\phi_m) + \boldsymbol{D}_{kl}^{AB} \cos(\phi_m)$.

A general constraint can be a combination of the above two cases, based on the type of joint and the stiffness matrix can be formed accordingly with the help of Lagrange multiplier. All the constraints at every joint are then assembled in the global constraint matrix $\boldsymbol{C}$ in Eq. (9.20). This concludes the description of the flexible multibody structural framework for small strain problems. We next apply this framework to fluid-structure interaction problems by considering different coupling techniques, viz., line-to-surface/surface-to-line and radial basis function based node-to-node mapping.

9.4 Fluid-Structure Coupling Based on Line-to-Surface and Surface-to-Line Mapping

If we consider one-dimensional structural components such as beam and cable in the multibody system, one challenge for a three-dimensional fluid-structure interaction solver is the coupling at the interface. Here, we discuss such kind of coupling by transferring structural displacements onto the fluid mesh via line-to-surface coupling and by transferring the fluid forces onto the structure via surface-to-line coupling.

9.4.1 Line-to-Surface Coupling

The structural displacements from the structural side of the interface are transferred to the fluid side by satisfying the ALE compatibility condition and the velocity continuity at the fluid-structure interface of the multibody component Γ_i^{fs}. To accomplish

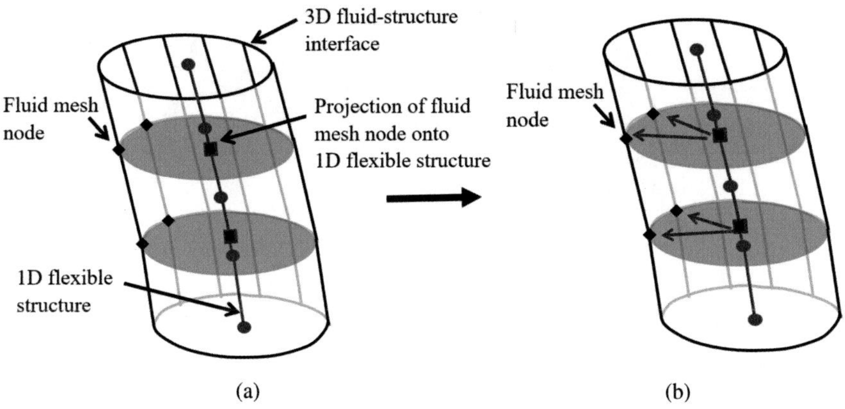

Fig. 9.3 Schematic of line-to-surface projection of displacements and velocity: **a** search for the corresponding structural line node for the fluid mesh nodes and **b** projection of the variables to the fluid mesh nodes

this, a nodal projection scheme is utilized, as shown in Fig. 9.3 which is carried out in the following manner:

First, the fluid mesh nodes are projected on the one-dimensional structural line element (Fig. 9.3a). Then, with the help of shape functions of the structural discretization, the displacement and velocities at the projected point on the line element are interpolated. The interpolated values are then assigned to the corresponding fluid node as

$$\boldsymbol{u}_k^{\mathrm{f,n+1}}\left(\boldsymbol{x}_k^{\mathrm{f,n+1}}\right) = \sum_{j=1}^{ns} \boldsymbol{\phi}_j^{\mathrm{s}}\left(\mathbb{P}\left(\boldsymbol{x}_k^{\mathrm{f,n+1}}\right)\right)\left(\boldsymbol{v}_{n+1}^{\mathrm{s},-}\right)_j \quad \forall \boldsymbol{x}_k^{\mathrm{f,n+1}} \in \Gamma_i^{\mathrm{fs}}(t^{\mathrm{n+1}}), \tag{9.32}$$

where $\boldsymbol{u}_k^{\mathrm{f,n+1}}$ and $\boldsymbol{x}_k^{\mathrm{f,n+1}}$ represent the fluid mesh displacement and position of the kth node at time $t^{\mathrm{n+1}}$ respectively, the function $\mathbb{P}\left(\boldsymbol{x}_k^{\mathrm{f,n+1}}\right)$ is the projection of the fluid mesh node $k \in \Gamma_i^{\mathrm{fs}}(t^{\mathrm{n+1}})$ onto the structural line elements, $\boldsymbol{\phi}_j^{\mathrm{s}}$ denotes the structural shape function for the 1D flexible multibody mesh node j and ns is the number of 1D structural mesh nodes.

The velocity continuity at the interface is satisfied by equating the fluid velocity with the mesh velocity for all the fluid nodes on the interface $\Gamma_i^{\mathrm{fs}}(t^{\mathrm{n}+\alpha^{\mathrm{f}}})$, i.e.

$$\boldsymbol{v}_k^{\mathrm{f,n}+\alpha^{\mathrm{f}}} = \boldsymbol{v}_k^{\mathrm{m,n}+\alpha^{\mathrm{f}}} = \frac{\boldsymbol{u}_k^{\mathrm{f,n+1}} - \boldsymbol{u}_k^{\mathrm{f,n}}}{\Delta t}. \tag{9.33}$$

This completes the description of the line-to-surface coupling for transferring the structural displacements across the fluid-structure interface.

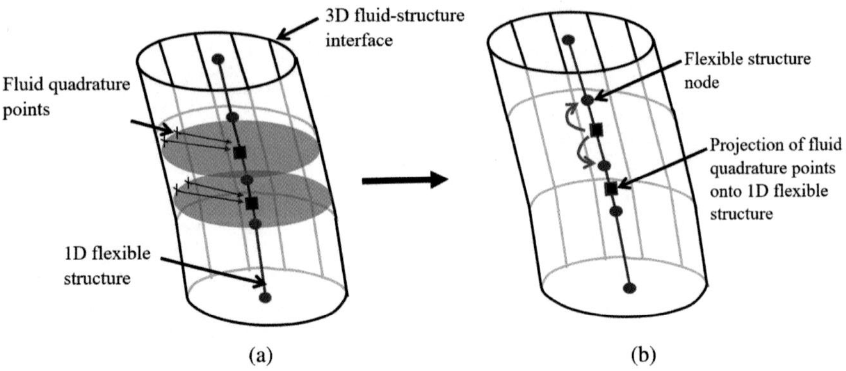

Fig. 9.4 Schematic of surface-to-line projection of traction: **a** quadrature projection of the fluid tractions from the fluid surface mesh to the corresponding structural line element and **b** evaluation of the fluid tractions at the structural nodes using the shape functions of the line element

9.4.2 Surface-to-Line Coupling

The fluid forces from the three-dimensional fluid mesh elements are transferred to the 1D flexible multibody line component via a quadrature projection technique where the fluid tractions at the quadrature points on the fluid surface mesh are projected onto the structural element, as shown in Fig. 9.4. First, the fluid surface mesh quadrature points are projected onto the structural line elements (Fig. 9.4a). These projected points are then associated with the elemental fluid tractions ($\boldsymbol{t}_q^{\mathrm{f}}$) evaluated at their corresponding quadrature points. Consequently, the fluid traction on the jth node of the 1D multibody mesh is computed with the help of the shape function ($\boldsymbol{\phi}_j^{\mathrm{s}}$) of the line element as

$$\boldsymbol{t}_j^{\mathrm{s}} = \sum_{k=1}^{nf} \sum_{q=1}^{nq} \int_{\gamma} \boldsymbol{\phi}_j^{\mathrm{s}} \left(\mathbb{P}\left(\boldsymbol{x}_q^{\mathrm{f}}, t\right)\right) \boldsymbol{t}_q^{\mathrm{f}} d\Gamma, \tag{9.34}$$

where $\boldsymbol{x}_q^{\mathrm{f}}$ represents the position of the fluid mesh quadrature point, γ is the 1D flexible multibody line element onto which quadrature point is projected, nq and nf are the number of quadrature points and nodes on the 1D flexible multibody mesh, respectively. It can be shown that this transfer of load satisfies the conservation property by the summation property of the shape functions.

9.4.3 Implementation Details

The flexible multibody structural framework is coupled with the incompressible flow equations with turbulence modeling via a partitioned iterative coupling. The nonlinear iterative force correction (NIFC) discussed in Sect. 7.3.2 is utilized to correct the fluid forces for stability in low structure-to-fluid mass ratio regimes. The search for the corresponding elements during the projection technique from line-to-surface and surface-to-line mapping is carried out by a split-tree algorithm. The partitioned algorithm is similar to that discussed in Sect. 8.3.2 neglecting the two-phase modeling equation. The details about the hybrid parallelism are also given in Sect. 8.3.3.

9.4.4 Validation of Vortex-Induced Vibration of Flexible Cylinder

We conduct validation study of the multibody fluid-structure framework by considering a uniform fluid flow across a long flexible cylinder or an offshore riser, similar to Sect. 7.4.3. The experiment conducted in [1] of a pinned-pinned riser subjected to uniform current of 0.2 m/s is considered for comparison of the results.

The schematic representation of the computational domain is shown in Fig. 9.5. The diameter of the riser is D and it spans a length of $481.5D$ in the Z-direction. The inlet and outlet of the domain are at a distance of $10D$ and $25D$ from the center of the riser respectively. The side walls are $20D$ from each other. While slip boundary condition is satisfied at the side walls and the planes perpendicular to the axis of the riser, no-slip condition is imposed on the riser wall. A uniform flow velocity is given at the inlet boundary. The non-dimensional parameters for the set-up are given in Table 9.1, where K_B, K_T, Re and ρ^* represent the non-dimensional riser bending rigidity, pre-tension force, Reynolds number and density ratio, respectively and E, I and P denote the Young's modulus of the riser material, the second moment of area of the riser cross-section and the pre-tension applied on the riser, respectively.

The computational domain can be categorized into the fluid domain and the structural domain. The discretization in the fluid domain consists of approximately 3.5 million nodes with unstructured eight-node hexahedron elements. On the other hand, the structural domain is discretized into 200 nonlinear beam elements. A boundary layer and wake regions are constructed around the riser cross-section to maintain sufficient resolution and the spanwise discretization consists of 200 layers. The boundary layer thickness is $0.25D$ with a wall-normal discretization such that $y^+ < 1$. The non-dimensional time step size ($\Delta tU/D$) is selected as 0.1 for the present case.

The root mean square (rms) amplitudes of the riser response in both the in-line (IL) and cross-flow (CF) directions are depicted in Fig. 9.6. A good agreement with the experimental data for the CF response is obtained with a difference of $< 1\%$. However, there is some over-prediction in the IL response and the difference is $\sim 10\%$. This may be due to the geometric imperfections of the riser surface and complex dynamics of the flow separation. The results obtained from the linear modal analysis of the riser (Sect. 10.3.3 and [113]) are also compared with the current nonlinear multibody structural model in Fig. 9.6. The temporal variation of the riser CF response at the spanwise position of $z/L = 0.55$ is compared with that of the experiment in Fig. 9.7 where a good agreement is observed.

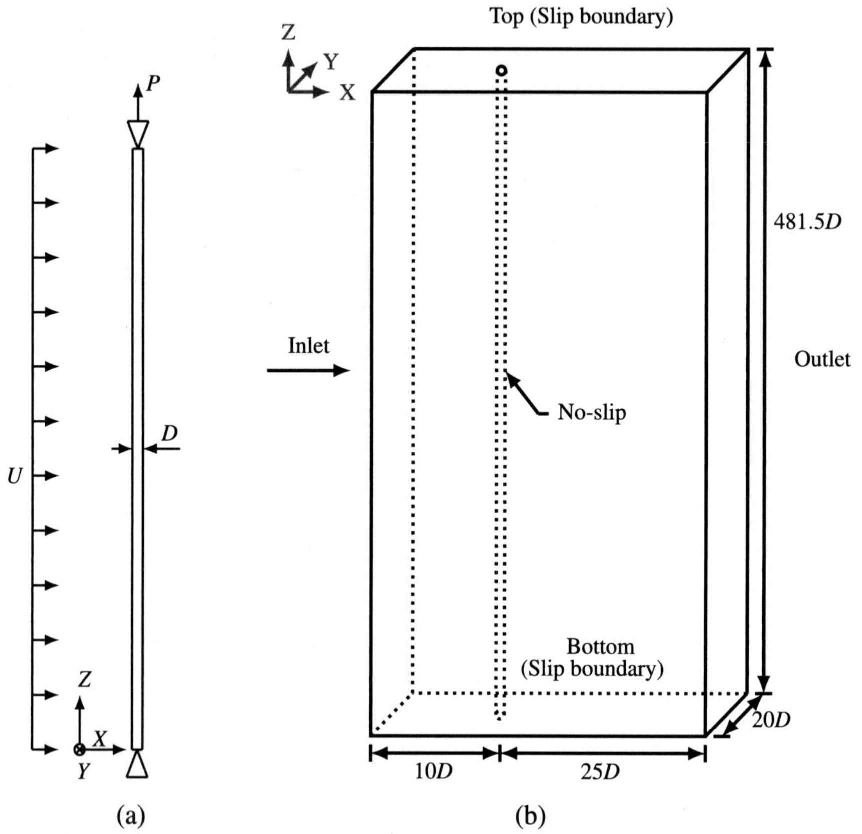

Fig. 9.5 A long flexible riser model in a uniform current flow along the Z-axis: **a** pinned-pinned tensioned riser with uniform flow, **b** schematic illustration of the computational setup and boundary conditions

Table 9.1 Non-dimensional parameters used in riser VIV simulation

Parameters	Value
$K_B \left(= \frac{EI}{\rho^f U^2 D^4}\right)$	2.1158×10^7
$K_T \left(= \frac{P}{\rho^f U^2 D^2}\right)$	5.10625×10^4
$Re \left(= \frac{\rho^f U D}{\mu^f}\right)$	4000
$\rho^* \left(= \frac{\rho^s}{\rho^f}\right)$	2.23

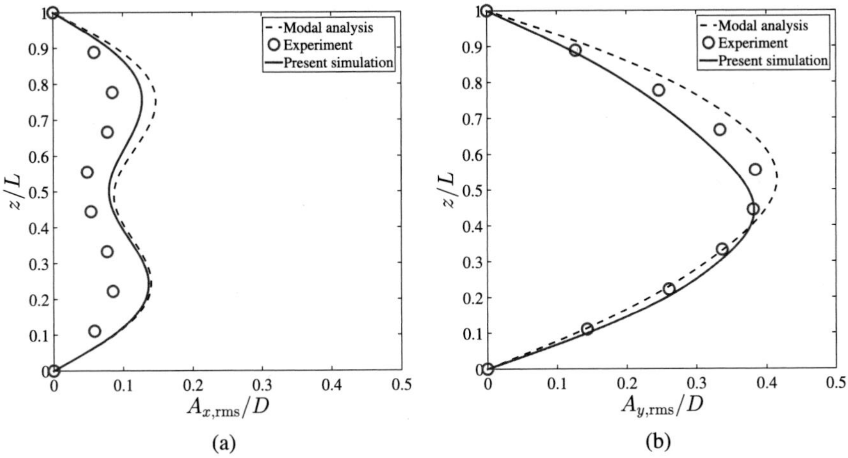

Fig. 9.6 Rms-amplitudes of displacements for uniform current flow past a flexible riser modeled using nonlinear beam at $(Re;m^*) = (4000; 2.23)$: **a** in-line and **b** cross-flow directions. (–) Modal analysis [113], (∘) Experiment [1] and (—) Present simulation. The riser is vibrating in the fundamental mode in the CF and the second mode for the IL directions

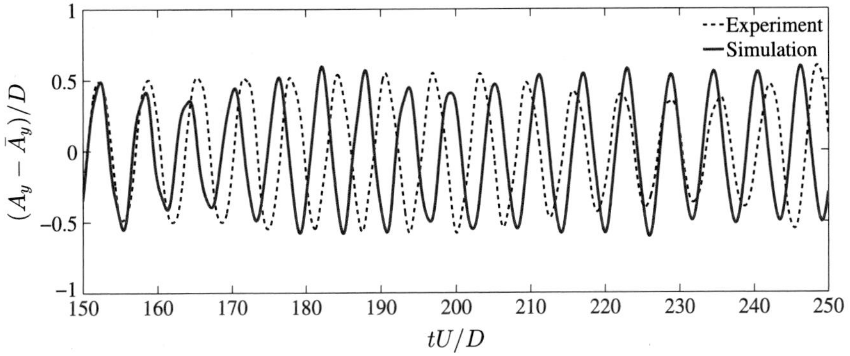

Fig. 9.7 Uniform current flow past a flexible riser at $Re = 4000$: Comparison of the cross-flow response at $z/L = 0.55$ with that of the experimental data for identical parameters

The flow visualization in the wake of the riser is depicted by plotting the vortical structures using two-dimensional vortices and iso-surface of Q-criterion in Fig. 9.8 for $tU/D = 250$. It can be observed that the vortex shedding process is quite complex with a 2S mode of shedding at most of the locations. A wider 2S mode of shedding is present at the locations where the amplitude of vibration is large. The iso-surfaces of the 3D vortices can be divided into the upper half and lower half along the spanwise direction in Fig. 9.8b. They form a tube-like shape around the anti-nodes of the vibration ($z/L = 0.25$ and 0.75) and smaller vortices are formed at the node of the vibration ($z/L = 0.5$) and near the supports. These observations are similar to those made in [113, 233].

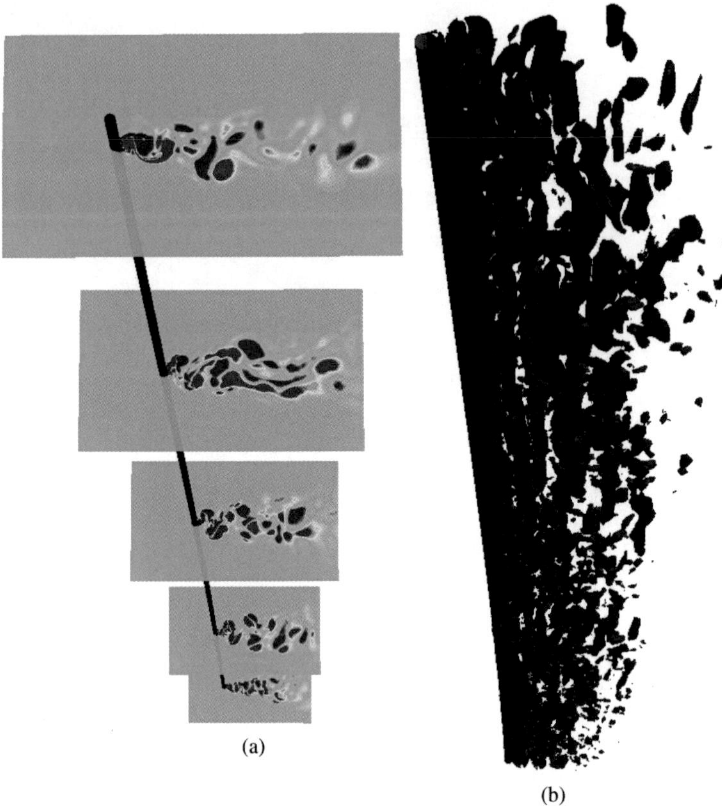

Fig. 9.8 Vortex patterns (at $tU/D = 250$) formed due to the flow induced vibration of flexible riser modeled using nonlinear beam: **a** Z-vorticity contours in various spanwise sections, red and blue color indicate the positive and negative vorticity respectively, and **b** instantaneous iso-surfaces of vortical structures

9.4.5 Application to Coupled Floater-Mooring-Riser System

After validation of the multibody framework with the surface-to-line and line-to-surface coupling at the fluid-structure interface, a demonstration on a practical application of coupled floater-mooring-riser system is carried out in this section. In ocean and offshore engineering, floaters are bodies of large mass which are generally rigid and can undergo vortex-induced motion in high velocity ocean currents. Mooring lines are long cables utilized for station-keeping of the floating platform with its one end connected to the floater via joints/constraints and the other end anchored to the ocean floor. On the other hand, risers are long pipelines modeled as beams that connect the floater to a drilling rig and/or subsea system. A schematic illustration of this offshore floater-mooring-riser system is shown in Fig. 9.9. As can be observed, the presence of multibody system consisting of rigid body, beam and cables can make

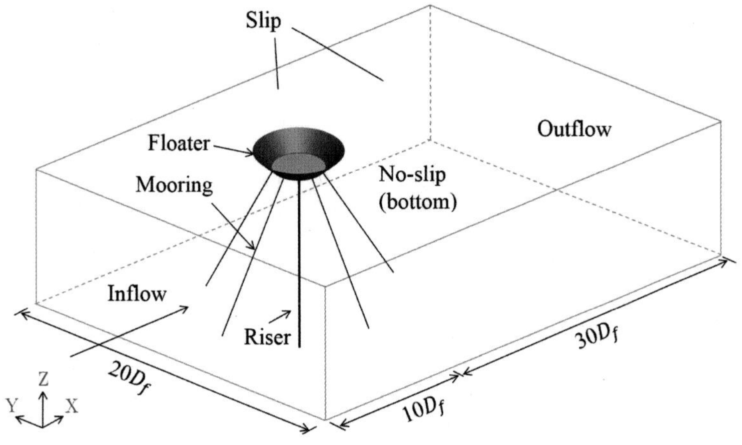

Fig. 9.9 Schematic illustration of the computational domain of the floater-mooring-riser system subjected to uniform current flow in the X direction

this system quite complex when it is exposed to environmental forces consisting of ocean currents. The floater can have vortex-induced motion and the moorings and riser can have their own local vortex-induced vibrations, thus forming a very complex coupled system.

In the present case, the floater-mooring-riser system consists of a rigid funnel-shaped floater, four taut mooring lines (modeled as cables) and a long riser (modeled as a beam) (see Fig. 9.9). The floater geometry corresponds to a typical Arctic hull with a downward ice-breaking slope with a maximum diameter of $D_f = 50D_r$ at the water-plane, D_r being the diameter of the riser. The floater has a draft of $20D_f$. The mooring lines have a length of $538.33D_r$ and diameter $0.1D_r$ and are equally spaced in a radial direction with azimuth angles of 90 deg, facing the current direction. The mooring cables are clamped at the bottom edge of the floater and extend till the seabed which is at a vertical distance of $481.5D_r$ from the bottom of the floater. Thus, the riser has a length of $481.5D_r$ and spans the Z-direction from the bottom of the floater to the seabed. The moorings and riser are clamped to the floater such that the top nodes of the moorings/riser move with the motion of the floater. The moorings and the riser are pinned at the seabed. The dimensions for the demonstration are chosen such that there is no contact between the flexible bodies undergoing motion.

The computational domain has a size of $[40D_f \times 20D_f \times 501.5D_r]$. The inlet and outlet boundaries are at a distance of $10D_f$ and $30D_f$ from the center of the floater respectively. The side walls are $10D_f$ from the floater center. The top surface of the domain is assumed to be a free-surface where slip boundary condition is satisfied, whereas a no-slip condition is imposed at the bottom boundary assuming it to be seabed. A traction-free condition is satisfied at the outlet and no-slip condition is imposed at the surface of each component of the multibody system, viz., floater, mooring lines and riser.

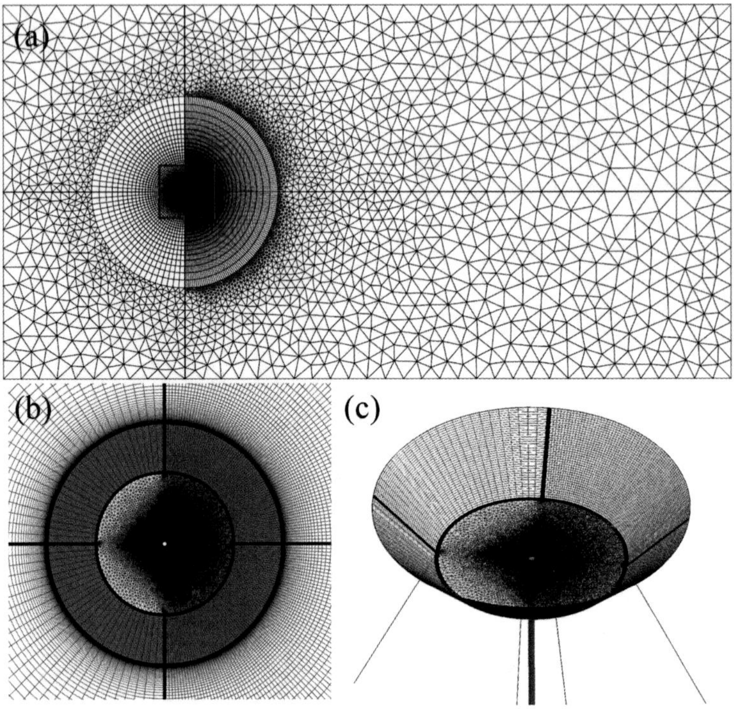

Fig. 9.10 Finite element discretization of the computational domain: **a** top-view of the fluid domain mesh, **b** close-up view of the mesh around the floater (inset box in **a**) and **c** isometric view of the mesh consisting of the floater, riser and the moorings

The domain can be decomposed into the fluid and the structural subdomains. The fluid domain is discretized into approximately 5 million nodes consisting of unstructured six-node wedge and eight-node hexahedron elements. On the other hand, the riser and the mooring lines are discretized into 100 nonlinear beam and cable elements respectively. The mesh closer to the boundary layer region and the wake of each multibody component is refined to get a reasonable resolution to capture the vortices, as shown in Fig. 9.10. The boundary layer thickness for all the components are selected as $0.25D$, D being the diameter of the multibody component, i.e., moorings, riser, with the stretching ratio, $\Delta y_{j+1}/\Delta y_j$ of 1.15 such that $y^+ < 1$. The time step size is taken as $\Delta t U/D_r = 0.1$. The dimensionless parameters for the problem set-up are given in Table 9.2, where E is the Young's modulus, I is the second moment of area of the cross-section, A is the cross-sectional area, K_A is the non-dimensional axial stiffness, m is the mass of the component of the multibody system and V_f is the volume of the displaced fluid by the floater.

The amplitude response envelope along the riser span is shown for the in-line (IL) and cross-flow (CF) directions in Fig. 9.11. A third dominant mode is observed in the IL direction, whereas the riser oscillates at the second dominant mode in the CF

Table 9.2 Dimensionless parameters used in floater-mooring-riser simulation

Dimensionless parameters	Value
$Re = \frac{\rho^f U D_r}{\mu^f}$	4000
Floater:	
$m^* = \frac{m}{V_f \rho^f}$	0.98
Riser:	
$K_B = \frac{EI}{\rho^f U^2 D^4}$	2.1158×10^7
$m^* = \frac{m}{\frac{\pi}{4} D^2 L \rho^f}$	2.23
Mooring:	
$K_A = \frac{EA}{\rho^f U^2 D^2}$	1.5708×10^{11}
$m^* = \frac{m}{\frac{\pi}{4} D^2 L \rho^f}$	8.0

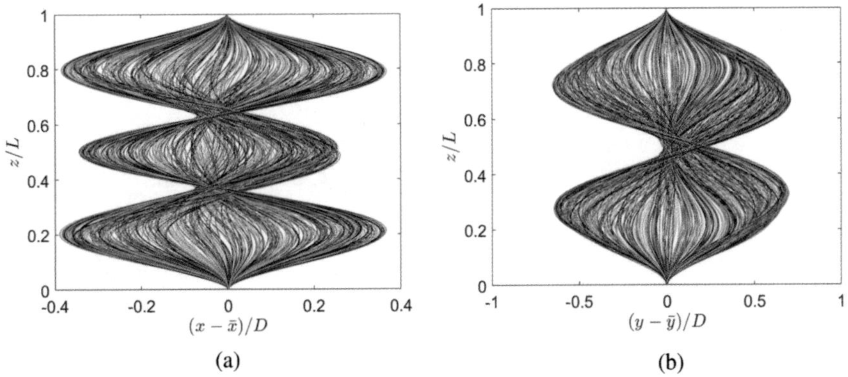

Fig. 9.11 Displacement response envelop for riser component in floater-mooring-riser system subjected to uniform current flow: **a** in-line and **b** cross-flow directions. Higher modes, i.e., third mode in the in-line (IL) direction and a second mode in the cross-flow (CF) direction is observed in comparison to the pinned-pinned riser case

direction. This is in contrast to the riser validation case in the previous section where second and first modes are observed in the IL and CF directions, respectively (see Fig. 9.6). Furthermore, a relatively higher amplitude of vibration is observed in both the directions compared to the VIV of a sole offshore riser. This may be attributed to the induced vibration due to the floater motion in the present case. Although the motion of the floater is small, $\mathcal{O}(10^{-3})$, it has significant effect on the VIV dynamics of the riser due to the coupled interaction. The response of the riser with time is plotted in Fig. 9.12, where a standing wave pattern is observed for both IL and CF directions.

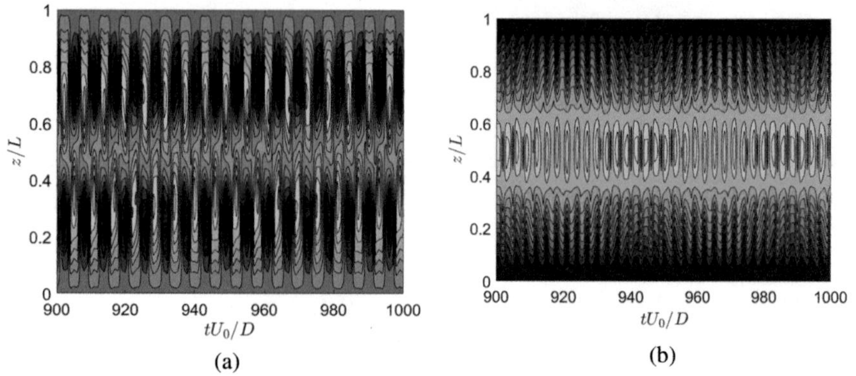

Fig. 9.12 Standing wave response of the flexible riser component in floater-mooring-riser system subjected to uniform current flow: **a** cross-flow and **b** in-line directions

Fig. 9.13 Instantaneous Z-vorticity behind the floater and the riser systems for the fully-coupled floater-mooring-riser system in a uniform current flow

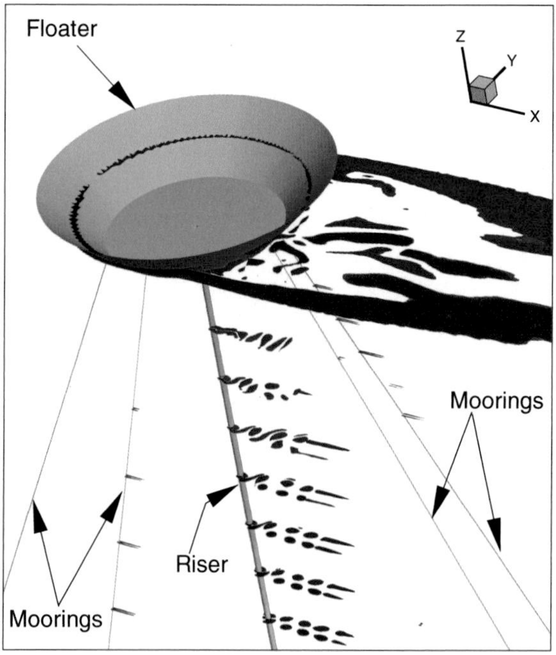

The vortex shedding process in the wake of the multibody system consisting of floater, mooring lines and riser is shown in Fig. 9.13 where spanwise Z-vorticity contour are shown along the riser and the mooring lines. A predominantly 2S mode of shedding is observed for most of the locations along the riser, similar to the riser VIV validation. Large vortices are observed in the wake of the floater due to its large dimension. These large vortices tend to either merge or destroy the small vortices which are generated by the riser and the mooring lines. The flow physics is captured with reasonable accuracy. Therefore, this practical demonstration shows the functionality and usability of the surface-to-line and line-to-surface coupling at the fluid-structure interface consisting of multiple bodies connected to each other and interacting with the surrounding fluid.

9.5 Fluid-Structure Coupling Based on Radial Basis Function Mapping

Next, we look into another type of coupling at the fluid-structure interface employing the flexible multibody structural framework which is referred to as point-to-point mapping via radial basis functions (RBF), discussed in Sect. 7.2.1. The convergence of the RBF mapping has already been discussed in Sect. 7.2.1.3. Here, mesh convergence and validation tests are performed considering a two-dimensional shell element in the flexible multibody framework by modeling flow across a pitching plate and a flexible flapping wing. Finally, the multibody system consisting of beams and shells connected to each other by joints describing the multibody components of a bat wing is considered to understand the flapping dynamics of bio-inspired wing designs.

9.5.1 Convergence and Validation Tests: Flow Across a Pitching Plate

In this section, we start by considering the flow across a pitching plate with a serration of 45° at the trailing edge as a numerical test to verify and validate the coupling between the flexible multibody framework and the data transfer at the fluid-structure interface via the RBF mapping. The problem set-up with the computational domain is shown in Fig. 9.14. The plate has a width, thickness and mean chord of $b = 0.1$ m, $h = 2.54 \times 10^{-3}$ m and $c = S/b = 0.1$ m, respectively. The surface area of the plate, thus is $S = 0.01$ m^2. The inflow and outflow boundaries are at a distance of $40c$ from the leading edge of the plate. All the other sides are around $40c$ from the middle of the plate. A freestream velocity of $U_\infty = 0.1$ m/s is given at the inflow boundary and stress-free condition is satisfied at the outflow. Slip condition is imposed at the side walls while a no-slip condition is satisfied on the surface of the plate. For the pitching motion, a prescribed motion of amplitude $\theta = \theta_{\max} \sin(2\pi f_0 t)$ is given at the

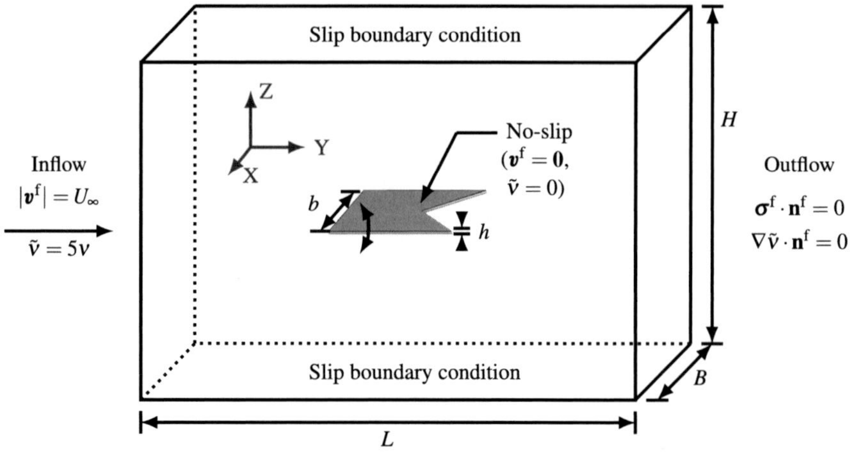

Fig. 9.14 Flow across a pitching plate: Schematic of the flow past a three-dimensional pitching plate. The computational setup and boundary conditions are shown for the turbulent Navier-Stokes equations. Here, $\boldsymbol{v}^{\mathrm{f}}$ denotes the fluid velocity, $L = 80c$, $B = 80c$, $H = 80c$ are the length, breadth and height of the computational domain respectively, and b and h are the span and thickness of the plate respectively

leading edge, where $\theta_{\max}$ is the maximum pitching amplitude and f_0 is the frequency of pitching. A hybrid RANS/LES type of model is considered for modeling the turbulence in the flow, which is discussed in Chapter 10.

The computational domain is divided into two subdomains, each corresponding to fluid and structural domains. A three-dimensional mesh is constructed for the fluid domain using eight-node hexahedron elements while a two-dimensional mesh is formed for the structural domain with the help of four-node quadrilaterals (pertaining to shell elements). A boundary layer mesh around the plate surface exists such that the first layer in the wall normal direction has $y^+ \leq 1$. Different mesh refinements are considered for the mesh convergence study for the two subdomains, thus resulting in a non-matching mesh at the fluid-structure interface, shown in Fig. 9.15.

The non-dimensional parameters for the problem of the flow across a pitching plate are the Reynolds number Re, Aeroelastic number Ae, mass ratio m^*, Poisson's ratio ν^{s} and the Strouhal number St, defined as follows:

$$\begin{aligned} Re &= \frac{\rho^{\mathrm{f}} U_\infty c}{\mu^{\mathrm{f}}} = 10,000, \; Ae = \frac{Eh}{(1/2)\rho^{\mathrm{f}} U_\infty^2 \, S} = 1.575 \times 10^8, \\ m^* &= \frac{\rho^{\mathrm{s}} h}{\rho^{\mathrm{f}} c} = 0.03, \qquad \nu^{\mathrm{s}} = 0.37, \qquad St = \frac{f_0 a}{U_\infty}, \end{aligned} \tag{9.35}$$

where $E = 3.1 \times 10^9$ N/m^2 is the Young's modulus of the structural material of the plate and $a = 2c\sin(\theta)$ is the characteristic width of the wake. For all the computations, the Strouhal number St is varied while the thrust coefficient (C_T) and the propulsive efficiency (η) of the pitching motion are quantified, which are defined as

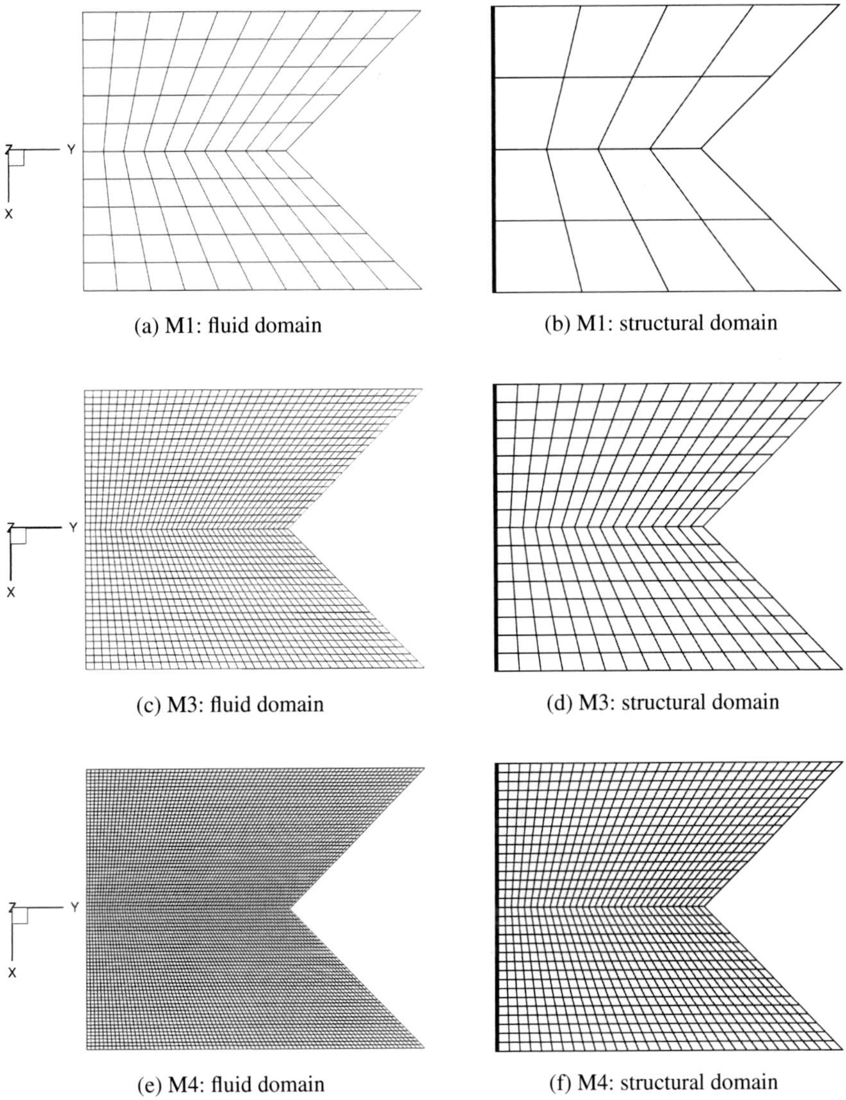

Fig. 9.15 Flow across a pitching plate: Non-matching mesh at the fluid-structure interface for the mesh convergence study

Table 9.3 Mesh characteristics and results for interface convergence study

Mesh	Structure (2D)		Fluid (3D)		$\overline{C}_T$	η
	Nodes	Elements	Nodes	Elements		
IC1	25	16	777,508	756,147	0.5641(1.79%)	0.0947(1.17%)
IC2	81	64	777,508	756,147	0.5558(0.29%)	0.0936(0%)
IC3	289	256	777,508	756,147	0.5547(0.09%)	0.0936(0%)
IC4	1089	1024	777,508	756,147	0.5543(0.02%)	0.0936(0%)
IC5	4225	4096	777,508	756,147	0.5542	0.0936

Table 9.4 Mesh characteristics and results for mesh convergence study

Mesh	Structure (2D)		Fluid (3D)		$\overline{C}_T$	η
	Nodes	Elements	Nodes	Elements		
M1	25	16	275,068	264,447	0.51397(4.39%)	0.10416(0.18%)
M2	81	64	415,348	401,547	0.50744(3.06%)	0.10394(1.59%)
M3	289	256	777,508	756,147	0.49740(1.02%)	0.10191(0.39%)
M4	1089	1024	1,828,228	1,786,947	0.49236	0.10231

$$C_T = \frac{T}{(1/2)\rho^{\mathrm{f}} U_\infty^2 S}, \qquad \eta = \frac{\overline{T} U_\infty}{2\tau_{\max}\theta_{\max} f_0}, \tag{9.36}$$

where T is the thrust generated by the pitching plate, $\overline{T}$ is the mean thrust and $\tau_{\max}$ is the maximum value of the spanwise torque.

Interface Convergence: First, interface convergence study is performed to assess the refinement at the fluid-structure interface for the non-matching meshes in the two fluid and structural domains. For a fixed three-dimensional mesh in the fluid domain, the structural mesh is refined systematically to eliminate the effect of the interface discretization errors due to the non-matching mesh. The fluid mesh consists of 777,508 eight-node hexahedron elements and 756,147 nodes. The refinement details of the structural mesh are provided in Table 9.3. A Strouhal number of $St = 0.595$ is selected for this study and the results are summarized in Table 9.3. The percentage error compared to the finest structural mesh (IC5) is quantified and the mesh IC3 is found to be sufficiently converged with error $< 0.1\%$.

Mesh Convergence: Next, after determining the required mesh refinement ratio between the structural and the fluid mesh at the fluid-structure interface, a mesh convergence study is conducted by refining both the fluid and the structural domains simultaneously while keeping the ratio similar at the interface. The mesh refinement details are given in Table 9.4. Compared to the finest mesh M4, it is found that M3 is sufficiently converged with error $< 1\%$ and therefore, it is selected for the validation study. Mesh M3 consists of 256 four-node quadrilaterals with 289 nodes in the structural domain.

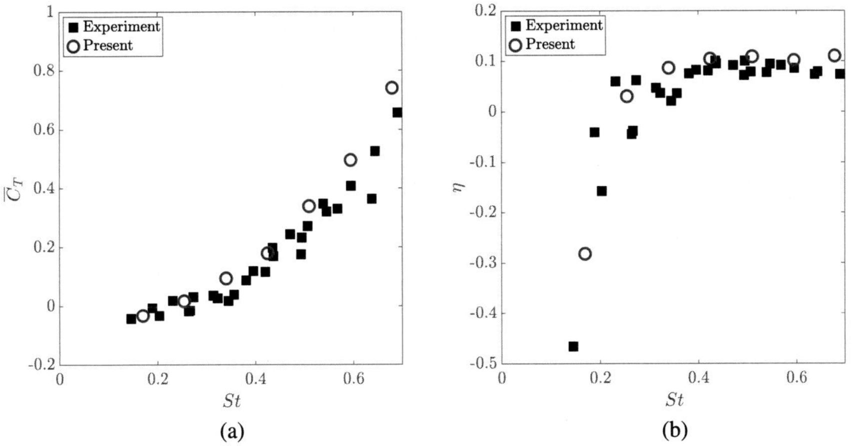

Fig. 9.16 Flow across a pitching plate: Variation with Strouhal number (St) of the **a** mean thrust coefficient $\overline{C}_T$, and **b** thrust efficiency η. The results are compared with that of the experimental study conducted in [219]

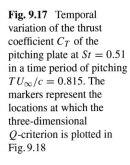

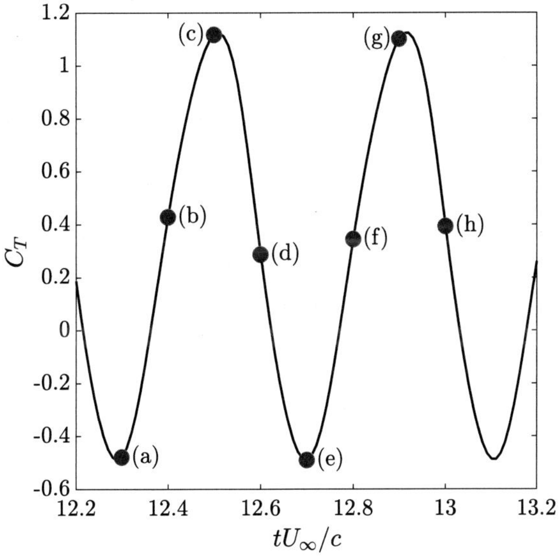

Fig. 9.17 Temporal variation of the thrust coefficient C_T of the pitching plate at $St = 0.51$ in a time period of pitching $TU_\infty/c = 0.815$. The markers represent the locations at which the three-dimensional Q-criterion is plotted in Fig. 9.18

Validation: Selecting the interface and mesh converged meshes, computations are carried out for varying Strouhal numbers in the range $St \in [0.17, 0.68]$. The thrust coefficient and propulsive efficiency are quantified and are shown in Fig. 9.16. The trend shows a good agreement with the experimental studies in [219]. Figure 9.17 shows the temporal variation of the thrust coefficient for a pitching cycle for $St = 0.51$. The frequency of the thrust oscillation is observed to be twice that of the pitching frequency. The vortices generated due to the pitching movement of the plate

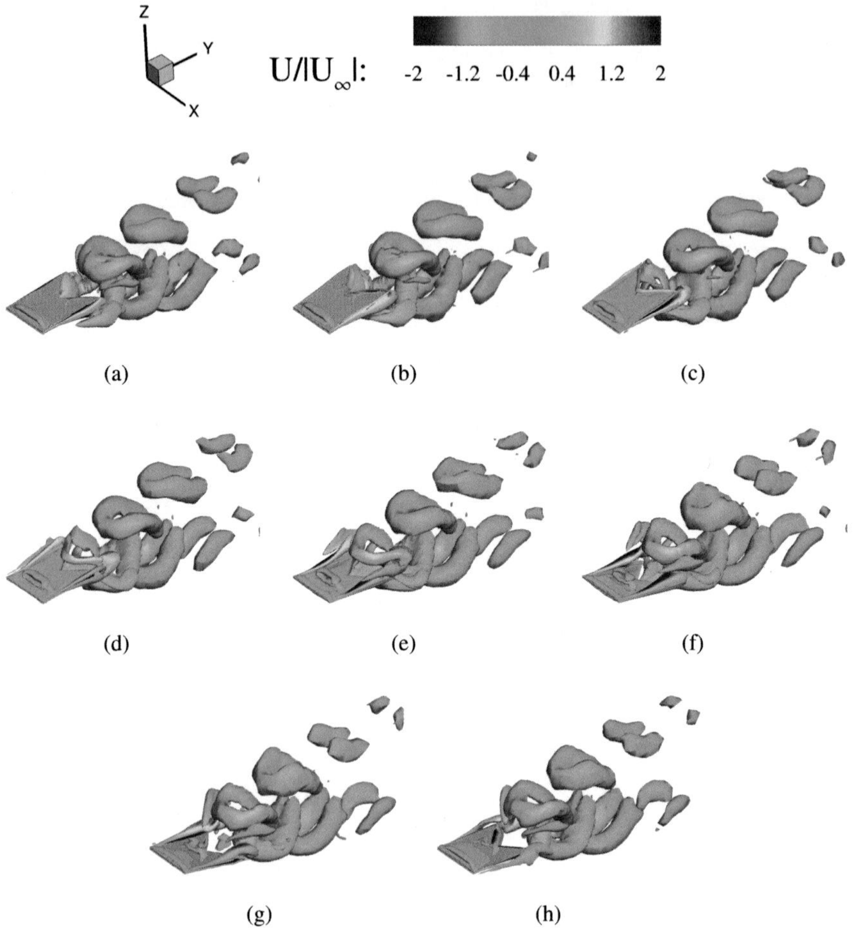

Fig. 9.18 Flow visualization with the help of Q-criterion colored by the free-stream velocity at $St = 0.51$ for the pitching plate at tU_∞/c: **a** 12.3, **b** 12.4, **c** 12.5, **d** 12.6, **e** 12.7, **f** 12.8, **g** 12.9 and **h** 13.0. These time locations correspond to the markers in Fig. 9.17 in one pitching cycle of the plate

in a pitching cycle are visualized using the iso-contour of Q-criterion colored by the streamwise velocity in Fig. 9.18. The different time instances correspond to the markers in Fig. 9.17. Horseshoe-like vortices created due to the 45° serration at the trailing edge are observed. These vortices merge with the tip vortices as they convect downstream due to the incoming flow.

9.5.2 *Flow Across a Flexible Flapping Wing*

We next validate the flexible multibody framework with the experimental work of [240] by considering a flexible flapping wing consisting of anisotropic material properties along the wing.

9.5.2.1 Mesh Convergence of an Isotropic Wing

We begin by conducting a mesh convergence study on an isotropic wing. The wing geometry is adopted from the experiment and its properties are given in Table 9.5. A Zimmerman shaped isotropic Aluminium wing is considered to study the aerodynamic characteristics under the flapping flight condition. The schematic of the wing and the computational domain are shown in Fig. 9.19. The root chord length C and the span of the wing are 25 mm and 75 mm, respectively. Both the inlet and outlet boundaries are at a distance of $30C$ from the leading edge root of the wing. The top and bottom walls are $60C$ from each other and the side walls have $80C$ distance between them. A freestream velocity at the inlet is set to zero and the reference velocity is the velocity at the wing tip. Slip boundary conditions are satisfied on the top, bottom and side walls, while a traction-free condition is imposed at the outlet boundary. A no-slip condition is applied on the flapping wing, which is undergoing hovering motion at 21° amplitude and 10 Hz flapping frequency.

The discretized computational domain can be decomposed into a fluid mesh and a structural mesh. The fluid mesh consists of unstructured eight-node hexahedron finite elements and the structural domain is discretized by two-dimensional four-node quadrilateral elements. The mesh details for the different mesh resolutions are

Table 9.5 Aeroelastic parameters for an isotropic aluminum wing under hovering motion

Parameters	Value
Semi-span at quarter chord	0.075 m
Chord length at wing root	0.025 m
Structural thickness	0.0004 m
Poisson's ratio	0.3
Material density	2700 kg/m^3
Young's modulus of material	70.0 GPa
Reference flow velocity (hovering)	1.0995 m/s
Air density	1.209 kg/m^3
Mean chord-based Reynolds number	2605
Flapping frequency	10 Hz
Flapping amplitude	21°
Aspect ratio	7.65

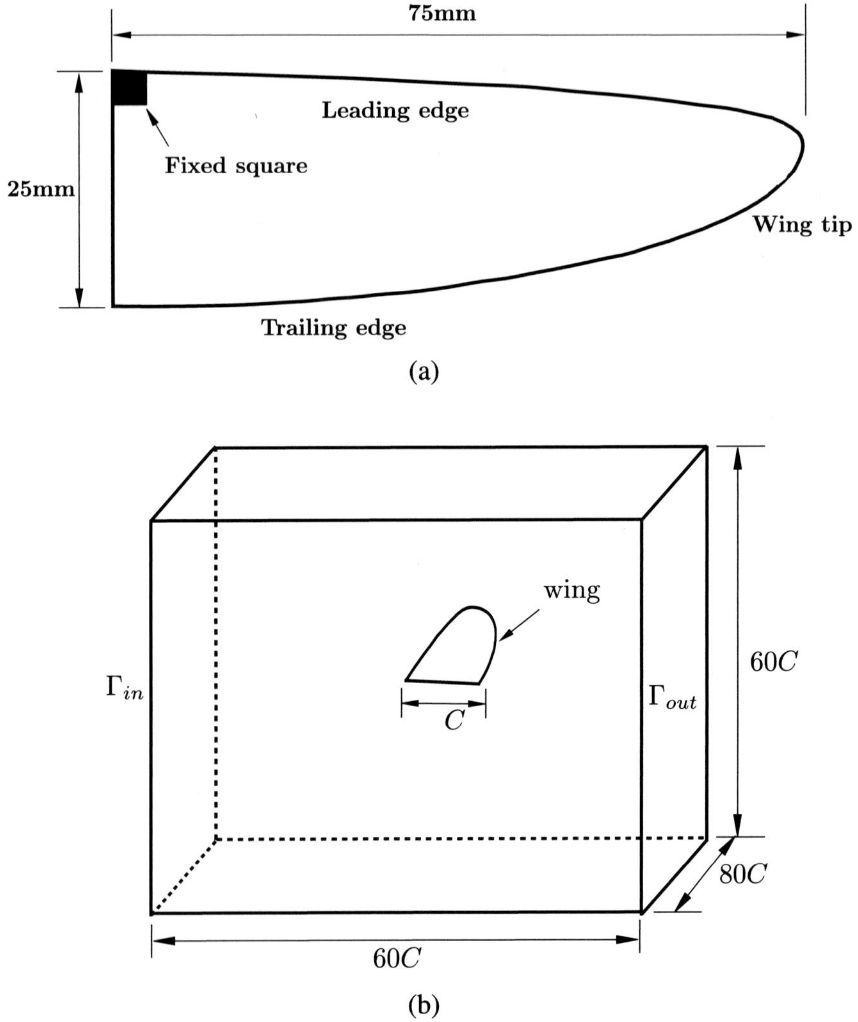

Fig. 9.19 Flow past a flapping wing: **a** geometry of isotropic wing, **b** schematic of computational setup

summarized in Table 9.6. A boundary layer is constructed around the wing such that $y^+ \approx 0.5$ in the direction normal to the wall. The mesh size in the chord-wise and span-wise directions is varied along with the size outside the boundary layer region. The computational time step size is selected as $\Delta t U_{ref}/C = 0.01$. A typical mesh (M2) is shown in Fig. 9.20.

Mesh convergence is conducted by comparing the amplitude of the lift coefficient for the various mesh refinements in Table 9.7. The result from the literature [11] is also provided for comparison. The error in the results compared to the finest mesh

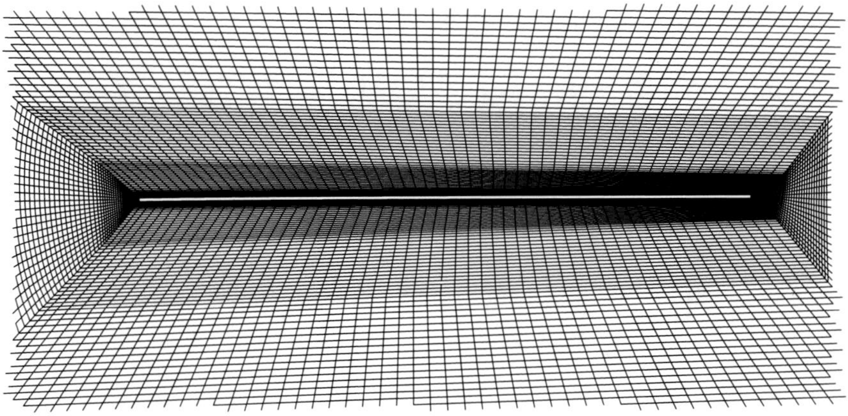

(a) Fluid space mesh

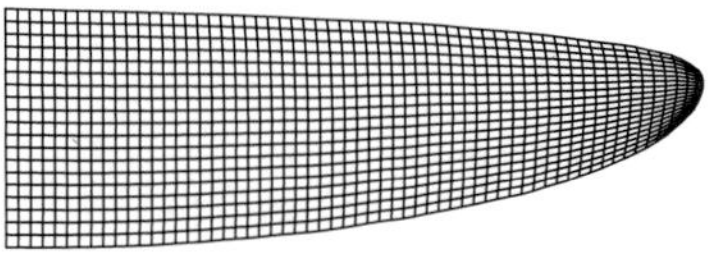

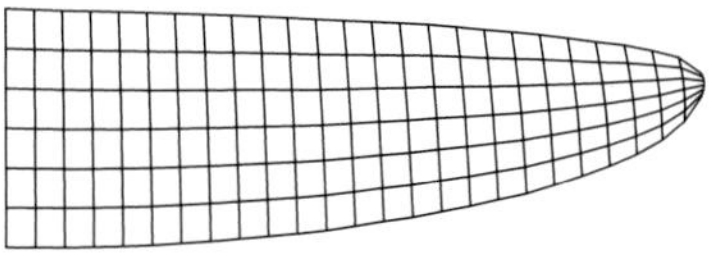

(b) Fluid surface mesh (c) Structural surface mesh

Fig. 9.20 Schematic of mesh characteristics: **a** in space for flapping wing in the fluid domain (M2), **b** on surface for flapping wing in the fluid domain (M2), **c** on surface for flapping wing in the structural domain (M2)

Table 9.6 Mesh statistics for an isotropic wing under hovering motion

Mesh	Fluid nodes	Fluid elements	Structural nodes	Structural elements
M1	509,082	492,630	114	90
M2	816,312	795,614	182	150
M3	1,311,120	1,284,226	506	450

M3 is given in the parentheses. The error percentage for M2 is around 0.15% and is selected for comparison with the literature. The variation of the lift coefficient with time is compared with the experimental data for one cycle in Fig. 9.21a, where a good agreement is observed between the obtained results and the literature. The normalized wing tip displacement is shown in Fig. 9.21b where it is observed to have a good match with the experimental observations in [240].

The normalized velocity magnitude contour, X-vorticity contour at quarter position along the chord and Y-vorticity contour at mid position along the span at two time instances $t/T = 0.3$ and $t/T = 0.48$ are shown in Fig. 9.22. It is observed that a pair of main vortices is generated at the leading edge and trailing edge during

Table 9.7 Mesh convergence study for the lift coefficient C_L

Results	Amplitude of lift coefficient C_L
Present (M1)	6.65 (0.61%)
Present (M2)	6.6 (−0.15%)
Present (M3)	6.61
Numerical simulation [11]	6.24

The percentage differences are computed based on M3 result

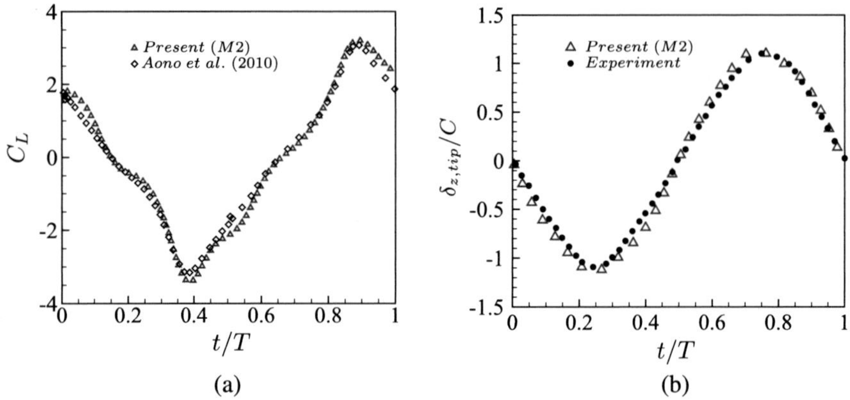

Fig. 9.21 Comparisons of time traces of the isotropic flapping wing: **a** lift coefficient: (△) Present simulation, (◇) Numerical simulation [11], **b** normalized displacement at wing tip: (△) Present simulation, (•) Experiment [240]

the hovering motion of the flexible wing. A comparison of the normalized velocity magnitude distributions along the vertical direction of the wing at the wing tip and mid-span for $t/T = 0.3$ and $t/T = 0.48$ with the observations from the literature is carried out in Fig. 9.23. The experimental results are plotted with 95% errorbars of the instantaneous values. A good qualitative agreement is found in the comparison of the numerical results with the literature.

The three-dimensional wake dynamics of the flapping wing is visualized at different time instances in Fig. 9.24 with the help of iso-surfaces of Q-criterion where the vortex structures are colored by the normalized velocity magnitude during the hovering motion.

9.5.2.2 Validation of an Anisotropic Wing

Here, we consider a wing designed to mimic a real hummingbird wing based on Zimmerman planform. The wing consists of several skeletons of unidirectional carbon fibres supporting a membrane of Capran material. The schematic of the anisotropic wing is depicted in Fig. 9.25a. A rigid triangle is used to mount the wing for the

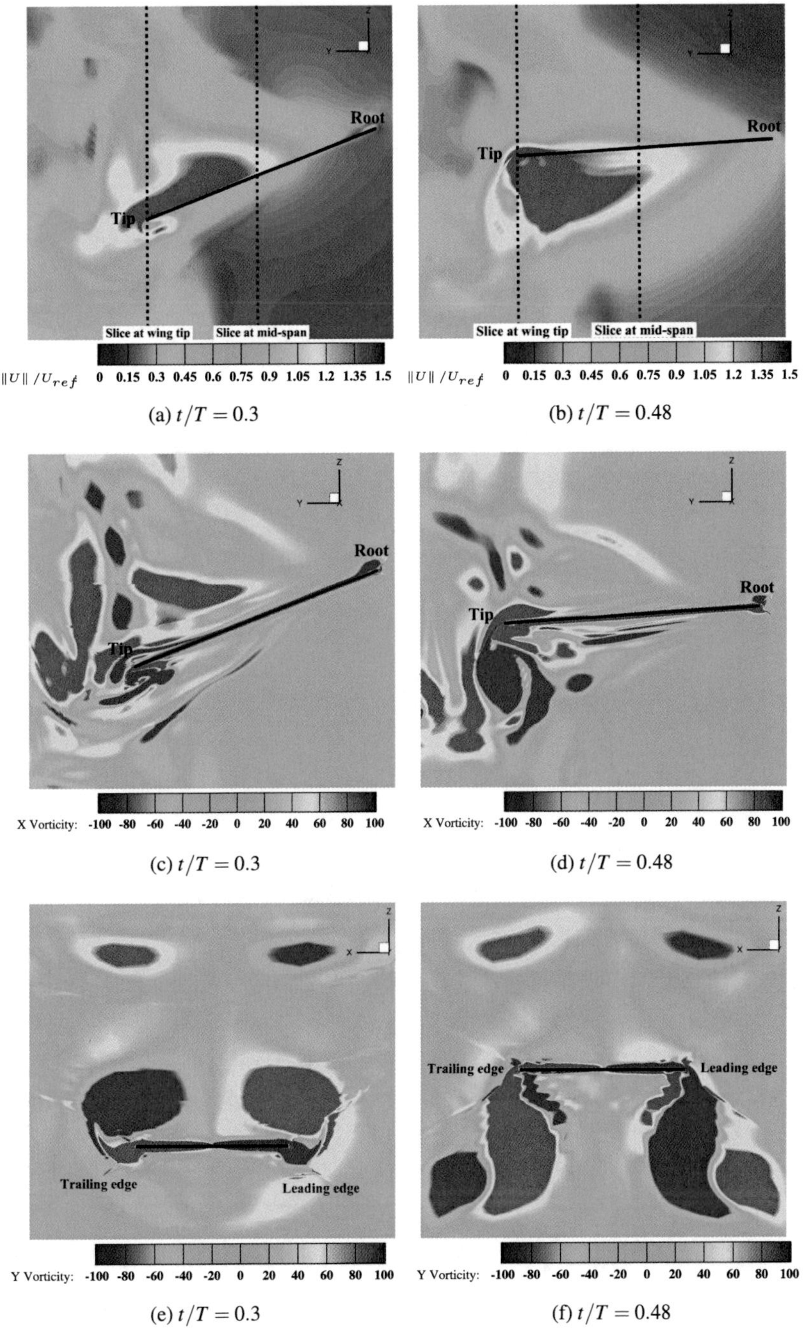

Fig. 9.22 Flow fields around the isotropic flapping wing in uniform flow: normalized velocity magnitude at **a** $t/T = 0.3$, **b** $t/T = 0.48$, X-vorticity at **c** $t/T = 0.3$, **d** $t/T = 0.48$, and Y-vorticity contours at **e** $t/T = 0.3$, **f** $t/T = 0.48$

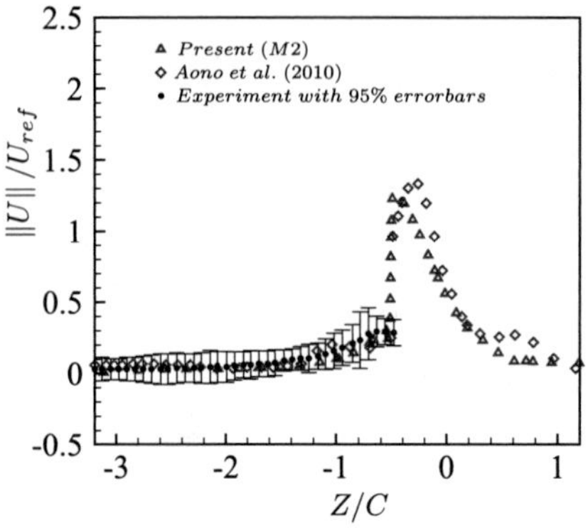

(a) $t/T = 0.3$ at mid-span

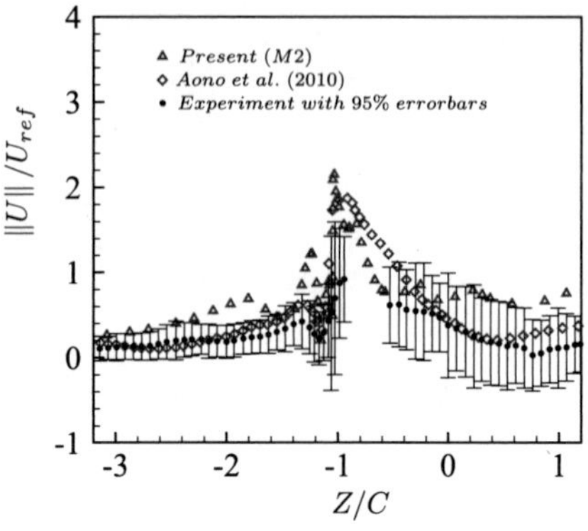

(b) $t/T = 0.3$ at wing tip

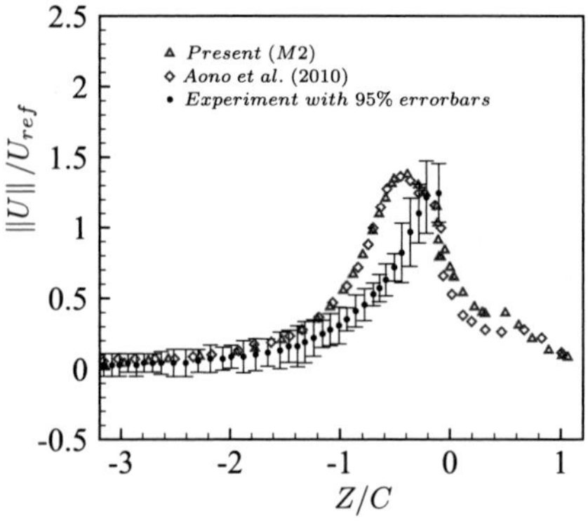

(c) $t/T = 0.48$ at mid-span

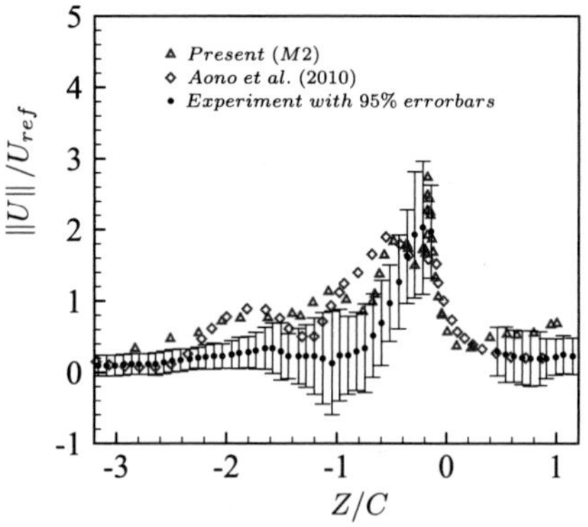

(d) $t/T = 0.48$ at wing tip

Fig. 9.23 Flow past an isotropic flapping wing: Comparison of instantaneous normalized velocity magnitude for isotropic wing at various time instants: **a** $t/T = 0.3$ at mid-span, **b** $t/T = 0.3$ at wing tip, **c** $t/T = 0.48$ at mid-span, **d** $t/T = 0.48$ at wing tip. ($\triangle$) Present simulation, ($\diamond$) Numerical simulation [11], ($\bullet$) experiment with 95% errorbars [240]

flapping motion at the corner of the wing root and the leading edge. The topological layout of the wing is shown in Fig. 9.25c. The wing is reinforced with different layers of carbon fibre strips of 0.8 mm width at the leading edge, the wing root and the surface of the membrane. Different layouts of the wing are termed as LiBj, L and B representing the leading edge and batten respectively, and i, j denoting the number of layers in the leading edge and batten, respectively. The wing root is reinforced with 2 carbon fibre layers for all the wings. Here, we consider L2B1 and L3B1 models for validation. The material properties of the skeleton and the wing membrane for the anisotropic wing are given in Table 9.8. Some structural material parameters are adjusted to satisfy the structural mode frequencies and shapes obtained from experiments according to the work in [79].

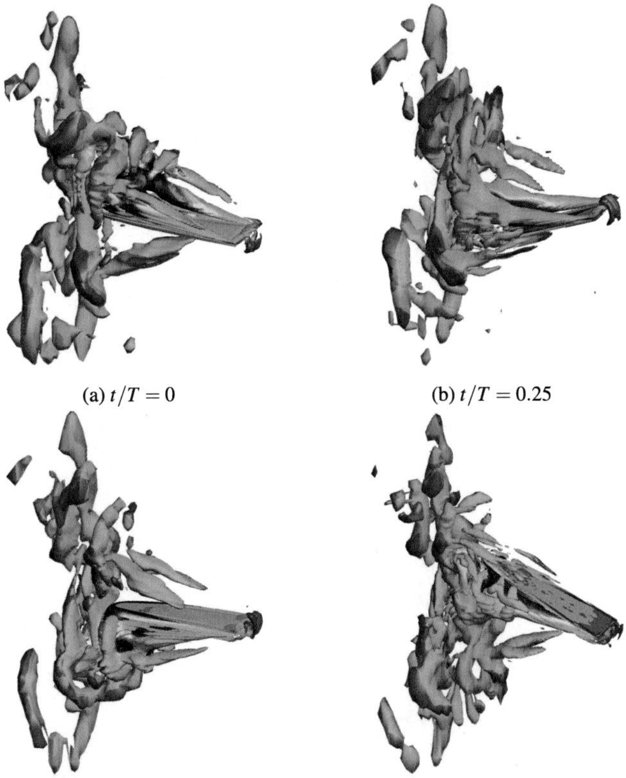

(a) $t/T = 0$ (b) $t/T = 0.25$

(c) $t/T = 0.5$ (d) $t/T = 0.75$

Fig. 9.24 Flow past an isotropic flapping wing: Wake structures based on the instantaneous iso-surfaces of $Q(= -\frac{1}{2}\frac{\partial u_i^f}{\partial x_j}\frac{\partial u_j^f}{\partial x_i})$ value at **a** $t/T = 0$, **b** $t/T = 0.25$, **c** $t/T = 0.5$, **d** $t/T = 0.75$. Iso-surfaces of non-dimensional $Q^+ \equiv Q(C/U_{ref})^2 = 1$ are colored by the normalized velocity magnitude $\|U\| / U_{ref}$

The anisotropic wing is discretized using co-rotational shell elements with anisotropic material properties for the skeleton and shell elements with isotropic properties for the wing membrane. Kinematic constraints are considered for connections between adjacent elements. The discretized mesh consists of 315 structured four-node quadrilateral elements and is shown in Fig. 9.25b.

The first six natural frequencies for the two models are analyzed in Table 9.9 where a good agreement is found with the experiment [240] and the literature [79]. A prescribed rotation with flapping amplitude of 35° and frequency 25 Hz is given at the mounted rigid triangle along the root of the wing in a vacuum in the experiment. The structural response at the wing tip is compared to the experimental results in [240]. The actual location in the vertical direction of the wing tip is given by $\delta_{z,tip}$ while $\delta_{w,tip}$ denotes the distance between the deformed actual wing tip and the

Table 9.8 Material properties of the composite skeleton and wing membrane for the anisotropic wing (L2B1 and L3B1 models) [79]

Component	Material properties
Carbon fiber prepreg (properties of one layer)	$E_{11} = 233$ GPa
	$E_{22} = 23.7$ GPa
	$G_{12} = 10.5$ GPa
	$G_{23} = G_{31}$=1.7 GPa
	$\upsilon_{12} = 0.05$
	$\upsilon_{23} = \upsilon_{31}$=0.32
	$\rho = 1740$ kg/m^3
	Thickness = 0.1 mm
Capran membrane (from experiments)	$E = 2.76$ GPa
	$\upsilon_{12} = 0.489$
	$\rho = 1384$ kg/m^3
	Thickness = 0.015 mm

Table 9.9 Comparison of the natural frequencies for anisotropic wing

Model	Result	Natural frequencies (Hz)					
		1st	2nd	3rd	4th	5th	6th
L2B1	Present	45.03	73.16	76.39	94.45	106.35	116.31
	Gogulapati A., 2011	47.00	72.00	76.50	88.00	109.00	118.80
	Experiment	42.00			84.00		126.00
L3B1	Present	62.71	76.13	79.55	106.31	111.00	119.39
	Gogulapati A.	65.00	75.50	76.80	107.00	109.00	120.00
	Experiment	59.00			104.00		138.00

undeformed reference plane. The initial reference plane, undeformed reference plane and deformed actual plane is shown in Fig. 9.25d.

The temporal evolution of the normalized location in the vertical direction and the normalized displacement at the wing tip for the two models undergoing rotational motion in vacuum are shown in Fig. 9.26. The results show a good agreement with the experimental measurements. The structural response is averaged over several cycles based on the periodic assumption considered in the experiments. The instantaneous structural displacement for the L2B1 model for different time instances is given in Fig. 9.27. A relatively large elastic deformation and the wing twist due to the high flexibility of the wing is observed.

Flapping of the anisotropic wing in air condition is also carried out to investigate the aeroelastic effects. Here, we consider the L3B1 model under hovering motion with zero freestream velocity for the validation purpose. The material properties across the wing are summarized in Table 9.8. The parameters for the numerical com-

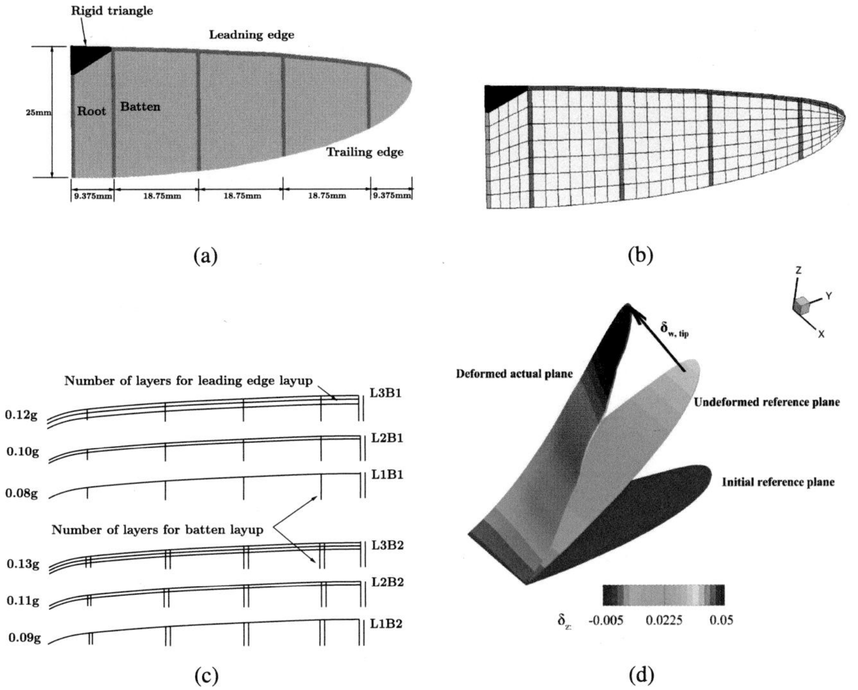

Fig. 9.25 Problem set-up for anisotropic wing configuration: **a** geometry information, **b** finite element representation, **c** the topological layout [240], **d** initial reference plane, undeformed reference plane and deformed actual plane

putation are given in Table 9.10. The computational domain employed is similar to that considered for the isotropic wing, as shown in Fig. 9.19b along with identical boundary conditions and similar three-dimensional fluid mesh discretization consisting of 492,630 unstructured finite elements. The mesh is constructed such that $y^+ \approx 0.5$ in the boundary layer region. The structural mesh is shown in Fig. 9.25b and consists of 315 structured elements. The time step size for the computation is selected as $\Delta t U_{ref}/C = 0.0018$.

The normalized location in the vertical direction and the normalized displacement at the tip of the wing are compared with the experimental results in Fig. 9.28, where a good qualitative agreement is found. The vortex structures are visualized by isosurfaces of Q-criterion colored by the normalized velocity magnitude in Fig. 9.29 for the L3B1 model, where the evolution of the turbulent wake structures during the flapping motion is observed.

This completes the convergence and validation study for the flexible multibody framework consisting of the shell elements coupled with the fluid domain via point-to-point radial basis function mapping. In the next section, this framework is demonstrated on a three-dimensional application of flapping dynamics of bio-inspired bat wing design.

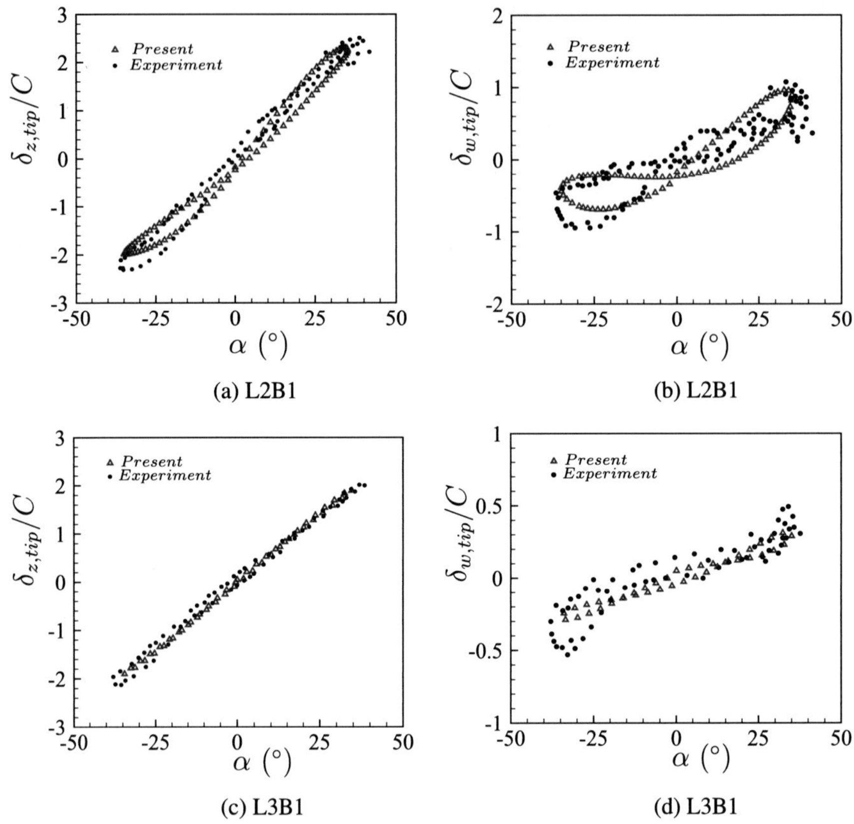

Fig. 9.26 Comparison of structural responses for anisotropic wing: **a** normalized location in vertical direction of L2B1 model at wing tip, **b** normalized displacement of L2B1 model at wing tip, **c** normalized location of L3B1 model in vertical direction at wing tip, **d** normalized displacement of L3B1 model at wing tip. (△) Present simulation, (•) Experiment [240]

Table 9.10 Aeroelastic parameters for an anisotropic wing under hovering motion

Parameters	Value
Reference velocity (hovering)	4.58 m/s
Air density	1.206 kg/m^3
Reynolds number	7304
Flapping frequency	25 Hz
Flapping amplitude	35°

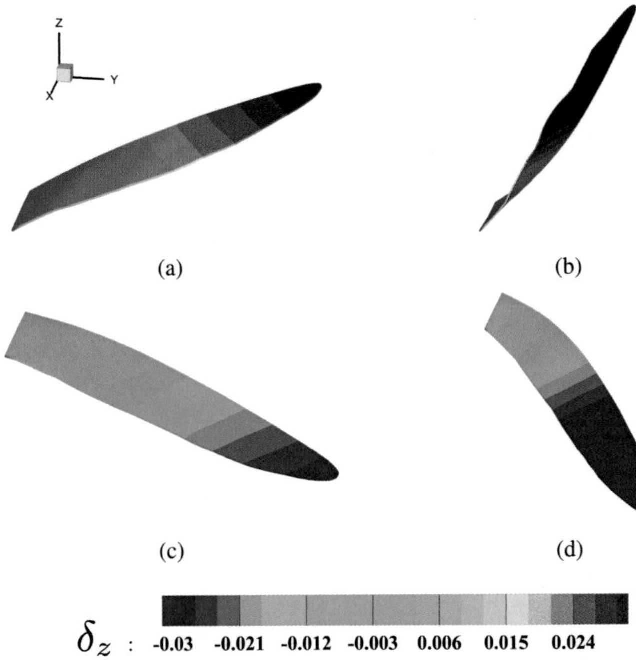

Fig. 9.27 Structural displacement contours of L2B1 model at various time instants: **a** $t/T = 0.19$ with flapping angle 32.54°, **b** $t/T = 0.34$ with flapping angle 29.55°, **c** $t/T = 0.55$ with flapping angle $-10.50°$, **d** $t/T = 0.69$ with flapping angle $-32.82°$

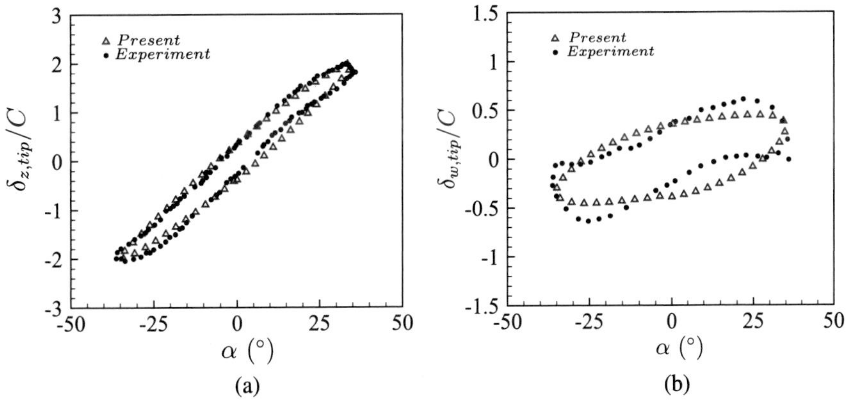

Fig. 9.28 Comparison of coupled aeroelastic responses for anisotropic flapping wing: **a** normalized location of L3B1 model in vertical direction at wing tip, **b** normalized displacement of L3B1 model at wing tip. (△) Present simulation, (•) experiment [240]

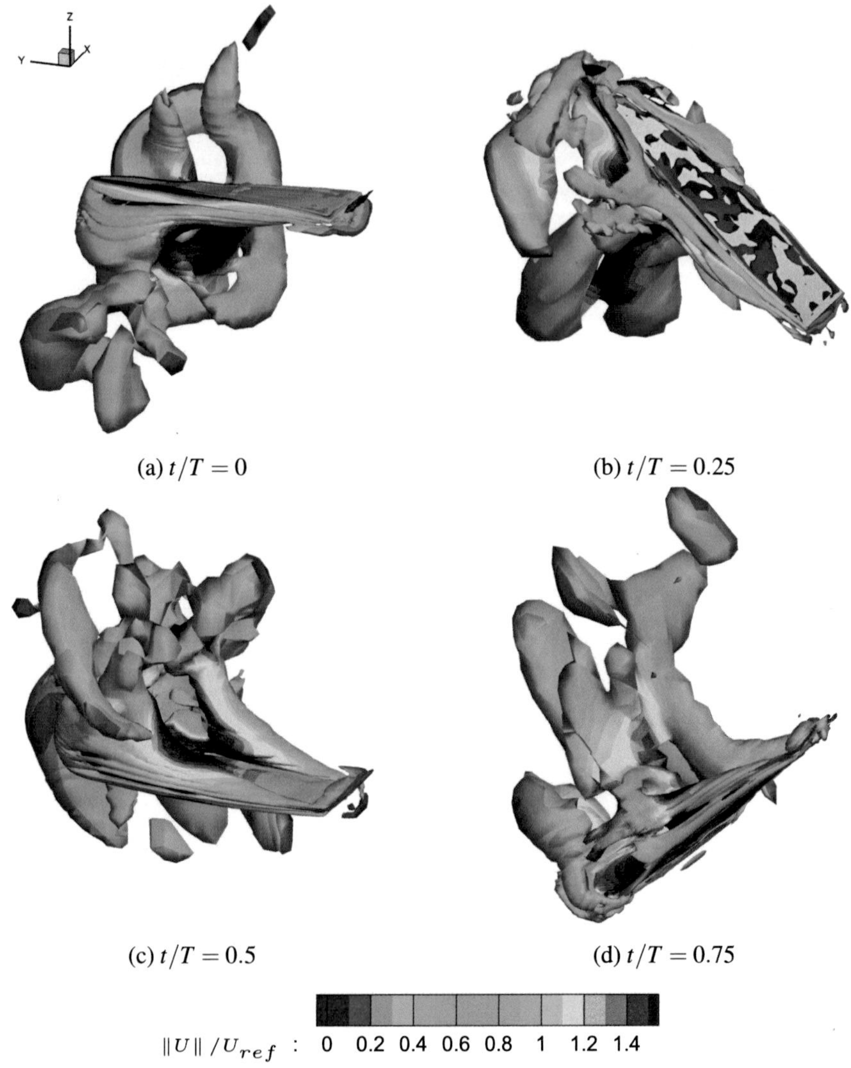

Fig. 9.29 Flow past an anisotropic flapping wing: Wake structures of L3B1 model based on the instantaneous iso-surfaces of $Q(= -\frac{1}{2}\frac{\partial u_i^f}{\partial x_j}\frac{\partial u_j^f}{\partial x_i})$ value at **a** $t/T = 0$, **b** $t/T = 0.25$, **c** $t/T = 0.5$, **d** $t/T = 0.75$. Iso-surfaces of non-dimensional $Q^+ \equiv Q(C/U_{ref})^2 = 1$ are colored by the normalized velocity magnitude $\|U\| / U_{ref}$

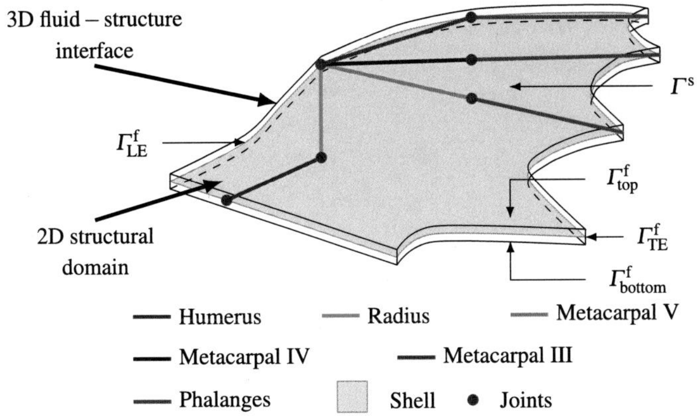

Fig. 9.30 The structural model of the bat wing where the bone fingers and membranes are represented by multibody components such as beams (lines) and thin shells (surfaces) connected by joints (dots). The structural boundary of the interface $\Gamma^s = \Gamma^s_{Beam} \cup \Gamma^s_{Shell} \cup \Gamma^s_{Joints}$ and the fluid boundary is $\Gamma^f = \Gamma^f_{top} \cup \Gamma^f_{bottom} \cup \Gamma^f_{LE} \cup \Gamma^f_{TE}$

9.5.3 *Three-Dimensional Flapping Dynamics of a Bat at $Re = 12{,}000$*

The wing anatomy of a bat can be considered as a flexible multibody system which consists of bone fingers connected to each other by joints of varying degrees of freedom. The skeleton formed by these bones forms the support for the flexible membrane of the wing as shown in Fig. 9.30. The different types of bones on a bat wing are humerus, radius, metacarpals and phalanges. The joints (shown as dots in the figure) can have multiple degrees of freedom and can allow active as well as passive movements of the wing. The complex kinematics of the movements between the bone fingers during a bat flight makes its computational modeling very challenging. Furthermore, the varying structural properties of the bones and the flexible thin membranes in the spanwise and chordwise directions adds up to the challenge.

Here, the flapping dynamics of such a multibody bat wing is demonstrated. The bone fingers and the flexible membranes of the wing are modeled as Euler-Bernoulli beam and thin shell structures, respectively. Although there are many joints in a real bat wing, we just consider the major 6 joints modeled as revolute joints in the bat wing model for simplicity, as shown in Fig. 9.30. Passive movement of the joints is assumed for simplicity. The geometry of the wing is based on the Pallas' long tongued bat species *Glossophaga soricina* for which the geometrical parameters are given in Table 9.11.

Table 9.11 Physical parameters of the geometry of the Pallas' long tongued bat *Glossophaga soricina*

Parameter	Value
Wing span (b)	0.2369 m
Chord length at root (c_{root})	0.0384 m
Wing surface area (S)	8.3217×10^{-3} m^2
Mean chord length ($c = S/b$)	0.03512 m

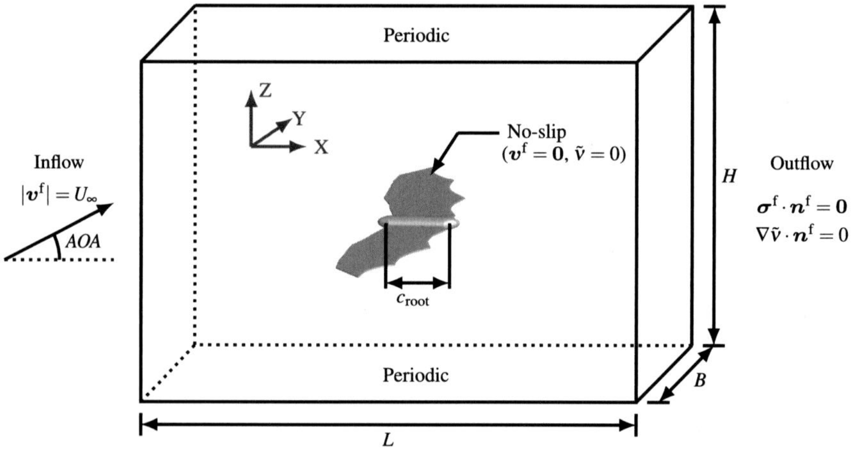

Fig. 9.31 A schematic of the flow past a bat. The computational setup and boundary conditions are shown for the turbulent Navier-Stokes equations. Here, $\boldsymbol{v}^{\text{f}}$ denotes the fluid velocity, θ is the angle of attack at the inflow boundary, L, B, H are the length, breadth and height of the computational domain respectively, and c_{root} is the chord at the root of the bat wing

9.5.3.1 Computational Domain and Key Fluid-Structure Parameters

The computational domain for the demonstration, shown in Fig. 9.31, consists of a cuboidal domain $L \times B \times H$, where $L \approx 65c$, $B \approx 68c$ and $H \approx 68c$, with c being the mean chord of the bat wing. The inflow and outflow boundaries are at a distance of about $30c$ and $35c$ from the center of the bat body respectively. A freestream velocity of $\boldsymbol{v}^{\text{f}} = (U_\infty \cos(AOA), 0, U_\infty \sin(AOA))$, where $AOA = 10°$ is the angle of attack, is given at the inflow boundary. On the other hand, a stress-free condition is satisfied at the outflow boundary. A periodic condition is allowed on the top and bottom boundaries whereas slip condition is imposed on the side walls parallel to the X-Z plane. On the surfaces of the bat wing and the bat body, no-slip condition is satisfied. A hybrid RANS/LES type of model is considered for modeling the turbulence in the flow.

The flapping motion of the wing is realized by prescribing a sinusoidal rotational motion along the X-axis at the revolute joint connecting the humerus bone with the bat body. The amplitude of the prescribed rotation is $\theta_{\max} = 25°$ and frequency of 1 Hz.

Table 9.12 Anisotropic elastic properties for the bone fingers (modeled as beam) in the flexible wing

Component	Flexural rigidity (EI) (Nm^2)	Length (l) (cm)
Humerus	1.56×10^{-3}	1.55
Radius	1.31×10^{-3}	3.00
Metacarpal V	0.12×10^{-3}	2.05
Metacarpal IV	0.98×10^{-3}	2.84
Metacarpal III	0.23×10^{-3}	4.62
Phalanx (digit V)	0.04×10^{-3}	3.29
Phalanx (digit IV)	0.04×10^{-3}	2.74
Phalanx (digit III)	0.04×10^{-3}	2.67

Table 9.13 Anisotropic elastic properties for the flexible membrane (modeled as thin shells) in the flexible wing

Component	Young's modulus (E) (N/m^2)	Thickness (h) (cm)	Density (ρ^s) (kg/m^3)	Poisson's ratio (ν^s)
Membrane	7×10^5	0.046	373.05	0.33

For the non-dimensionalization of the parameters, the reference velocity is selected as the magnitude of the inflow velocity, $U_{ref} = U_\infty$. Thus, the non-dimensional rotational frequency is $fc/U_{ref} = 0.03512$ resulting in a non-dimensional time period of rotation of $T_w U_{ref}/c = 28.5$, where T_w is the flapping time period.

The Reynolds number for the demonstration is defined as $Re = \rho^f U_{ref} c/\mu^f = 12,000$. The bone fingers which are modeled as Euler-Bernoulli beams have varying flexural rigidity EI and length l, E and I being the Young's modulus and second moment of the cross-sectional area of the beam, respectively. The beam is assumed to have a circular cross-section of diameter 0.5 cm and its material density is taken as $\rho^s = 2200$ kg/m^3. The material properties of the different bone fingers are given in Table 9.12 based on the experimental estimates in [205]. Furthermore, the thin flexible membranes are modeled as shell elements with properties given in Table 9.13.

The following aerodynamic coefficients are quantified by the computation which are evaluated by non-dimensionalizing the forces by the respective characteristic quantities:

$$C_L = \frac{1}{\frac{1}{2}\rho^f U_{ref}^2 S} \int_\Gamma (\boldsymbol{\sigma}^f \cdot \boldsymbol{n}) \cdot \boldsymbol{n}_z d\Gamma, \tag{9.37}$$

$$C_D = \frac{1}{\frac{1}{2}\rho^f U_{ref}^2 S} \int_\Gamma (\boldsymbol{\sigma}^f \cdot \boldsymbol{n}) \cdot \boldsymbol{n}_x d\Gamma, \tag{9.38}$$

where $\boldsymbol{n}$ is the normal to the surface of the bat. $\overline{X}$, $\overline{X}_{,max}$ and $\overline{X}_{,rms}$ represent the mean, maximum and the root mean square from the mean of the coefficient X.

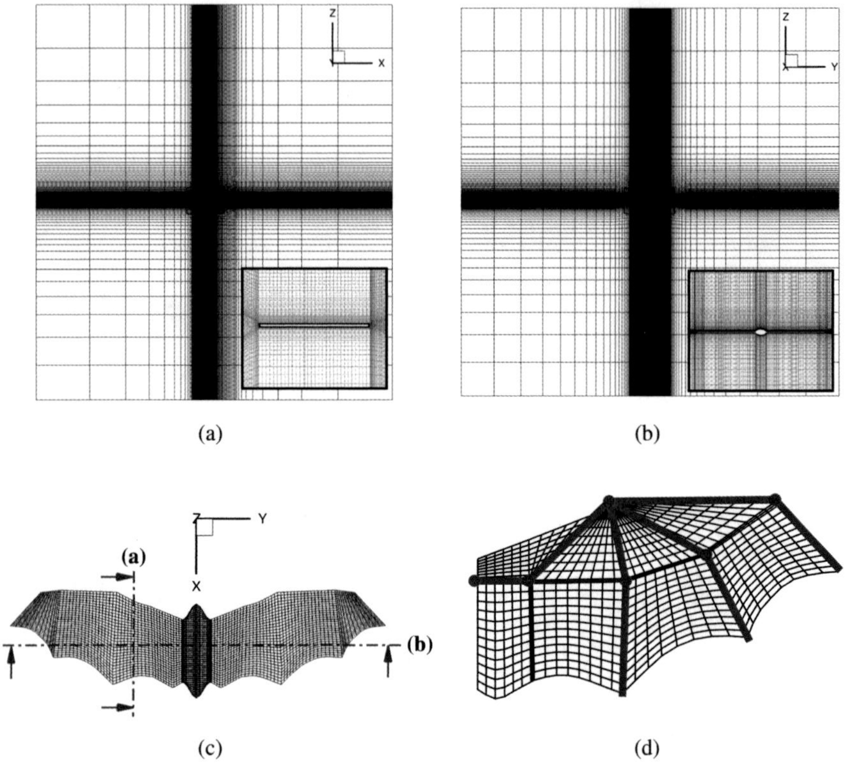

Fig. 9.32 Forward flight of a bat: Three-dimensional fluid mesh in the **a** $X-Z$ plane, **b** $Y-Z$ plane, **c** bat body; **d** two-dimensional structural mesh consisting of multibody components as beams, shells and revolute joints for the right wing

9.5.3.2 Discretization Characteristics

The computational domain is discretized using three-dimensional mesh for the fluid subdomain and two-dimensional mesh for the structural subdomain. A typical mesh is shown in Fig. 9.32. A boundary layer mesh is constructed enveloping the two wings to capture the boundary layer flow physics. The mesh consists of 4.04 million nodes with 3.96 million eight-node hexahedron elements in the fluid subdomain. The elements increase in size progressively from the center of the bat to the outer computational domain. On the other hand, the two-dimensional structural mesh consists of 792 four-node quadrilateral elements for the shell along with 74 two-node beam elements for each wing (shown in Fig. 9.32d). The non-dimensional time step size for the computation is chosen as $\Delta t U_{\text{ref}}/c = 0.071$ or $\Delta t/T_w = 2.5 \times 10^{-3}$. For this complex fluid-structure interaction problem, we utilize radial basis function mapping across the non-matching meshes at the fluid-structure interface. This mapping is illustrated in Fig. 9.33 for the bat wing.

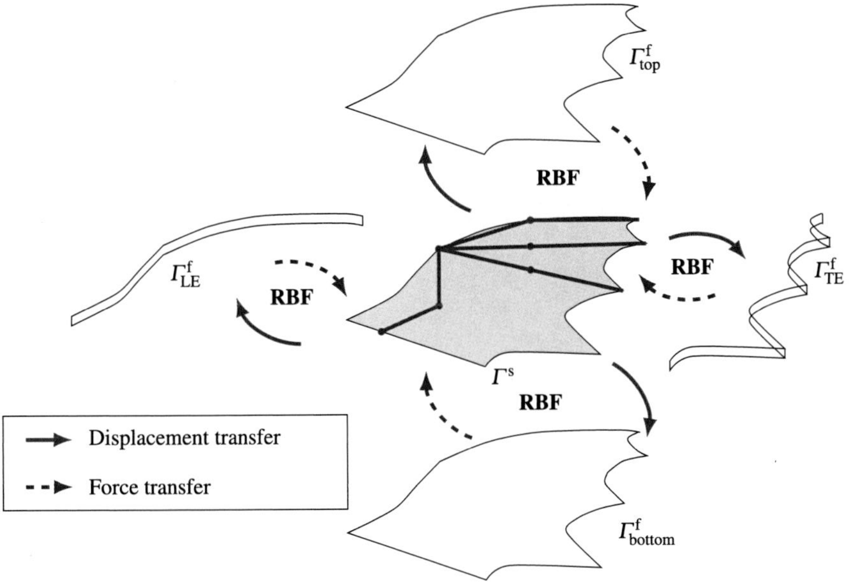

Fig. 9.33 The structural displacements of the discretized structural nodes is mapped onto the fluid mesh (which may be non-matching) by the radial basis function (RBF) mapping and the fluid forces from the fluid nodes to the structural nodes are also transferred in the same manner

9.5.3.3 Numerical Predictions

For the current demonstration, the amplitude response, aerodynamic coefficients and wake dynamics of the flapping bat are investigated at 10° angle of attack. The amplitude response of the tip of the right wing and the time history of the lift and drag coefficients on the surface of the wings in a flapping cycle are plotted in Figs. 9.34a and 9.34b, respectively. The figures show that major lift is generated during the downstroke, as observed in the literature. The high lift during the downstroke is a result of the larger effective angle of attack. The high frequency oscillations in the aerodynamic coefficients may be a result of the different modes of vibration along the flexible membrane. In the figure, the markers F1, F2, F3 and F4 represent the locations in a flapping cycle chosen for further analysis and post-processing. The deformation along the wing showing the approximate locations of the bone fingers and the joints is depicted in Fig. 9.35.

The statistics of the aerodynamic quantities is provided in Table 9.14. A rigid non-flapping case of the bat wing, similar to a fixed-wing flying vehicle is also considered for comparison. The data shows an improvement in both the mean lift coefficient (8 %) and the mean lift-to-drag coefficient ratio (3.5 %). A higher value of the unsteady lift (120 %) for the flexible wing is particularly interesting, which corresponds to the basic mechanism for lift generation for natural flyers.

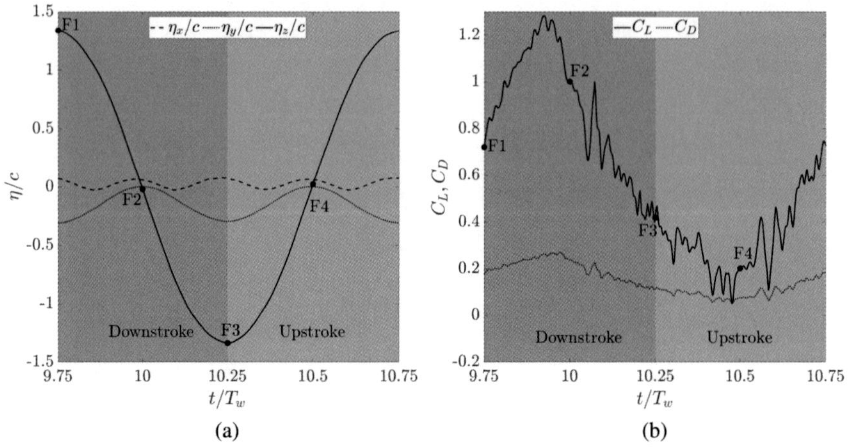

Fig. 9.34 Time histories of **a** normalized displacement at the right wing tip, and **b** lift and drag coefficients for a time period of oscillation at angle of attack of 10°. F1, F2, F3 and F4 indicate the points at one-quarter times of a period of flapping which have been utilized for further analysis

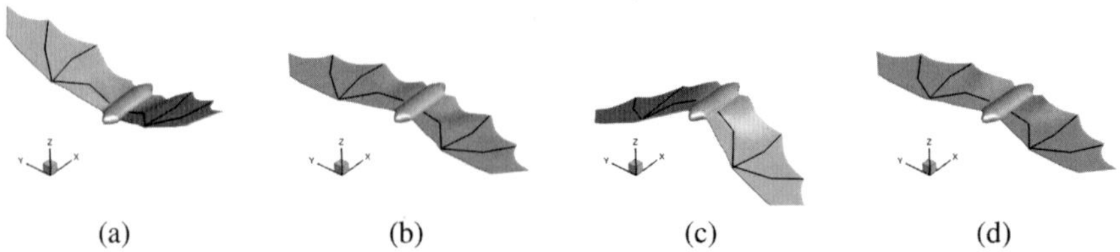

Fig. 9.35 The deformation of the wing with approximate locations of the joints and bone fingers for the flexible flapping wings at: **a** $t/T_w = 9.75$ (F1), **b** $t/T_w = 10$ (F2), **c** $t/T_w = 10.25$ (F3) and **d** $t/T_w = 10.5$ (F4)

Table 9.14 Aerodynamic coefficients for the different wing configurations at $AOA = 10°$

Parameter	Rigid non-flapping wing	Flexible flapping wing
$\overline{C_L}$	0.5623	0.6059
$\overline{C_D}$	0.1462	0.1522
$\overline{C_{L,\text{rms}}}$	0.0069	0.3446
$\overline{C_{D,\text{rms}}}$	0	0.0623
$C_{L,\text{max}}$	0.5822	1.2838
$\overline{C_L}/\overline{C_D}$	3.8452	3.9805

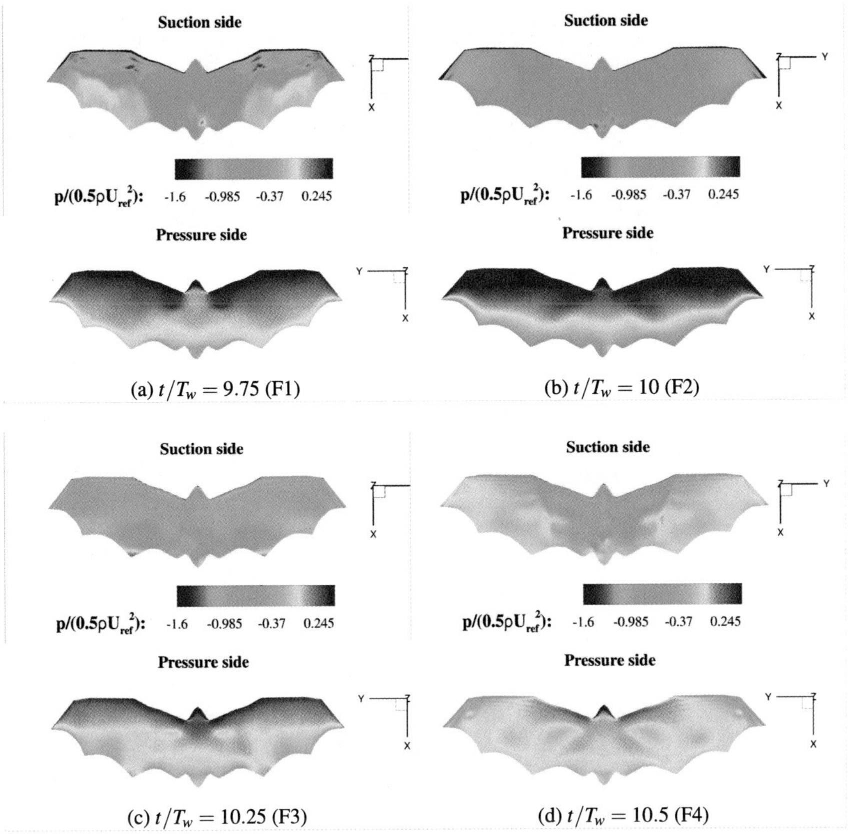

Fig. 9.36 Pressure distribution on the pressure and suction sides of the bat wing at $AOA = 10^{\circ}$

To further analyse the higher unsteady lift, pressure distribution on the suction and the pressure sides of the wing are depicted in Fig. 9.36 for F1, F2, F3 and F4 locations of the flapping period. A larger pressure differential across the wings is observed during the downstroke (F1 and F2) compared to the upstroke (F3 and F4).

The three-dimensional wake of the flapping bat is shown in Fig. 9.37 by plotting the iso-contours of the Q-criterion colored by the streamwise velocity. Similar to the pattern observed in the wake of the pitching plate in Fig. 9.18, horseshoe-like vortices, observed in the near- and far-wake, are generated due to the inward serration of the trailing edge of the wing. A complex interaction between the tip and leading-edge vortices can be seen. Moreover, the Y-vorticity contours at different cross-sections along the span of the right wing are shown in Fig. 9.38. The wing deforms passively pertaining to the varying structural properties along the cross-section due to the presence of the bone fingers and the membrane. In reality, the prime bone fingers (humerus, radius and metacarpals) are controlled explicitly by the bat, leading to active kinematics of the wing, more deformation and rich vortex dynamics. The investigation of such complex interaction is beyond the scope of this book.

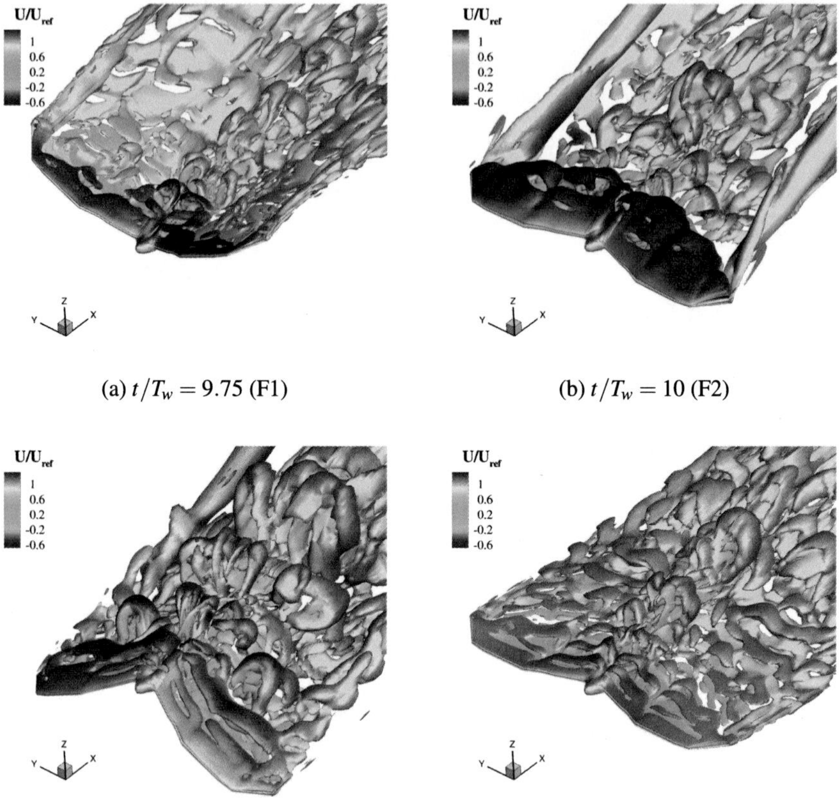

(a) $t/T_w = 9.75$ (F1)

(b) $t/T_w = 10$ (F2)

(c) $t/T_w = 10.25$ (F3)

(d) $t/T_w = 10.5$ (F4)

Fig. 9.37 The three-dimensional vorticity patterns visualized by the iso-surface of Q-criterion at $Q^+ = 100$ colored by the normalized velocity magnitude for the flexible flapping wings

To summarize, the flexible multibody structural framework consisting of various components connected by joints and constraints is presented, validated and demonstrated for an offshore application of VIV of a riser and flapping dynamics of a bio-inspired bat wing design in the present chapter.

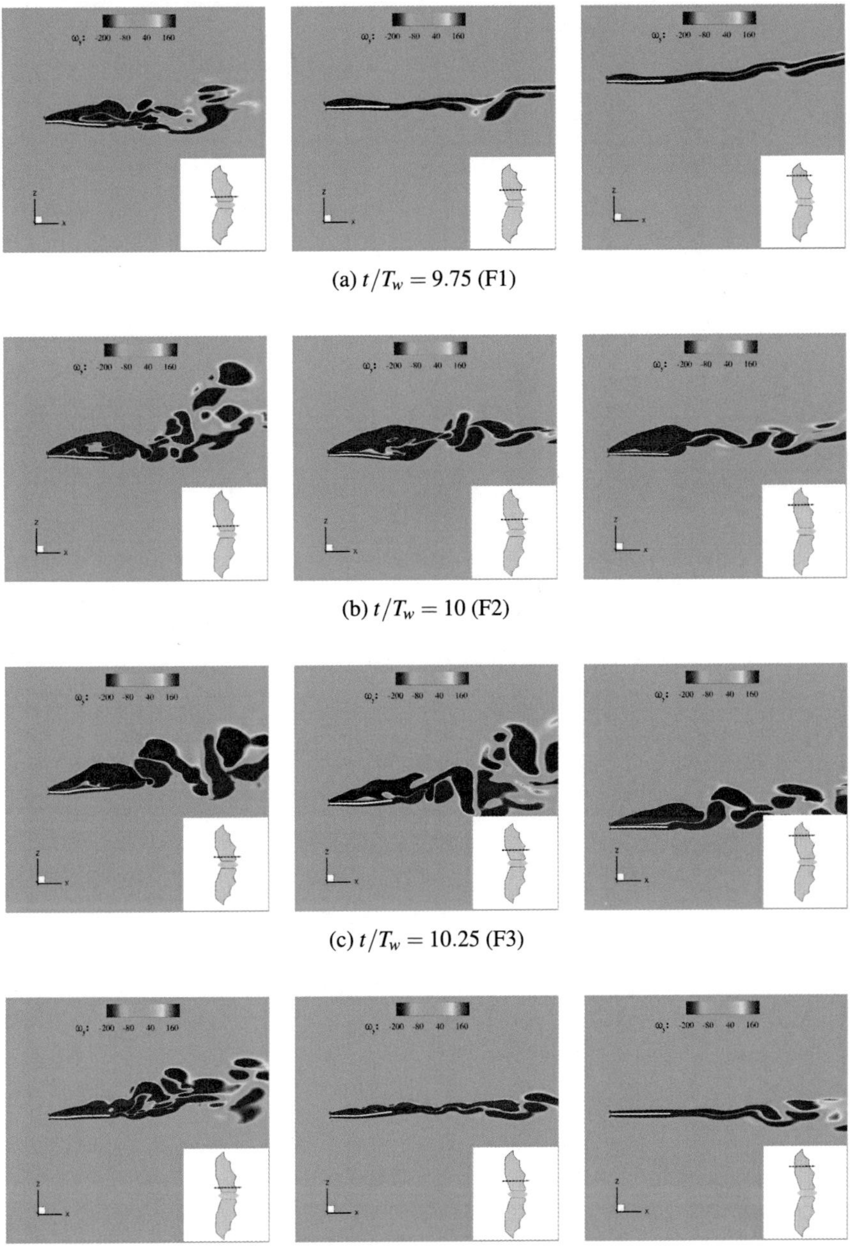

(a) $t/T_w = 9.75$ (F1)

(b) $t/T_w = 10$ (F2)

(c) $t/T_w = 10.25$ (F3)

(d) $t/T_w = 10.5$ (F4)

Fig. 9.38 The Y-vorticity shown at different slices of the right wing (0.51c, 1.02c and 2.05c from the body center) for the flexible flapping wings

Chapter 10
Turbulence Modeling in Fluid-Structure Interaction

10.1 Introduction

Continuing the journey of the partitioned-type of coupling techniques, we focus on the modeling of turbulent effects in this chapter. These effects are predominant at high Reynolds numbers and are inherently chaotic and complex to capture by a numerical simulation. The closure problem for the turbulent flow can be broadly solved by three approaches, viz., direct numerical simulation (DNS), unsteady Reynolds averaged Navier-Stokes (RANS), and large eddy simulation (LES). The DNS resolves all the physical turbulent effects and eddies completely, thus it requires the computational mesh refinement to be extremely fine at high Reynolds numbers, leading to a very high computational cost. On the other hand, RANS or unsteady RANS only models the turbulent effects close to the boundary layer, thus being ineffective in capturing separated flow structures in the turbulent wake. LES models the massively separated flow structures, provided the boundary layer is sufficiently refined. For large-scale applications, both DNS and LES are considered beyond the capacity of current computer hardware and URANS is not reliable for separated flows [53].

In this chapter, we first discuss an explicit LES filtering technique which provides a mechanism of resolving the large-scale flow structures and modeling the subgrid-scale stresses via the Smagorinsky-based functional modeling [187]. The technique only models the impact of the subgrid scales on the evolution of the resolved scales by the filtered Navier-Stokes equations [187]. It works on the eddy-viscosity assumption pertaining to the analogy between the related turbulence mechanism and the dissipation by molecular motion. The classical Smagorinsky model suffers from excessive damping of the large coherent flow structures, erroneous behavior in wall bounded and transitional flows, inaptness to accomodate backscatter, etc. A dynamic approach, to resolve the above drawbacks, based on the Germano identity [76] is presented for the fluid-structure interaction solver. In this approach, adaptive computation of the subgrid-scale coefficient which takes into account the difference between the subgrid stresses on the computational mesh and on a test-filter scale, resolves the inconsistency in the classical Smagorinsky model. The turbulence model is val-

R. K. Jaiman and V. Joshi, *Computational Mechanics of Fluid-Structure Interaction*,
https://doi.org/10.1007/978-981-16-5355-1_10

idated by performing a numerical test of a stationary and vibrating square cylinder at moderate Reynolds number and then demonstrating an application to a large-scale semi-submersible floater. These applications involve flow past bluff bodies at high Reynolds number where turbulent fluctuations are nonlinearly superimposed on the periodic vortex shedding motion which increases complexity in the numerical simulation.

Then, a hybrid LES/RANS approach, known as delayed detached eddy simulation (DDES) is discussed which gives a practical approach for large-scale problems. In this hybrid technique, the boundary layer turbulent effects are modeled through the Spalart-Allmaras (S-A) URANS equation, and the model acts as LES-type of filtering in the separated flow regions. This, in turn, combines the advantages of both RANS and LES, and reduces the computational cost by a large scale compared to DNS and LES. The underlying S-A equation for the DDES is phenomenological and poses several challenges with discretization in finite element methodology, such as (a) inclusion of the reaction terms, (b) production and destruction effects, and (c) oscillations in the numerical solution near high gradient layers. The turbulent viscosity variable of the S-A equation should exhibit non-negative values for the positive boundary and initial conditions. Numerical discretization may result in oscillations near the regions of large gradients of the solution. Note that this phenomenon is similar to that discussed in Chap. 4 for convection-diffusion-reaction equation. Several methods have been proposed to circumvent this issue in the S-A model [7, 142, 153, 198]. We apply the positivity preserving variational (PPV) method discussed in Chap. 4 to the DDES transport equation to obtain a positivity preserving and bounded solution for the turbulent viscosity. The variational methodology is validated by considering flow across a stationary circular cylinder at Reynolds numbers of 3900 and 140,000. We then demonstrate the hybrid RANS/LES technique on a large-scale problem of an offshore riser subjected to uniform ocean current at Reynolds number of 4000.

For turbulence modeling consisting of either LES or hybrid RANS/LES, the Navier-Stokes equations can be written after spatial filtering or Reynolds averaging as

$$\rho^{\mathrm{f}}\frac{\partial \overline{\boldsymbol{v}}^{\mathrm{f}}}{\partial t} + \rho^{\mathrm{f}}\overline{\boldsymbol{v}}^{\mathrm{f}}\cdot\nabla\overline{\boldsymbol{v}}^{\mathrm{f}} = \nabla\cdot\overline{\boldsymbol{\sigma}}^{\mathrm{f}} + \nabla\cdot\boldsymbol{\sigma}^{\mathrm{turb}} + \rho^{\mathrm{f}}\boldsymbol{b}^{\mathrm{f}}, \quad \text{on } \Omega^{\mathrm{f}}(t)\times[0,T], \tag{10.1}$$

$$\nabla\cdot\overline{\boldsymbol{v}}^{\mathrm{f}} = 0, \quad \text{on } \Omega^{\mathrm{f}}(t)\times[0,T], \tag{10.2}$$

where all the primitive variables (fluid velocity and pressure) are filtered or Reynolds averaged (depending on either LES or hybrid RANS/LES model) and $\boldsymbol{\sigma}^{\mathrm{turb}}$ is the turbulent stress tensor. For LES, the expression for the turbulent stress tensor is given by $\boldsymbol{\sigma}^{\mathrm{sgs}}$ and for DDES, it is written as $\boldsymbol{\sigma}^{\mathrm{ddes}} = \mu_T^{\mathrm{f}}(\nabla\overline{\boldsymbol{v}}^{\mathrm{f}} + (\nabla\overline{\boldsymbol{v}}^{\mathrm{f}})^T)$, with μ_T^{f} representing the turbulent dynamic viscosity. The closure problem for computing the turbulent dynamic viscosity in DDES is resolved by solving the Spalart-Allmaras one-equation turbulence model for the turbulent kinematic viscosity, $\nu_T^{\mathrm{f}} = \mu_T^{\mathrm{f}}/\rho^{\mathrm{f}}$ in DDES. We now discuss the formulation of the different turbulence models and their applications.

10.2 Dynamic Subgrid-Scale Model

The mixed SGS-based model relies on the standard explicit filtering based on scale separation that reduces the computational cost compared to the direct numerical simulation [187]. The dynamic subgrid scale model is based on modeling the scales smaller than the spatial resolution Δ after spatial filtering of the incompressible Navier-Stokes equations. The spatial filtering introduces an extra subgrid-scale stress term, $\sigma_{ij}^{\text{sgs}} = \overline{u_i^{\text{f}} u_j^{\text{f}}} - \overline{u}_i \overline{u}_j^{\text{f}}$ in the filtered Navier-Stokes equations (Eq. (10.1)), where $\overline{u_i}^{\text{f}}$ denotes the component of the filtered fluid velocity $\overline{\boldsymbol{v}}^{\text{f}}$. The stress needs to be modeled as the term $\overline{u}_i^{\text{f}} \overline{u}_j^{\text{f}}$ consists of unknown quantities u_i^{f} and u_j^{f}. The nonlinear SGS stress tensor can be expressed as [71, 232]

$$\sigma_{ij}^{\text{sgs}} - \frac{\delta_{ij}}{3}\sigma_{kk}^{\text{sgs}} \approx -2\mu_t \overline{S}_{ij} - C_{NL} 6\mu_t^2/\sigma_{kk}^{\text{sgs}} \left(\overline{S}_{ik}\overline{\Omega}_{kj} + \overline{S}_{jk}\overline{\Omega}_{ki} - 2\overline{S}_{ik}\overline{S}_{kj} + \frac{2}{3}\overline{S}_{nk}\overline{S}_{kn}\delta_{ij} \right), \tag{10.3}$$

where $\mu_t = \rho^{\text{f}} (C_s \overline{\Delta})^2 |\overline{S}|$ denotes the dynamic eddy viscosity, $\overline{S}_{ij} \equiv \frac{1}{2}\left(\frac{\partial \overline{u}_i^{\text{f}}}{\partial x_j} + \frac{\partial \overline{u}_j^{\text{f}}}{\partial x_i} \right)$ is the resolved strain-rate tensor with the norm $|\overline{S}| \equiv (2\overline{S}_{ij}\overline{S}_{ij})^{1/2}$, and $\overline{\Omega}_{ij} \equiv \frac{1}{2}\left(\frac{\partial \overline{u}_i^{\text{f}}}{\partial x_j} - \frac{\partial \overline{u}_j^{\text{f}}}{\partial x_i} \right)$ represents the filtered rate-of-rotation tensor. Assuming a local equilibrium between the dissipation and production of energy, a theoretical value of $C_s \approx 0.17$ can be derived for the Smagorinsky model [194]. Equation (10.3) recovers to the standard linear SGS stress model for $C_{NL} = 0$.

In the dynamic SGS model [76], two filters are then defined: the grid filter and the test filter with scale dimension Δ and $\widehat{\Delta}$ respectively. The former is naturally provided by the resolution of the grid and the latter (denoted by $\widehat{()}$) can be any coarser level filter [187]. Subgrid-scale stresses generated by different filters are related by the following [76]:

$$L_{ij} = T_{ij} - \sigma_{ij}^{\text{sgs}}, \tag{10.4}$$

where L_{ij} is the Leonard tensor generated by performing the test filter on the grid filtered data as

$$L_{ij} = \widehat{\overline{u}_i^{\text{f}} \overline{u}_j^{\text{f}}} - \widehat{\overline{u}}_i^{\text{f}} \widehat{\overline{u}}_j^{\text{f}} \tag{10.5}$$

and T_{ij} is the stress at the test level,

$$T_{ij} = \widehat{\overline{u_i^{\text{f}} u_j^{\text{f}}}} - \widehat{\overline{u}}_i^{\text{f}} \widehat{\overline{u}}_j^{\text{f}}. \tag{10.6}$$

As there is no assumption regarding the modeling of the turbulent stresses in Eqs. (10.5) and (10.6), any dynamic procedure can be developed for evaluating the subrid-scale stresses.

The relation for the Leonard stress can be written, by considering the Smagorinsky eddy-viscosity model for the unknown stresses $\boldsymbol{\sigma}^{\text{sgs}}$ and $\boldsymbol{T}$, as

$$L_{ij} = -2C_s^2 \left(\widehat{\Delta}^2 |\widehat{\overline{S}}| \widehat{\overline{S}}_{ij} - \Delta^2 |\overline{S}| \overline{S}_{ij} \right) \tag{10.7}$$

The Lilly determination [137] of the Smagorinsky constant leads to the closure

$$C_{ds} = \frac{1}{2}\frac{\langle M_{ij}L_{ij}\rangle}{\langle M_{lk}M_{lk}\rangle} \tag{10.8}$$

where $M_{ij} \equiv \left(\widehat{\Delta}^2|\widehat{\overline{S}}|\widehat{\overline{S}}_{ij} - \Delta^2|\overline{S}|\overline{S}_{ij}\right)$ and $\langle\cdot\rangle$ indicates some type of smoothing process such as averaging, planar averaging or Lagrangian averaging and C_{ds} denotes the dynamic coefficient as opposed to a constant C_s in Eq. (10.3). For $C_{ds} > 0$, we take $C_s = \sqrt{C_{ds}}$, or otherwise $C_s = 0$.

For the computing of the element residuals and the Jacobian matrices, the eddy viscosity values at the nodes are interpolated to the quadrature points. Additional test-filtered data for $\widehat{\overline{S}}$ and $\widehat{|\overline{S}|\overline{S}_{ij}}$ is required for the Smagorinsky model. Therefore, L_2-projection of the required data from the quadrature to the nodal points is carried out. For example, the L_2-projection of the filtered quantity ψ at the nodal point a (i.e., the test-filtered data) is given by

$$\widehat{\psi}_a = \frac{\sum_e \int_{\Omega_e} N_a \psi \, d\Omega}{\sum_e \int_{\Omega_e} N_a \, d\Omega} \tag{10.9}$$

The above projection is carried out using the lumped mass matrix technique [95]. We loop through the elements and quadrature points, evaluate the variables, assemble them in a global vector and scale them by a lumped mass matrix, to construct a smooth filtered set of variables. To ensure a reasonable behavior along the wall due to the inability of the LES to resolve the viscous sublayer, a simple algebraic eddy viscosity model [13, 171] is utilized: $\frac{\mu_t}{\mu^{\mathrm{f}}} = \kappa y_w^+ (1 - e^{-y_w^+}/A)^2$, where $y^+ = y_w u_\tau/\nu$ is the non-dimensional distance to the wall based on the instantaneous friction velocity u_τ, κ is the model coefficient and the constant $A = 19$. Further details can be found in [108].

A variational multiscale (VMS) framework was introduced in [94, 100] as an alternative to the dynamic Smagorinsky model. The VMS technique relies on the separation of the resolved scales into a large- and small-scales and rationalizes the stabilization methods to circumvent numerical difficulties related to the standard Galerkin technique. This multiscale LES involves the decomposition of the turbulent stress tensor into large- and small-scale parts. Various variants of the VMS method are assessed and found to be accurate as the standard dynamic model for the turbulent channel problem in [230]. Although the VMS technique provides flexibility and general framework for practical geometries and complex flows using finite element [145], isogeometric [22] and discontinuous Galerkin methods, an extensive assessment and comparison of the dynamic LES with the VMS method is beyond the scope of the current text.

10.2.1 Stationary and Vibrating Square Cylinder at Moderate Reynolds Number

In this section, we validate the dynamic subgrid-scale LES model mentioned in the previous section. To accomplish this, we consider benchmark problems of vortex-shedding phenomenon for an incompressible flow across a square cylinder at high Reynolds number. This configuration constitutes the scenarios of hydrodynamic loading and vibration behavior of the structures found in offshore engineering. We present stationary and the vibrating cases for three-dimensional interaction of the incompressible flow with a square cylinder.

10.2.1.1 LES Validation for Stationary Square Cylinder

We begin with the validation of the dynamic subgrid-scale model by considering the flow past a stationary square cylinder at the Reynolds number $Re = 22{,}000$ against the experiment [143] and other numerical techniques.

Figure 10.1a depicts the computational domain of the problem set-up. The problem consists of a square cylinder of diameter D (cylinder axis parallel to the Z-axis) with the inlet and outlet boundaries at a distance of $20D$ and $40D$ from the center of the cylinder respectively. A blockage of 2.5% is considered as the side walls are at a distance of $40D$ from each other and are equidistant from the center of the cylinder. A uniform freestream velocity of U is specified at the inlet along the X-axis. Slip condition is satisfied on the side walls while a no-slip condition is applied on the cylinder wall. The computational domain is discretized using unstructured finite elements consisting of 1,060,818 nodes and 3,664,114 six-node wedge elements. The mesh nodes are distributed uniformly along the span of the cylinder with a distance of $\Delta z = 0.156D$ between two points. The mesh is shown in Fig. 10.1b where the square cylinder is surrounded by a finer square-shaped grid to capture the vortex shedding and a $y^+ < 1$ is maintained in the boundary layer. The selection of the time step Δt is crucial for the capturing of forces and turbulent statistics. For this case, a $\Delta t = 0.025$ is selected.

The turbulent flow over the square cylinder is modeled with the help of the dynamic SGS model discussed in Sect. 10.2. The hydrodynamic coefficients and other data obtained from the results are compared with that of the experiment and other numerical studies in Table 10.1. It is observed that the quantities such as Strouhal number St, the mean and fluctuating lift and drag coefficients and the mean recirculation l_R agree well with the experimental results. Moreover, the exclusion of the dynamic SGS for the other methods resulted in poor prediction of the Strouhal frequency St, due to the neglection of the energy fluctuation of the high frequency subgrid-scale vortices.

The capability of the SGS model to capture the fine details of the turbulence in the flow can be observed in Fig. 10.2. The Z-vorticity distribution in the mid-plane is shown in Fig. 10.2a indicating the size of the wake generated by the stationary

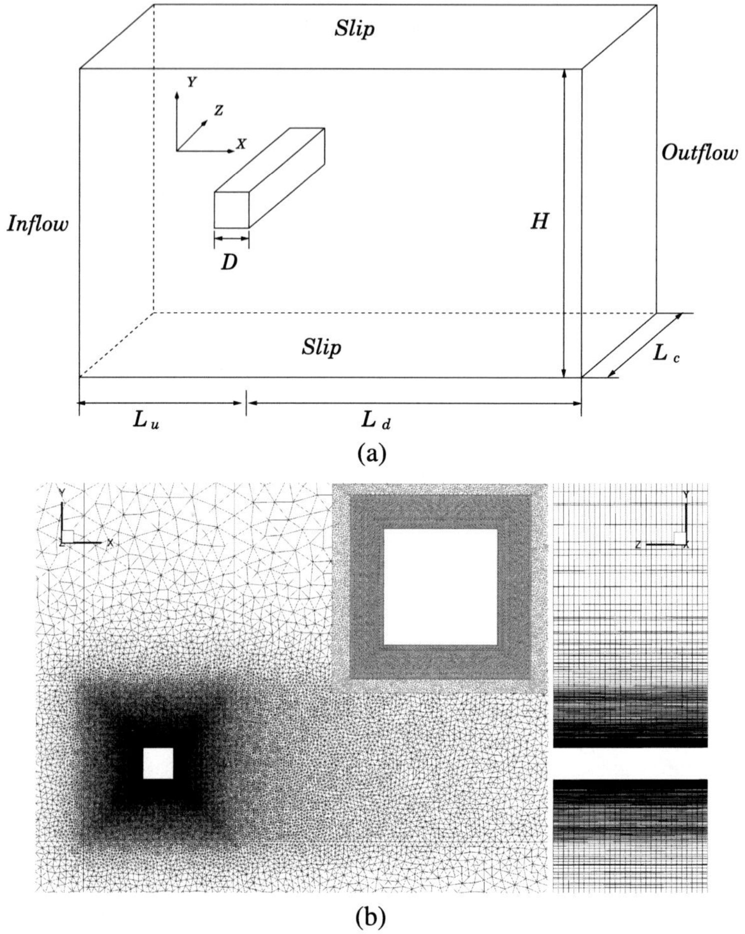

Fig. 10.1 Stationary square cylinder configuration for dynamic SGS validation: **a** schematic representation of the fluid domain (top and bottom planes are given periodic boundary condition), **b** mesh in a cross-sectional plane with a close-up view of the boundary layer and the sectional view of the mesh in the spanwise direction (right side)

cylinder. The dynamic model is seen to be clearly resolving the opposite signed vortex islands within the traditional von-Karman vortices. The full three-dimensional vortical structures of the flow in the wake can be observed by visualizing the Q-criterion in Fig. 10.2b. The structures are observed to be consistently distributed in the streamwise direction. In the spanwise direction, structures of the order of cylinder diameter are sufficiently resolved along the spanwise length. It is well known that the turbulence consists of a wide range of spatial and temporal scales. The fluctuation of energy is generated at the larger eddies associated with low wave numbers. These large eddies are transformed into smaller eddies through vortex

Table 10.1 Comparison of the present simulation results with other numerical and experimental studies for flow around stationary square cylinder at $Re = 22{,}000$ (NLES—Non-linear LES, AVM^4 - algebraic variational multiscale-multigrid-multifractal method)

Studies	St	C_D^{mean}	C_D^{rms}	C_L^{rms}	l_R
Experimental [143]	0.132	1.9–2.2		1.38	1.68
DNS [226]	0.13	2.09	0.18	1.45	
Smagorinsky [180]	0.13	2.30	0.14	1.15	1.46
Dynamic SGS [234]	0.13	2.17	0.18	1.29	1.20
2D LES [28]	0.134	2.18		1.62	
NLES [144]	0.115	2.1627	0.259	1.14	1.10
AVM^4 [175]	0.15	2.28	0.15	1.17	1.32
Present dynamic SGS	0.122	2.0766	0.190	1.460	1.46

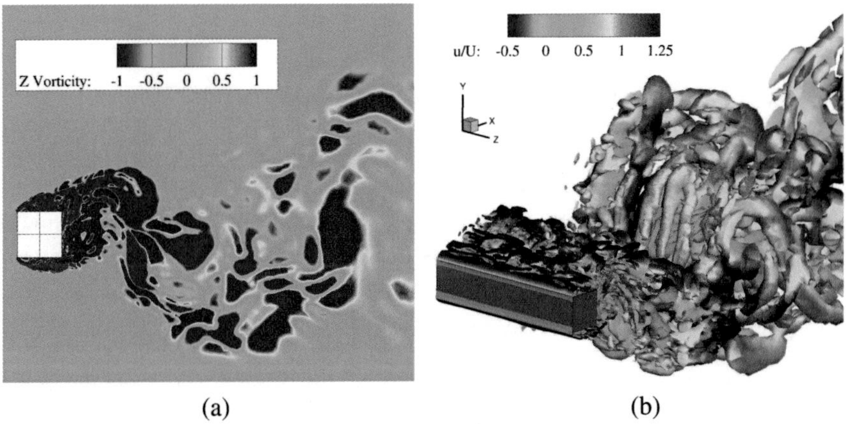

Fig. 10.2 Representative results of dynamic SGS for $Re = 22{,}000$ flow over 5D span of a square cylinder: **a** vorticity distribution of the center XY plane **b** wake structures (iso-surfaces of $Q^+ = 0.25$ colored by normalized streamwise velocity u/U)

stretching mechanism and thus, the energy cascades down to higher wave numbers in the inertial spectrum. In this range, the slope of the energy spectrum remains constant and is found to be a function of the wave number. The single-sided spectra of the transverse velocity fluctuations in several downstream locations along the centerline of the wake of the cylinder is plotted in Fig. 10.3. The vortex-shedding frequency is observed to be $f/f_{vs} = 1.0$ and the inertial region shows a very good agreement with the Kolmogorov's $-5/3$ rule.

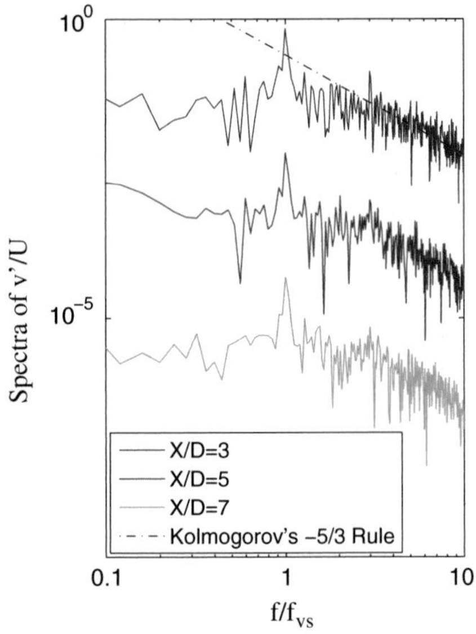

Fig. 10.3 Energy spectra of the transverse velocity fluctuations for three downstream locations (X/D=3 is not shifted. Other spectra are shifted for clarity)

10.2.1.2 Vibrating Square Cylinder at Moderate Reynolds Number

Continuing the validation study for the dynamic SGS model, we consider the coupling between the incompressible fluid flow and the spring-mounted rigid square cylindrical structure in this section. The square cylinder is mounted on a spring and is allowed to vibrate in the transverse direction. The investigation is carried out for two cases reported in the experiments [247]: (a) zero incidence angle, $Re = 11182$, $U_r = 14.26$ and $m^* = 2.64$, (b) 45° incidence angle, $Re = 2572$, $U_r = 3.28$ and $m^* = 2.64$, where U_r and m^* are the reduced velocity and mass ratio, respectively. The low mass ratio regime in these cases require the non-linear iterative force correction (NIFC) algorithm (discussed in Chap. 7) for the stability in the fluid-structure interaction problem. The computational domain is similar to that in the previous section (Fig. 10.1a) and the same discretized mesh is utilized for the 0° incidence case. For the 45° incidence, a circular cylindrical refined region instead of a square one surrounds the square cylinder.

The time histories and frequency spectra of the fluid forces and the transverse displacement of the square cylinder are compared with that of the experiment for the 0° incidence in Fig. 10.4 where a good agreement is observed. The transverse displacement is found to have a single dominant frequency while the transverse force has two dominating frequencies, as observed in the experiment. The computation has also captured the high frequency small amplitude local minima and maxima in C_L due to a sampling rate of 40 Hz compared to the 4 Hz measurements in the experiment.

The flow structures in the wake for the two cases at three locations corresponding to the peak positive, mean and peak negative positions of the vibrating cylinder are

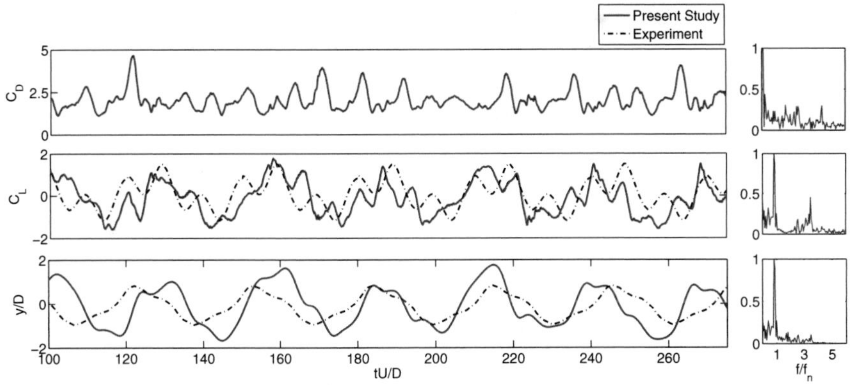

Fig. 10.4 Comparison of the time histories (left) and single-sided spectra of the vibration responses (right) with experimental results [247]. Response corresponds to 1-DOF vibration of square cylinder in a flow with Re = 11,182, $m^* = 2.64$, $U_r = 14.26$ at zero incidence angle

shown in Fig. 10.5. The streamwise vortex ribs are observed to be pointing somewhat vertical for 0° incidence but are stretched horizontally for the 45° incidence. The streamwise vortices are more aligned in the 45° incidence compared to the zero incidence case. The formation and release of a hairpin-like vortical structure in the near-wake region for the 0° incidence case is observed which is found to be similar to the observations in the near-wall turbulence experiments and DNS studies in the literature. These vortices can be associated with the continuous interaction of the streamwise vortices with the free shear layers of the vibrating cylinder. The vortex streets for the two cases are shown in Fig. 10.6 where a 2S vortex shedding mode is observed which is consistent with that of the experiment.

The hydrodynamic coefficients and other relevant quantities consisting of Strouhal frequency, the force coefficients and the non-dimensional vibrational amplitude are quantified in Table 10.2 and compared to the results from the literature. The results agree well with the experimental data for both 0° (galloping dominated) and 45° (VIV dominated) incidences. The correct predictions of these complex fluid-elastic instabilities help in the assessment of the reliability and accuracy of the dynamic SGS turbulence model.

10.2.2 Application to Semi-Submersible Floater

Next, we demonstrate the dynamic SGS turbulence model by considering a fluid-structure interaction application to the problem of flow-induced vibrations of a three-dimensional multicolumn semi-submersible floater subjected to a uniform flow at Re = 20,000. The geometry parameters of the semi-submersible model are adopted from [231]. The computational domain and the floater geometry are shown in Fig. 10.7a

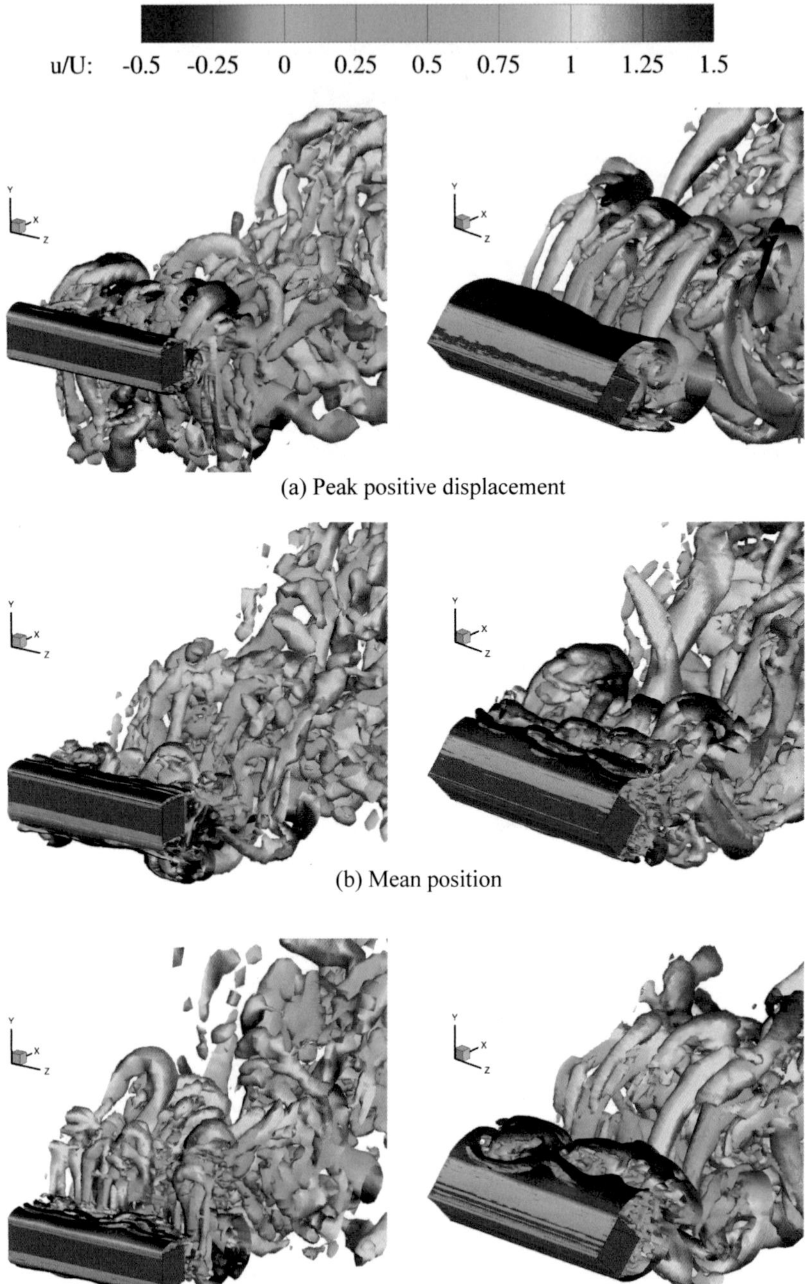

(a) Peak positive displacement

(b) Mean position

(c) Peak negative displacement

Fig. 10.5 Vortical wake structures for zero incidence (left) and 45° incidence (right). Iso-surfaces are of $Q^+ = 0.25$ and colored by u/U). Unsteady hairpin structure can be clearly observed for zero incidence and peak negative displacement

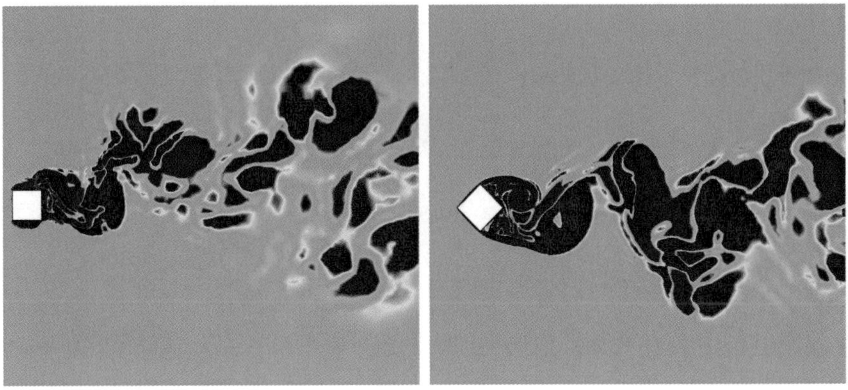

(a) (b)

Fig. 10.6 Spanwise Z-vorticity contours of the vortex wake modes: **a** zero incidence **b** 45° incidence case

Table 10.2 Comparison of transversely vibrating cylinder simulation results with experimental data

Quantity	0° incidence at $Re = 11{,}182, m^* = 2.64, U_r = 14.26$		45° incidence at $Re = 2572, m^* = 2.64, U_r = 3.28$	
	Experiment [247]	Present	Experiment [247]	Present
C_L^{rms}	0.762	0.816	1.98	1.738
C_D^{max}	4.00	4.663	–	2.0558
St	0.79	0.814	0.95	1.1719
A_y/D	0.96	1.0329	0.40	0.409

and b, respectively. The computational model is scaled to 1:70 to the real operational semi-submersible with the sharp cornered columns. It consists of four columns with four pontoons in a full-load and deep draft condition. The structure is elastically mounted and is free to vibrate in both in-line and transverse directions. The test model in [231] consisted of 3 low friction air bearings that slid along a horizontal plate mounted to the carriage. In the demonstration, detailed free decay tests are conducted in sway, surge and yaw directions to ensure that the friction of the air bearing is less than 1% of the critical damping. A vertical pretension is applied to adjust the model into the water, without restricting the horizontal motions.

The computational domain consists of columns of characteristic diameter D. The inlet and outlet boundaries are $20D$ and $40D$ from the center of the semi-submersible respectively. The side walls are at a distance of $40D$ from each other and are equidistant from the center of the semi-submersible. The bottom boundary is $20D$ from the top water level. At the inlet, a freestream velocity is applied, while slip boundary conditions are imposed at the top, bottom and side boundaries. A stress-free boundary is satisfied at the outlet and a no-slip condition is applied on the

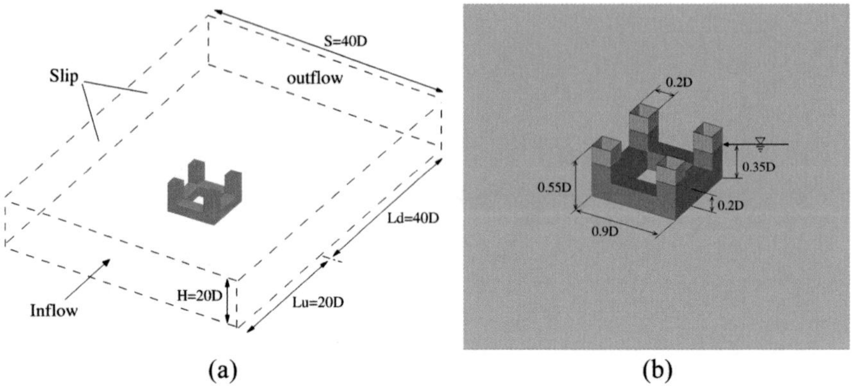

Fig. 10.7 Flow past a deep draft semi-submersible floater **a** full domain (not to scale) **b** detailed dimensions of column and pontoon. The water level is shown in the right figure

surfaces of the semi-submersible structure. The Reynolds number is based on D, the kinematic viscosity of the fluid and the freestream velocity, U. As shown in Fig. 10.8, a three-dimensional unstructured mesh consisting of 1.7 million nodes and 10 million linear tetrahedral elements, is constructed using the open-source mesh generator [77]. The mesh is formed such that a reasonable resolution and isotropic LES-type grid is maintained around the column and pontoon surfaces. The distance between the column surface and the closest first grid line is $y_w = 0.03D$. The edge of the two-dimensional cross-section of the column consists of 30 grid points and there are 25 and 30 layers of grid resolution along the column and pontoon length, respectively. A time step size of $\Delta t = 0.10$ is chosen for the demonstration. Reynolds number of $Re = 20,000$ with mass ratio $m^* = 0.83$ at zero incidence and the damping ratio $\zeta = 0, 0.01$ for varying reduced velocity U_r are considered for the computation.

The instantaneous streamwise velocity and the pressure fields across different cross-sections of the pontoons and the columns of the semi-submersible are shown in Fig. 10.9. The flow patterns seem symmetrical about the centerline and the three-dimensionality and shedding vortices convey fresh fluid into the wake region, magnifying the flow entrainment between the core and the wake regions of the semi-submersible structure.

The instantaneous spanwise vorticity field at horizontal and vertical cross-sectional planes is shown in Fig. 10.10. Concentrated vorticity layers are shed from the corner of the square columns, with negative and positive vorticities colored by blue and red respectively. For the VIV mode, the spanwise vorticity at the pontoon level shows some symmetry along the gap axis, as expected. Vortex streets generated by the four columns and their interaction is depicted in Fig. 10.10a (right). The streamwise vorticity at the front vertical and the mid-plane section is shown in Fig. 10.10b.

The response statistics from the transient data are quantified as: (a) nominal response

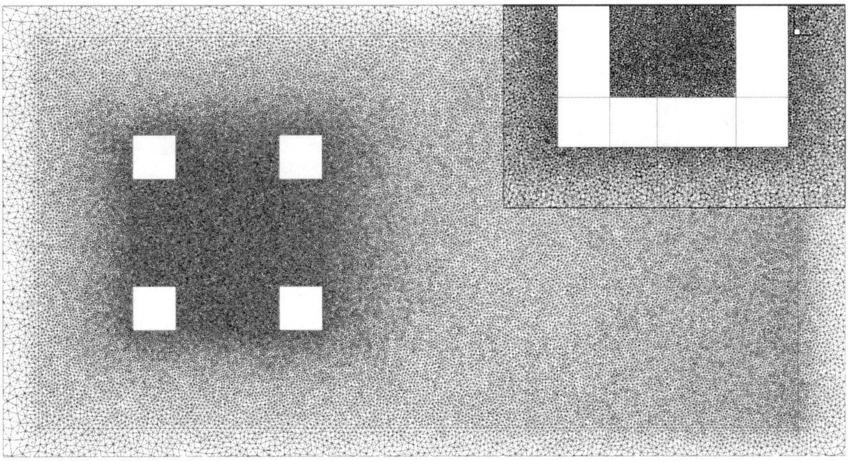

Fig. 10.8 3D unstructured mesh for semi-submersible platform with a close-up view in wake region and cross-sectional view at the mid-plane is shown in the right corner inset

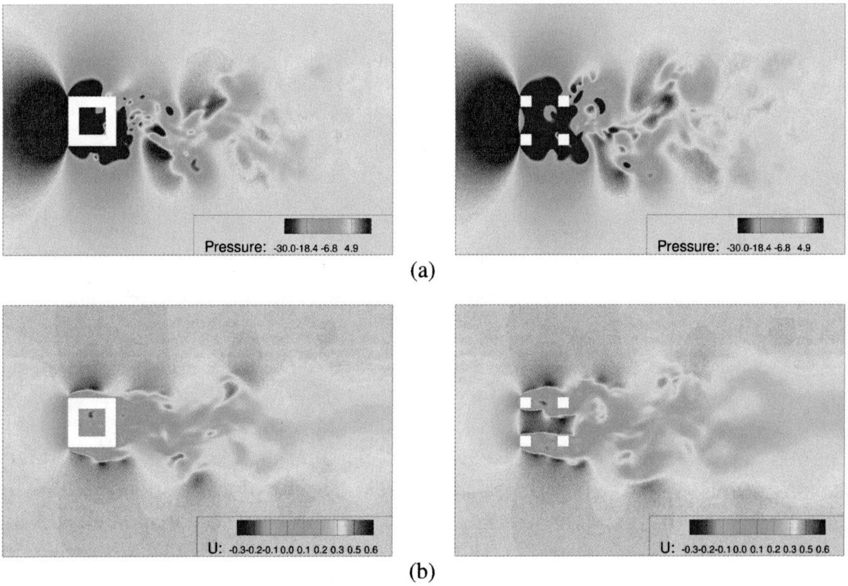

Fig. 10.9 Instantaneous flow fields at pontoon and columns levels: **a** pressure field **b** streamwise velocity

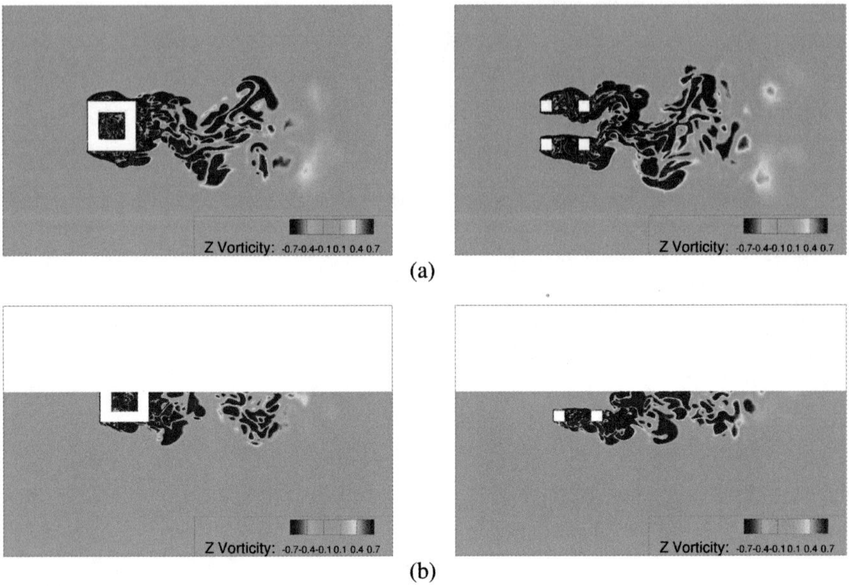

Fig. 10.10 Instantaneous spanwise vorticity fields for the semi-submersible at cross-sectional planes of pontoons and columns: **a** horizontal (spanwise) sections of the pontoon (left) and top column (right); **b** vertical sections at the front (left) and mid-section (right)

$$\left(\frac{A}{D}\right)_{\text{nomimal}} = \frac{\sqrt{2} \times \text{std}(Y(t))}{D}, \tag{10.10}$$

where std denotes the standard deviation of $Y(t)$; and (b) maximum response

$$\left(\frac{A}{D}\right)_{\text{maximum}} = \frac{\max(Y(t)) - \min(Y(t))}{2D}. \tag{10.11}$$

The nominal response shows the averaged response amplitude which can be useful for the fatigue analysis of the moorings and the risers. The maximum response deals with the highest maximum and the lowest minimum excursion. The statistical results are compared to that of the experimental data in Table 10.3. The results agree well with the experimental observations. Compared to the experimental data, the difference is about 6% and 1% for the maximum transverse response and Strouhal frequency respectively.

The relationship between the reduced velocity and the transverse response is shown in Fig. 10.11. The "lock-in" region occurs between $U_r \in [7, 15]$, leading to a change to the galloping mode as the reduced velocity is increased further. As can be seen, the maximum transverse amplitudes agree well with the experimental data. The significance of yaw motion in the galloping behavior of deep draft floaters at high

Table 10.3 Dependence of damping on response amplitudes and frequency at $Re = 20{,}000$, $m^* = 0.83$ and $U_r = 6$

Damping coefficient ζ	A_y/D_{maximum}	A_y/D_{nominal}	St
$\zeta = 0$	0.2351	0.1924	0.1724
$\zeta = 0.01$	0.2239	0.1905	0.1620
$\zeta \approx 0.01$ (experimental)	0.21	–	0.16

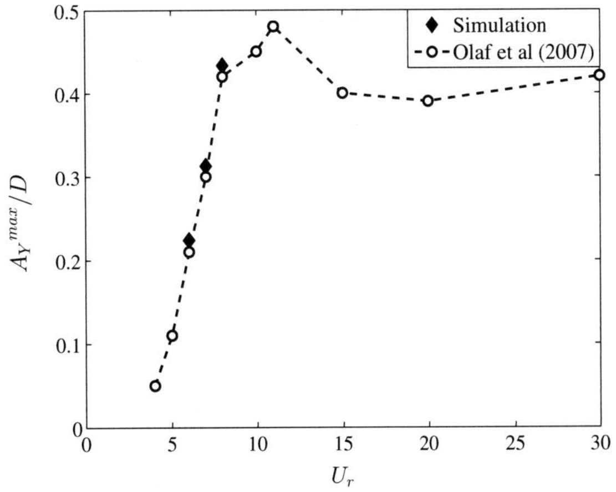

Fig. 10.11 Comparison of transverse amplitude versus reduced velocity between current dynamic SGS simulation and MARIN experiment

reduced velocity was noted in [231], however, such complex coupling between the transverse and yaw motions has not been considered in the present demonstration.

The temporal variation in the force coefficients and the response amplitude are plotted in Figs. 10.12a, b respectively. The frequency of C_L and A_Y and the $X - Y$ trajectory of the floater are shown in Figs. 10.12c, d respectively. The particular case represents a VIV mode at $U_r = 8$. The vortex shedding frequency is synchronized with the natural frequency of the structure. Moreover, the transverse vibration amplitude remains in-phase with the lift coefficient. Therefore, the dynamic SGS turbulence model has been demonstrated for a large-scale offshore floating platform subjected to an incoming ocean current at high Reynolds number.

10.3 Spalart-Allmaras Based Delayed Detached Eddy Simulation

As mentioned before, LES is useful in resolving the separated flow regions. The boundary layer region has to be refined sufficiently to capture the boundary layer effects while utilizing LES modeling. This could be computational expensive.

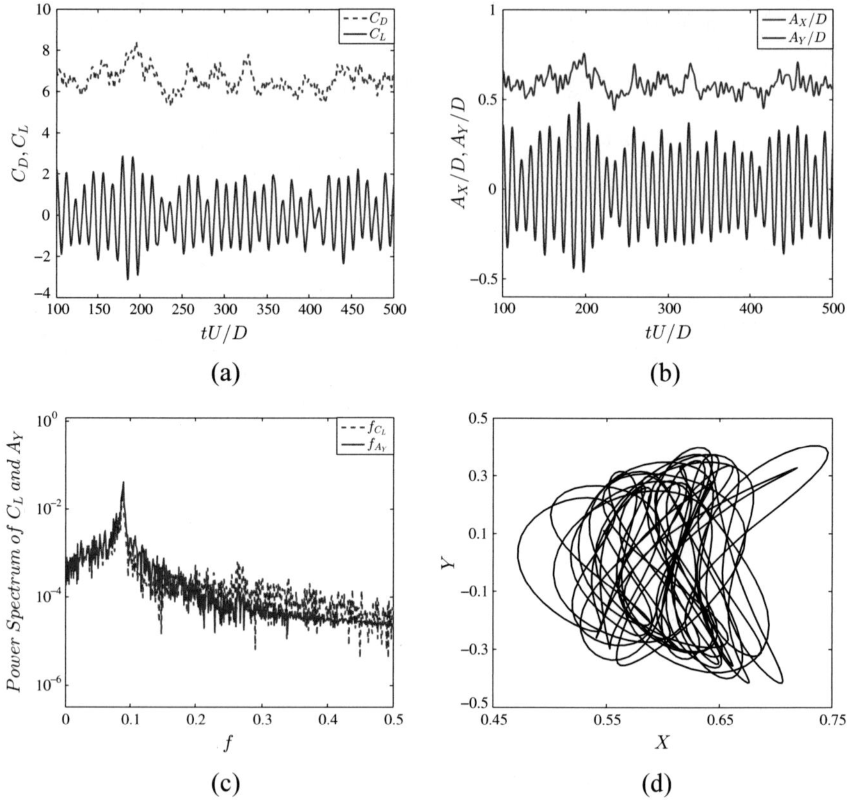

Fig. 10.12 Response characteristics of semi-submersible floater at $U_r = 8$ for $m^* = 0.83$ and $\zeta = 0.01$

Another alternative is the category of hybrid RANS/LES modeling where the model acts as RANS near the boundary layer region and as LES-type filter in the separated flow regions, thus giving the advantages of both RANS and LES and reducing the overall computing cost. In this section, we formulate a hybrid RANS/LES model known as delayed detached eddy simulation based on the Spalart-Allmaras (S-A) one-equation model. We first formulate the equation and briefly discuss its variational form. Then, the coupled equations are validated by flow across a stationary circular cylinder at Re = 3900 and Re = 140,000 followed by a demonstration to a large-scale application of uniform flow across an offshore riser leading to vortex-induced vibration.

10.3.1 Formulation

The challenges in the one-equation S-A model have been discussed in the introduction. We write the equation to be solved for the closure of turbulence modeling and then apply the positivity preserving variational technique discussed in Chap. 4 to reduce the oscillations that one may find in the numerical solution, thus maintaining positivity and boundedness in the solution.

10.3.1.1 Strong Differential Form

Consider the physical domain $\Omega^{\mathrm{f}}(t)$ with the spatial and temporal coordinates denoted by $\boldsymbol{x}^{\mathrm{f}}$ and t respectively. The S-A one-equation which solves for the eddy viscosity $\tilde{\nu}$ is given as

$$\left.\frac{\partial \tilde{\nu}}{\partial t}\right|_{\chi} + (\overline{\boldsymbol{v}}^{\mathrm{f}} - \boldsymbol{w}) \cdot \nabla \tilde{\nu} = P - D + \frac{1}{\tilde{\sigma}}\big[\nabla \cdot ((\nu^{\mathrm{f}} + \tilde{\nu})\nabla \tilde{\nu}) + c_{b2}(\nabla \tilde{\nu})^2\big], \quad \text{on } \Omega^{\mathrm{f}}(t) \times [0, T], \tag{10.12}$$

$$\tilde{\nu} = \tilde{\nu}_D, \quad \text{on } \Gamma_D^{\tilde{\nu}}(t) \times [0, T], \tag{10.13}$$

$$\nabla \tilde{\nu} \cdot \boldsymbol{n}^{\tilde{\nu}} = \tilde{\nu}_N, \quad \text{on } \Gamma_N^{\tilde{\nu}}(t), \tag{10.14}$$

$$\tilde{\nu} = \tilde{\nu}_0, \quad \text{on } \Omega^{\mathrm{f}}(0), \tag{10.15}$$

where $P = c_{b1}\tilde{S}\tilde{\nu}$ and $D = c_{w1} f_w (\tilde{\nu}/\tilde{d})^2$ are the production and destruction terms, $\tilde{S} = S + (\tilde{\nu}/(\kappa^2 \tilde{d}^2)) f_{v2}$, S being the magnitude of vorticity, $c_{b1}, c_{b2}, \tilde{\sigma}, \kappa, c_{w1}$ and c_{v1} are the constants defined for the turbulence model in [199], $\tilde{d} = d - f_d \max(0, d - C_{DES}\Delta)$, d is the distance to the closest wall, C_{DES} is the DES coefficient and Δ is based on the largest dimension of the grid element. Here, $\Gamma_D^{\tilde{\nu}}(t)$ and $\Gamma_N^{\tilde{\nu}}(t)$ denote the Dirichlet and Neumann boundaries for the eddy viscosity respectively. $\tilde{\nu}_D$ and $\tilde{\nu}_N$ denote the Dirichlet and Neumann conditions on the eddy viscosity respectively, $\boldsymbol{n}^{\tilde{\nu}}$ denotes the unit normal to the Neumann boundary and $\tilde{\nu}_0$ represents the initial condition for the eddy viscosity. Definitions of all the parameters can be found in [113, 199].

The turbulent kinematic viscosity, then can be computed as

$$\nu_T^{\mathrm{f}} = \tilde{\nu} f_{v1}, \qquad f_{v1} = \frac{\tilde{\chi}^3}{\tilde{\chi}^3 + c_{v1}^3}, \qquad \tilde{\chi} = \frac{\tilde{\nu}}{\nu^{\mathrm{f}}}, \tag{10.16}$$

where $\nu^{\mathrm{f}} = \mu^{\mathrm{f}}/\rho^{\mathrm{f}}$ is the molecular kinematic viscosity of the fluid. This turbulent viscosity gives the closure to the turbulence modeling stress term in Eq. (10.1) as $\boldsymbol{\sigma}^{\mathrm{ddes}} = \mu_T^{\mathrm{f}}(\nabla \overline{\boldsymbol{v}}^{\mathrm{f}} + (\nabla \overline{\boldsymbol{v}}^{\mathrm{f}})^T)$.

10.3.1.2 Semi-Discrete Variational Form

We write the variational formulation for the turbulence transport equation under the positivity preserving variational framework discussed in Chap. 4 to reduce the spurious oscillations in the solution corresponding to the eddy viscosity. Before proceeding to the presentation of the spatial PPV-based variational formulation, we first discretize the equation in the time domain via generalied-α time integration [44, 111]. Let $\partial_t \tilde{\nu}^{\mathrm{n}+\alpha_{\mathrm{m}}^{\mathrm{f}}}$ be the derivative of $\tilde{\nu}$ with respect to time at $t^{\mathrm{n}+\alpha^{\mathrm{f}}}$. The expressions for the temporal discretization can be written as:

$$\tilde{\nu}^{\mathrm{n}+1} = \tilde{\nu}^{\mathrm{n}} + \Delta t \partial_t \tilde{\nu}^{\mathrm{n}} + \gamma^{\mathrm{f}} \Delta t (\partial_t \tilde{\nu}^{\mathrm{n}+1} - \partial_t \tilde{\nu}^{\mathrm{n}}), \tag{10.17}$$

$$\partial_t \tilde{\nu}^{\mathrm{n}+\alpha_{\mathrm{m}}^{\mathrm{f}}} = \partial_t \tilde{\nu}^{\mathrm{n}} + \alpha_{\mathrm{m}}^{\mathrm{f}} (\partial_t \tilde{\nu}^{\mathrm{n}+1} - \partial_t \tilde{\nu}^{\mathrm{n}}), \tag{10.18}$$

$$\tilde{\nu}^{\mathrm{n}+\alpha^{\mathrm{f}}} = \tilde{\nu}^{\mathrm{n}} + \alpha^{\mathrm{f}} (\tilde{\nu}^{\mathrm{n}+1} - \tilde{\nu}^{\mathrm{n}}), \tag{10.19}$$

where Δt is the time step size, $\alpha_{\mathrm{m}}^{\mathrm{f}}, \alpha^{\mathrm{f}}$ and γ^{f} are the generalized-α parameters defined in Eq. (5.23). The time discretized turbulence transport equation is therefore,

$$\partial_t \tilde{\nu}^{\mathrm{n}+\alpha_{\mathrm{m}}^{\mathrm{f}}} + (\overline{\boldsymbol{v}}^{\mathrm{f}} - \boldsymbol{w}) \cdot \nabla \tilde{\nu}^{\mathrm{n}+\alpha^{\mathrm{f}}} - c_{b1} \tilde{S} \tilde{\nu}^{\mathrm{n}+\alpha^{\mathrm{f}}} + c_{w1} f_w \frac{(\tilde{\nu}^{\mathrm{n}+\alpha^{\mathrm{f}}})^2}{\tilde{d}^2}$$
$$- \frac{c_{b2}}{\tilde{\sigma}} (\nabla \tilde{\nu}^{\mathrm{n}+\alpha^{\mathrm{f}}})^2 - \frac{1}{\tilde{\sigma}} \nabla \cdot (\nu^{\mathrm{f}} + \tilde{\nu}^{\mathrm{n}+\alpha^{\mathrm{f}}}) \nabla \tilde{\nu}^{\mathrm{n}+\alpha^{\mathrm{f}}} = 0. \tag{10.20}$$

Equation (10.20) can be expressed in the form of a convection-diffusion-reaction equation as follows:

$$\partial_t \tilde{\nu}^{\mathrm{n}+\alpha_{\mathrm{m}}^{\mathrm{f}}} + \tilde{\boldsymbol{v}} \cdot \nabla \tilde{\nu}^{\mathrm{n}+\alpha^{\mathrm{f}}} - \nabla \cdot (\tilde{k} \nabla \tilde{\nu}^{\mathrm{n}+\alpha^{\mathrm{f}}}) + \tilde{s} \tilde{\nu}^{\mathrm{n}+\alpha^{\mathrm{f}}} = 0, \text{ on } \Omega^{\mathrm{f}}(t), \tag{10.21}$$

where $\tilde{\boldsymbol{v}}$, $\tilde{k}$ and $\tilde{s}$ are the modified convection velocity, diffusion coefficient and reaction coefficient respectively, which are given as

$$\tilde{\boldsymbol{v}} = \overline{\boldsymbol{v}}^{\mathrm{f}} - \boldsymbol{w} - \frac{c_{b2}}{\tilde{\sigma}} \nabla \tilde{\nu}^{\mathrm{n}+\alpha^{\mathrm{f}}}, \qquad \tilde{k} = \frac{1}{\tilde{\sigma}} (\nu^{\mathrm{f}} + \tilde{\nu}^{\mathrm{n}+\alpha^{\mathrm{f}}}), \qquad \tilde{s} = -c_{b1} \tilde{S} + c_{w1} f_w \frac{\tilde{\nu}^{\mathrm{n}+\alpha^{\mathrm{f}}}}{\tilde{d}^2}. \tag{10.22}$$

Defining the space of trial solution as $\mathcal{S}_{\tilde{\nu}}^{\mathrm{h}}$ and that of the test function as $\mathcal{V}_{\tilde{\nu}}^{\mathrm{h}}$ such that

$$\mathcal{S}_{\tilde{\nu}}^{\mathrm{h}} = \left\{ \tilde{\nu}_{\mathrm{h}} \mid \tilde{\nu}_{\mathrm{h}} \in H^1(\Omega^{\mathrm{f}}(t)),\ \tilde{\nu}_{\mathrm{h}} = \tilde{\nu}_D \text{ on } \Gamma_D^{\tilde{\nu}}(t) \right\}, \tag{10.23}$$

$$\mathcal{V}_{\tilde{\nu}}^{\mathrm{h}} = \left\{ \tilde{w}_{\mathrm{h}} \mid \tilde{w}_{\mathrm{h}} \in H^1(\Omega^{\mathrm{f}}(t)),\ \tilde{w}_{\mathrm{h}} = 0 \text{ on } \Gamma_D^{\tilde{\nu}}(t) \right\}, \tag{10.24}$$

the variational statement for the turbulence transport equation employing the positivity preserving variational technique is given as: find $\tilde{\nu}_{\mathrm{h}}(\boldsymbol{x}^{\mathrm{f}}, t^{\mathrm{n}+\alpha^{\mathrm{f}}}) \in \mathcal{S}_{\tilde{\nu}}^{\mathrm{h}}$ such that $\forall \tilde{w}_{\mathrm{h}} \in \mathcal{V}_{\tilde{\nu}}^{\mathrm{h}}$,

$$\begin{aligned}
&\int_{\Omega^{\mathrm{f}}(t)} \left(\tilde{w}_{\mathrm{h}} \partial_t \tilde{\nu}_{\mathrm{h}} + \tilde{w}_{\mathrm{h}} (\tilde{\boldsymbol{v}} \cdot \nabla \tilde{\nu}_{\mathrm{h}}) + \nabla \tilde{w}_{\mathrm{h}} \cdot (\tilde{k} \nabla \tilde{\nu}_{\mathrm{h}}) + \tilde{w}_{\mathrm{h}} \tilde{s} \tilde{\nu}_{\mathrm{h}} \right) d\Omega \\
&+ \sum_{\mathrm{e}=1}^{\mathrm{n_{el}}} \int_{\Omega^{\mathrm{e}}} \Big(\big(\tilde{\boldsymbol{v}} \cdot \nabla \tilde{w}_{\mathrm{h}} + |\tilde{s}| \tilde{w}_{\mathrm{h}} \big) \tau_{\mathrm{t}} \big(\partial_t \tilde{\nu}_{\mathrm{h}} + \tilde{\boldsymbol{v}} \cdot \nabla \tilde{\nu}_{\mathrm{h}} - \nabla \cdot (\tilde{k} \nabla \tilde{\nu}_{\mathrm{h}}) + \tilde{s} \tilde{\nu}_{\mathrm{h}} \big) \Big) d\Omega^{\mathrm{e}} \\
&+ \sum_{\mathrm{e}=1}^{\mathrm{n_{el}}} \int_{\Omega^{\mathrm{e}}} \chi \frac{|\mathcal{R}(\tilde{\nu}_{\mathrm{h}})|}{|\nabla \tilde{\nu}_{\mathrm{h}}|} k_s^{\mathrm{add}} \nabla \tilde{w}_{\mathrm{h}} \cdot \left(\frac{\tilde{\boldsymbol{v}} \otimes \tilde{\boldsymbol{v}}}{|\tilde{\boldsymbol{v}}|^2} \right) \cdot \nabla \tilde{\nu}_{\mathrm{h}} d\Omega^{\mathrm{e}} \\
&+ \sum_{\mathrm{e}=1}^{\mathrm{n_{el}}} \int_{\Omega^{\mathrm{e}}} \chi \frac{|\mathcal{R}(\tilde{\nu}_{\mathrm{h}})|}{|\nabla \tilde{\nu}_{\mathrm{h}}|} k_c^{\mathrm{add}} \nabla \tilde{w}_{\mathrm{h}} \cdot \left(\boldsymbol{I} - \frac{\tilde{\boldsymbol{v}} \otimes \tilde{\boldsymbol{v}}}{|\tilde{\boldsymbol{v}}|^2} \right) \cdot \nabla \tilde{\nu}_{\mathrm{h}} d\Omega^{\mathrm{e}} = \int_{\Gamma_N^{\tilde{\nu}}(t)} \tilde{w}_{\mathrm{h}} \tilde{\nu}_N d\Gamma,
\end{aligned} \tag{10.25}$$

where the first and the second lines represent the Galerkin and linear stabilization terms respectively. The linear stabilization of PPV behaves as Galerkin/least-squares when $\tilde{s} > 0$ and as subgrid-scale when $\tilde{s} < 0$. The third and fourth lines denote the nonlinear PPV stabilization terms in the streamline and crosswind directions respectively. The residual of the equation is given by

$$\mathcal{R}(\tilde{\nu}_{\mathrm{h}}) = \partial_t \tilde{\nu}_{\mathrm{h}} + \tilde{\boldsymbol{v}} \cdot \nabla \tilde{\nu}_{\mathrm{h}} - \nabla \cdot (\tilde{k} \nabla \tilde{\nu}_{\mathrm{h}}) + \tilde{s} \tilde{\nu}_{\mathrm{h}}. \tag{10.26}$$

The linear stabilization parameter τ_{t} is expressed as:

$$\tau_{\mathrm{t}} = \left[\left(\frac{2}{\Delta t} \right)^2 + \tilde{\boldsymbol{v}} \cdot \boldsymbol{G} \tilde{\boldsymbol{v}} + C_I \tilde{k}^2 \boldsymbol{G} : \boldsymbol{G} + \tilde{s}^2 \right]^{-1/2}, \tag{10.27}$$

where C_I and $\boldsymbol{G}$ are the constant obtained from the inverse estimate and the contravariant tensor defined in Eq. (5.40), respectively.

The PPV parameters for the current context of the turbulence transport equation are given as:

$$\chi = \frac{2}{|\tilde{s}|h + 2|\tilde{\boldsymbol{v}}|}, \tag{10.28}$$

$$k_{\mathrm{s}}^{\mathrm{add}} = \max\left\{\frac{||\tilde{\boldsymbol{v}}| - \tau_{\mathrm{t}}|\tilde{\boldsymbol{v}}|\tilde{s} + \tau_{\mathrm{t}}|\tilde{\boldsymbol{v}}||\tilde{s}||h}{2} - (\tilde{k} + \tau_{\mathrm{t}}|\tilde{\boldsymbol{v}}|^2) + \frac{(\tilde{s} + \tau_{\mathrm{t}}\tilde{s}|\tilde{s}|)h^2}{6}, 0\right\}, \tag{10.29}$$

$$k_{\mathrm{c}}^{\mathrm{add}} = \max\left\{\frac{||\tilde{\boldsymbol{v}}| + \tau_{\mathrm{t}}|\tilde{\boldsymbol{v}}||\tilde{s}||h}{2} - \tilde{k} + \frac{(\tilde{s} + \tau_{\mathrm{t}}\tilde{s}|\tilde{s}|)h^2}{6}, 0\right\}, \tag{10.30}$$

where $|\tilde{\boldsymbol{v}}|$ is the magnitude of the convection velocity of the transport equation. The characteristic element length h is selected as the streamline element length.

The turbulence transport equation for the DDES modeling is then coupled in a partitioned manner with the Navier-Stokes equations for the closure. This partitioning procedure along with the nonlinear iterative force correction (NIFC) (Section 7.3.2) scheme provides stability for low structure-to-fluid mass ratios in the fluid-structure interaction problems. More details about the PPV implementation and the partitioned coupling to the DDES modeling can be found in [113].

10.3.2 Flow Across Circular Cylinder at Re =3900 and Re = 140,000

The hybrid RANS/LES type of turbulence model discussed in the previous section is validated considering flow across a stationary cylinder at subcritical Reynolds numbers of 3,900 (laminar separation) and 140,000 (turbulent separation). The computational domain for the problem is shown in Fig. 10.13 and consists of a circular cylinder of diameter D. The inlet and outlet boundaries of the domain are at a distance of $10D$ and $20D$ from the center of the cylinder respectively. The side walls are at a distance of $30D$ from each other and are equidistant from the cylinder center, giving a blockage of 3.3%. The span of the cylinder is kept at $3D$ such that the three-dimensional turbulence structures are captured. A freestream velocity U_∞ parallel to the X-axis is applied to the inlet boundary with $\tilde{\nu} = 0$ for the laminar separation case and $\tilde{\nu} = 5\nu$ for Re = 140,000, where ν is the molecular viscosity of the fluid. A no-slip boundary condition and $\tilde{\nu} = 0$ is satisfied on the cylinder wall. Slip condition is imposed on the side walls and a periodic condition is applied on the boundaries along the Z-axis. A non-zero initial condition for $\tilde{\nu} = 5\nu$ is selected.

It is crucial to maintain the quality of the mesh and isotropic element size in the separated flow region to get the benefits of the DDES [197]. The mesh adopted in this validation was similar to that in [213]. The computational domain is discretized using unstructured finite elements consisting of 1,457,595 nodes and 1,421,820 eight-node hexahedron elements, as shown in Fig. 10.14a. The boundary layer thickness is $0.25D$ along with the stretching ratio of $\Delta y_{j+1}/\Delta y_j = 1.25$ and number of divisions in the wall-normal direction such that $y^+ \approx 0.12$ for Re = 140,000 and $y^+ \approx 0.004$ for Re = 3900 cases. An isotropic grid is maintained in the LES focus region (Fig. 10.14b).

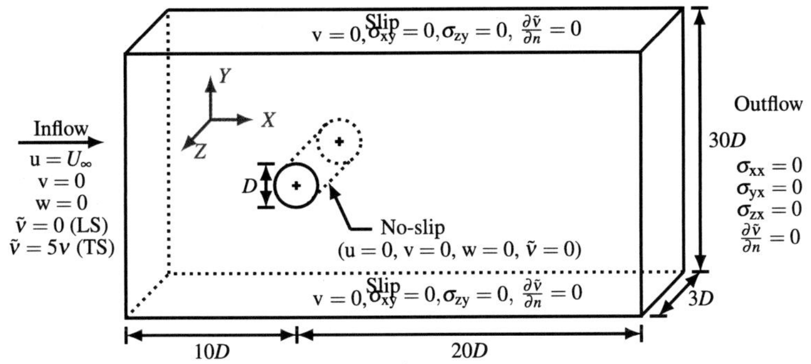

Fig. 10.13 A schematic of the flow past a three-dimensional circular cylinder at high Reynolds number. The computational setup and boundary conditions are shown for the Navier-Stokes and DDES equations. Here, $\boldsymbol{v}^{\mathrm{f}} = (\mathrm{u, v, w})$ denotes the components of the fluid velocity, LS and TS denote the laminar separation and turbulent separation cases respectively

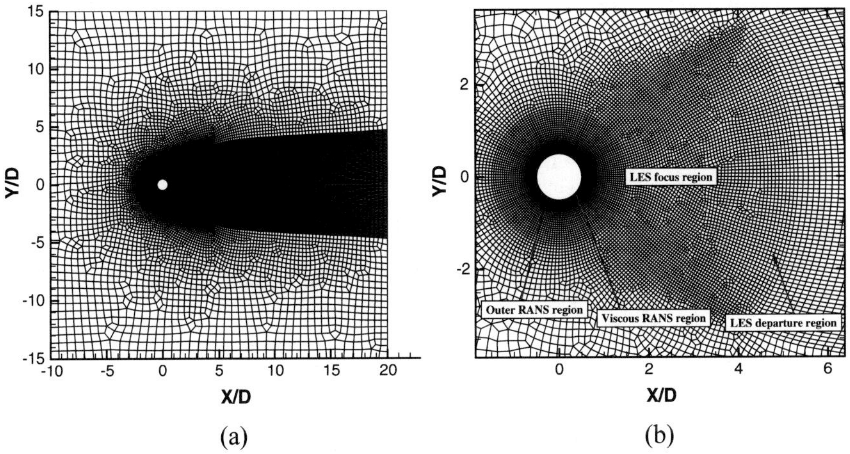

Fig. 10.14 Mesh characteristics for the flow past a circular cylinder: **a** mesh in the full computational domain in the cross-sectional X-Y plane, **b** various mesh regions around the cylinder for the well-designed DES mesh. While the elements in the LES focus region are isotropic and well resolved for capturing the wake dynamics, the viscous RANS region maintains $y^+ < 1$ close to the cylinder surface

The discretization in the spanwise direction is $\Delta z = 0.05D$ consisting of 60 layers. For ensuring the Courant condition of $Co = |\boldsymbol{v}^{\mathrm{f}}|\Delta t/\Delta \leq 1$ in the LES focus region (Δ being the largest dimension of the grid element), the time step size is chosen as $\Delta t = 0.05$.

The statistical results pertaining to the hydrodynamic coefficients are evaluated from $tU_\infty/D = 100 - 250$ for *Re* = 140,000 case and from $tU_\infty/D = 100 - 450$ for *Re* = 3900 case, neglecting the initial transient period of up to $tU_\infty/D = 100$. Thus, the number of vortex shedding cycles that are captured for *Re* = 3900 and *Re*

Table 10.4 Comparison of the hydrodynamic parameters pertaining to the flow across a circular stationary cylinder at Re = 3900 (Laminar separation)

Study	$\overline{C}_D$	C_L^{rms}	St	L_r/D	$-C_{pb}$
Present	1.0069	0.1715	0.2075	1.462	0.8998
DES [249]	1.0246	0.1897	0.2166	1.4909	–
Experiment [37, 158, 166]	0.99±0.05	–	0.215±0.005	1.4±0.1	0.88±0.05

= 140,000 are 70 and 45 respectively. The hydrodynamic coefficients and parameters for the two cases are compared with the experimental and numerical observations in Tables 10.4 and 10.5. We observe that the various quantities are in good agreement with the literature. Notice that the experiment for the turbulent separation case can be performed by inducing turbulence in the boundary layer by tripping. Therefore, the compared values are from the experiments performed at very high Reynolds number for the turbulent separation case. In Tables 10.4 and 10.5, St is the Strouhal number and L_r/D is the measured length from the base of the cylinder to the point of null mean streamline velocity along the centerline behind the cylinder. The force coefficients, viz., the lift (C_L) and drag (C_D) coefficients are evaluated by integrating the surface traction over the first layer of the elements on the cylindrical surface, as

$$C_L = \frac{1}{\frac{1}{2}\rho^{\text{f}} U_\infty^2 D} \int_\Gamma (\boldsymbol{\sigma}^{\text{f}} \cdot \boldsymbol{n}) \cdot \boldsymbol{n}_y d\Gamma, \tag{10.31}$$

$$C_D = \frac{1}{\frac{1}{2}\rho^{\text{f}} U_\infty^2 D} \int_\Gamma (\boldsymbol{\sigma}^{\text{f}} \cdot \boldsymbol{n}) \cdot \boldsymbol{n}_x d\Gamma, \tag{10.32}$$

where $\boldsymbol{n}_x$ and $\boldsymbol{n}_y$ are the unit normal $\boldsymbol{n}$ to the cylindrical surface and $\boldsymbol{\sigma}^{\text{f}}$ is the fluid stress tensor with Γ being the cylindrical surface. The pressure coefficient is defined as

$$C_p = \frac{p - p_\infty}{\frac{1}{2}\rho^{\text{f}} U_\infty^2}, \tag{10.33}$$

where p and p_∞ are the pressure at the concerned point and pressure at far-field respectively. The base pressure coefficient (C_{pb}) is obtained by replacing p with the base pressure of the cylinder.

The temporal variation of the force coefficients and the frequency spectrum of the lift coefficient are shown in Figs. 10.15 and 10.16 respectively for the two Reynolds number cases.

Furthermore, the variation of the mean pressure coefficient along the circumference of the cylinder is also plotted in Fig. 10.17 where it is compared with the results from the literature. For both the cases, a good agreement in the variation is found. The effect of the positivity preserving variational (PPV) technique can be observed

Table 10.5 Comparison of the hydrodynamic parameters pertaining to the flow across a circular stationary cylinder at Re = 140,000 (turbulent separation)

Study	$\overline{C}_D$	C_L^{rms}	St	L_r/D	$-C_{pb}$
Present	0.5782	0.0957	0.2930	1.113	0.6991
DES (TS1) [213]	0.57	0.08	0.30	1.1	0.65
DES (TS2) [213]	0.59	0.06	0.31	1.2	0.67
Experiment [82]	0.59	–	0.29	–	0.72
Experiment [182]	0.62–0.74	–	0.27	–	0.5–0.9

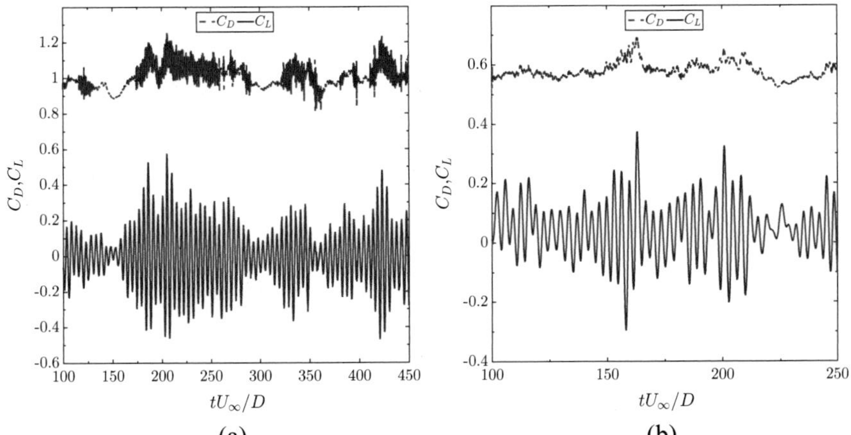

Fig. 10.15 Time histories of the hydrodynamic force coefficients for the flow past a circular stationary cylinder at two subcritical Reynolds numbers: **a** Re = 3900, and **b** Re = 140,000

in Fig. 10.17a where we compare the results with the method not employing the nonlinear PPV terms. Absence of the nonlinear terms does not give good prediction of the pressure distribution in the leeward side of the cylinder. Thus, PPV method is essential in capturing high gradients in the solution in the shear layer dominated regions in the near-wake of the cylinder.

The flow structures in the wake of the cylinder are visualized by plotting the iso-surfaces of Q-criterion colored by the normalized streamwise velocity for the two cases in Figs. 10.18a, c. A narrow wake for Re = 140,000 with the shift of the separation point to the back side of the cylinder is clearly visible. Large rolls of spanwise vorticity along with the streamwise ribs connecting the rolls through the braid region can be seen in Figs. 10.18b, d. More fine eddies and small turbulent structures can be observed in the braid region depicting the resolution of small scales in the hybrid RANS/LES turbulence modeling.

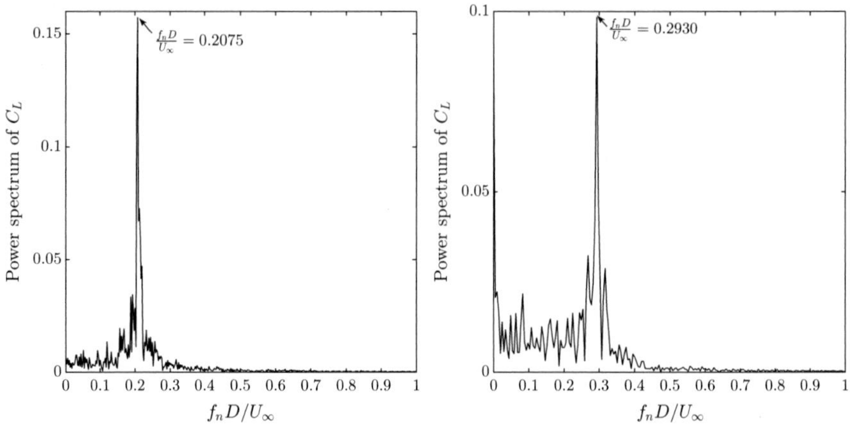

(a) (b)

Fig. 10.16 Spectrum of the lift coefficient C_L for the stationary circular cylinder at: **a** Re = 3900, **b** Re = 140,000

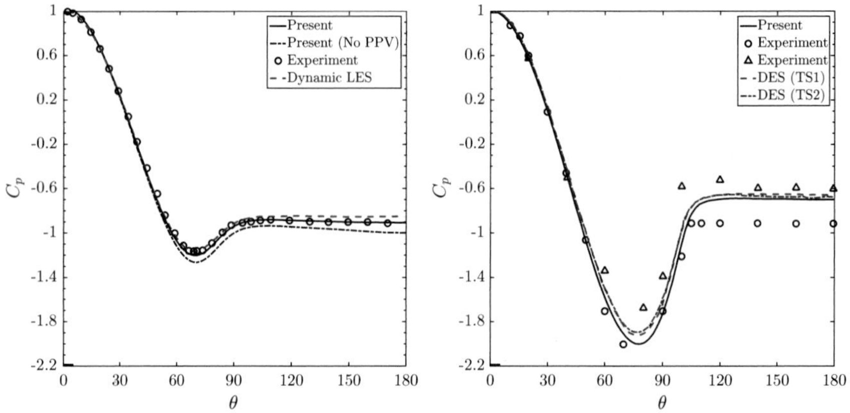

(a) (b)

Fig. 10.17 Flow past a stationary circular cylinder: Variation of C_p along the circumference with angle θ at **a** Re = 3900: (—) Present simulation, (–) Present (no PPV method), (○) Experiment [158], (- -) Dynamic LES [178]; **b** Re = 140,000: (—) Present simulation, (○) Experiment at $Re = 8.5 \times 10^6$ [182], (△) Experiment at $Re = 7.6 \times 10^6$ [221], (- -) DES [213], (-·-) DES [213]. Here the angle $\theta = 0$ denotes the stagnation point

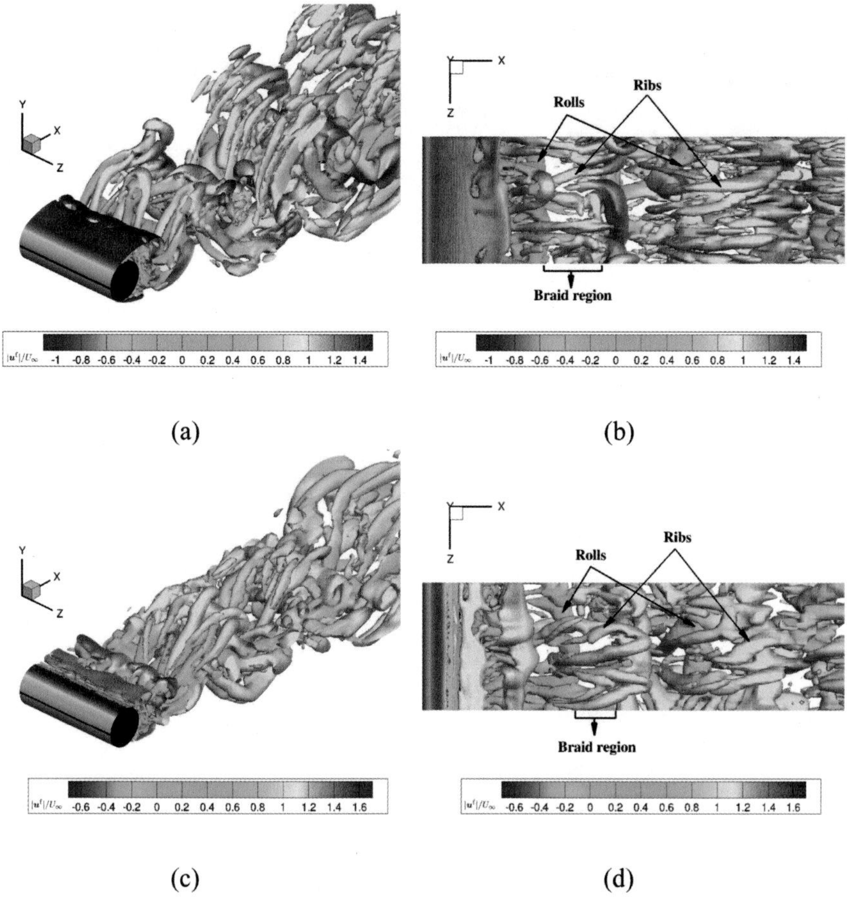

Fig. 10.18 Flow past a stationary circular cylinder at subcritical Reynolds numbers: Vortical wake structures of iso-surfaces of $Q^+ = 0.25$ colored by $|\boldsymbol{v}^{\mathrm{f}}|/U_\infty$ at $tU_\infty/D = 150$: Three-dimensional view at **a** Re = 3900 and **c** Re = 140,000. Top-view- The rolls with the braid region consisting of small-scale structures and the ribs can be observed at **b** Re = 3900 and **d** Re = 140,000

As mentioned earlier, turbulence consists of multiscales where energy is generated by the large eddies. Energy is lost by these eddies by vortex stretching mechanism, thus breaking into smaller eddies, which finally lose their energy to the molecular viscosity. The slope of the energy spectrum with the wavenumber/frequency remains constant at $-5/3$ in the inertial range of this energy cascade, obeying the Kolmogorov's rule. This slope can be observed for the transverse velocity fluctuations at various downstream locations along the centerline of the cylinder wake for the different Reynolds number cases in Figs. 10.19a, b.

Further comparison with the results from the literature is carried out for the mean streamwise and transverse velocities at various locations in the wake of the cylinder

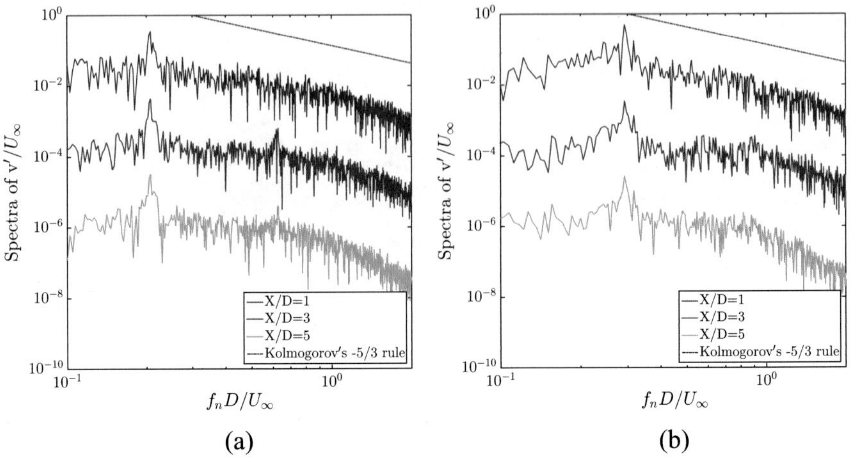

Fig. 10.19 Flow past a stationary circular cylinder: Spectra of the transverse velocity fluctuations and the Kolmogorov's rule at **a** Re = 3900, **b** Re = 140,000

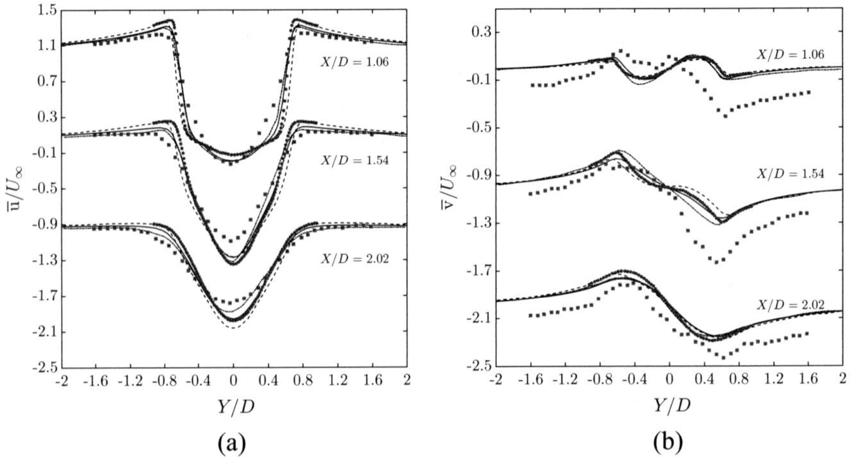

Fig. 10.20 Mean velocity at three locations in the wake of the stationary circular cylinder at Re = 3900: **a** Streamwise velocity; **b** Transverse velocity. (—) Present simulation, (- -) LES [169], (-··-) B-Spline simulation [128], (•) PIV experiment [169], (■) Experiment [133]

for Re = 3900 in Fig. 10.20. It can be observed that the turbulent characteristics of the flow are successfully captured by the hybrid RANS/LES based delayed detached eddy simulation.

10.3.3 Application to VIV of an Offshore Riser

We demonstrate the hybrid RANS/LES turbulence model with the coupled fluid-structure interaction problem of vortex-induced vibration of an offshore riser in this section. Excitation of several modes along the slender riser due to the vortex shedding process makes the study of its flow dynamics challenging pertaining to the multi-modal and multi-frequency response. Here, we consider a flexible riser subjected to a uniform current of 0.2 m/s at Reynolds number Re = 4000 to understand its flow dynamics and compare the results with that of the experiment (case no. 1103 in [1]).

The riser of diameter D subjected to a uniform current U_∞ is modeled as Euler-Bernoulli beam under an axial tension P and neglecting any damping and shear effects (Fig. 10.21a). The structural equations are solved in a modal space to predict the displacement which can be expressed as a linear combination of the different modes. The details of this linear modal analysis can be found in [113]. The key non-dimensional parameters for the computation are:

$$Re = \frac{\rho^{\mathrm{f}} U_\infty D}{\mu^{\mathrm{f}}} = 4000, \qquad m^* = \frac{m^{\mathrm{s}}}{\pi D^2 L \rho^{\mathrm{f}}/4} = 2.23, \qquad U_r = \frac{U_\infty}{f_1 D} = 5.6,$$
$$EI^* = \frac{EI}{\rho^{\mathrm{f}} U_\infty^2 D^4} = 2.1158 \times 10^7, \qquad P^* = \frac{P}{\rho^{\mathrm{f}} U_\infty^2 D^2} = 5.10625 \times 10^4, \tag{10.34}$$

where Re, m^*, U_r, EI^* and P^* are the Reynolds number, mass ratio, nominal reduced velocity, flexural rigidity and axial tension respectively. The nominal reduced velocity is evaluated at the first modal natural frequency of the structure f_1.

The computational domain is shown in Fig. 10.21b. The inlet and outlet boundaries are $10D$ and $25D$ away from the center of the riser respectively. The side walls are equidistant ($10D$) from the riser center giving a blockage of 5%. The riser spans a length of $481.5D$ in the Z-direction. A uniform freestream velocity is given at the inlet boundary with $\tilde{\nu} = 0$. Slip conditions are satisfied at the side walls and boundaries perpendicular to the axis of the riser, while no-slip conditions are imposed with $\tilde{\nu} = 0$ at the riser surface.

The domain is discretized using three-dimensional unstructured finite elements. For such large-scale and complex problem, mesh resolution is very crucial to capture the boundary layer, near- and far-wake characteristics accurately. A relatively coarser grid is employed with the cross-sectional mesh shown in Fig. 10.22 to maintain a reasonable level of boundary layer and wake resolution. The number of divisions along the circumference of the riser is 160. The boundary layer is constructed with a thickness of $0.25D$ and stretching ratio of $\Delta y_{j+1}/\Delta y_j = 1.15$ such that $y^+ < 1$ in the wall-normal direction. The spanwise discretization varies for different mesh resolutions in the study, which are summarized in Table 10.6. The non-dimensional time step size is selected as $\Delta t U_\infty / D = 0.1$.

The temporal variation of the amplitude response of the location $z/L = 0.55$ of the riser is shown in Fig. 10.23 for the finest mesh M3. The total number of time steps

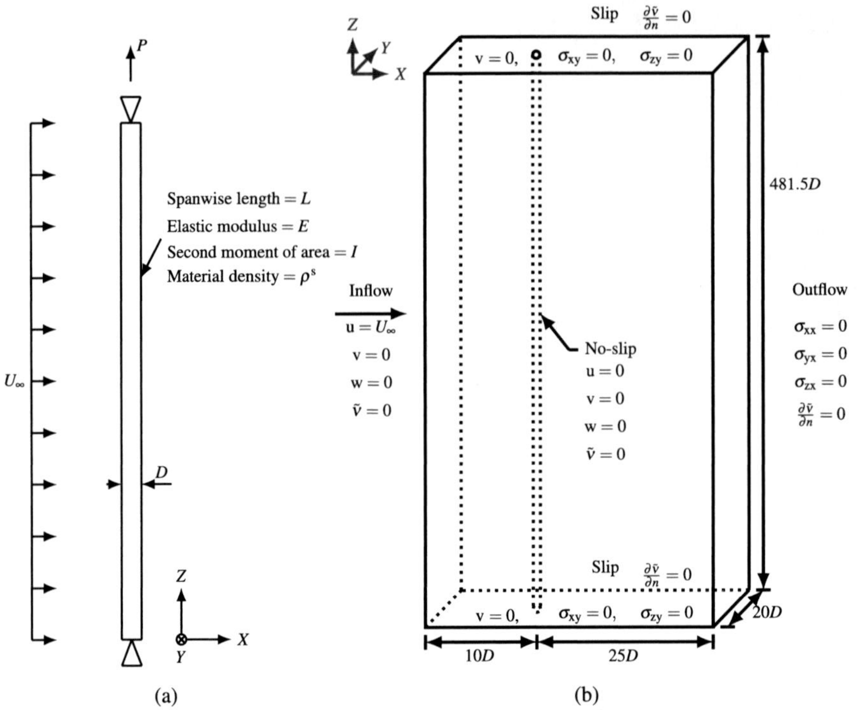

Fig. 10.21 A long flexible riser model in uniform current flow along the Z-axis: **a** pinned-pinned tensioned riser with uniform flow, **b** sketch to illustrate the computational setup and boundary conditions for the PPV-based DDES on flow past a flexible riser. Here, $\boldsymbol{v}^{\mathrm{f}} = (\mathrm{u}, \mathrm{v}, \mathrm{w})$ denotes the fluid velocity components. Consistent with the experimental measurement, a tensioned beam pinned at both ends and free to oscillate in the IL (X) and CF (Y) directions is employed in the numerical simulation

Table 10.6 Mesh details for flexible riser with $L/D = 481.5$ in a uniform current flow at Re = 4000

Mesh	Number of nodes	Number of elements	Spanwise layers
M1	3,247,155	3,200,800	200
M2	6,478,155	6,401,600	400
M3	12,940,155	12,803,200	800

for the computation was 8000. The statistics of the computation are only shown for $tU_\infty/D \in [400, 800]$ to avoid any initial transient effects. The mesh convergence study shows that the variation in the in-line response for meshes M1 and M2 is large, but it reduced for the finest mesh M3. On the other hand, the cross-flow response remains similar across the meshes. A comparison of the root mean square values of the in-line and cross-flow amplitude response along the riser across the meshes is plotted in Fig. 10.24. A_x and A_y denote the displacement amplitude at a point along

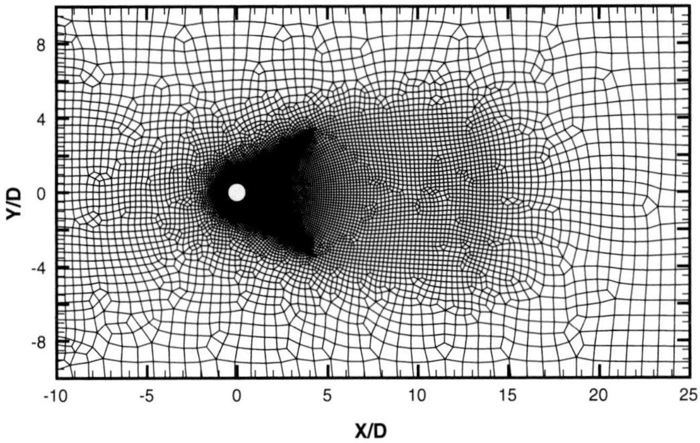

Fig. 10.22 Uniform flow past a long flexible riser model: Two-dimensional mesh in the X-Y plane employed in the computation of flexible riser VIV. This mesh is extruded in the third dimension to obtain the three-dimensional mesh

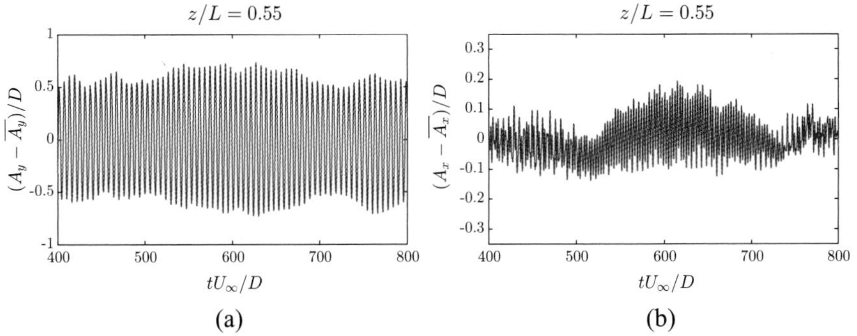

Fig. 10.23 Uniform current flow past a flexible riser model $L/D = 481.5$ at $(Re, m^*, U_r) = (4000, 2.23, 5.6)$: Dependence of riser response at $z/L = 0.55$ with spanwise resolution for 800 layers: **a** cross-flow response and **b** in-line response

the riser; $\overline{A_x}$ and $\overline{A_y}$ represent the time averaged values of the in-line and cross-flow displacements at a particular spanwise location, respectively.

The root mean square values of the amplitude can be written as:

$$A_{x,\text{rms}} = \sqrt{\frac{1}{N}\sum_{i=1}^{N}(A_{x,i} - \overline{A_x})^2}, \qquad A_{y,\text{rms}} = \sqrt{\frac{1}{N}\sum_{i=1}^{N}(A_{y,i} - \overline{A_y})^2}, \qquad (10.35)$$

where N represents the number of samples taken across time (4000 in this case). The comparison of the results with the experimental observations shows a good agreement for the cross-flow response, however, some over-prediction for the in-line response is observed. The peak in-line displacement for the numerical and experimen-

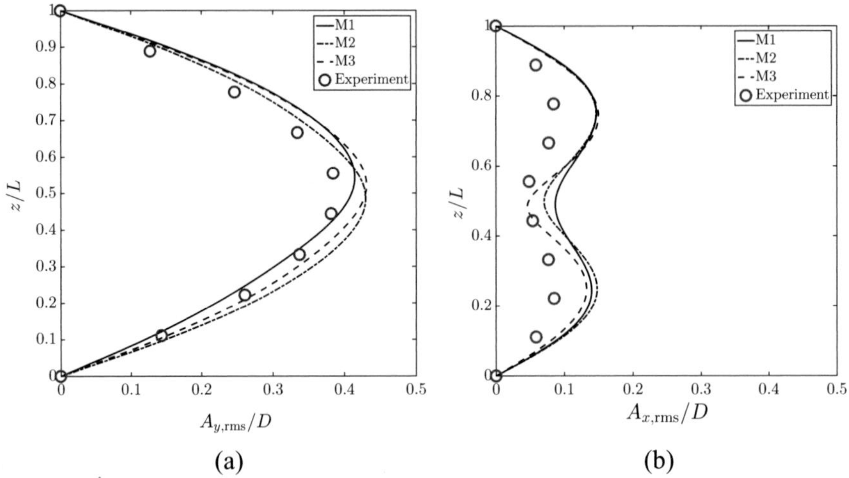

Fig. 10.24 Uniform current flow past a flexible riser model $L/D = 481.5$ at $(Re, m^*, U_r) = (4000, 2.23, 5.6)$: Variation of the root mean square amplitude along the span of the riser for different employed meshes: **a** cross-flow amplitude; **b** in-line amplitude. The first mode (fundamental) response for the CF and the second mode for the IL can be seen

tal observations differ by $< 15\%$ with respect to the total peak rms amplitude A/D, where $A = \sqrt{(A^2_{x,\mathrm{rms}} + A^2_{y,\mathrm{rms}})}$. Small numerical and experimental uncertainties in the complex physical phenomena (for example, the lock-in range and boundary-layer separation) can have a huge impact in the in-line response and drag force due to these sensitivities. Pertaining to the best performance, mesh M3 is selected for further analysis.

The riser response envelope in a uniform current flow is illustrated in Fig. 10.25. The riser vibrates with a dominant second mode and first mode in the in-line (IL) and cross-flow (CF) directions, respectively. The dominant frequency of the in-line response is generally twice that of the cross-flow response, which is reflected by the dominant modes of the riser. The riser response along the span is plotted as a function of time in Fig. 10.26. A frequency twice that of the cross-flow response is observed in the in-line direction. Furthermore, a standing wave pattern is observed for the riser response, agreeing with a recent observation in [52] that a pure standing wave response is manifested by a single mode. A detailed discussion about the physical understanding of these response patterns can be found in [113].

As a result of spectral analyses, the frequencies along the locations of the riser along with the orbital trajectories are shown in Figs. 10.27 and 10.28 for the two different time windows. Alternating trajectories of the different cross-sections along the riser is also observed which change with time. Thus, we focus on two time windows: $tU_\infty/D \in [400, 470]$ where the lower part of the riser ($0 < z/L < 0.5$) is undergoing counter-clockwise trajectory and $tU_\infty/D \in [600, 670]$ where the upper part ($0.5 < z/L < 1$) has a counter-clockwise trajectory to the incoming flow. For

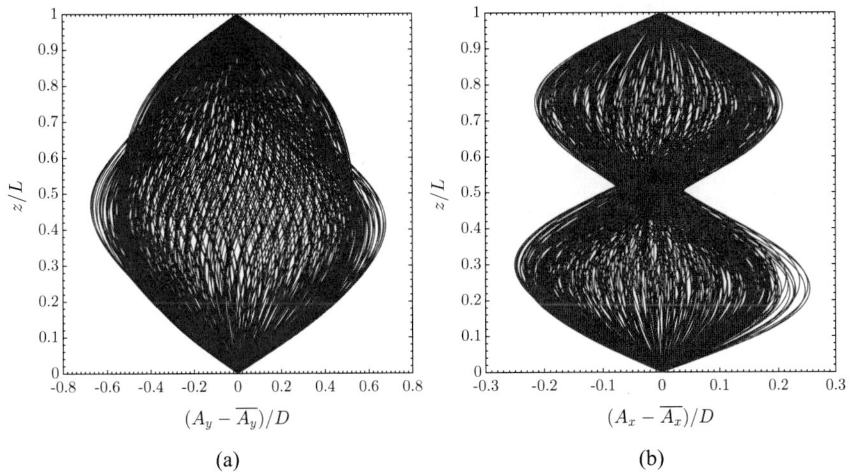

Fig. 10.25 Riser response envelope with spanwise resolution of 800 layers for uniform current flow past a flexible riser model at $(Re, m^*, U_r) = (4000, 2.23, 5.6)$: **a** cross-flow and **b** in-line directions

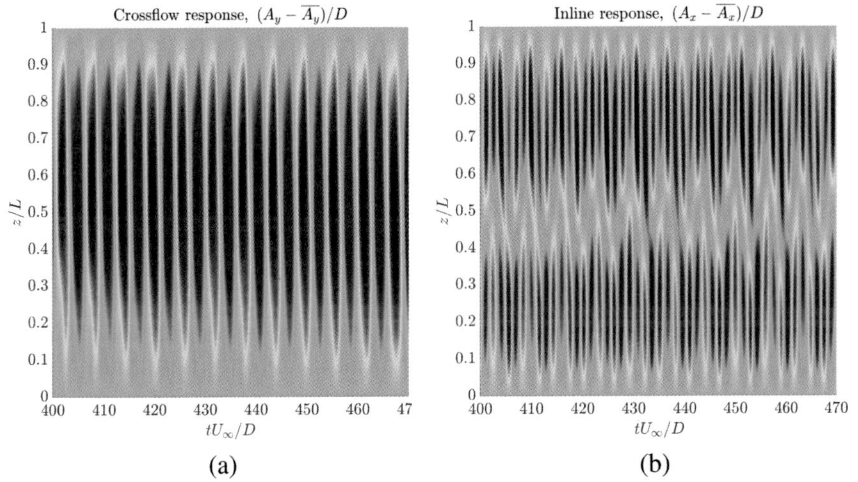

Fig. 10.26 Standing wave response of flexible riser VIV with spanwise resolution of 800 layers for uniform current flow past a flexible riser model at $(Re, m^*, U_r) = (4000, 2.23, 5.6)$: **a** cross-flow and **b** in-line directions

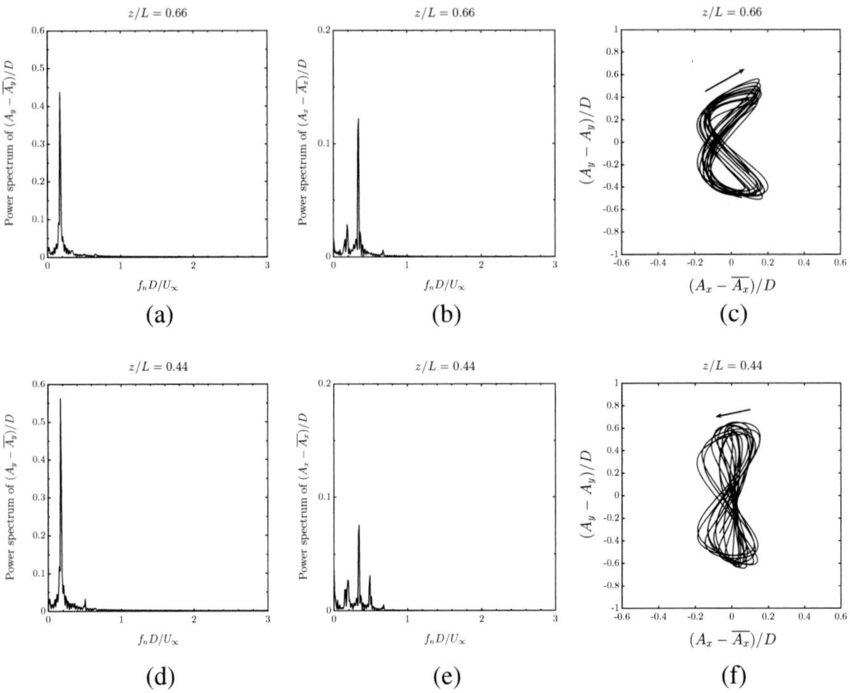

Fig. 10.27 Oscillation frequencies and orbital trajectory at different locations $z/L \in [0.44, 0.66]$ along the span of the riser in the time window $tU_\infty/D \in [400, 470]$. The horizontal row corresponds to the same location of the riser with the first and second column giving the power spectra of cross-flow and in-line amplitudes respectively

every location along the riser, a dominant frequency (fD/U_∞) of 0.166 and 0.342 is observed for the CF and IL response respectively. These values of the frequency are much closer to the experimental values than the numerical tests performed in [233]. The in-line response spectra exhibits higher modes of frequency twice that of the dominant frequency across most of the locations.

With respect to the trajectories, the figure-8 configuration is observed for most of the locations. A transition in the trajectory pattern along the riser and with time is also observed. These observations are consistent with the experiment [1]. The contours of the vibration amplitude and the spanwise Z-vorticity patterns at different locations along the riser ($z/L \in [0.11, 0.88]$) are shown in Fig. 10.29 at $tU_\infty/D = 460$ and $tU_\infty/D = 650$. A higher amplitude response is observed at the bottom side of the riser in Fig. 10.29a as the lower side is undergoing counter-clockwise motion. This trajectory is opposed to the incoming flow and produces more shear as the fluid does net work on the structure [54]. This pattern reverses for $tU_\infty/D = 650$ in Fig. 10.29b. Moreover, the vortex shedding pattern is very complex where the 2S shedding mode is observed in most of the locations. A wider 2S with two rows configuration is observed near the location of large amplitude response and 2P mode is also observed

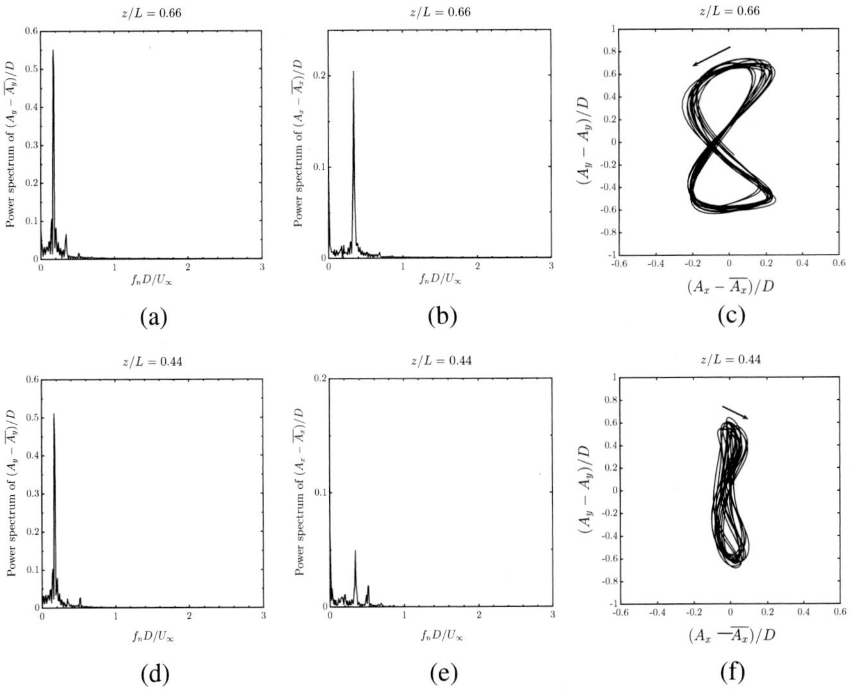

Fig. 10.28 Oscillation frequencies and orbital trajectory at different locations $z/L \in [0.44, 0.66]$ along the span of the riser in the time window $tU_\infty/D \in [600, 670]$. The horizontal row corresponds to the same location of the riser with the first and second column giving the power spectra of cross-flow and in-line amplitudes respectively

at some locations. To summarize, we demonstrated the hybrid RANS/LES turbulence model for a large-scale problem of VIV of an offshore riser, while comparing the results with the experiment and drawing parallels from the literature. Further analysis is required to better assess the vortex dynamics in detail.

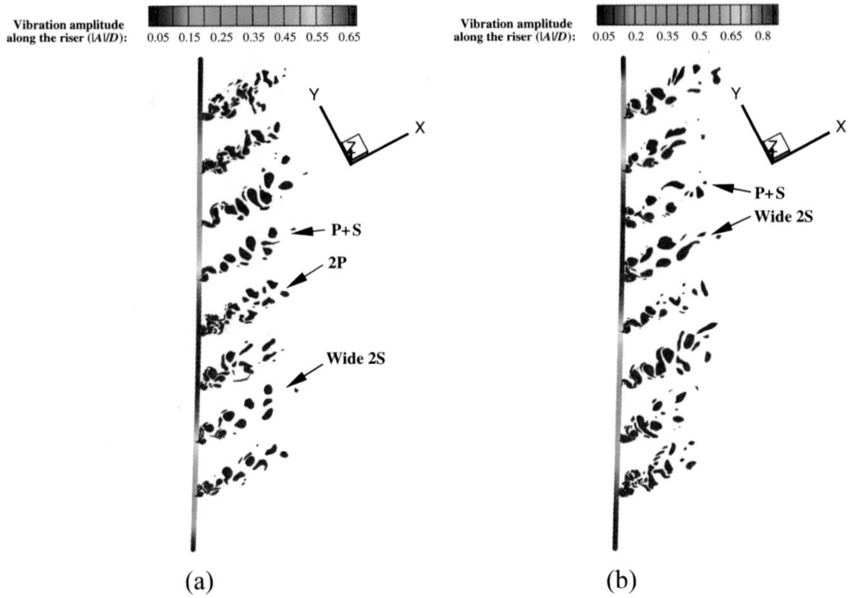

Fig. 10.29 Flow visualization along the flexible riser undergoing VIV at $(Re, m^*, U_r) = (4000, 2.23, 5.6)$: Vibration amplitude along the riser with Z-vorticity contours in various sections-red and blue color indicate the positive and negative vorticity respectively: **a** $tU_\infty/D = 460$ and **b** $tU_\infty/D = 650$. The vorticity contours are plotted for different locations along the riser corresponding to $z/L \in [0.11, 0.88]$

References

1. Vortex induced vibration data repository—Datasets from ExxonMobil (Test case 1103) (http://web.mit.edu/towtank/www/vivdr/downloadpage.html)
2. MPI webpage. Technical Report. www.mpi-formum.org (2009)
3. Akkerman, I., Bazilevs, Y., Benson, D.J., Farthing, M.W., Kees, C.E.: Free-surface flow and fluid-object interaction modeling with emphasis on ship hydrodynamics. J. Appl. Mech. **79**(1), 10905 (2012)
4. Akkerman, I., Bazilevs, Y., Calo, V.M., Hughes, T.J.R., Hulshoff, S.: The role of continuity in residual-based variational multiscale modeling of turbulence. Comput. Mech. **41**(3), 371–378 (2007, 2008)
5. Akkerman, I., Bazilevs, Y., Kees, C.E., Farthing, M.W.: Isogeometric analysis of free-surface flow. J. Comput. Phys. **230**(11), 4137 – 4152 (2011). Special issue High Order Methods for CFD Problems
6. Allen, S.M., Cahn, J.W.: A microscopic theory for antiphase boundary motion and its application to antiphase domain coarsening. Acta Metallurgica **27**(6), 1085–1095 (1979)
7. Allmaras, S.R., Johnson, F.T., Spalart, P.R.: Modifications and clarifications for the implementation of the Spalart-Allmaras turbulence model. In: 7th Conference on Computational Fluid Dynamics, Big Island, Hawaii (2012)
8. Anderson, D.M., McFadden, G.B.: A diffuse-interface description of fluid systems. In: NIST IR 5887. National Institute of Standards and Technology (1996)
9. Anderson, D.M., McFadden, G.B., Wheeler, A.A.: Diffuse-interface methods in fluid mechanics. Annu. Rev. Fluid Mech. **30**(1), 139–165 (1998)
10. Antman, S.S.: Nonlinear Problems of Elasticity. Springer (1995)
11. Aono, H., Chimakurthi, S.K., Wu, P., Sällström, E., Stanford, B.K., Cesnik, C.E., Ifju, P., Ukeiley, L., Shyy, W.: A computational and experimental study of flexible flapping wing aerodynamics. In: 48th AIAA Aerospace Sciences Meeting Including the New Horizons Forum and Aerospace Exposition pp. 4–7 (2010)
12. Bottasso, C.L., Borri, M.: Energy preserving/decaying schemes for non-linear beam dynamics using the helicoidal approximation. Comput. Methods Appl. Mech. Eng. **143**(3), 393–415 (1997)
13. Balaras, E., Benocci, C., Piomelli, U.: Two layer approximate boundary conditions for large-eddy simulations. AIAA J. **34**, 1111–19 (1996)
14. Bank, R.E., Smith, R.K.: A posteriori error estimates based on hierarchical bases. SIAM J. Numer. Anal. **30**(4), 921–935 (1993)

R. K. Jaiman and V. Joshi, *Computational Mechanics of Fluid-Structure Interaction*,
https://doi.org/10.1007/978-981-16-5355-1

15. Bartels, S.: A posteriori error analysis for time-dependent Ginzburg-Landau type equations. Numerische Mathematik **99**(4), 557–583 (2005)
16. Bartels, S., Müller, R., Ortner, C.: Robust a priori and a posteriori error analysis for the approximation of Allen-Cahn and Ginzburg-Landau equations past topological changes. SIAM J. Numer. Anal. **49**(1), 110–134 (2011)
17. Bauchau, O.: A self-stabilized algorithm for enforcing constraints in multibody systems. Int. J. Solids Struct. **40**(13–14), 3253–3271 (2003)
18. Bauchau, O.: Dymore User's Manual. Georgia Institute of Technology, Atlanta (2007)
19. Bauchau, O.: Flexible Multibody Dynamics. Springer (2010)
20. Bauchau, O., Bottasso, C.: On the design of energy preserving and decaying schemes for flexible, nonlinear multi-body systems. Comput. Methods App. Mech. Eng. **169**(1), 61–79 (1999)
21. Bauchau, O., Theron, N.: Energy decaying scheme for non-linear beam models. Comput. Methods App. Mech. Eng. **134**(1), 37–56 (1996)
22. Bazilevs, Y., Calo, V.M., Cottrell, J.A., Hughes, T.J.R., Reali, A., Scovazzi, G.: Variational multiscale residual-based turbulence modeling for large eddy simulation of incompressible flows. Comput. Methods App. Mech. Eng. **197**(1), 173–201 (2007)
23. Bazilevs, Y., Calo, V.M., Hughes, T.J.R., Zhang, Y.: Isogeometric fluid-structure interaction: theory, algorithms, and computations. Comput. Mech. **43**(1), 3–37 (2008)
24. Beckert, A.: Coupling fluid (CFD) and structural (FE) models using finite interpolation elements. Aerosp. Sci. Technol. **4**(1), 13–22 (2000)
25. Blevins, R.D.: Flow-Induced vibration. Van Nostrand Reinhold, New York (1990)
26. Boffi, D., Gastaldi, L.: Stability and geometric conservation laws for ALE formulations, pp. 4717–4739
27. Boris, J.P., Book, D.L.: Flux-corrected transport. I. SHASTA, a fluid transport algorithm that works. J. Comput. Phys. **11**, 38–69 (1973)
28. Bouris, D., Bergeles, G.: 2D LES of vortex shedding from a square cylinder. J. Wind Eng. Indu. Aerodyn. **80**(1), 31–46 (1999)
29. Brackbill, J.U., Kothe, D.B., Zemach, C.: A continuum method for modeling surface tension. J. Comput. Phys. **100**(2), 335–354 (1992)
30. Brassel, M., Bretin, E.: A modified phase field approximation for mean curvature flow with conservation of the volume. Math. Methods Appl. Sci. **34**(10), 1157–1180 (2011)
31. Brezinski, C.: Projection Methods for Systems of Equations, vol. 7. North-Holland, Amsterdam (1997)
32. Brezinski, C.: A classification of quasi-newton methods. Numer. Algor. **33**(1), 123–135 (2003)
33. Bronsard, L., Stoth, B.: Volume-preserving mean curvature flow as a limit of a nonlocal Ginzburg-Landau equation. SIAM J. Math. Anal. **28**(4), 769–839 (1997). Copyright—Copyright © 1997. Soc. Ind. Appl. Math. Last updated 29 Feb 2012
34. Brooks, A.N., Hughes, T.J.R.: Streamline upwind/Petrov-Galerkin formulations for convection dominated flows with particular emphasis on the incompressible Navier-Stokes equations. Compu. Methods Appl. Mech. Eng. **32**(1), 199–259 (1982)
35. Cahn, J.W., Hilliard, J.E.: Free energy of a nonuniform system. I. Interfacial free energy. J. Chem. Phys. **28**(2) (1958)
36. Calderer, A., Kang, S., Sotiropoulos, F.: Level set immersed boundary method for coupled simulation of air/water interaction with complex floating structures. J. Comput. Phys. **277**, 201–227 (2014)
37. Cardell, G.S.: Flow past a circular cylinder with a permeable splitter plate. Ph.D. thesis, California Institute of Technology (1993)
38. Casey, M., Wintergerste, T.: Best practice guidelines for industrial computational fluid dynamics. In: ERCOFTAC (2000)
39. Causin, P., Gerbeau, J.F., Nobile, F.: Added-mass effect in the design of partitioned algorithms for fluid-structure problems, pp 4506–4527
40. Cebral, J.R., Lohner, R.: Conservative load projection and tracking for fluid-structure problems. AIAA J. **35**(4), 687–692 (1997)

41. Chadwick, P.: Continuum Mechanics: Concise Theory and Problems. Dover (1999)
42. Chakrabarti, S.K.: Hydrodynamics of Offshore Structures. Springer (1987)
43. Chen, L., Zhang, C.: A coarsening algorithm on adaptive grids by newest vertex bisection and its applications. Journal of Computational Mathematics **28**(6), 767–789 (2010)
44. Chung, J., Hulbert, G.M.: A time integration algorithm for structural dynamics with improved numerical dissipation: the generalized-α method. J. App. Mech. **60**(2), 371–375 (1993)
45. Codina, R.: A discontinuity-capturing crosswind-dissipation for the finite element solution of the convection-diffusion equation. Comput. Methods Appl. Mech. Eng. **110**, 325–342 (1993)
46. Codina, R.: A finite element formulation for the numerical solution of the convection-diffusion equation. In: CIMNE, Barcelona (1993)
47. Codina, R.: Comparison of some finite element methods for solving the diffusion-convection-reaction equation. Comput. Methods Appl. Mech. Eng. **156**(1), 185–210 (1998)
48. Codina, R.: On stabilized finite element methods for linear systems of convection-diffusion-reaction equations. Comput. Methods Appl. Mech. Eng. **188**(1), 61–82 (2000)
49. Codina, R., Oñate, E., Cervera, M.: The intrinsic time for the streamline upwind/ Petrov-Galerkin formulation using quadratic elements. Comput. Methods Appl. Mech. Eng. **94**, 239–262 (1992)
50. Codina, R., Soto, O.: Approximation of the incompressible Navier-Stokes equations using orthogonal subscale stabilization and pressure segregation on anisotropic finite element meshes. Comput. Methods Appl. Mech. Eng. **193**(15–16), 1403–1419 (2004)
51. Connell, B.S.H., Yue, D.K.P.: Flapping dynamics of a flag in a uniform stream, pp 33–67
52. Constantinides, Y., Oakley, O.H.: Numerical simulation of cylinder VIV focusing on high harmonics. In: ASME 28th International Conference on Ocean, Offshore and Arctic Engineering. Honolulu, USA (2009)
53. Corson, D., Jaiman, R., Shakib, F.: Industrial application of rans modelling: capabilities and needs. Int. J. Comput. Fluid Dyn. **23**(4), 337–347 (2009)
54. Dahl, J.M., Hover, F.S., Triantafyllou, M.S.: Resonant vibrations of bluff bodies cause multivortex shedding and high frequency force. Phys. Revi. Lett. **99**(144503), 1–4 (2007)
55. Donea, J.: Arbitrary Lagrangian-Eulerian finite element methods, pp 4195–4215
56. Donea, J., Huerta, A.: Finite Element Methods for Flow Problems. Wiley (2003)
57. Du, Q., Nicolaides, R.A.: Numerical analysis of a continuum model of phase transition. SIAM J. Numer. Anal. **28**(5), 1310–1322 (1991)
58. Dutra do Carmo, E.G., Alvarez, G.B.: A new stabilized finite element formulation for scalar convection-diffusion problems: the streamline and approximate upwind/Petrov-Galerkin method. Comput. Methods Appl. Mech. Eng. **192**, 3379–3396 (2003)
59. Elias, R.N., Coutinho, A.L.G.A.: Stabilized edge-based finite element simulation of free-surface flows. Int. J. Numer. Methods Fluids **54**(6–8), 965–993 (2007)
60. Enright, D., Fedkiw, R., Ferziger, J., Mitchell, I.: A hybrid particle level set method for improved interface capturing. J. Comput. Phys. **183**(1), 83–116 (2002)
61. Enright, D., Losasso, F., Fedkiw, R.: A fast and accurate semi-Lagrangian particle level set method. Comput. Struct. **83**(6–7), 479–490 (2005)
62. Farhat, C., Lesoinne, M., Tallec, P.L.: Load and motion transfer algorithms for fluid/structure interaction problems with non-matching discrete interfaces: Momentum and energy conservation, optimal discretization and application to aeroelasticity. Comput. Methods Appl. Mech. Eng. **157**(1), 95–114 (1998)
63. Felippa, C.A., Park, K.C.: Staggered transient analysis procedures for coupled mechanical systems. Formulation 61–111
64. Felippa, C.A., Park, K.C., Farhat, C.: Partitioned analysis of coupled mechanical systems. Comput. Methods Appl. Mech. Eng. **190**(24), 3247 – 3270 (2001). Advances in Computational Methods for Fluid-Structure Interaction
65. Feng, X., Wu, H.: A posteriori error estimates and an adaptive finite element method for the Allen-Cahn equation and the mean curvature flow. J. Sci. Comput. **24**(2), 121–146 (2005)
66. Förster, C., Wall, W.A., Ramm, E.: Artificial added mass instabilities in sequential staggered coupling of nonlinear structures and incompressible viscous flows, pp. 1278–1293

67. Franca, L., Frey, S.: Stabilized finite element methods: II. The incompressible Navier-Stokes equations. Comput. Methods Appl. Mech. Eng. **99**, 209–233 (1992)
68. Franca, L.P., Dutra do Carmo, E.G.: The Galerkin gradient least-squares method. Comput. Methods Appl. Mech. Eng. **74**, 41–54 (1989)
69. Franca, L.P., Valentin, F.: On an improved unusual stabilized finite element method for the advective-reactive-diffusive equation. Comput. Methods Appl. Mech. Eng. **190**, 1785–1800 (2001)
70. Galeão, A.C., Dutra do Carmo, E.G.: A consistent approximate upwind Petrov-Galerkin method for convection-dominated problems. Comput. Methods Appl. Mech. Eng. **68**(1), 83–95 (1988)
71. Gatski, T., Speziale, C.: On explicit algebraic stress models for complex turbulent flows. J. Fluid Mech. **254**, 59 (1993)
72. Gee, M.W., Küttler, U., Wall, W.A.: Truly monolithic algebraic multigrid for fluid-structure interaction. Int. J. Numer. Methods Eng. **85**(8), 987–1016
73. Georgoulis, E.H., Makridakis, C.: On a posteriori error control for the Allen-Cahn problem. Math. Methods App. Sci. **37**(2), 173–179 (2014)
74. Geradin, M., Cardona, A.: Kinematics and dynamics of rigid and flexible mechanisms using finite elements and quaternion algebra. Comput. Mech. **4**, 115–135 (1989)
75. Geradin, M., Cardona, A.: Flexible Multibody Dynamics. A Finite Element Approach. Wiley (2001)
76. Germano, M., Piomelli, U., Moin, P., Cabot, W.: A dynamic subgrid-scale eddy viscosity model. Phys. Fluids **3**, 1760–1765 (1991)
77. Geuzaine, C., Remacle, J.F.: Gmsh: a 3-D finite element mesh generator with built-in pre- and post-processing facilities. In. J. Numer. Methods Eng. **79**(11), 1309–1331 (2009)
78. Gibou, F., Min, C.: Efficient symmetric positive definite second-order accurate monolithic solver for fluid/solid interactions. J. Comput. Phys. **231**(8), 3246–3263 (2012)
79. Gogulapati, A.: Nonlinear approximate aeroelastic analysis of flapping wings in Hover and Forward flight. Ph.D thesis, University of Michigan (2011)
80. Gonsalez, Stuart, A.M.: A First Course in Continuum Mechanics. Academic Press, New York (1977)
81. Gurtin, M.: An Introduction to Continuum Mechanics. Academic Press (1981)
82. Hansen, R.P., Forsythe, J.R.: Large and detached Eddy simulations of a circular cylinder using unstructured grids. In: AIAA 2003-0775 (2003)
83. Harari, I., Hughes, T.J.R.: What are C and h?: Inequalities for the analysis and design of finite element methods. Int. J. Numer. Methods Eng. **97**(2), 157–192 (1992)
84. Harari, I., Hughes, T.J.R.: Stabilized finite element methods for steady advection-diffusion with production. Int. J. Numer. Methods Eng. **115**, 165–191 (1994)
85. Harten, A.: High resolution schemes for hyperbolic conservation laws. J. Comput. Phys. **49**, 357–393 (1983)
86. Hauke, G.: A simple subgrid scale stabilized method for the advection-diffusion-reaction equation. Int. J. Numer. Methods Eng. **191**, 2925–2947 (2002)
87. Hauke, G., Garcia-Olivares, A.: Variational subgrid scale formulations for the advection-diffusion-reaction equation. Int. J. Numer. Methods Eng. **190**, 6847–6865 (2001)
88. Heil, M., Hazel, A.L., Boyle, J.: Solvers for large-displacement fluid-structure interaction problems: segregated versus monolithic approaches. Comput. Mech. **43**, 91–101 (2008)
89. Hounjet, M.H.L., Meijer, J.J.: Evaluation of elastomechanical and aerodynamic data transfer methods for non-planar configurations in computational aeroelastic analysis. ICAS-Publication, 10.1–10.25 (1994)
90. Hron, J., Turek, S.: A monolithic FEM/multigrid solver for an ALE formulation of fluid-structure interaction with applications in biomechanics. In: Bungartz, H-J., Schäfer, M. (eds.) Fluid-Structure Interaction, pp. 146–170. Springer, Berlin, Heidelberg
91. Hu, H., Patankar, N., Zhu, M.: Direct numerical simulations of fluid-solid systems using the arbitrary Lagrangian-Eulerian technique, pp. 427–462
92. Hughes, T.: The Finite Element Method. Prentice-Hall, Englewood Cliffs, NJ (1987)

93. Hughes, T.J.R.: Finite Element Methods for Convection Dominated Flows, vol. 34. ASME, New York (1979)
94. Hughes, T.J.R.: Multiscale phenomena: Green's functions, the Dirichlet-to-Neumann formulation, subgrid scale models, bubbles and the origins of stabilized methods. Comput. Methods Appl. Mech. Eng. **127**(1), 387–401 (1995)
95. Hughes, T.J.R.: The Finite Element Method: Linear Static and Dynamic Finite Element Analysis. Dover Publications (2000)
96. Hughes, T.J.R., Feijóo, G.R., Mazzei, L., Quincy, J.B.: The variational multiscale method-a paradigm for computational mechanics. Comput. Methods Appl. Mech. Eng. **166**, 3–24 166
97. Hughes, T.J.R., Franca, L.P., Hulbert, G.M.: A new finite element formulation for computational fluid dynamics: VIII. The Galerkin/least-squares method for advective-diffusive equations. Comput. Methods Appl. Mech. Eng. **73**(2), 173–189 (1989)
98. Hughes, T.J.R., Liu, W., Zimmerman, T.: Lagrangian-Eulerian finite element formulation for incompressible visous flows, pp. 329–349
99. Hughes, T.J.R., Mallet, M., Mizukami, A. A new finite element formulation for computational fluid dynamics: II. Beyond SUPG. Comput. Methods Appl. Mech. Eng. **54**(3), 341–355 (1986)
100. Hughes, T.J.R., Mazzei, L., Oberai, A.A., Wray, A.A.: The multiscale formulation of large eddy simulation: decay of homogeneous isotropic turbulence. Phys. Fluids **13**, 505–512 (2001)
101. Idelsohn, S.R., Heinrich, J.C., Oñate, E.: Petrov-Galerkin methods for the transient advective-diffusive equation with sharp gradients. Int. J. Numer. Methods Eng. **39**(9), 1455–1473 (1996)
102. Idelsohn, S.R., Nigro, N., Storti, M., Buscaglia, G.: A Petrov-Galerkin formulation for advective-reaction-diffusion problems. Comput. Methods Appl. Mech. Eng. **136**, 27–46 (1996)
103. Issa, R., Violeau, D.: Test-case 2, 3D dambreaking, Release 1.1. Technical Report, Electricité De France, Laboratoire National d'Hydraulique et Environnement. http://app.spheric-sph.org/sites/spheric/tests/test-2 (2006)
104. Ito, S.: Study of the transient heave oscillation of a floating cylinder. Ph.D. thesis, Massachusetts Institute of Technology (1971)
105. Jaiman, R., Geubelle, P., Loth, E., Jiao, X.: Stable and accurate loosely-coupled scheme for unsteady fluid-structure interaction. In: AIAA 2007-334 CP (2007)
106. Jaiman, R., Jiao, X., Geubelle, P., Loth, E.: Conservative load transfer along curved fluid-solid interface with non-matching meshes. J. Comput. Phys. **218**(1), 372–397 (2006)
107. Jaiman, R., Sen, S., Gurugubelli, P.: A fully implicit combined field scheme for freely vibrating square cylinders with sharp and rounded corners. Comput. Fluids **112**, 1–18 (2015)
108. Jaiman, R.K., Guan, M.Z., Miyanawala, T.P.: Partitioned iterative and dynamic subgrid-scale methods for freely vibrating square-section structures at subcritical Reynolds number. Comput. Fluids **133**, 68–89 (2016)
109. Jaiman, R.K., Jiao, X., Geubelle, P.H., Loth, E.: Assessment of conservative load transfer for fluid-solid interface with non-matching meshes. Int. J. Numer. Methods Eng. **64**(15), 2014–2038 (2005)
110. Jaiman, R.K., Pillalamarri, N.R., Guan, M.Z.: A stable second-order partitioned iterative scheme for freely vibrating low-mass bluff bodies in a uniform flow. Comput. Methods Appl. Mech. Eng. **301**, 187–215 (2016)
111. Jansen, K.E., Whiting, C.H., Hulbert, G.M.: A generalized-α method for integrating the filtered Navier-Stokes equations with a stabilized finite element method. Comput. Methods Appl. Mech. Eng. **190**(3–4), 305–319 (2000)
112. Joshi, V., Jaiman, R.K.: A positivity preserving variational method for multi-dimensional convection-diffusion-reaction equation. J. Comput. Phys. **339**, 247–284 (2017)
113. Joshi, V., Jaiman, R.K.: A variationally bounded scheme for delayed detached eddy simulation: application to vortex-induced vibration of offshore riser. Comput. Fluids **157**, 84–111 (2017)
114. Joshi, V., Jaiman, R.K.: An adaptive variational procedure for the conservative and positivity preserving Allen-Cahn phase-field model. J. Comput. Phys. (Accepted for publication) (2018). https://arxiv.org/abs/1710.10406v2

115. Joshi, V., Jaiman, R.K.: A positivity preserving and conservative variational scheme for phase-field modeling of two-phase flows. J. Comput. Phys. **360**, 137–166 (2018)
116. Joshi, V., Jaiman, R.K.: A hybrid variational Allen-Cahn/ALE scheme for the coupled analysis of two-phase fluid-structure interaction. Int. J. Numer. Methods Eng. **117**(4), 405–429 (2019)
117. Jung, K.H., Chang, K.A., Jo, H.J.: Viscous effect on the roll motion of a rectangular structure. J. Eng. Mech. **132**(2), 190–200 (2006)
118. Karypis, G., Kumar, V.: Metis: a software package for partitioning unstructured graphs, partitioning meshes, and computing fill-reducing orderings of sparse matrices, version 4.0. Technical Report. http://glaros.dtc.umn.edu/gkhome/metis/metis/download
119. Kessler, D., Nochetto, R.H., Schmidt, A.: A posteriori error control for the Allen–Cahn problem: Circumventing Gronwall's inequality. ESAIM: Math. Model. Numer. Anal. **38**(1), 129–142 (2004)
120. Kim, J.: A continuous surface tension force formulation for diffuse-interface models. J. Comput. Phys. **204**(2), 784–804 (2005)
121. Kim, J.: Phase-field models for multi-component fluid flows. Commun. Comput. Phys. **12**(3), 613–661 (2012)
122. Kim, J., Jaiman, R., Cosgrove, S., O'Sullivan, J.: Numerical wave tank analysis of wave run-up on a truncated vertical cylinder. In: ASME 30th International Conference on Ocean, Offshore and Arctic Engineering (OMAE), Rotterdam, The Netherlands (2011)
123. Kim, J., Lee, S., Choi, Y.: A conservative Allen-Cahn equation with a space-time dependent Lagrange multiplier. Int. J. Eng. Sci. **84**, 11–17 (2014)
124. Kirk, B.S., Peterson, J.W., Stogner, R.H., Carey, G.F.: libMesh: a C++ library for parallel adaptive mesh refinement/coarsening simulations. Eng. Comput. **22**(3), 237–254 (2006)
125. Kleefsman, K.M.T., Fekken, G., Veldman, A.E.P., Iwanowski, B., Buchner, B.: A volume-of-fluid based simulation method for wave impact problems. J. Comput. Phys. **206**(1), 363–393 (2005)
126. Kleefsman, T.: Water impact loading on offshore structures. Ph.D. thesis, University of Groningen, The Netherlands (2005)
127. Koh, C.G., Gao, M., Luo, C.: A new particle method for simulation of incompressible free surface flow problems. Int. J. Num. Methods Eng. **89**(12), 1582–1604 (2012)
128. Kravchenko, A.G., Moin, P.: Numerical studies of flow over a circular cylinder at $Re_D = 3900$. Phys. Fluids **12**(2), 403–417 (2000)
129. Kuhl, E., Hulshoff, S., de Borst, R.: An arbitrary Lagrangian Eulerian finite-element approach for fluid-structure interaction phenomena. Int. J. Numer. Methods Eng. **57**(1), 117–142 (2003)
130. Kuzmin, D.: A Guide to Numerical Methods for Transport Equations. Friedrich-Alexander-Universität, Erlangen-Nü
131. Kuzmin, D., Löhner, R., Turek, S.: Flux-corrected Transport: Principles, Algorithms and Applications. Springer (2005)
132. Formaggia, L., A.M. Quarteroni, Veneziani, A.: Cardiovascular Mathematics: Modeling and Simulation of the Circulatory System. Springer, Milano (2009)
133. Laurenco, L.M., Shih, C.: Characteristics of the plane turbulent near wake of a circular cylinder. A particle image velocimetry study. Private communication (data taken from "B-Spline methods and zonal grids for numerical simulations of turbulent flows", Kravchenko, A. G., 1998) (1993)
134. Lee, H.G.: High-order and mass conservative methods for the conservative Allen-Cahn equation. Comput. Math. Appl. **72**(3), 620–631 (2016)
135. LeVeque, R.J.: Numerical Methods for Conservation Laws. Birkhäuser Verlag, Basel; Boston (1992)
136. Lew, A., Marsden, J.E., Ortiz, M., West, M.: Variational time integrators. Int. J. Numer. Methods Eng. **60**(1), 153–212 (2004)
137. Lilly, D.: A proposed modification of the Germano subgrid-scale closure method. Phys. Fluids **4**, 633–635 (1992)
138. Liu, J.: One field formulation and a simple explicit scheme for fluid structural interaction

139. Liu, J., Jaiman, R.K., Gurugubelli, P.S.: A stable second-order scheme for fluid-structure interaction with strong added-mass effects, pp. 687–710
140. Liu, J., Jaiman, R.K., Gurugubelli, P.S.: A stable second-order scheme for fluid-structure interaction with strong added-mass effects. J. Comput. Phys. **270**, 687–710 (2014)
141. Löhner, R.: Applied Computational Fluid Dynamics Techniques, pp. 269–297. Wiley (2008)
142. Lorin, E., Ben Haj Ali, A., Soulaimani, A.: A positivity preserving finite element-finite volume solver for the Spalart-Allmaras turbulence model. Comput. Methods Appl. Mech. Eng. **196**, 2097–2116 (2007)
143. Lyn, D.A., Einav, S., Rodi, W., Park, J.-H.: A laser-doppler velocimetry study of ensemble-averaged characteristics of the turbulent near wake of a square cylinder. J. Fluid Mech. **304**, 285–319 (1995)
144. Mankbadi, M.R., Georgiadis, N.J.: Examination of parameters affecting large-eddy simulations of flow past a square cylinder. AIAA J. **53**(6), 1706–1712 (2015)
145. Masud, A., Calderer, R.: A variational multiscale method for incompressible turbulent flows: bubble functions and fine scale fields, pp. 2577–2593
146. Masud, A., Hughes, T.J.R.: A space-time Galerkin/least-square finite element formulation of the navier-stokes equations for moving domain problems, pp. 91–126
147. Matthies, H.G., Niekamp, R., Steindorf, J.: Algorithms for strong coupling procedures, pp. 2028–2049
148. Meng, F., Banks, J., Henshaw, W., Schwendeman, D.: A stable and accurate partitioned algorithm for conjugate heat transfer. J. Comput. Phys. **344**, 51–85 (2017)
149. Milenkovic, V.: Coordinates suitable for angular motion synthesis in robots. In: Proceedings of the Robot VI Conference, Detroit, Michigan (1982)
150. Mittal, R., Iaccarino, G.: Immersed boundary methods. Annu. Rev. Fluid Mech. **37**(1), 239–261 (2005)
151. Mittal, S.: On the performance of high aspect ratio elements for incompressible flows. Comput. Methods Appl. Mech. Eng. **188**, 269–287 (2000)
152. Mizukami, A., Hughes, T.J.R.: A Petrov-Galerkin finite element method for convection-dominated flows: An accurate upwinding technique for satisfying the maximum principle. Comput. Methods Appl. Mech. Eng. **50**(2), 181–193 (1985)
153. Moro, D., Nguyen, N.C., Peraire, J.: Navier-Stokes solution using hybridizable discontinuous Galerkin methods. In: AIAA-2011-3407 (2011)
154. Morris-Thomas, M.: An investigation into wave run-up on vertical surface piercing cylinders in monochromatic waves. Ph.D. thesis, The University of Western Australia (2003)
155. Nadukandi, P., Oñate, E., Garcia, J.: A high-resolution Petrov-Galerkin method for the 1D convection-diffusion-reaction problem. Comput. Methods Appl. Mech. Eng. **199**, 525–546 (2010)
156. Nadukandi, P., Oñate, E., Garcia, J.: A high-resolution Petrov-Galerkin method for the convection-diffusion-reaction problem. Part II-A multidimensional extension. Comput. Methods Appl. Mech. Eng. **213–216**, 327–352 (2012)
157. Nogueira, E., Cotta, R.M.: Heat transfer solutions in laminar co-current flow of immiscible liquids. Wärme - und Stoffübertragung **25**(6), 361–367 (1990)
158. Norberg, C.: Effects of Reynolds number and a low-intensity free-stream turbulence on the flow around a circular cylinder. Publication 87/2. Department of Applied Thermodynamics and Fluid Mechanics. Chalmers University of Technology, Sweden (1987)
159. Ogden, R.: Non-Linear Elastic Deformations. Dover (1997)
160. Olsson, E., Kreiss, G.: A conservative level set method for two phase flow. J. Comput. Phys. **210**(1), 225–246 (2005)
161. Olsson, E., Kreiss, G., Zahedi, S.: A conservative level set method for two phase flow II. J. Comput. Phys. **225**(1), 785–807 (2007)
162. Oñate, E.: Derivation of stabilized equations for numerical solution of advective-diffusive transport and fluid flow problems. Comput. Methods Appl. Mech. Eng. **151**, 233–265 (1998)
163. Oñate, E., Miquel, J., Hauke, G.: Stabilized formulation for the advection-diffusion-absorption equation using finite calculus and linear finite elements. Comput. Methods Appl. Mech. Eng. **195**, 3926–3946 (2006)

164. Oñate, E., Miquel, J., Hauke, G.: Stabilized formulation for the advection-diffusion-absorption equation using finite calculus and linear finite elements. Comput. Methods Appl. Mech. Eng. **195**, 3926–3946 (2006)
165. Oñate, E., Zarate, F., Idelsohn, S.R.: Finite element formulation for the convective-diffusive problems with sharp gradients using finite calculus. Comput. Methods Appl. Mech. Eng. **195**, 1793–1825 (2006)
166. Ong, L., Wallace, J.: The velocity field of the turbulent very near wake of a circular cylinder. Exp. Fluids **20**(6), 441–453 (1996)
167. OpenMp: https://computing.llnl.gov/tutorials/openmp/s
168. Paidoussis, M.P., Price, S.J., de Langre, E.: Fluid-Structure Interactions: Cross-Flow-Induced Instabilities. Cambridge University Press (2010)
169. Parnaudeau, P., Carlier, J., Heitz, D., Lamballais, E.: Experimental and numerical studies of the flow over a circular cylinder at Reynolds number 3900. Phys. Fluids **20**(8), 85101 (2008)
170. Peng, D., Merriman, B., Osher, S., Zhao, H., Kang, M.: A PDE-based fast local level set method. J. Comput. Phys. **155**(2), 410–438 (1999)
171. Piomelli, U., Balaras, E.: Wall-layer models for large-eddy simulations. Annu. Rev. Fluid Mech. **34**, 349 (2002)
172. Piperno, S., Farhat, C., Larrouturou, B.: Partitioned procedures for the transient solution of coupled aeroelastic problems—Part II: energy transfer analysis and three-dimensional applications, pp. 3147–3170
173. Principe, J., Codina, R.: On the stabilization parameter in the subgrid scale approximation of scalar convection-diffusion-reaction equations on distorted meshes. Comput. Methods Appl. Mech. Eng. **199**(21), 1386–1402 (2010)
174. Prosperetti, A., Tryggvason, G.: Computational Methods for Multiphase Flow. Cambridge University Press (2007)
175. Rasthofer, U.: Computational multiscale methods for turbulent single and two-phase flows. Ph.D. thesis, München, Technische Universität München, Dissertation, 2015 (2015)
176. Rice, J.G., Schnipke, R.J.: A monotone streamline upwind finite element method for convection-dominated flows. Comput. Methods Appl. Mech. Eng. **48**(3), 313–327 (1985)
177. Rider, W.J., Kothe, D.B.: Reconstructing volume tracking. J. Comput. Phys. **141**(2), 112–152 (1998)
178. Rizzetta, D.P., Visbal, M.R., Blaisdell, G.A.: A time-implicit high-order compact differencing and filtering scheme for large-eddy simulation. Int. J. Numer. Methods Fluids **42**(6), 665–693 (2003)
179. Robinson-Mosher, A., Schroeder, C., Fedkiw, R.: A symmetric positive definite formulation for monolithic fluid structure interaction. J. Comput. Phys. **230**(4), 1547–1566 (2011)
180. Rodi, W., Ferziger, J., Breuer, M., Pourquie, M.: Status of large eddy simulation: results of a workshop. Trans.-Am. Soc. Mech. Eng. J. Fluids Eng. **119**, 248–262 (1997)
181. Roe, B., Jaiman, R., Haselbacher, A., Geubelle, P.H.: Combined interface boundary condition method for coupled thermal simulations. Int. J. Numer. Methods Fluids **57**(3), 329–354 (2008)
182. Roshko, A.: Experiments on the flow past a circular cylinder at very high Reynolds number. J. Fluid Mech. **10**(2), 345–356 (1961)
183. Rubinstein, J., Sternberg, P.: Nonlocal reaction-diffusion equations and nucleation. IMA J. Appl. Math. **48**(3), 249–264 (1992)
184. Russo, G., Smereka, P.: A remark on computing distance functions. J. Comput. Phys. **163**(1), 51–67 (2000)
185. Saad, Y.: Iterative Methods for Sparse Linear Systems. Society of Industrial and Applied Mathematics (2003)
186. Saad, Y., Schultz, M.H.: GMRES: a generalized minimal residual algorithm for solving non-symmetric linear systems. SIAM J. Sci. Stat. Comput. **7**(3), 856–869 (1986)
187. Sagaut, P.: Large eddy simulation for incompressible flows. In: Scientific Computing. Springer (2006)
188. Sauerland, H., Fries, T.-P.: The extended finite element method for two-phase and free-surface flows: a systematic study. J. Comput. Phys. **230**(9), 3369–3390 (2011)

189. Scardovelli, R., Zaleski, S.: Direct numerical simulation of free-surface and interfacial flow. Annu. Rev. Fluid Mech. **31**(1), 567–603 (1999)
190. Schmidt, A., Siebert, K.G.: Design of Adaptive Finite Element Software. Springer (2005)
191. Sethian, J.A., Smereka, P.: Level set methods for fluid interfaces. Annu. Rev. Fluid Mech. **35**(341) (2003). Copyright—Copyright Annual Reviews, Inc. Last updated 18 May 2014
192. Shabana, A.: Dynamics of Multibody Systems. Wiley (1998)
193. Shakib, F., Hughes, T.J.R., Johan, Z.: A new finite element formulation for computational fluid dynamics: X. The compressible Euler and Navier-Stokes equations. Comput. Methods Appl. Mech. Eng. **89**, 141–219 (1991)
194. Smagorinsky, J.: General circulation experiments with the primitive equations, I. The basic experiment. Monthly Weather Rev. **91**, 99–164 (1963)
195. Sorensen, R.M.: Basic Wave Mechanics: For Coastal and Ocean Engineers. Wiley (1993)
196. Sotiropoulos, F., Yang, X.: Immersed boundary methods for simulating fluid-structure interaction. Progr. Aerosp. Sci. **65**, 1–21 (2014)
197. Spalart, P.R.: Young-person's guide to Detached-Eddy Simulation grids. In: NASA/ CR-2001-211032 (2001)
198. Spalart, P.R., Allmaras, S.R.: A one-equation turbulence model for aerodynamic flows. In: AIAA-92-0439 (1992)
199. Spalart, P.R., Allmaras, S.R.: A one-equation turbulence model for aerodynamic flows. La Rech. Aé rospatiale **1**, 5–21 (1994)
200. Stein, K., Tezduyar, T., Benney, R.: Mesh moving techniques for fluid-structure interactions with large displacements, pp. 58–63
201. Stokes, G.G.: On the Theory of Oscillatory Waves, vol. 1 of Cambridge Library Collection—Mathematics, pp. 197–229. Cambridge University Press (2009)
202. Stynes, M.: Steady-state convection-diffusion problems. Acta Numerica **14**, 445–508 (2005)
203. Sussman, M., Fatemi, E.: An efficient, interface-preserving level set redistancing algorithm and its application to interfacial incompressible fluid flow. SIAM J. Sci. Comput. **20**(4), 1165–27. Copyright—Copyright © (1999) Society for Industrial and Applied Mathematics. Last updated 15 Feb 2012
204. Sussman, M., Puckett, E.G.: A coupled level set and volume-of-fluid method for computing 3D and axisymmetric incompressible two-phase flows. J. Comput. Phys. **162**(2), 301–337 (2000)
205. Swartz, S.M., Middleton, K.M.: Biomechanics of the bat limb skeleton: scaling, material properties and mechanics. Cells Tissues Organs **187**, 59–84 (2008)
206. Tan, Z., Lim, K.M., Khoo, B.C.: An adaptive mesh redistribution method for the incompressible mixture flows using phase-field model. J. Comput. Phys. **225**, 1137–1158 (2007)
207. Temam, R.: Navier-Stokes Equations. Theory and Numerical Analysis. AMS Chelsea Publishing (2001)
208. Temam, R.: Mathematical Modeling in Continuum Mechanic. Cambridge Publishers (2005)
209. Tezduyar, T.E., Mittal, S., Ray, S., Shih, R.: Incompressible flow computations with stabilized bilinear and linear equal-order interpolation velocity-pressure elements. Comput. Methods Appl. Mech. Eng. **95**, 221–242 (1992)
210. Tezduyar, T.E., Park, Y.J.: Discontinuity-capturing finite element formulations for nonlinear convection-diffusion-reaction equations. Comput. Methods Appl. Mech. Eng. **59**, 307–325 (1986)
211. Thiagarajan, K.P., Repalle, N.: Wave run-up on columns of deepwater platforms. Proc. Inst. Mech. Eng. Part M: J. Eng. Maritime Environ. **227**(3), 256–265 (2013)
212. Tierra, G., Guillén-González, F.: Numerical methods for solving the Cahn-Hilliard equation and its applicability to related energy-based models. Arch. Comput. Methods Eng. **22**(2), 269–289 (2015)
213. Travin, A., Shur, M., Strelets, M., Spalart, P.: Detached-eddy simulations past a circular cylinder. Flow Turbul. Combust. **63**(1), 293–313 (2000)
214. Truesdell, C.: The Elements of Continuum Mechanics, vol. 6. Springer, NewYork (1966)

215. Truesdell, C.: A First Course in Rational Continuum Mechanics. Academic Press, New York (1977)
216. Turek, S., Hron, J.: Proposal for numerical benchmarking of fluid-structure interaction between an elastic object and laminar incompressible flow. In: Fluid-Structure Interaction, pp. 371–385. Springer (2006)
217. Unverdi, S.O., Tryggvason, G.: A front-tracking method for viscous, incompressible, multi-fluid flows. J. Comput. Phys. **100**(1), 25–37 (1992)
218. van Brummelen, E.H., van der Zee, K.G., Garg, V.V., Prudhomme, S.S.: Flux evaluation in primal and dual boundary-coupled problems
219. Van Buren, T., Floryan, D., Brunner, D., Senturk, U., Smits, A.J.: Impact of trailing edge shape on the wake and propulsive performance of pitching panels. Phys. Rev. Fluids **2**, 014702 (2017)
220. van der Vorst, H.A.: Iterative Krylov Methods for Large Linear Systems. Cambridge University Press (2003)
221. van Nunen, J.W.G.: Pressure and forces on a circular cylinder in a cross flow at high Reynolds numbers. In: Naudascher, E. (ed.) Flow Induced Structural Vibrations, pp. 748–754. Springer, Berlin (1974)
222. Vasconcelos, D.F.M., Rossa, A.L., Coutinho, A.L.G.A.: A residual-based Allen-Cahn phase field model for the mixture of incompressible fluid flows. Int. J. Numer. Methods Fluids **75**(9), 645–667 (2014)
223. Verfürth, R.: Adaptive finite element methods. Lecture notes, Fakultät für Mathematik, Ruhr-Universität, Bochum, Germany
224. Verfürth, R.: A posteriori error estimation and adaptive mesh-refinement techniques. J. Comput. Appl. Math. **50**(1–3), 67–83 (1994)
225. Verfürth, R.: A Review of a Posteriori Error Estimation and Adaptive Mesh-Refinement Techniques. Wiley (1996)
226. Verstappen, R., Veldman, A.: Direct numerical simulation of turbulence at lower costs. J. Eng. Math. **32**(2–3), 143–159 (1997)
227. Veubeke, B.F.D.: The dynamics of flexible bodies. Int. J. Eng. Sci. **14**(10), 895–913 (1976)
228. Vignjevic, R., Campbell, J.: Review of Development of the Smooth Particle Hydrodynamics (SPH) Method, pp. 367–396. Springer, Boston, MA, USA (2009)
229. Vist, S., Pettersen, J.: Two-phase flow distribution in compact heat exchanger manifolds. Exp. Thermal Fluid Sci. **28**(2), 209–215 (2004). The International Symposium on Compact Heat Exchangers
230. Vreman, A.: The filtering analog of the variational multiscale method in large-eddy simulation. Phys. Fluids **15**, L61 (2003)
231. Waals, O., Phadke, A., Bultema, S.: Flow-induced motions of multi-column floaters. In: ASME Offshore Mechanics and Arctic Engineering OMAE07-29539 CP (2007)
232. Wang, B.C., Bergstrom, D.J.: A dynamic nonlinear subgrid-scale stress model. Phys. Fluids **17**, 035109 (2005)
233. Wang, E., Xiao, Q.: Numerical simulation of vortex-induced vibration of a vertical riser in uniform and linearly sheared currents. Ocean Eng. **121**, 492–515 (2016)
234. Wang, G., Vanka, S.: Large-eddy simulations of high Reynolds number turbulent flow over a square cylinder. Department of Mechanical and Industrial Engineering Report No. CFD, pp. 96–102
235. Wang, Z., Yang, J., Koo, B., Stern, F.: A coupled level set and volume-of-fluid method for sharp interface simulation of plunging breaking waves. Int. J. Multiphase Flow **35**(3), 227–246 (2009)
236. Warming, R.F., Hyett, B.J.: The modified equation approach to the stability and accuracy analysis of finite-difference methods. J. Comput. Phys. **14**, 159–179 (1974)
237. Wendland, H.: Piecewise polynomial, positive definite and compactly supported radial functions of minimal degree. Adv. Comput. Math. **4**, 389–396 (1995)
238. Wendland, H., Rieger, C.: Approximate interpolation with applications to selecting smoothing parameters. Numerische Mathematik **101**, 729–748 (2005)

239. Wiener, T.F.: Theoretical analysis of Gimballess inertial reference equipment using delta-modulated instruments. Ph.D. thesis, Massachusetts Institute of Technology (1962)
240. Wu, P.: Experimental characterization, design, analysis and optimization of flexible flapping wings for micro air vehicles. Ph.D. thesis, University of Florida (2010)
241. Wu, Z.: Compactly supported positive definite radial functions. Adv. Comput. Math. **4**, 283–292 (1995)
242. Yang, X., James, A.J., Lowengrub, J., Zheng, X., Cristini, V.: An adaptive coupled level-set/volume-of-fluid interface capturing method for unstructured triangular grids. J. Comput. Phys. **217**(2), 364–394 (2006)
243. Yenduri, A., Ghoshal, R., Jaiman, R.: A new partitioned staggered scheme for flexible multi-body interactions with strong inertial effects. Compu. Methods Appl. Mech. Eng. **315**, 316–347 (2017)
244. Zalesak, S.T.: Fully multidimensional flux-corrected transport algorithms for fluids. J. Comput. Phys. **31**, 335–362 (1979)
245. Zhang, J., Du, Q.: Numerical studies of discrete approximations to the Allen-Cahn equation in the sharp interface limit. SIAM J. Scie. Comput. **31**(4), 3042–3063 (2009)
246. Zhang, Z., Tang, H.: An adaptive phase field method for the mixture of two incompressible fluids. Comput. Fluids **36**(8), 1307–1318 (2007)
247. Zhao, J., Leontini, J., Jacono, D., Sheridan, J.: Fluid-structure interaction of a square cylinder at different angles of attack. J. Fluid Mech. **747**, 688–721 (2014)
248. Zhao, L., Bai, X., Li, T., Williams, J.J.R.: Improved conservative level set method. Int. J. Numer. Methods Fluids **75**(8), 575–590 (2014). FLD-13-0333.R1
249. Zhao, R., Liu, J., Yan, C.: Detailed, investigation of detached-eddy simulation for the flow past a circular cylinder at Re=3900. In: Progress in Hybrid RANS-LES Modelling: Papers Contributed to the 4th Symposium on Hybrid RANS-LES Methods, Beijing, China, vol. 2012, pp. 401–412. Springer, Berlin, Heidelberg (2011)
250. Zienkiewicz, O.C., Zhu, J.Z.: A simple error estimator and adaptive procedure for practical engineerng analysis. Int. J. Numer. Methods Eng. **24**(2), 337–357 (1987)